Studies in Mechanobiology, Tissue Engineering and Biomaterials

Volume 9

Series Editor
Amit Gefen, Ramat Aviv, Israel

Further volumes of this series can be found on our homepage:
http://www.springer.com/series/8415

Amit Gefen
Editor

Patient-Specific Modeling in Tomorrow's Medicine

Amit Gefen
Department of Biomedical Engineering
Faculty of Engineering
Tel Aviv University
Tel Aviv
Israel

ISSN 1868-2006
ISBN 978-3-642-24617-3
DOI 10.1007/978-3-642-24618-0
Springer Heidelberg Dordrecht London New York

e-ISSN 1868-2014
e-ISBN 978-3-642-24618-0

Library of Congress Control Number: 2011943311

© Springer-Verlag Berlin Heidelberg 2012
This work is subject to copyright. All rights are reserved, whether the whole or part of the material is concerned, specifically the rights of translation, reprinting, reuse of illustrations, recitation, broadcasting, reproduction on microfilm or in any other way, and storage in data banks. Duplication of this publication or parts thereof is permitted only under the provisions of the German Copyright Law of September 9, 1965, in its current version, and permission for use must always be obtained from Springer. Violations are liable to prosecution under the German Copyright Law.
The use of general descriptive names, registered names, trademarks, etc. in this publication does not imply, even in the absence of a specific statement, that such names are exempt from the relevant protective laws and regulations and therefore free for general use.

Cover design: WMXDesign GmbH, Heidelberg

Printed on acid-free paper

Springer is part of Springer Science+Business Media (www.springer.com)

Preface

Patient-specific modeling (PSM) is an emerging field in biomedical engineering. PSM is aimed at implementing the powerful modeling tools and techniques developed over the years by biomedical engineers and medical physicists for the benefit of patients. This is typically done by first creating a three-dimensional computational reconstruction of the anatomy of the tissues or a mathematical model of the organ of interest in the individual patient, based on imaging scans or other individualized parameters, and then by using the model for calculations that provide a diagnosis, prognosis or prediction of treatment outcomes. Hence, a particularly challenging aspect of PSM is that it requires integration of expertise from various technological and bioengineering subdisciplines, such as biosolid and biofluid mechanics, biomass transport, medical imaging, constitutive tissue modeling, numerical simulations and computer visualization to solve actual medical problems.

The PSM approach is currently being put into practice to assist in managing a wide range of different medical conditions, such as in orthopaedics, cardiology, neurology, oncology and ophthalmology. Critical issues in making PSM standard and routine are the ease of use and interpretation of data by medical staff, as well as successful validation of the predicted outcome measures, which are all significant barriers for PSM technologies to become clinically acceptable. Substantial research efforts are underway worldwide to resolve these issues, and it does appear that PSM technologies will eventually become an integral part of modern medicine, as PSM can be naturally combined with common imaging examinations such as MRI, CT or ultrasound scans. This book reviews the frontier of research and clinical applications of PSM, and provides a comprehensive and rigorous update as well as perspectives on future directions in this exciting field.

The frontier in PSM research is presented in this volume through contributions of internationally leading groups in this field, from the US, Australia, New Zealand, Israel and five different European countries. The book is useful for medical physicists, computer scientists, biomedical engineers and other engineers who are interested in the science and technology aspects of PSM, as well as for medical specialists such as radiologists, orthopaedists, cardiologists and others who wish to be updated about the state of implementation. Academics, medical doctors and students alike can use this book to learn about the state-of-the-art and current achievements in PSM as well as on the challenges that will need to be addressed in the near future to ensure that the great promises that PSM brings are indeed put into clinical practice.

Amit Gefen

Contents

Part I Bones

Reliable Patient-Specific Simulations of the Femur 3
Zohar Yosibash and Nir Trabelsi

**Patient-Specific Diagnosis and Visualization
of Bone Micro-Structures** 27
L. Podshivalov, A. Fischer and P. Z. Bar-Yoseph

Patient Specific Modeling of Musculoskeletal Fractures 53
Eran Peleg

Part II Spine

**Advancements in Spine FE Mesh Development:
Toward Patient-Specific Models** 75
Nicole A. Kallemeyn, Kiran H. Shivanna, Nicole A. DeVries,
Swathi Kode, Anup A. Gandhi, Douglas C. Fredericks,
Joseph D. Smucker and Nicole M. Grosland

Patient-Specific Modeling of Scoliosis 103
J. Paige Little and Clayton J. Adam

**Subject-Specific Models of the Spine for the Analysis
of Vertebroplasty** 133
Alison C. Jones, Vithanage N. Wijayathunga, Sarrawat Rehman
and Ruth K. Wilcox

Part III Heart and Circulation

Morphological and Functional Modeling of the Heart Valves and Chambers 157
Razvan Ioan Ionasec, Dime Vitanovski and Dorin Comaniciu

Patient-Specific Analysis of Blood Flow and Mass Transport in Small and Large Arteries 189
X. Y. Xu, N. Sun, D. Liu and N. B. Wood

Patient-Specific Modeling of Leg Compression in the Treatment of Venous Deficiency 217
Stéphane Avril, Pierre Badel, Laura Dubuis, Pierre-Yves Rohan, Johan Debayle, Serge Couzan and Jean-François Pouget

Part IV Airways

Scan-Based Flow Modelling in Human Upper Airways 241
Perumal Nithiarasu, Igor Sazonov and Si Yong Yeo

A Decission Support System for Endoprosthetic Patient-Specific Surgery of the Human Trachea 281
Olfa Trabelsi, Angel Ginel, Jose L. López-Villalobos, Miguel A González-Ballester, Amaya Pérez-del Palomar and Manuel Doblaré

Part V Brain

Image-Based Computational Fluid Dynamics for Patient-Specific Therapy Design and Personalized Medicine 337
Andreas A. Linninger

Patient-Specific Modeling and Simulation of Deep Brain Stimulation 357
Karin Wårdell, Elin Diczfalusy and Mattias Åström

Contents

Part VI Patients-Specific Modeling in Diagnostics, Surgical Planning and Rehabilitation

Patient-Specific Modeling of Breast Biomechanics with Applications to Breast Cancer Detection and Treatment.................... 379
Thiranja P. Babarenda Gamage, Vijayaraghavan Rajagopal, Poul M. F. Nielsen and Martyn P. Nash

Soft-Tissue Simulation for Cranio-Maxillofacial Surgery: Clinical Needs and Technical Aspects........................ 413
Hyungmin Kim, Philipp Jürgens and Mauricio Reyes

Patient-Specific Modeling of Subjects with a Lower Limb Amputation 441
Sigal Portnoy and Amit Gefen

Patient-Specific Modeling of the Cornea..................... 461
Roy Asher, Amit Gefen and David Varssano

Part VII Systems Biology Approaches in Patients-Specific Modeling

Computational Modeling of Gene Translation and its Potential Applications in Individualized Medicine...................... 487
Tamir Tuller and Hadas Zur

Neural Network Modeling Approaches for Patient Specific Glycemic Forecasting 505
Scott M. Pappada and Brent D. Cameron

Author Index 531

Part I
Bones

Reliable Patient-Specific Simulations of the Femur

Zohar Yosibash and Nir Trabelsi

Abstract Valuable simulations aimed at diagnosis and optimal treatment in clinical orthopedic practice should demonstrate that their results are verified, validated by experimental observations and obtained on a patient-specific level in a short time-scale. Such verified and validated simulations, based on CT-scans, were recently introduced using high-order finite element methods (p-FEMs), demonstrating an unprecedented prediction capability. We describe herein the methods used for creating p-FEM models of patient-specific femurs (including assignment of inhomogeneous material properties) and the large set of in-vitro experiments used to assess the validity of the simulation results. We thereafter extend the simulation capabilities for the analysis of bone fixations by metallic inserts, demonstrating again the high quality results. Such reliable patient-specific simulations are in an advanced stage to be used on a daily basis in clinical practice.

1 Introduction

Although bone simulations by finite element methods (FEMs) have been performed for more than 25 years, these have had very little impact on current clinical practice. This may be due to improper use of verification methods that assess the accuracy of the numerical approximations, imprecise knowledge on the three main ingredients of such simulations (precise geometry, material properties and

---

Z. Yosibash (✉) · N. Trabelsi
Department of Mechanical Engineering,
Computational Mechanics Laboratory,
Ben-Gurion University of the Negev,
Beer Sheva, Israel
e-mail: zohary@bgu.ac.il

constitutive laws and precise boundary conditions) and due to a lack of proper validation protocols for comparison to in-vitro experimental observations which allow in a following step case-studies and in-vivo validation. At the same time, new and improved technological methods (as quantitative computer tomography, qCT) and advanced computational mechanics algorithms (p-version FEMs) have emerged which in conjunction with experimental procedures, allow a *reliable* finite element simulation of the femur that account for several ingredients:

(a) An accurate description of femur's geometry,
(b) an accurate description of the material inhomogeneous properties,
(c) accurate and as close as possible to physiologic boundary conditions (loads and constrains),
(d) a verification process assuring that the numerical errors are bounded by a specified tolerance,
(e) a validation process in which the FE results are compared against all measurable data (strains and displacements) recorded during simplified in-vitro tests on a variety of femurs (with a wide spread in age, gender and weight), see e.g. [20].

These ingredients are mandatory (and not-sufficient), but unfortunately they are ignored by many of the publications on the subject. Non-the-less, recently several works are worth of mentioning in the context of reliable FE simulation of the human femurs [40, 4, 30, 28, 37, 38] following the pioneering works by Keyak that started in the early 90s with [23]. Utilizing qCT scans in conjunction with *p*-FEMs [33] we herein provide a systematic method for performing reliable FE simulations of the human femur, validated by in-vitro tests on the largest (so far reported) cohort of fresh-frozen femurs.

In classical *h*-FEMs the inhomogeneous distribution of material properties is usually attained by assigning constant distinct values to distinct elements (see e.g. [35] and references therein), thus the material properties become mesh dependent. Furthermore, the bone's surface is approximated by piecewise flat tessellation or piecewise parabolic tessellation, introducing un-smoothness of the surface and inaccuracies in the surface strains. Instead, *p*-FEMs have many advantages over conventional *h*-FEMs: *p*-FEMs accurately represent the bone's surfaces by using blending-function techniques, keep the mesh unchanged and only increases the polynomial degree p of the shape functions to achieve convergence and allow naturally functional variation of the material properties within each element [33]. In addition, *p*-elements are much larger, may be far more distorted and have large aspect ratio and yet produce considerable faster convergence rates compared to their *h*-FEM counterparts.

The first step in a reliable patient-specific simulation is a fast automatic algorithm for the generation of the *p*-FE femur's model from clinical qCT scans having inhomogeneous continuous isotropic or even orthotropic material properties assigned based on the Hounsfield Units (HUs) described in Sect. 2.1. Most researchers assign inhomogeneous isotropic material properties in their FE models because of lack of knowledge on the orthotropic inhomogeneous properties and the material trajectories from qCT scans. In this case, the Young's modulus E is determined by empirical relationships to a densitometric measure, and the "best" relationship found to well represent the whole-organ mechanical response is discussed. The isotropic assumption is justified if simplified loading conditions are applied to the femur. Micro-mechanical approaches can be used instead to assign orthotropic properties to the whole organ [19, 41, 36]. The difference in the predicted mechanical response when isotropic or orthotropic properties are assigned to the p-FE models is demonstrated by considering four femurs with a variety of loading cases.

Once the p-FE analyses are finalized (the entire process for one patient-specific femur lasts less than 3 h) and the results are verified so the numerical error is assured to be controlled, we validate the results by a set of in-vitro experiments on seventeen human fresh-frozen femurs. Five experiments are performed in-house and twelve others are performed by a different group so a "blinded" non-biased comparison can be performed. The different experiments are detailed in Sect. 3, and both measured strains and displacements are compared. The validity of the presented methods is clearly demonstrated by the excellent match between experimental and the predicted results—both strains and displacements. We also discuss the comparison "norm"—most previous publications report the slope and the linear regression coefficient of the experiment vis FE-prediction data, however, as we shall demonstrate this kind of comparison may sometimes be a bit misleading, so instead the Bland-Altman plot is suggested [5].

Finally, we discuss and demonstrate some potential applications of patient-specific p-FE analyses in daily clinical practice (like arthoplasty) and suggest further research activities necessary to bring the methods to a mature clinical-usage stage.

2 p-FE Analysis of the Femur

2.1 Generation of Femur's p-FE Model from qCT-Scans

All DICoM (Digital Imaging and Communication in Medicine) format qCT scans are *automatically* manipulated by in-house Matlab programs. First the scans are transformed into binary images in which nonzero pixels belong to the bone and the value 0 is assigned to pixels representing the background. The proximal femur bone's axis is aligned with the z axis. No exact HU exists that distinguishes between the cortical and trabecular regions. Following [1, 13, 16, 2] we associated voxel values of $HU > 600 (\rho_{ash} > 0.6 \, g/cm^3)$ with the cortical bone and values of $HU \leq 600$ to the trabecular bone. Exterior, interface and interior boundaries are traced and x-y arrays are generated, each representing different boundaries of a given slice. These arrays are manipulated by a 3-D smoothing algorithm that generates smooth raw data arrays by a 3-D spherical filter, 2-D averaging filter and

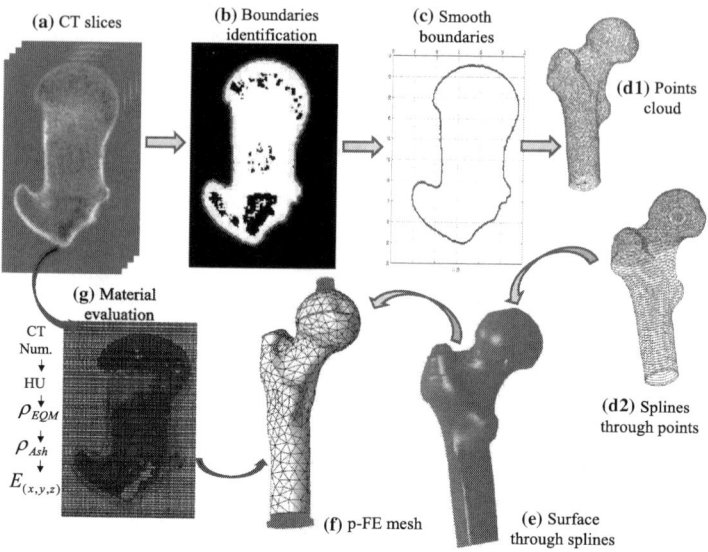

Fig. 1 Schematic flowchart describing the generation of the p-FE model from qCT scans. **a** Typical CT-slice, **b** Contour identification, **c** Smoothing boundary points, **d1** Points cloud representing the bone surface. **d2** Close splines for all slices, **e** Bone surface, **f** p-FE mesh and **g** Material evaluation from CT data. (Figure from [36])

spline interpolation (3-D spherical filter calculates the new location of each point in a specific slice using data from other slices around it). The smooth edges, using cubic spline interpolation, are saved as new arrays that are read into the CAD package SolidWorks-2010 (SolidWorks Corporation, MA, USA) and manipulated to generate a surface representation of the femur and subsequently a solid model. Large surfaces are generated, which are essential for the *p*-mesh generator, in contrast with many small patches typically obtained by other methods. Separating by a surface the cortical and trabecular regions allows an easy assignment of different material properties to each region. The resulting 3D solid is imported into the p-FE StressCheck[1] code. An auto-mesher is thereafter applied that generates tetrahedral high-order elements. Blending function method [33] maps the elements to the standard element so the surfaces are accurately represented in the FE simulation. The entire algorithm (qCT to FE) is schematically illustrated in Fig. 1.

Remark 1 The numerical error associated with the use of tetrahedron *p*-elements, compared to hexahedral *p*-elements, was investigated and was confirmed to be of the same order of accuracy when the polynomial degree is increased [42]. The "over-stiff" behavior of tetrahedron *h*-elements does not occur for the p-version of the FEM.

---

[1] StressCheck is trademark of Engineering Software Research and Development, Inc, St. Louis, MO, USA.

2.2 Assigning Isotropic Material Properties to the FE Model, Determined by Empirical Correlation

Many empirical relations between Young's modulus and bone density, with a constant Poisson's ratio were suggested, see e.g. [7, 8, 24, 22, 25]. In [42, 37] we found that p-FE analyses with the relationships in [24] (the cortical connections are based on [22]) provide the closest results to in-vitro experiments on the proximal femur:

$$\rho_{EQM} = 10^{-3}(a \times HU - b) \quad [g/cm^3] \tag{1}$$

$$\rho_{ash} = (1.22 \times \rho_{EQM} + 0.0523) \quad [g/cm^3] \tag{2}$$

$$E_{Cort} = 10200 \times \rho_{ash}^{2.01} \quad [MPa] \quad \rho_{ash} > 0.6 \tag{3}$$

$$E_{Trab} = 5307 \times \rho_{ash} + 469 \quad [MPa] \quad 0.27 < \rho_{ash} \leq 0.6 \tag{4}$$

$$E_{Trab} = 33900 \times \rho_{ash}^{2.20} \quad [MPa] \quad \rho_{ash} \leq 0.27 \tag{5}$$

where ρ_{EQM} is the equivalent mineral density, ρ_{ash} is the ash density, E_{Cort}, E_{Trab} are the Young's modulii in the cortical and trabecular regions and the parameters a and b are determined by the K_2HPO_4 phantoms in the CT-scan. Constant Poisson ratio $v = 0.3$ was assigned to the entire bone. According to a sensitivity analysis in [40, 42] the influence of v on the results is very small. These $E - \rho_{ash}$ relations are shown graphically in Fig. 2.

Remark 2 Numerical experiments show that the use of (4) for the entire range $0 < \rho_{ash} \leq 0.6 \, g/cm^3$ results in almost identical results as if using (5) in the range $\rho_{ash} \leq 0.27 \, g/cm^3$. Thus in all our p-FE analyses the relationship (5) is disregarded.

Determination of $E(x,y,z)$ at each integration point (Gauss points, 512 for a tetrahedral element) in the FE model is performed as follows. First a moving average algorithm is applied to average the HU data in each voxel based on a pre-defined cubic volume of $3 \times 3 \times 3 \, mm^3$ surrounding it (cubic volumes of 27,64,343 mm^3 showed similar results in [40]). HU averaged data is subsequently converted to an equivalent mineral density ρ_{EQM} by (1) which is determined by the calibration phantom, i.e. five burettes containing different concentrations of K_2HPO_4 ranging from 0 to 300 $[mg/cm^3]$—see for details [6, 40]. During CT scans, the phantom was placed as close as possible to the bone to minimize errors introduced by none-uniformity of the CT numbers within the scan filed. Under different CT conditions the HU to ρ_{EQM} relation may be also different (see Fig. 3). Using a calibration phantom as a correlation factor is necessary to obtain accurate results and assure to provide same material evaluation for the same scanned object. An excellent linear correlation between HUs and corresponding concentrations of K_2HPO_4 is always established.

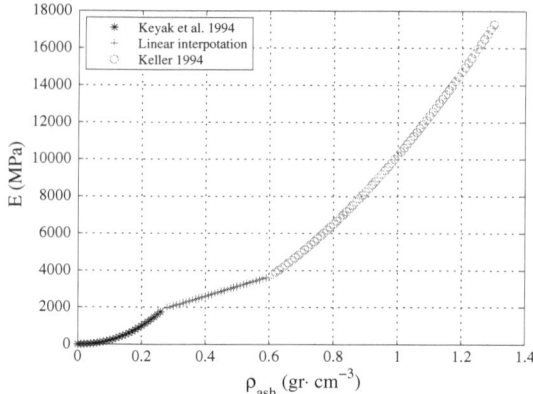

Fig. 2 $E(HU)$ connections in (3–5) according to [24]

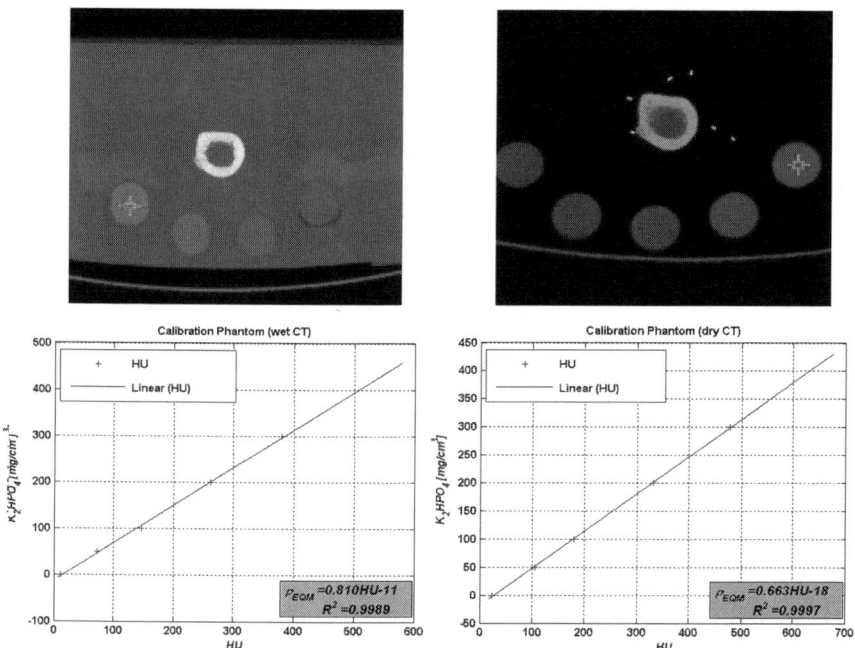

Fig. 3 ρ_{EQM} to HU relation. *Left*: Wet CT scan of bone's shaft and K_2HPO_4 calibration phantoms and the linear correlation; *right*: Dry CT scan. (Figure from [37])

The E value at every Gauss point is computed using a weighted point average (WPA) method from the qCT file: eight vertices of a cube in which the Gauss point is located are identified (see Fig. 4) and the value at the Gauss point is computed by its relative distance from vertices. For CT-scans with high enough resolution

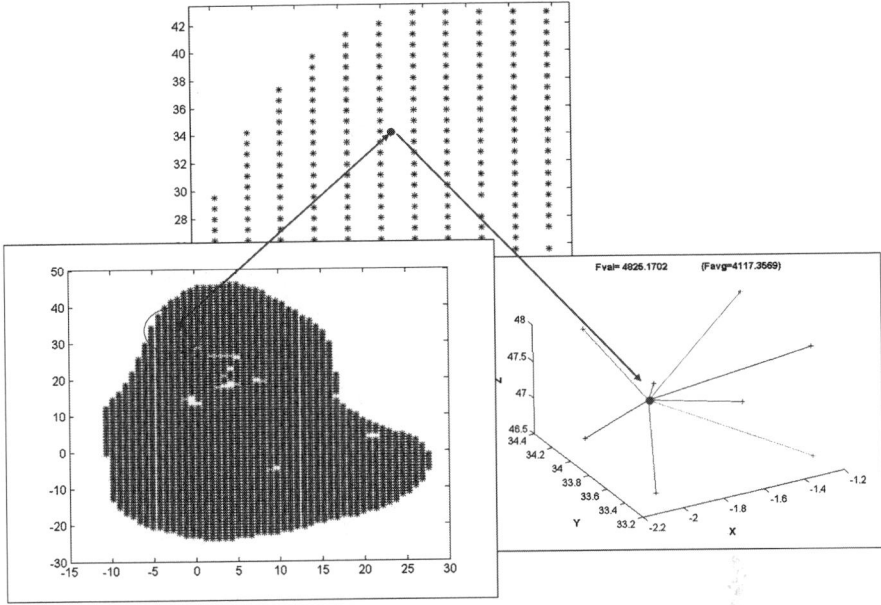

Fig. 4 Weighted point average: Identifying the point in the 3D array, Gauss point value is set as the average according to its surrounding. (Figure from [37])

the algorithm is more efficient if the value at the closest Gauss point in the CT scan is chosen instead, with very little influence on the results.

2.3 Micro-Mechanics-Based Orthotropic Material Properties

Since the bone-tissue is definitely not-isotropic on one hand, and a clinical qCT scan provides only the HU at a point (one data point) on the other hand (which does not allow to determine the 9 material properties for an orthotropic material), several assumptions and a more sophisticated approach has to be applied. A continuum micro-mechanics-based (MM-based) model (see also [41, 19]) may be applied on the qCT scans to determine (non-empirical) relations between orthotropic elasticity tensor components and HU. It is based on two consecutive steps [15, 19]:

- Based on voxel average rules for the attenuation coefficients, we assign to each voxel the volume fraction occupied by water (marrow) and that occupied by solid bone matrix. The volume fraction is identical to the vascular porosity, as given by:

$$\phi(x) = \begin{cases} \frac{HU_{BM}-HU(x)}{HU_{BM}} & \forall HU \leq 1600 \\ 0 & \text{otherwise} \end{cases} \qquad (6)$$

We denote by x the position of the individual voxel with $HU = 0$ representing pure water, and $HU_{BM} \geq 1600$ represents a "perfectly compact" bone in a human femur (see [41] for details). The lower HU values refer to very porous trabecular bone, with a vascular porosity (ϕ) close to 100%. At the upper end of the HU, values are identified as associated with vanishing vascular porosity, $\phi \sim 0$.

- By means of a MM model for bone based on mechanical properties of solid bone matrix and of water, we convert ϕ into voxel-specific orthotropic material tensor components. The model, cast in the framework of random homogenization theory [43], is of the Mori-Tanaka type, so that the effective stiffness tensor $\mathbb{C}_{eff}$ of the bone at position x is given by:

$$\mathbb{C}_{eff} = \{\phi \mathbb{C}_{H_2O} : [\mathbb{I} + \mathbb{P}_{cyl} : (\mathbb{C}_{H_2O} - \mathbb{C}_{BM})]^{-1} + (1-\phi)\mathbb{C}_{BM}\} :$$
$$: \{\phi[\mathbb{I} + \mathbb{P}_{cyl} : (\mathbb{C}_{H_2O} - \mathbb{C}_{BM})]^{-1} + (1-\phi)\mathbb{I}\}^{-1} \qquad (7)$$

where $\mathbb{I}$ is the fourth-order identity tensor (9), $\mathbb{P}_{cyl}$ is the fourth-order Hill tensor (10) accounting for the cylindrical pore shape in a bone matrix of stiffness $\mathbb{C}_{BM}$ (15), $\mathbb{C}_{H_2O}$ (8) is the bulk elastic stiffness, ":" denotes the second-order tensor contraction and:

$$\mathbb{C}_{H_2O} = 3 \cdot k_{H_2O} \mathbb{J} \text{ where } J_{ijkl} = \frac{1}{3}\delta_{ij}\delta_{kl} \text{ and } k_{H_2O} = 2.3 [GPa] \qquad (8)$$

$$I_{ijkl} = (\delta_{ik}\delta_{jl} + \delta_{il}\delta_{jk})/2 \qquad (9)$$

$$\mathbb{P}_{cyl} = \frac{1}{2\pi}\int_0^{2\pi} \mathbb{B} d\Phi \qquad (10)$$

$$B_{ijkl} = \frac{1}{4}(\xi_i \bar{G}_{jk}\xi_l + \xi_j \bar{G}_{ik}\xi_l + \xi_i \bar{G}_{jl}\xi_k + \xi_j \bar{G}_{il}\xi_k) \qquad (11)$$

$$\bar{G} = K^{-1} \qquad (12)$$

$$K = \xi \cdot \mathbb{C}_{BM} \cdot \xi, \qquad K_{jk} = \xi_i C_{BM,ijkl} \xi_l \qquad (13)$$

$$\xi = \cos\Phi \hat{e}_1 + \sin\Phi \hat{e}_2 + \hat{e}_3 \qquad (14)$$

where "$\cdot$" denotes the first-order tensor contraction (also called inner product). The bone-matrix material tensor obtained at the micro level by nanoindentation techniques [14] was used very successfully in [36] for the analysis of the proximal femur:

$$\mathbb{C}_{BM} = \begin{pmatrix} 20.8 & 10.4 & 16.6 & 0 & 0 & 0 \\ 10.4 & 22.1 & 16.5 & 0 & 0 & 0 \\ 16.6 & 16.5 & 41.9 & 0 & 0 & 0 \\ 0 & 0 & 0 & 15.4 & 0 & 0 \\ 0 & 0 & 0 & 0 & 11.2 & 0 \\ 0 & 0 & 0 & 0 & 0 & 9.4 \end{pmatrix} [GPa] \quad (15)$$

The resulted inhomogeneous orthotropic material properties are along material trajectories, which are determined in the following subsection.

2.3.1 Comparison of Empirical-Based Isotropic and MM-based Orthotropic Material Properties

The MM-based orthotropic constants are compared to the empirical isotropic Young's modulus given by (3)–(4) in Fig. 5 (empirical shear modulus is computed with ($v = 0.3$)). In the longitudinal direction (E_3) is higher for $HU < 1400$ while in the transverse detections (E_1 and E_2) are smaller compared to the isotropic empirical E, whereas the Poisson ratio is clearly nonconstant. Similar shear modulus are obtained for $HU < 600$. One can notices that $(0.9 < E_1/E_2 < 1)$ and $(0.3 < E_2/E_3 < 0.6)$, supporting our observation that the longitudinal direction is the "stiff" one and that the transverse direction is rather isotropic.

2.3.2 Determination of Material Trajectories for the MM-based Orthotropic Properties

The (vectorial) material trajectories cannot be determined clearly from a clinical CT scan thus additional information like the characteristic density distributions within the bony organ may be used. Herein, we take an assumption that the inhomogeneous (voxel-specific) material trajectories are aligned with the principal strains (see e.g. [27]). The longitudinal direction (axis-3) is considered as the "stiff" material direction (having the largest Young's modulus) and the other two transverse directions as weak (having the smaller Young's modulus). The transverse plane is rather isotropic (E_1 is close in value to E_2).

The main assumption is that bone tissue orientation correlates well to principal strain direction [21]. The stance position loading (the magnitude of load does not influence the principal directions but only the magnitude of the strains) is used for approximating the principal strain directions assuming a homogeneous isotropic material. This approach is also motivated by the recent experimental results in [11], which show that the principal strain directions on femur's surface is almost independent on the loading conditions on femur's head, covering the range of directions spanned by the hip joint force. Although the principal strains magnitude varied greatly between loading configurations,

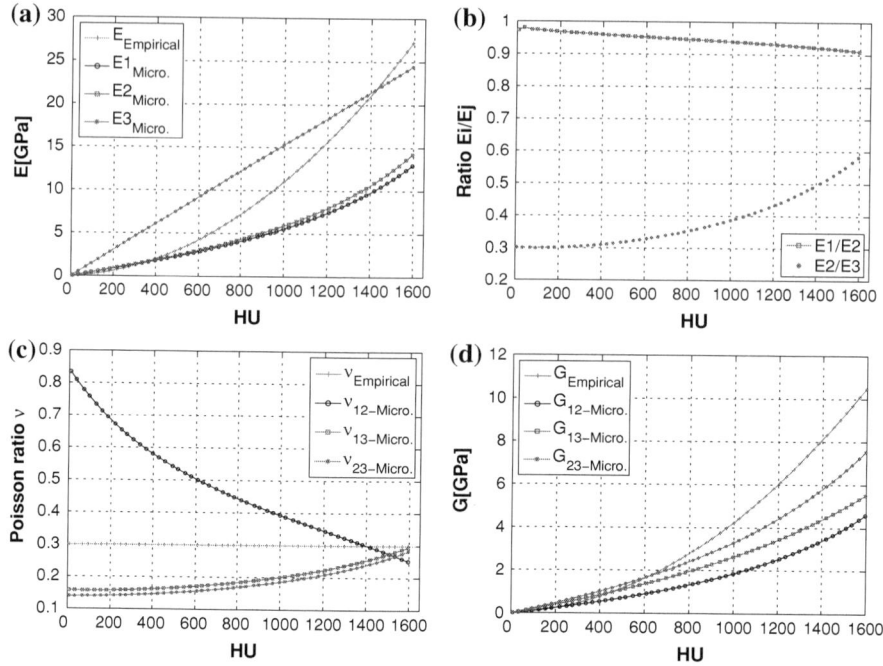

Fig. 5 a $E(HU)$ relation for orthotropic MM-based and isotropic empirical-based models (3)–(4). **b** Ratio of Young's modulus in different directions. **c** Poisson ratio dependence on HU for MM-based model ($v = 0.3$ for empirically based model). **d** Shear moduli relation to HU for MM-based and empirically-based models. (Figure from [36])

the principal strain direction varied very little. This suggests that the anatomy and the distribution of anisotropic material properties in the proximal femur is probably not strongly affected by the various loading directions, and can be determined from the direction of the principal strains resulting from the stance loading configuration.

At each point of interest the "stiffest" direction 3 is associated with the largest absolute value of the principal strain. For each x, the coordinate system along principal strain directions is denoted by $X_\alpha^m, \alpha = 1, 2, 3$ (3 is the "stiffest"). Since the material tensor $\mathbb{C}^m$ is assumed to be along X_α^m it is transformed into the femur's coordinate system by [32]:

$$C_{ijkl} = C_{\alpha\beta\gamma\delta}^m \ell_{\alpha i} \ell_{\beta j} \ell_{\gamma k} \ell_{\delta l} \tag{16}$$

where $\ell_{\alpha i} \equiv cos(X_\alpha^m, x_i)$.

For example, in most of the shaft region the "stiffest" trajectory is orientated along femur's x_3 axis and is in agreement with other studies [31]. At locations

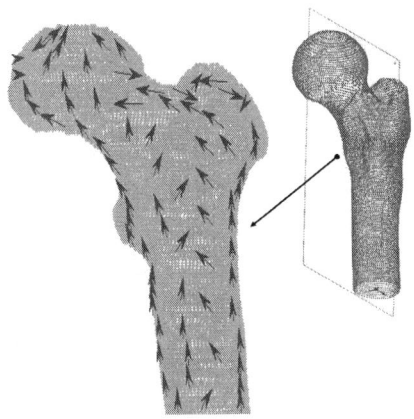

Fig. 6 Axial material trajectories throughout one of the analyzed FF femurs

close to femur's outer boundary in the head and neck regions, the directions are also coinciding with the outer boundary directions which is to be expected. In the central femoral's head and neck, some resemblance to "Wolf's law" is evident (see Fig. 6).

2.4 Verification of p-FE Results and Sensitivity Analyses

p-FE results are being verified so to ensure that the numerical error is under a specific tolerance. Convergence is realized by keeping a fixed mesh (with relatively large elements) and increasing the polynomial degree of the approximated solution p. To this end, the polynomial degree over the elements is increased until the relative error in energy norm is small, and the strains at the points of interest converge to a given value. Such a verification, for example for FF3, when increasing p from 1 to 5 is presented in Fig. 7. Each of the femurs's FE model consists between 3500 to 4500 elements ($\sim 150,000$ degrees of freedom (DOFs) at $p = 4$ and $\sim 300,000$ DOFs at $p = 5$). The proper material properties are assigned to each integration point. For an orthotropic material the material directions are also determined at each integration point.

2.4.1 Sensitivity Studies

To ensure the reliability of the FE analyses sensitivity studies were performed to ensure that the obtained results are not very sensitive to the changes: (a) several constant Poisson ratios $v = 0.01, 0.1, 0.3, 0.4$ were applied to the entire isotropic model, as in [8, 34, 40], (b) the distal face residing in PMMA was either clamped

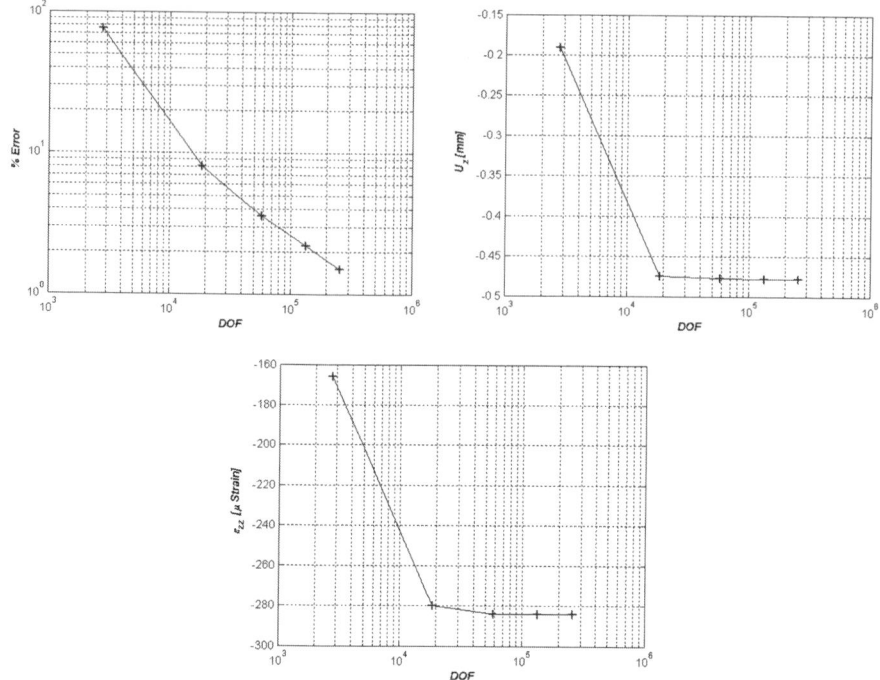

Fig. 7 Convergence in energy norm, head displacement and ε_{zz} at a representative point of interest in FF3. (Figure from [37])

or modeled as attached to a distributed spring, (c) strains at SGs locations were obtained in $\pm 5°$ offset orientations, (d) strains were obtained either as averaged over an element or as maximum or minimum values.

For example, the sensitivity to the loading was checked by applying it in the FE model in several ways so to mimic the in-vitro experiments (described in Sect. 3): (a) a pressure on a planar face that trimmed femur's head at two heights having 1 mm difference, (b) tractions on a circular surface determined by the interface of the cone and femur's head. In both cases the head was free to move perpendicular to the load (see Fig. 8).

3 In-Vitro Experiments

Biomechanical experiments on seventeen fresh-frosen human cadaver femurs were conducted to validate the FEA. Two sets of experiments were used for validation, one set includes five femurs (denoted FF1-FF5) for which experiments were performed in-house, and one includes six pairs of femurs (denoted by 1–6) tested

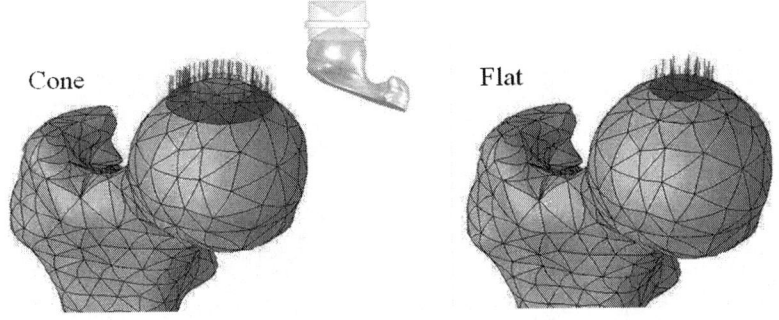

Fig. 8 Load configurations applied at an angle of $0°$ to the shaft axis

Table 1 Data of femurs and CT scan resolution

Donor label	Side	Age (years)	Height (cm)	Weight (kg)	Gender	Slice thickness (mm)	Pixel size (mm)	Load rate (mm/sec)
1	L & R	59	180	96	Female	1.00	0.547	1/6
2	L & R	53	193	98	Male	1.00	0.488	1/6
3	L & R	48	170	55	Male	1.00	0.488	1/6
4	L & R	64	168	136	Female	1.00	0.488	1/6
5	L & R	54	178	161	Male	1.00	0.547	1/6
6	L & R	58	185	86	Male	1.00	0.547	1/6
FF1	L	30	N/A	N/A	Male	0.75	0.78	1/600–1/30
FF2	R	20	N/A	N/A	Female	1.5	0.73	1/600, 1/120, 1/6
FF3	L	54	N/A	N/A	Female	1.25	0.52	1/2
FF4	R	63	N/A	N/A	Male	1.25	0.195	1/60, 1/6, 1
FF5	R	56	N/A	N/A	Male	1.25	0.26	1/2

by a research institute in Germany and results were unknown until analyses were completed to avoid any bias. Table 1 summarizes the data on femurs and CT scan resolutions.

FF1-FF5 femurs were scanned by a Phillips Brilliance 16 CT (Einhoven, The Netherlands) with following parameters: 120–140 kVp, 250 mAs, 0.75–1.5 mm slice thickness, axial scan without overlap, with pixel size of 0.2–0.7 mm. Thereafter, and within one day since defrosting experiments were conducted to mimic a simple stance position configuration in which the femurs were loaded through their head while inclined at different inclination angles (0, 7, 15 and 20 degrees) as shown in Fig. 9. We measured the vertical and horizontal displacements of femur's head, the strains at the inferior and superior parts of the neck, and on the medial and lateral femoral shaft. Between 5 and 10 strain-gauges were bonded on each of the tested femurs. In all experiments a linear response between force and displacements and strains was observed beyond 200 N preload. The experimental error is within a $\pm 5\%$ range. This range was estimated by the measurements error (calibration to known displacements/loads/strains), deviation

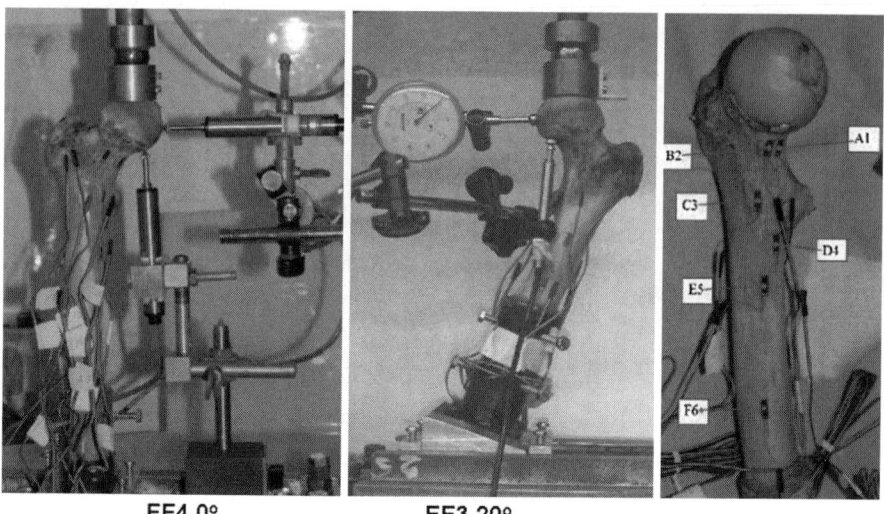

Fig. 9 Typical experiments on FF1-FF5 at $0°$ and $20°$ inclination angles. Right - Representative strain gauges location at the neck and shaft regions

between consecutive measurements and estimation of the linear response (details are provided in [42, 37, 36]).

The six pairs 1–6 were qCT-scaned by a GE Healthcare machine (Lightspeed VCT, WI, USA) including a mineral density calibration phantom (HEAD CT Calibration Phantom, Mindways, Austin, TX, USA) to determine the relation between Houndsfield units (HU) and bone mineral density (BMD). Five uniaxial strain-gauges (SGs) were bonded to the surface of the femurs at the superior neck, the inferior neck, the lesser trochanter, the medial shaft and the lateral shaft (Fig. 10). Optical markers were distributed over the specimen, the adapter of the testing machine, and the cardan joint. The axes of the SGs were aligned with the femoral neck axis or the femoral shaft axis, respectively to measure compressive or tensile strains during axial compression. The distal end of each femur was potted with casting resin in an aluminum case that fitted into a cardan joint so that the line of force went through the center of the femoral head and the center of the epicondyles. The femoral head was potted in a hemisphere of casting resin that fitted the proximal adapter of the test setup (Fig. 10b). Femurs were then loaded at a rate of 10 mm/min by a servo-electric testing machine in a single leg standing position configuration [12]. Three cycles from 0 to 250 N, three cycles from 0 to 500 N, three cycles from 0 to 750 N and three cycles from 0 to 1000 N were applied on each femur. Local displacements at points on the bone were measured by an optical sensor at the optical markers. Eight to nine optical markers per specimen were distributed over the frontal plane of the bones (Fig. 10). The mean value of strains were calculated per load step. These mean values were used for the later comparison with the FEA. The total displacements $u_{tot} = \sqrt{u_x^2 + u_y^2 + u_z^2}$ of

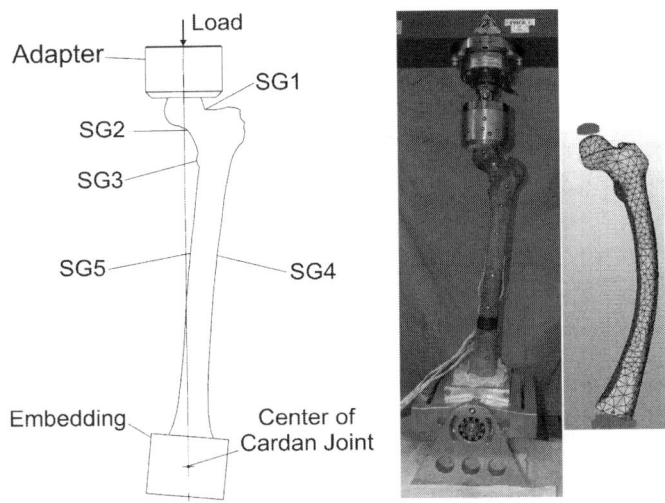

Fig. 10 a (*left*) Sketch of the frontal plane of an embedded and instrumented left femur. The adapter applied the load by the testing machine to the specimen. The proximal embedding (hidden in adapter) builds a ball-joint with the adapter. Strain gauges (SG1-SG5) are applied to specific anatomic sites. SG1 - located at the middle of the superior neck. SG2 - located opposite to SG1 at the inferior neck. SG3 - located next to the most prominent part of the lesser trochanter. SG4 - located 100 mm distally to SG3 at the medial side of the shaft. SG5 - located opposite to SG4 at the lateral side of the shaft. **b** (*right*) Experimental setup with the optical markers on an instrumented left femur and its corresponding deformed (magnified) FE model (Figure from [38])

the optical markers on the bone surface were also calculated. Details on these experiments are provided in [38].

Remark 3 Note that in all experiments no difference in femur's mechanical response was noticed for an applied load rate of $1/600 - 1\text{mm}/\text{sec}$.

4 Validation: Comparison Between the *p*-FE Results and Experimental Observations

4.1 Isotropicic p-FE Models

The verified FE analyses (with inhomogeneous isotropic properties) that mimic the in-vitro experiments are used for validation purposes, i.e. to ensure that these indeed represent the biomechanical response. For each FE analysis the strains at the location of the strain-gauges (SGs) were averaged over a small area representing the area over which the gauge or the SG measured the strains. Because uni-axial SGs were used in our experiments, we considered the FE-strain

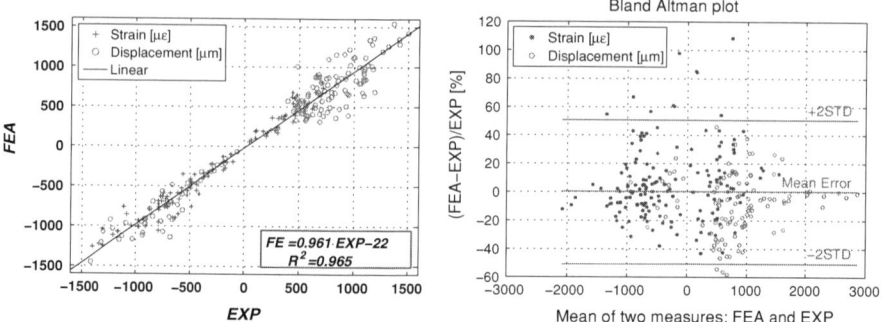

Fig. 11 Comparison of the computed strains +, ∗ and displacements ○ to the experimental observations normalized to 1000 N load. Inhomogeneous isotropic material properties (*Left*) Linear regression, (*Right*) Bland-Altman plot

component in the direction coinciding with the SG direction (in most cases the SGs are aligned along the principle strain directions). A total of 102 displacements and 161 strains on the 17 femurs were used to assess the validity of the p-FE simulations with isotropic material properties. In Fig. 11 the pooled FE strains and displacements are compared to the the experimental observations—this comparison is demonstrated by two different statistical measures: By a linear regression plot, inspecting the slope, intersection and R^2 of the linear regression between the experimental observations and FE predictions, and by a Bland-Altman plot [5] emphasising the errors between the two, and the 95% confidence limits.

Remark 4 Note that for twelve of the seventeen femurs a blinded comparison was performed, i.e. the group that performed the experiments did not know the FE results, and vice-versa, the experimental results were not known by the group that performed the analysis.

One may notice the unprecedented match between the predicted and measured data for femurs under stance position loading: the slope and R^2 of the linear regression are very close to 1, and the average error in the Bland-Altman is zero. It is evident from the Bland-Altman plot that the large discrepancies between experiments and analyses occur at locations where the strains are small. This is because minor changes in the strains result in large relative errors.

In Fig. 12 we consider the predicted strains match alone and predicted displacements match alone. As expected, the strains prediction is highly accurate, but not less important, the displacements are also well predicted.

The validated FE results may be also utilized to investigate the internal state of strains within the femur—for example, Fig. 13 shows the maximum (tensile) and minimum (compression) principal strains at a cutting plane within FF5.

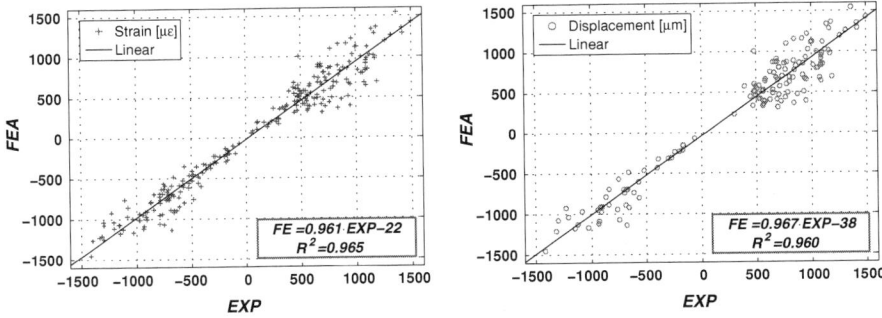

Fig. 12 Comparison of the computed strains alone (*Left*) and displacements alone (*Right*) to the experimental observations. Inhomogeneous isotropic material properties

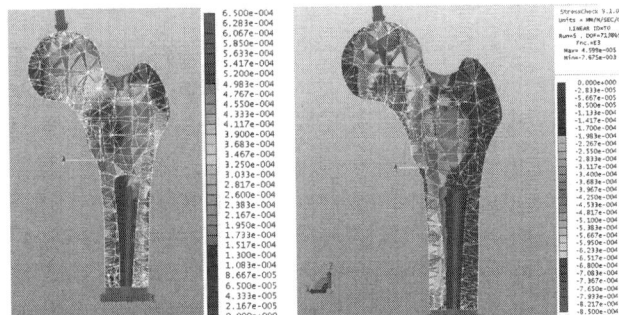

Fig. 13 Principle maximum strains (*left*) and minimum strains (*right*) in FF5 loaded at 7°

4.2 Orthotropic p-FE Models

A simplified uniaxial loading condition was applied on the femur's head in the in-vitro experiments thus an isotropic inhomogeneous assumption represents well the experimental observations, all being along the principal direction. A more realistic (physiological) and complex load scenario is expected to require an orthotropic model to well simulate the bone response.

We demonstrate that *p*-FE models of femurs *FF1-FF4* with orthotropic material properties under the simple loading conditions yield results in accordance with the experimental observations, and thus are similar to the FE models having isotropic material properties. In Table 2 we compare the linear regression parameters (orthotropic vis experiments and isotropic vis experiments). Separate comparison for different inclination angles ($0°$ and $20°$) is also presented. The results are divided into three main groups of interest: strains at the femur's neck region, strains at the femur's shaft region and displacements of femur's head. To quantify and emphasize the differences between the isotropic versus the orthotropic model, a comparison between the FE results based on the different material models is presented in Table 3.

Table 2 Summary of linear regression: FE results compared to experimental observations

Region and Tilt	Material model	Slope	R^2
All results	ISO.	1.01	0.96
Tilt 0°	ORTH.	1.09	0.96
All results	ISO.	0.88	0.91
Tilt 20°	ORTH.	1.04	0.88
All results	ISO.	0.98	0.95
Tilt 0°, 20°	ORTH.	1.08	0.94
All SGs	ISO.	1.02	0.96
Tilt 0°	ORTH.	1.10	0.96
All SGs	ISO.	0.87	0.90
Tilt 20°	ORTH.	1.03	0.87
All SGs	ISO.	0.98	0.94
Tilt 0°, 20°	ORTH.	1.09	0.93
Displacements	ISO.	0.99	0.98
Tilt 0°, 20°	ORTH.	1.0	0.99
Shaft region	ISO.	1.1	0.94
Tilt 0°, 20°	ORTH.	1.06	0.93
Neck region	ISO.	0.88	0.96
Tilt 0°, 20°	ORTH.	1.09	0.94

* All Results = BOTH strains and displacements, SGs- strain gauges results

Table 3 Linear regression: isotropic vs. orthotropic FE results (Tilt 0° & 20°)

Region of interest	Slope	R^2
All results	1.08	0.97
All SGs	1.09	0.97
Displacements	1.01	0.99
Shaft region	0.97	0.99
Neck region	1.24	0.97

To assess if the orthotropic and isotropic models yield similar results for a more complex state of loading we applied on the same FE models a compression load on femur's head in addition to a torsional load (resulting in a moment along the z axis). Because experiments with complex loading configurations on fresh-frozen human femurs are unavailable at this time, we only address the numerical results. Comparing the FE predicted strains by the different material models show a significant differences between them ($Slope = 1.35, R^2 = 0.96$, and $Slope = 1.19, R^2 = 0.99$ for displacements). This numerical study demonstrates that when a more complex loading condition is applied, the FE predicted mechanical response of the femur is considerably different (the orthotropic model is significantly less stiff). Nevertheless, since no experimental observations for such loading is available, it is impossible to assess at this time which of the two models better represents the reality.

5 Discussion, Clinical Applications for Prostheses and Implants and Further Required Research

New algorithms that use patient-specific qCT-scans in conjunction with p-FE methods, together with advanced methods for assigning inhomogeneous isotropic or orthotropic material properties to the FE models are presented. Due to the semi-automatic procedures, the required timeframe to manipulate the qCT-scans and complete the FE analysis of a femur with isotropic inhomogeneous material properties is less than three hours. The methods allow to keep the discretization errors under control, enabling to focus the attention on the idealization errors. The validation of the FE-results was performed on 17 fresh frozen femurs demonstrating an excellent agreement between the FE results and experiments, which to the best of the authors' knowledge are more accurate as reported in the literature when strains are addressed, and moreover a very good agreement is noticed in the displacements also (not reported in studies by other authors to the best of our knowledge). Thanks to the double-blinded validation performed on 12 femurs [38], this validation process is also bias-free.

The differences between the isotropic model and the orthotropic MM model are in the range of the experimental errors so unequivocal conclusions cannot be drawn regarding which model is better. Both models seem to provide very good correlation to experimental observations under the simplified loading condition.

The mechanical response of the fresh-frozen femurs was proven to be insensitive to the strain rate when loading was applied at a rate of 0.0017–1 mm/sec. The yield/failure onset, on the other hand, is quite sensitive to the applied rate (strain rate), and evidence is available that at high loading rates as 10–20 mm/sec, the global mechanical response is almost linear elastic up to fracture.

As a potential clinical application, we briefly present herein a recent study that investigates the strain/stress state in a femur that undergo arthoplasty so that a cemented-prosthesis is inserted. Of-course that we validate our p-FE results by performing an in-vitro experiment on FF5 after being fractured and the prosthesis inserted. The fresh frozen bone FF5, has been qCT scanned, loaded and the p-FE model was demonstrated to very well predict the mechanical response in an in-vivo test as shown in Fig. 14. After mechanical tests were completed, the femur was loaded until fracture occurring at the femur's neck. The fractured femur was "fixated" by a "total hip replacement" procedure, during which a metallic prosthesis with cement was placed in the remaining part of the femur. Four strain-gauges were placed on the prosthesis in addition to the existing strain gauges on femur's surface. The fixated bone was qCT scanned (where the metallic prosthesis smeared the qCT scan so we had to manipulate the scans to identify more precisely the prosthesis location and cement), after which a p-FE analysis and an in-vitro experiment was performed on the fixated bone as shown in Fig. 15. Again, excellent correlation was observed between the experimental observations and the p-FE results. The p-FE model can be used to investigate the stress/strain field in the bone after prosthesis is inserted to identify if stress-shielding occurs and to optimize the size/location/type of

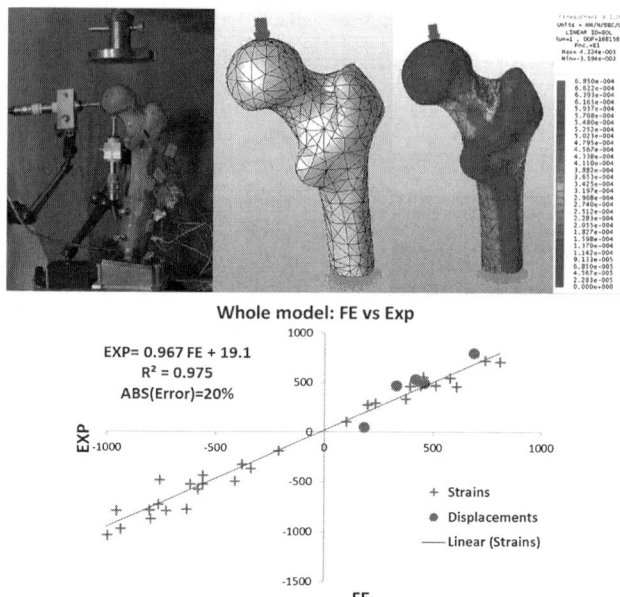

Fig. 14 p-FE model and principle stress of intact FF5 femur validated by an in-vitro experiment

such prostheses prior to surgical intervenience. This case study is provided to demonstrate how such analyses may be used in clinical practice to identify an optimal position and shape of the prosthesis to reduce stress-shielding.

The following future research activities are required to further corroborate the simulation results:

(a) Validation of the p-FE models for femurs under multiple and more complex loading conditions is required especially for the orthotropic ones. We investigated by in-vitro tests only a stance position loading where axial load and bending moment were excited in the bone. Future in-vitro experiments resulting in a more complex state of stresses (a more realistic physiological load) is necessary to support our conclusions. Such a future study may examine if indeed the inhomogeneous orthotropic material properties are necessary (as opposed to a simplified isotropic assumption) for a reliable simulation of the mechanical response, especially in the head and neck locations of the proximal femur. This is of major importance to failure analysis of osteoporotic bones, and the possibility to predict such failures as a function of bones' density and local geometry. "More physiological" loadings may have a large influence on femur's mechanical response [11] thus p-FE models loaded according to [3, 17, 18] must be compared to to in-vitro experiments.

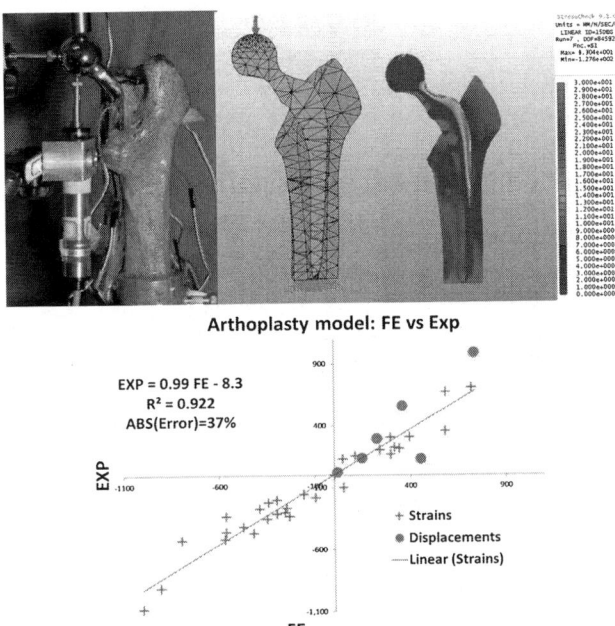

Fig. 15 p-FE model and principle stress of the fractured FF5 femur with the prosthesis inserted validated by an in-vitro experiment. A view through a surface in the middle of the bone

(b) The orthotropic MM model is based on a clinical CT scan and does not accurately represent the trabecular bone morphology. A more detailed analysis of the femur, especially in the trabecular zone, may be achieved, for example, by using μCT-scans, which are of resolutions of an order of 20 μm and may provide an accurate description at the lamella-level [29]. Such μFEs are applied to the femur in [39] demonstrating that the continuum FE model results in same strain accuracy as the μFE models. The μCT-scans are unfortunately unapplicable in clinical practice due to the extensive radiation capacity, and thus may only be used on cadaver specimens thus other clinical methods have to be sought instead.

(c) p-FE simulations must be demonstrated to perform well on a variety of clinical pathological scenarios to enhance their potential use in clinical practice.

(d) The $E - \rho$ relationships may be further investigated. A systematic and well accepted protocol for material properties evaluation is necessary. A clear flow of mathematical relations from the CT-data to HU and Bone Mineral Density (BMD) based calibration phantoms to bone ash or apparent density to material constants is still lacking. Anisotropic such relations are even less established. The use of the fabric tensor [9, 26, 10] may be the advocated method for achieving this aim. The fabric tensor is also associated with the determination of patient specific material trajectories—new technologies have to be investigated so to determine these in clinical practice.

(e) Although the mechanical response may be well determined, still a reliable failure criterion for bone tissue is lacking. This leads to very limited capabilities to predict fracture in the whole bone which requires a 3-D failure laws in an anisotropic setting. Such failure criteria for bone tissue and failure laws in a 3-D stress state setting are most probably rate dependent.
(f) A better communication path with the medical community is lacking. The mechanical simulation results must be presented in a manner which is fast and useful for clinical decision making. Especially, the use of measures as "bone stiffness" and "fixation stability" should be rephrased and more meaningful and practical measures should be used instead.

In spite of the aforemention still missing information, this study exemplifies that p-FEMs based on qCT scans are reliable enough and in an advanced stage to be used in clinical practice for patient-specific in-vivo validation studies.

Acknowledgements We would like to thank Prof. Charles Milgrom from the Hadassah Hospital, Jerusalem, for femurs supply and his help with the CT-scans and experiments. Special thanks are extended to Mr. Alon Katz, a graduate student under the supervision of the first author, for his help with FE analyses and experiments. The authors gratefully acknowledge the generous support of the Technische Universität München–Institute for Advanced Study and International Graduate School of Science and Engineering, funded by the German Excellence Initiative, which made parts of this research possible.

References

1. Alho, A., Husby, T., Hoiseth, A.: Bone mineral content and mechanical strength An ex-vivo study on human femora and autopsy. Clin. Orthop **227**, 292–297 (1988)
2. Bayraktar, H.H., Morgan, E.F., Niebur, G.L., Morris, G.E., Wong, E.K., Keaveny, M.: Comparison of the elastic and yield properties of human femoral trabecular and cortical bone tissue. J. Biomech. **37**, 27–35 (2004)
3. Bergmann, G., Deuretzbacher, G., Heller, M.O., Graichenm, F., Rohlmann, A., Strauss, J., Haas, N.P., Duda, G.N.: Hip contact forces and gait patterns from routine activities. J. Biomech. **34**, 859–871 (2001)
4. Bessho, M., Ohnishi, I., Matsuyama, J., Matsumoto, T., Imai, K., Nakamura, K.: Prediction of strength and strain of the proximal femur by a CT-based finite element method. J. Biomech. **40**(8), 1745–1753 (2007)
5. Bland, J.M., Altman, D.G.: Statistical methods for assessing agreement between two methods of clinical measurement. Lancet **1**(8476), 307–310 (1986)
6. Cann, C.E.: Quantitative CT for determination of bone mineral density: a review. Radiology **166**, 509–522 (1988)
7. Carter, D.R., Hayes, W.C.: The compressive behavior of bone as a two-phase porous structure. J. Bone Joint Surg. Am. **59**, 954–962 (1977)
8. Cody, D.D., Hou, F.J., Divine, G.W., Fyhrie, D.P.: Short term in vivo study of proximal femoral finite element modeling. Ann. Biomed. Eng. **28**, 408–414 (2000)
9. Cowin, S.C.: The relationship between the elasticity tensor and the fabric tensor. Mech. Mater. **4**, 137–147 (1985)
10. Cowin, S.C.: Anisotropic poroelasticity: fabric tensor formulation. Mech. Mater. **36**, 665–677 (2004)

11. Cristofolini, L., Juszczyk, M., Taddei, F., Viceconti, M.: Strain distribution in the proximal human femoral metaphysis. Prof. Inst. Mech. Eng. Part H: J. Eng. Med. **223**, 273–288 (2009)
12. Eberle, S., Gerber, C., von Oldenburg, G., Hungerer, S., Augat, P.: Type of hip fracture determines load share in intramedullary osteosynthesis. Clin. Orthop. Rela. Res. **467**, 1972–1980 (2009)
13. Esses, S.I., Lotz, J.C., Hayes, W.C.: Biomechanical properties of the proximal femur determined in vitro by single-energy quantitative computed tomography. J. Bone Miner. Res. **4**, 715–722 (1989)
14. Franzoso, G., Zysset, P.K.: Elastic anisotropy of human cortical bone secondary osteons measured by nanoindentation. ASME J. Biomech. Eng. 021001/1–11 (2009)
15. Fritsch, A., Hellmich, C.: Universal microstructural patterns in cortical and trabecular, extracellular and extravascular bone materials: micromechanics-based prediction of anisotropic elasticity. J. Theor. Biol. **244**, 597–620 (2007)
16. Heismann, B.J., Leppert, J., Stierstorfer, K.: Density and atomic number measurements with spectral x-ray attenuation method. J. App. Phys. **94**, 2073–2079 (2003)
17. Heller, M.O., Bergmann, G., Deuretzbacher, G., Durselen, L., Pohl, M., Claes, L., Haas, N.P., Duda, G.N.: Musculo-skeletal loading conditions at the hip during walking and stair climbing. J. Biomech. **34**, 883–893 (2001)
18. Heller, M.O., Bergmann, G., Kassi, J.-P., Claes, L., Haas, N.P., Duda, G.N.: Determination of muscle loading at the hip joint for use in pre-clinical testing. J. Biomech. **38**, 1155–1163 (2005)
19. Hellmich, C., Kober, C., Erdmann, B.: Micromechanics-based conversion of CT data into anisotropic elasticity tensors, applied to FE simulations of a mandible. Ann. Biomed. Eng. **36**, 108–122 (2008)
20. Henninger, H.B., Reese, S.P, Anderson, A.E., Weiss, J.A.: Validation of computational models in biomechanics. Proc. IMechE, Part H: Eng. Med. **224**(H7), 801–812 (2010)
21. Hert, J., Fiala, P., Petrtyl, M.: Osteon orientation of the diaphysis of the long bones in man. Bone **15**(3), 269–277 (1994)
22. Keller, T.S.: Predicting the compressive mechanical behavior of bone. J. Biomech. **27**, 1159–1168 (1994)
23. Keyak, J.H., Meagher, J.M., Skinner, H.B., Mote, C.D. Jr: Three-dimensional finite element modelling of bone: a new method. ASME J. Biomech. Eng. **12**, 389–397 (1990)
24. Keyak, J.H., Falkinstein, Y.: Comparison of in situ and in vitro CT scan-based finite element model predictions of proximal femoral fracture load. Med. Eng. Phys. **25**, 781–787 (2003)
25. Morgan, E.F., Bayraktar, H.H., Keaveny, T.M.: Trabecular bone modulus-density relationships depend on anatomic site. J. Biomech. **36**, 897–904 (2003)
26. Odgaard, A.: Three-dimensional methods for quantification of cancellous bone architecture. Bone **20**(4), 315–328 (1997)
27. Pietruszczak, S., Inglism, D., Pande, G.N.: A fabric-dependent fracture criterion for bone. J. Biomech. **32**(10), 1071–1079 (1999)
28. Pise, U.V., Bhatt, A.D., Srivastava, R.K., Warkedkar, R.: A B-spline based heterogeneous modeling and analysis of proximal femur with graded element. J. Biomech. **42**, 1981–1988 (2009)
29. Ruegsegger, P., Koller, B., Muller, R.: A microtomographic system for the nondestructive evaluation of bone architecture. Calcified Tissue Int. **58**(1), 24–29 (1996)
30. Schileo, E., DallAra, E., Taddei, F., Malandrino, A., Schotkamp, T., Baleani, M., Viceconti, M.: An accurate estimation of bone density improves the accuracy of subject-specific finite element models. J. Biomech. **41**, 2483–2491 (2008)
31. Schneider, R., Faust, G., Hindenlang, U., Helwig, P.: Inhomogeneous, orthotropic material model for the cortical structure of long bones modelled on the basis of clinical CT or density data. Comput. Meth. Appl. Mech. Engrg. **198**, 2167–2174 (2009)
32. Sokolnikoff, I.S.: Mathematical Theory of Elasticity. McGraw-Hill, New York (1956)
33. Szabó, B.A., Babuska, I.: Finite Element Analysis. John-Wiley, New York (1991)

34. Taddei, F., Cristofolini, L., Martelli, S., Gill, H.S., Viceconti, M.: Subject-specific finite element models of long bones: an in vitro evaluation of the overall accuracy. J. Biomech. **39**, 2457–2467 (2006)
35. Taddei, F., Schileo, E., Helgason, B., Cristofolini, L., Viceconti, M.: The material mapping strategy influences the accuracy of CT-based finite element models of bones: an evaluation against experimental measurements. Med. Eng. Phys. **29**(9), 973–979 (2007)
36. Trabelsi, N., Yosibash, Z.: Patient-specific fe analyses of the proximal femur with orthotropic material properties validated by experiments. ASME J. Biomech. Eng. **133**(6), 061001-1–061001-11 (2011)
37. Trabelsi, N., Yosibash, Z., Milgrom, C.: Validation of subject-specific automated p-FE analysis of the proximal femur. J. Biomech **42**, 234–241 (2009)
38. Trabelsi, N., Yosibash, Z., Wutteb, C., Augat, R., Eberle, S: Patient-specific finite element analysis of the human femur - a double-blinded biomechanical validation. J. Biomech. **44**, 1666–1672 (2011)
39. Verhulp, E., van Rietbergen, B., Huiskes, R.: Comparison of micro-level and continuum-level voxel models of the proximal femur. J. Biomech. **39**(16), 2951–2957 (2006)
40. Yosibash, Z., Padan, R., Joscowicz, L., Milgrom, C.: A CT-based high-order finite element analysis of the human proximal femur compared to in-vitro experiments. ASME J. Biomech. Eng. **129**(3), 297–309 (2007)
41. Yosibash, Z., Trabelsi, N., Hellmich, C.: Subject-specific p-FE analysis of the proximal femur utilizing micromechanics based material properties. Int. J. Multiscale Comput. Eng. **6**(5), 483–498 (2008)
42. Yosibash, Z., Trabelsi, N., Milgrom, C.: Reliable simulations of the human proximal femur by high-order finite element analysis validated by experimental observations. J. Biomech. **40**, 3688–3699 (2007)
43. Zaoui, A.: Continuum micromechanics: survey. J. Eng. Mech. (ASCE) **128**(8), 808–816 (2002)

Patient-Specific Diagnosis and Visualization of Bone Micro-Structures

L. Podshivalov, A. Fischer and P. Z. Bar-Yoseph

Abstract Bone is a hierarchical bio-material whose architecture differs at each level of hierarchy and whose mechanical properties can vary considerably, even on the same specimen, due to bone heterogeneity. Because of their complexity and large number of details, these models are considered to be large-scale models. Modeling, visualization and diagnosis of such models is challenging, since a large amount of data must be processed rapidly. Moreover, if physically based modeling is required, material properties are also included in the computational model in addition to geometrical data, making the task more difficult. Therefore, advanced technology and computational methods are required for efficient, reliable and robust visualization and diagnosis. In this chapter we describe state-of-the-art technologies and methods that facilitate the processing of bone structure at the micro-scale. Specifically, we relate to computational methods that enable structural analysis of this highly detailed structure for medical diagnosis.

1 Introduction

Visualization of bone micro-structures and diagnosis of bone diseases are two interconnected processes. Visualization techniques are driven by diagnostic needs, while the diagnosis process, which utilizes emerging medical imaging technology, dictates how the bone structure should be visualized. During the last two decades,

---

L. Podshivalov · A. Fischer (✉)
Laboratory for CAD and LCE, Technion, Haifa, Israel
e-mail: meranath@tx.technion.ac.il

L. Podshivalov · P. Z. Bar-Yoseph
Computational Biomechanics Laboratory, Technion, Haifa, Israel

visualization methods for bone-micro structure have developed considerably for a number of reasons: (a) Recently, medical imaging technology provides high-resolution scanning that capture micro-scale structures; (b) computation and graphics techniques have developed tremendously, enabling modeling and analysis of complex models; (c) our understanding of the medical and biological processes involved in modeling and remodeling of bone tissue has improved, thus leading to new algorithms and methods for geometric modeling, analysis and visualization. In this chapter, we review the evolution of the field of patient-specific visualization and diagnosis of bone micro-structure, focusing on medical imaging technology, geometric modeling, structural analysis and visualization.

1.1 Bone Hierarchical Structure

Bones are composed of hierarchical bio-composite materials characterized by complex multi-scale structural geometry and complex behavior. The outer shell of bone is known as *cortical bone* and is characterized by its strong and dense structure. The inner structure consists of small plates and rods (in Latin *trabecula*) arranged in complex three-dimensional structures. These structures, known as called *trabecular bone*, are characterized by their high strength-to-weight ratio. For many years, scientists have been interested in this complex structure and the processes that govern its formation, regeneration and destruction throughout life. In 1892, the anatomist and orthopaedic surgeon Julius Wolff published a work in which he defined "Wolff's Law" that predicted a relationship between the geometrical structure and mechanical properties of bone micro-structures [40]. It is known today that bone tissue structure can be classified into five structural levels, ranging from macro-scale to nano-components [60, 71]:

- Macro-structure (mm–μm): trabecular and cortical bone.
- Micro-structure (10–500 μm): osteons and trabeculae.
- Sub-microstructure (1–10 μm): lamellae and single trabecula.
- Nanostructure (100 nm–1 μm): fibrillar collagen and embedded mineral.
- Sub-nano-structure (< 100 nm): molecular structure of mineral, collagen, and non-collagenous organic proteins.

Figure 1 illustrates the hierarchical structure of bone tissue, ranging from organ-to sub-nano-scale. Depending upon the anatomical site, bone architecture differs significantly at the micro-structural level with respect to shape, thickness, element direction and size. For example, the rod-like structures at the vertebra differ significantly from the plate-like structure at the femoral head. Bone architecture can also vary at different locations on the same site as a result of functionality and locally applied forces characterized by magnitude and direction. Thus, each patient's bone structure is unique and is influenced by gender, age, and physical condition.

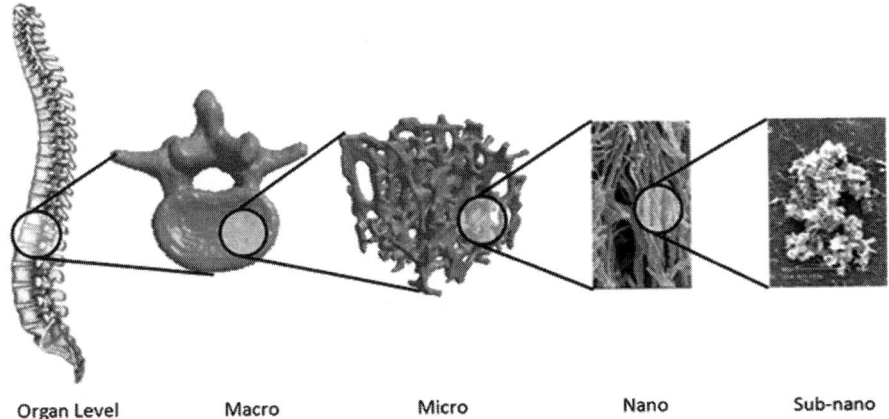

Fig. 1 Hierarchical structure of bone tissue (*from left to right*): spine, vertebra bone, trabecular bone structure, fibril array and bone crystals

As of today, due to technology limitations and limited understanding of the biological processes at other structural scales, macro-and micro-scales remain at present the only diagnostically relevant structures. Current research aims to define structure-mechanical relationships at the nano-scale [30] and to model the modeling and remodeling processes of the bone tissue at the cellular levels [41].

1.2 Diagnostic Approach

Diagnosis of bone micro-structures is especially relevant in cases of metabolic bone diseases. For example, metabolic bone diseases such as osteoporosis are characterized by micro-architectural deterioration of bone tissue, leading to micro fractures; therefore, early diagnosis at the micro-scale is a key to intervention. Current guidelines issued by the World Health Organization (WHO) define Bone Mineral Density (BMD) measurements as a default assessment tool for diagnosing osteoporosis. BMD results are reported as a T-score and a Z-score.

The T-score represents the number of standard deviations above or below the mean bone mineral density for sex and race matched to 30 year-olds. A Z-score compares the patient with a population adjusted for age, sex and race. In postmenopausal women, osteoporosis is defined as a BMD T-score of more than 2.5 standard deviations below the sex-adjusted mean for normal young adults at peak bone mass. The BMD result is used to classify patients into three categories: (a) normal; (b) osteopenic; and (c) osteoporotic. A fourth category is severe osteoporosis, which is not defined by BMD but rather by the presence of osteoporotic fractures. Patients with normal BMD values need no further therapy, osteopenic patients should be counseled and treated with preventive therapy, and

Table 1 Comparison of micro-scale medical imaging technologies

Technology	Method	Resolution	Advantages	Disadvantages
CT	pQCT	80 μm in vivo	Lower radiation than μCT method	Radiation is limited to peripheral sites
	μCT	6 μm in vitro	Highest spatial resolution for micro-structures	Unacceptably high radiation doses for human use limited field of view
		50 μm in vivo	Highest in vivo resolution	Radiation
MRI	μMRI	90 μm in vivo	No radiation exposure	Long acquisition time

should receive active therapy aimed at increasing bone density and decreasing fracture risk. The most widely used techniques for assessing bone mineral density are: (a) Dual-Energy X-ray Absorptiometry (DEXA); (b) Dual Photon Absorptiometry (DPA); (c) ultrasound; and (d) Quantitative Computerized Tomography (QCT), the only technology capable of providing 3D results. Although BMD testing is commonly used, it presents a number of limitations:

- The BMD result does not describe the complex 3D bone micro-structure, but rather only offers an indication with a single scalar value.
- BMD is not applied on 3D micro-architecture.
- BMD has only 70% reliability.
- BMD is difficult to measure accurately.
- No uniform threshold exists for all instruments and sites.

Therefore, it is required to develop 3D micro-scale scanning methods from which 3D models can be constructed and analyzed.

1.3 Micro-Scale 3D Scanning Methods

Constant improvements in medical imaging technology allow high resolution in vivo scanning of large specimens or even whole bone models. These methods are: (a) peripheral Quantitative Computed Tomography (pQCT); (b) micro Computed Tomography (μCT); and (c) micro Magnetic Resonance Imaging (μMRI). All are based on the common technology of CT and MRI. Table 1 compares the existing 3D imaging methods. While CT-based micro-scale imaging technology provides higher spatial resolution than μMRI technology, the major advantage of μMRI lies in the absence of ionizing radiation.

These scanning technologies have become popular due to the development of new 3D computerized methods for the reconstruction of trabecular structure from micro CT and MRI images [5, 13] as well as to the increase in computational resources that has led to the development of analytical tools for structural and mechanical analysis.

Prior to diagnosis, medical images must be processed and a geometric model must be reconstructed. In following section the geometric methods for model reconstruction are presented.

2 Volumetric Model Reconstruction from μCT/μMRI Images

The first step in processing μCT/μMRI images of scanned tissue for diagnosis is to reconstruct the 3D geometric model from the medical images. 3D model reconstruction can be divided into two main categories: volumetric reconstruction and surface reconstruction. Subsequently, even if the geometry of a reconstructed model is of high visualization quality, the issue of mesh quality is handled through remeshing that enables further analyses, such as finite element analysis.

2.1 Methods for Geometric Model Reconstruction from CT/MRI Imaging

Methods for model reconstruction from micro-CT/MRI imaging, which is used today, are mainly based on traditional methods for macro-scale models. The following describes the most common methods for reconstructing macro-scale medical models from CT/MRI imaging.

2.1.1 Surface Reconstruction Methods for Macro-Scale Models

Existing approaches for surface reconstruction can be classified into two main categories: (a) Marching Cubes method; and (b) Reconstruction from parallel cross-sections.

Marching Cubes method: The most well-known method for surface reconstruction of 3D models from medical images is the Marching Cubes method developed by Lorensen and Cline in 1987 [46]. The output of this method is a surface model with triangulated mesh. The triangulation is achieved via a look-up table that defines all possible surface-edge intersections resulting from the vertex value (tissue or void). Although this method was considered revolutionary for its time, it is prone to creating inner holes. Since then, numerous enhancements and variations to this method have been published and implemented [3, 50, 51]. The main disadvantage of this method is that is generates unsmooth and step-like surfaces.

Reconstruction from parallel cross-sections: Surface reconstruction from μCT/μMRI images, defined as parallel cross-sections, yields a 3D triangulation model. One of the early works in this area was published by Keppel in 1975 [44] and provided a solution for a limited case in which each slice contains only

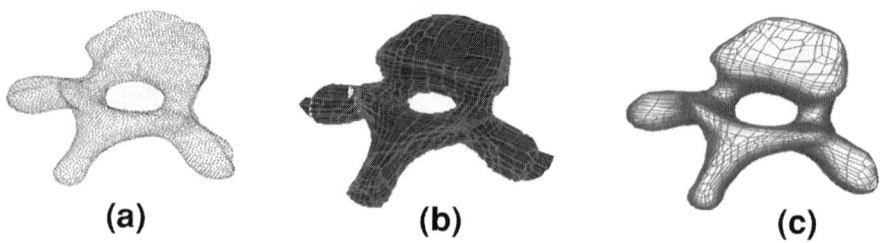

Fig. 2 Anisotropic grid-base volumetric reconstruction of vertebra. **a** Cloud of points retrieved from medical images. **b** Volumetric reconstruction using hexahedral elements. **c** Surface reconstruction using quads [5, 6]

one contour. In 1978, Christiansen and Sederberg [18] extended this method to dual contours. In the 1990s, increased use of CT and MRI technology for medical diagnosis gave this research area a boost. A large number of studies have been published that consider reconstruction of organs and systems with high geometrical complexity, such as the cardiac system and the brain [7, 17, 19]. One major contributor to research on surface reconstruction from parallel cross-sections is Prof. Barequet, who has published numerous works on this subject [9]. Recent research focuses on reconstruction from sparsely sampled parallel contours [53] and nested contours [8].

2.1.2 Volumetric Reconstruction Methods for Macro-Scale Models

Existing approaches to volumetric reconstruction can be classified into two main categories: (a) Grid-based method; and (b) Anisotropic grid-based method.

Grid-based method: This approach to 3D model reconstruction from medical images is based on direct conversion of pixels into hexahedron elements (voxels). This approach is widely used by the biomedical community for μFE analysis of bone micro-structure [4, 61, 67, 68], despite its disadvantage in the form of jagged edges on the envelope. In 2006, Boyd and Müller [13] published a smoothing algorithm for this type of elements, but the resulting surface still lacks natural smoothness. The grid-based method can be also used for creating a tetrahedral elements for finite element analysis, as in the work of Frey et al. [24]. This approach was used in early studies of μFE of bone micro-structure [66].

Anisotropic grid-based method: Another approach for volumetric reconstruction is based on anisotropic grid-based techniques [5, 48]. The main idea of this method is first to construct a geometric field, which is induced by the shape of the surface. This geometric field represents the natural directions and grid cell size for each point in R^3. Then, the imposed volumetric grid is deformed toward the object's shape according to the produced geometric field. This method produces models in which the basic volumetric unit can be either a hexahedron or a tetrahedron (Fig. 2a–c).

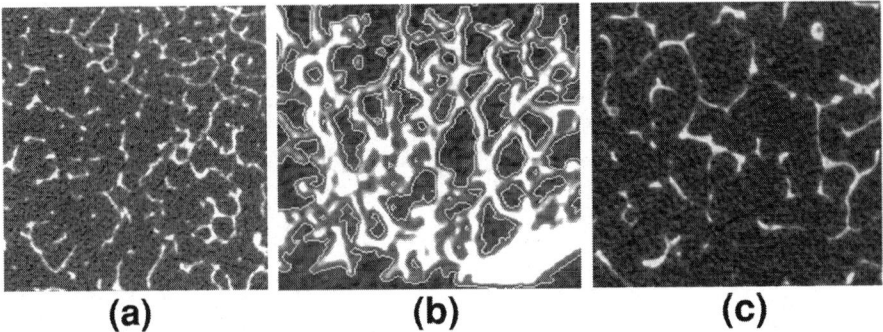

Fig. 3 Segmentation process of micro-CT images. **a** Original image. **b** Segmentation with incorrect thresholds. **c** Segmentation with correct thresholds

In the case of reconstruction from µCT/µMRI images, the above methods cannot be applied directly due to the large number of details and to the fact that high-resolution images are often noisy, as depicted in Fig. 3a. Therefore, more reliable filtering and segmentation methods are needed. Good segmentation requires choosing threshold values of the Hounsfield Units (HU) that optimally separate the bone micro-structure from the bone marrow. A threshold range that is too wide can lead to segmentation that includes bone marrow and interconnects between separated parts, as shown in Fig. 3b and c depicts segmentation results achieved with precise threshold values. In this case, curvature filter [47] and 3D bilateral filter [48] were applied to the original image prior to segmentation. Such filtering smoothes the images and eliminates the noisy background and preserves features.

2.2 Domain-Based Approach for Reconstruction from µCT/µMRI Imaging

To facilitate fast computational performance and obtain a mesh that can be easily handled for later optimization and visualization, a domain-based approach is desired for parallel computation of large-scale problems. The problem domain is subdivided into sub-domains on which the physical problem is computed in parallel. Each sub-domain contributes to the global solution of the problem. The subdivision into sub-domains can be grid-based, can utilize a load-balancing approach by means of graph partition, e.g. Metis [42] or can be subdivided into simple sweepable volumes using Voronoi graphs [64].

There are two main domain decomposition approaches for interaction between sub-domains: overlapping and non-overlapping methods. The first approach, known as the Schwarz Method, subdivides the space into overlapping sub-domains. This approach is practical for problems in which the global domain can be constructed from sub-domains with simple geometric shapes. The domains can

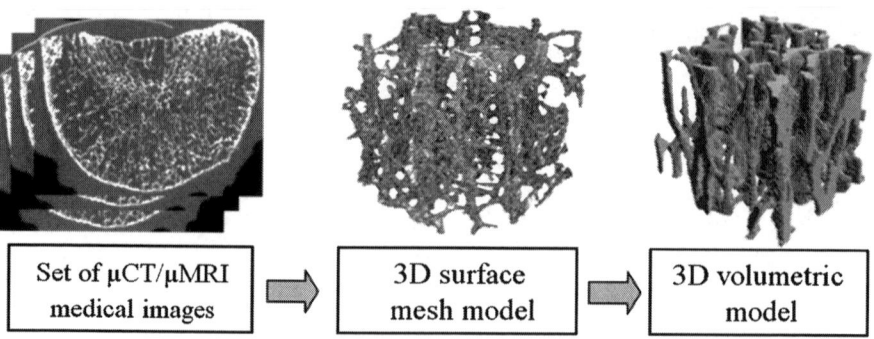

Fig. 4 Main stages in physically based multiscale modeling of large-scale porous structures

be solved in parallel, where the boundary values of the overlapped regions are updated after each iteration. The second approach subdivides the space into adjoining sub-domains, thus requiring that one sub-domain be solved prior to solving the next one or that the *ghost nodes approach* be used. Both types of domain decomposition methods facilitate the solution of complex problems by means of parallel computing. In case of processing of medical images, a non-overlapping method with grid-based subdivision can be utilized (Fig. 4).

Each sub-domain must capture the trabecular structure of the analyzed bone. In the field of image processing, the Nyquist sampling rate [28] defines the minimum spatial frequency required for correct sampling of the image. According to the Nyquist sampling rate, the sampled interval must be at least two times smaller than the size of the smallest detail to be captured. If the smallest detail is represented by the sum of morphological parameters, trabecular thickness (*Tb.Th*) and trabecular spacing (*Tb.Sp*) of the model, then the minimum size of sub-domain should be at least $2 \times (Tb.Th + Tb.Sp)$. According to Hildebrand et al. [32], the average values for *Tb.Th* and *Tb.Sp* are 0.3 mm and 0.8 mm, respectively. Thus, a sub-domain of 128^3 voxels (4.7 mm^3) in size with a spatial resolution of 37 μm, as used in this work, included on average $4 \times (Tb.Th + Tb.Sp)$.

2.3 From Surface Triangular Mesh to Volumetric Hexahedral Model

If the surface reconstruction approach is used, the surface model needs to be converted into a volumetric model. A voxelization technique can be utilized for this conversion. Voxelization can be carried out using different methods. Not all methods can deal with complex 3D geometry, so that structural information may be lost in the process. The voxelization method that provided the most optimal results is based on the flood-fill approach [49]. Flood-filling begins with a seed voxel, identified by stepping a short distance along the inward-pointing surface normal of a mesh triangle. This voxel is marked as an *internal voxel*. Then, a ray is cast recursively from each *internal voxel* to all its neighbors. If the ray does not cross the surface, the neighbor is marked as an internal voxel. If the ray does cross the surface, the neighbor is marked as a *border voxel*. The process is terminated when all *internal voxels* were handled.

2.4 Hierarchical Multiresolution Geometric Modeling

The resulted volumetric model can be used for structural analysis. However, such a model is still impractical for the majority of applications. For efficient mechanical analysis or even visualization, parallel processing is required. Such technology does not typically exist at medical centers, and thus an alternative approach is required. One of the recent approaches utilizes hierarchical multiresolution geometric modeling. It is useful for making storage, transmission, computation, and visualization of these models feasible and more efficient. Multiresolution modeling algorithms are also known as simplification algorithms [54]. They can be applied to surface meshes and grid-based volumetric structures. In the field of surface mesh representation, the *edge-collapse* approach proposed by Hoppe et al. [39] is the most common decimation operation. Edge contraction operates on a single edge and contracts the edge to a single vertex. *Edge collapse* is a reversible operation. The inverse of this operation is known as *vertex splitting*. Progressive Meshes (PM) [38] are built by considering a sequence of edge collapses that iteratively simplify a mesh. The original mesh M may be obtained from the base mesh M_0 through progressive refinement by applying all the vertex split updates in the sequence. If only part of the sequence is applied, a model at an intermediate level-of-detail (LOD) is obtained. Volumetric multiresolution algorithms have hierarchical characteristics that allow continuous bi-directional transition from a smooth macro-scale model to a porous micro-scale model [57]. The most prevalent 3D hierarchical multiresolution geometric data structure is the octree whose common property is the principle of recursive decomposition of space by a factor of eight [63], as depicted in Fig. 5a.

In octree, each node stores an explicit three-dimensional point, which is the center of the subdivision for that node. The root node of the octree represents a finite bounded space, so that the implicit centers are well-defined. A 3D hierarchical geometric model allows continuous transition from a smooth macro-scale model with low topological complexity to a trabecular micro-scale model with high topological complexity. The intermediate levels are equivalent to zooming from a dense trabecular structure to sub-group of trabeculae, and finally to a singular trabecula. The octree structure enables fast transition between different structural levels of bone model. Moreover, due to the domain-based approach, a high-resolution model can be straightforwardly presented at the *Volume of Interest* (*VOI*) while keeping the rest of the model at low resolution. The octree data structure and an example for macro-and micro scale vertebra models are depicted in Fig. 5b–e. Figure 6 shows the geometric differences at different levels of the

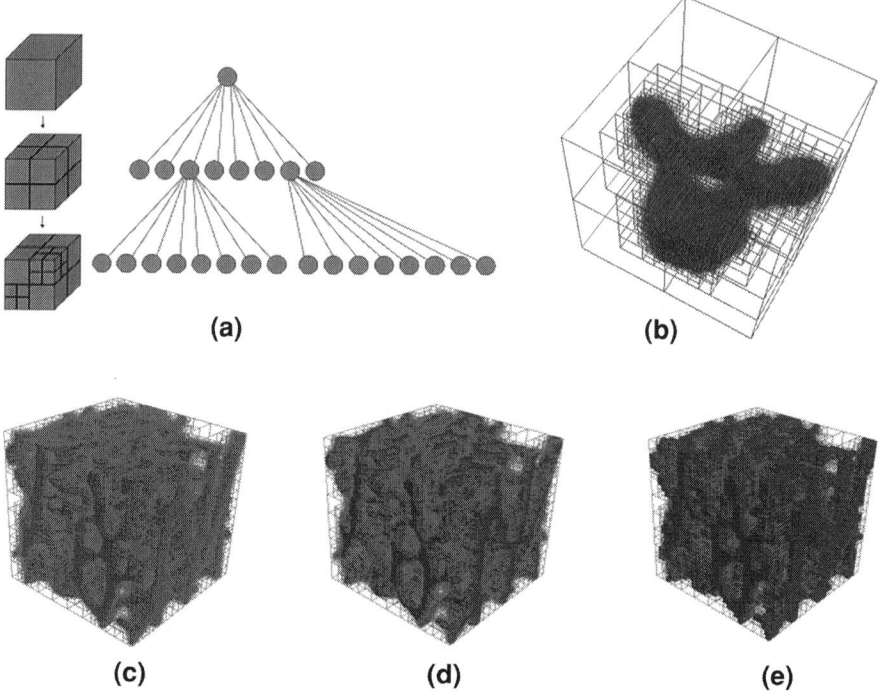

Fig. 5 Hierarchical modeling utilizing octree data-structure. **a** Recursive subdivision of a cube into octants and its corresponding octree. **b** The octree modeling of vertebra at the macro-scale. **c–e** Different resolutions of the octree representation of a sub-domain at the micro-scale level, 128^3, 64^3 and 32^3 resolutions respectively [58]

hierarchical octree representation. Although the global appearance of the model generally seems identical, detailed observation reveals distinctions, as seen in the zoomed regions. The dominant features of the structure have been maintained but at a higher resolution level, and minor features appear as well. These features are essential when an accurate representation of the bone micro-structure is desired.

It is important to emphasize that the multiresolution approach handles only the geometrical aspects of a model and does not take into account its physical properties, such as weight, density, elasticity and conductivity. Without these properties, the model loses its physical substance at the intermediate levels and cannot be used for mechanical analysis.

3 Structural Bone Analyses for Patient-Specific Diagnosis

Over the years, the challenge to understand how bone micro-structure influences the development and progress of metabolic bone diseases has attracted many researchers. Their research concentrated mainly on two areas: (a) development of

Fig. 6 : Multiresolution geometric modeling of trabecular bone with voxel size of (**a**), 37 μm and (**b**) 148 μm

structural parameters that can be retrieved initially from the medical images and later from the reconstructed volumetric models, and (b) structural mechanical analysis of bone utilizing finite element methods.

3.1 Extracting 3D Bone Structural and Topological Parameters

In the 1970s, X-rays and CT-based imaging technology allowed acquiring images on which micro-scale structure was recognizable. This led to the development of a set of structural parameters, some of which are used till today. Table 2 provides a partial list of these parameters. Two types of 3D parameters have been determined by commercial systems and in the literature: structural and topological. Structural parameters are based on geometric properties of the model alone, while topological parameters take into consideration bone connectivity and topology.

(a) *Basic parameters*: These basic parameters are used for estimating the surface (BS) and volume (BV) of bone tissue [23]. These parameters have characteristic values at each anatomical site and thus can be used for quantification of bone quantity.
(b) *Trabecular architecture descriptors*: This set of parameters describes the trabecular architecture of the bone, which can vary between two main structures: a rod-like structure characteristic of vertebrae, and a plate-like structure characteristic of femoral bone. These parameters are used for estimating trabecular thinning, measuring marrow cavity and evaluating number of plates per unit length. Originally these parameters were approximated by using a set of the basic parameters, assuming plate-like architecture [23], but currently their values can be calculated directly from a 3D model [32, 33].

Table 2 3D bone structural and topological parameters

Type	Group	Parameter	Notation
Structural parameters	Basic parameters	Bone surface	BS
		Bone volume	BV
	Trabecular architecture descriptors	Trabecular thickness	Tb.Th
		Trabecular separation	Tb.Sp
		Trabecular number	Tb.N
	Advanced structural descriptors	Marrow star volume	Ma.St.V
		Structure model index	SMI
		Percent plate	%Plate
	Anisotropy descriptors	Degree of anisotropy	DA
		Percent bone in load direction	%Bone$_{LD}$
Topological parameters	The connectivity descriptor	Connectivity density	ConnD
	Process descriptors	Surface-to-curve ratio	S/C
		Erosion index	EI

(c) *Advanced structural descriptors*: The Marrow Star Volume parameter (Ma.St.V) can be used to describe the "voids" within the trabecular structure. It is a sensitive descriptor for quantifying bone loss either through trabecular thinning or loss of entire trabecula [12]. Alternatives for the trabecular number parameter include a Structure Model Index (SMI) [34], which varies from 0 to 3 (an ideal plate structure and an ideal rod-like structure, respectively), and Percent Plate (%Plate), which allows quantitative estimation of the effect of bone resorption on the shape of the trabecula.

(d) *Anisotropy descriptors*: Degree of Anisotropy (DA) [29] and Percent Bone in Load Direction (%Bone$_{LD}$) [12] are used to describe the anisotropy of the trabecular bone structure. The DA parameter is used for defining the direction of the preferred orientation of trabeculae and is important for directionally dependent mechanical properties. The %Bone$_{LD}$ parameter allows quantitative estimation of the effect of bone resorption on the shape of the trabeculae.

(e) *The connectivity descriptor*: The Connectivity Density (ConnD) parameter [55] provides an estimate of the number of trabecular connections per mm^3. It is defined as the number of trabecular elements that may be removed without separating the network and is frequently referenced as the parameter mostly affected during the progression of osteoporosis [56].

(f) *Process descriptors*: Topologically based parameters for evaluation and characterization of bone micro-architecture have been defined [27, 69]. These parameters are used to provide detailed insight into the 3D trabecular network topology [70]. Estimation of these parameters is based on digital topological analysis that classifies each voxel according to its connectivity with the neighboring voxels. As a result, a number of powerful discriminators of different structural arrangements are defined. The first is the surface-to-curve ratio (S/C), which is expected to be a sensitive indicator for the conversion of plates to rods. The second is the Erosion Index (EI), defined as the ratio of the sum of parameters expected to increase upon osteoclastic resorption (edges), divided by the sum of parameters expected to decrease due to such processes (surface).

The above parameters have provided better understanding of bone microstructure; however, their estimation is limited to the available technology. Since digital image analysis was not initially available, most of the parameters were approximated from two basic parameters—BS and BV. Moreover, even the precision of these two parameters was dictated by the resolution of imaging technology. Nowadays, most of these parameters have already been integrated into commercial applications provided with μCT and μMRI scanners. However, they yield only implicit and partial descriptions of bone topology, structure and geometric properties (shape elements and orientation). Therefore, a complete 3D micro-structural mechanical analysis is preferable.

3.2 Finite Element Mechanical Analysis of Bone Micro-Structures

Along with parametric structural analysis, researchers began utilizing finite element mechanical analysis for estimating bone strength and predicting its condition. Mechanical analysis was initially restricted to the macro-scale [14, 16], and material properties were evaluated by applying the homogenization approach. The homogenization theory was developed to analyze the physics of micro-structured materials [11] and has been used extensively to analyze composite materials and predict optimal topology of micro-structured materials [10]. The method has also been used to predict the influence of trabecular bone architecture on effective stiffness and to estimate trabecular tissue stress and strain for arbitrary global loading [36, 37]. The homogenization methods allow replacing complex biological models with their respective simplified macro models that can be solved with ordinary computer resources. However, the asymptotic method is valid only when there is an order of magnitude difference between two consecutive levels, while the RVE method represents the complex geometric structure with a solid model. Averaging the micro-scale structure and replacing it with a solid model that has equivalent material properties lead to a loss of structural information that is important for precise diagnosis.

During the 1980 and 1990s, the growth in computational resources and in the availability of parallel computing has created two trends in the area of computational mechanics: (a) an increase in the number of elements in analyzed models and (b) a decrease in the size of these elements, leading to an increase in the overall number of degrees of freedom. The combination of these two trends has led to the development of large-scale computational models, known as micro Finite Element Analysis techniques (μFEA) [68]. Initially, the solution utilized the element-by-element precondition conjugate gradient (EBE-PCG) method [15, 26] and was executed on super-computers, e.g. Cray.

In the 2000s, modern algebraic multi-grid (AMG) linear solvers have been utilized for solving large-scale FE trabecular models with nearly half a billion degrees of freedom for a vertebra [2] and for the distal part of the radius of a human forearm [4], both with spatial resolution of 30 μm. The solution required state-of-the art parallel FE applications and thousands of processors. Indeed, analyses of high resolution models require considerable computational resources that are not commonly available at hospitals and medical centers. Thus, a multi-scale finite element analysis method was proposed.

3.3 Multiscale Finite Element Analysis of Bone Micro-Structure

Multiscale methods are an extension of the multiresolution approach, which integrates material properties with the geometrical method. In engineering, multiscale methods are used mainly for analysis, as they are intended to bridge between different structural levels. Yet state-of-the-art multiscale methods do not support continuous transition between consecutive structural methods and only allow discrete hierarchical modeling, assuming order of magnitude differences between these levels, e.g. macro-, micro-and nano levels. Most state-of-the-art multiscale methods utilize multi-step homogenization [25, 31, 52] or mesh superposition [43]. These methods assume order-of-magnitude scale separation between two consecutive levels of hierarchy. Therefore, continuous transition in terms of geometrical modeling and mechanical properties is not available.

Currently there are two main computational models in engineering: a homogenized macro-scale model in which the highly detailed structure is replaced with one that is homogeneous, and a high-resolution micro-scale model without intermediate models. The classic homogenization procedure is based on the asymptotic or Representative Volume Element (RVE) method [1, 72]. The homogenization theory was developed to analyze the physics of micro-structured materials [11] and has been used extensively to analyze composite materials and predict optimal topology of micro-structured materials [10]. Homogenization methods allow replacing complex biological models with their respective simplified macro models that can be solved with ordinary computer resources. However, the asymptotic method is valid only when there is an order of magnitude difference between two consecutive levels, while the RVE method represents the complex geometric structure with a homogeneous model. This approach has been used for physically-based modeling of deformable elastic inhomogeneous models for medical applications [45], surgery simulation and animation [22].

The proposed multiscale domain-based approach [59] enables using a "digital magnifying glass" for continuous transition between macro-and micro-scales, as illustrated in Fig. 7. This computational tool can be used by engineers for solving complex models on commonly available workstations rather than on

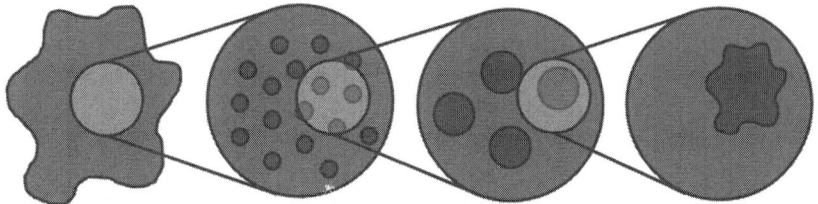

Fig. 7 Schematic representation of the continuous multiscale method

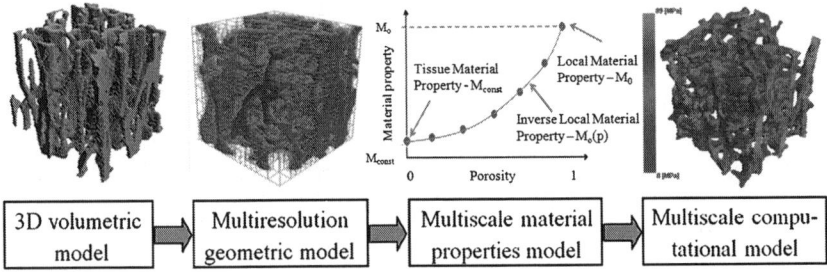

Fig. 8 Main stages of the presented physically based multiscale modeling for large-scale porous structures

super-computers or grid-based systems. The approach is based on the following stages: (a) creation of a volumetric geometric model from a surface model or set of image slices; (b) generation of a multiresolution geometric model using hierarchical data structure; (c) evaluation of a multiscale material properties model; and (d) integration between the geometric and material properties models. Figure 8 depicts these stages, as described below.

Realizing the proposed multiscale approach requires synergy between the multiresolution geometric model and the assignment of *local mechanical properties* at each intermediate scale. The local mechanical properties vary at each intermediate scale due to changes in the geometric model. The main aim at this stage is to preserve the *effective (global) material properties*. Once the effective property to be preserved is selected, the *local material properties* model can be established.

A trivial illustration of the proposed method would be to keep the weight (m) of a model as invariant. Then for an intermediate model of volume V_i the local density ρ_i of the material must be evaluated to guarantee that:

$$V_i \cdot \rho_i = m = const \tag{3.1}$$

A similar approach can be utilized for evaluation of different material properties. However, several effective material properties also depend on structural or topological information of the model, e.g. stiffness and conductivity. Thus, a preprocessing stage for estimating these properties is necessary. Following is a description of a method for estimating local material properties for intermediate models, assuming a linear elasticity model with small displacements [58]:

Stage 1: Estimate the effective material properties for a set of intermediate models without changing the local material properties. This set must include the model with the highest available resolution.

The elasticity model is represented by the generalized Hooke's law that defines the relationship between the dimensionless displacements applied to the model (strains-ε) and the forces per area (stresses-σ) developed in the material:

$$\sigma_{ij} = C_{ijkl}\varepsilon_{kl} \qquad (3.2)$$

where C_{ijkl} are fourth-degree tensor constants that construct the material stiffness matrix.

For homogenization of material properties, the RVE formulation can be used. Applying the RVE homogenization establishes a relation between the average values of stresses ($\bar{\sigma}$) and strains ($\bar{\varepsilon}$). For the representative volume V these values are defined as:

$$\bar{\varepsilon} = \frac{1}{V}\int_V \varepsilon dV \qquad \bar{\sigma} = \frac{1}{V}\int_V \sigma dV \qquad (3.3)$$

The boundary conditions are applied in terms of displacements u_i:

$$u_i = \varepsilon_{ij}^0 x_j \qquad (3.4)$$

where ε_{ij}^0 are constant strains and x_j are point coordinates.

The average strain values for the above boundary conditions are:

$$\bar{\varepsilon}_{ij} = \varepsilon_{ij}^0 \qquad (3.5)$$

Now the elastic stiffness matrix constants can be found using (3.2). In the case of fully anisotropic material, 21 independent stiffness matrix constants must be estimated. This calculation requires performing a finite element analysis on each sub-domain with six different boundary conditions. The first three sets of boundary conditions are equivalent to uniaxial compression tests, one for each major axis. The last three are equivalent to shear tests.

Before the algorithm proceeds to the next stage, the symmetry planes of the material are found and used for transforming the stiffness matrix to its orthotropic form. Cowin and Mehrabadi [20] showed that symmetry planes exist if the eigenvectors of the matrices (3.6) and (3.7) are the same and the eigenvalues are different.

$$C'_{ijkk} = \begin{bmatrix} C'_{11}+C'_{12}+C'_{13} & C'_{16}+C'_{26}+C'_{36} & C'_{15}+C'_{25}+C'_{35} \\ C'_{16}+C'_{26}+C'_{36} & C'_{12}+C'_{22}+C'_{23} & C'_{14}+C'_{24}+C'_{34} \\ C'_{15}+C'_{25}+C'_{35} & C'_{14}+C'_{24}+C'_{34} & C'_{13}+C'_{23}+C'_{33} \end{bmatrix} \qquad (3.6)$$

$$C'_{ikkj} = \begin{bmatrix} C'_{11} + C'_{55} + C'_{66} & C'_{16} + C'_{26} + C'_{45} & C'_{15} + C'_{46} + C'_{35} \\ C'_{16} + C'_{26} + C'_{45} & C'_{22} + C'_{44} + C'_{66} & C'_{24} + C'_{34} + C'_{56} \\ C'_{15} + C'_{46} + C'_{35} & C'_{24} + C'_{34} + C'_{56} & C'_{33} + C'_{44} + C'_{55} \end{bmatrix} \quad (3.7)$$

To find the eigenvectors of these matrices, the SVD decomposition is applied. The eigenvectors are further used as vectors of the rotation matrix Q for tensor transformation from anisotropic to orthotropic material tensor. After the tensor transformation is applied, the orthotropic material stiffness matrix C_{ijkl} is generated:

$$\bar{\sigma} = C \cdot \bar{\varepsilon} = \begin{bmatrix} \sigma_{11} \\ \sigma_{22} \\ \sigma_{33} \\ \sigma_{23} \\ \sigma_{31} \\ \sigma_{12} \end{bmatrix} = \begin{bmatrix} C_{11} & C_{12} & C_{13} & 0 & 0 & 0 \\ C_{21} & C_{22} & C_{23} & 0 & 0 & 0 \\ C_{31} & C_{32} & C_{33} & 0 & 0 & 0 \\ 0 & 0 & 0 & C_{44} & 0 & 0 \\ 0 & 0 & 0 & 0 & C_{55} & 0 \\ 0 & 0 & 0 & 0 & 0 & C_{66} \end{bmatrix} \cdot \begin{bmatrix} \varepsilon_{11} \\ \varepsilon_{22} \\ \varepsilon_{33} \\ \varepsilon_{23} \\ \varepsilon_{31} \\ \varepsilon_{12} \end{bmatrix} \quad (3.8)$$

Currently, nine independent material properties of the orthotropic material can be calculated using the generalized Hooke's law in stiffness form.

The calculated orthotropic material properties are:

$$E_{11} \quad E_{22} \quad E_{33} \quad G_{23} \quad G_{31} \quad G_{12} \quad v_{23} \quad v_{31} \quad v_{32} \quad (3.9)$$

where: E_{ii}-Young's modulus, G_{ij}-shear modulus and v_{ij}-Poisson's ratio.

Stage 2: Define a correlation between the geometric properties of the intermediate models (e.g. porosity) and their respective *effective material properties*. The correlation between the porosity (p) and the *effective material properties* (M) can be established in terms of a polynomial function. Coefficients of the polynomial (a_i) can be found straightforwardly by solving a set of linear equations. After computing all the coefficients, a common factor can be extracted from the equation and rewritten as follows:

$$M(p) = M_0 \cdot \sum_{i=n}^{0} \hat{a}_i p^i \quad (3.10)$$

where M_0 represents the initial *local material property*.

Stage 3: Find an inverse local material properties model that preserves the effective material properties at a constant value (M_{const}) for different intermediate models. According to the above, (3.10) can be rewritten as:

$$M_0(p) = M_{const} \cdot \left[\sum_{i=n}^{0} \hat{a}_i p^i \right]^{-1} \quad (3.11)$$

In this work, the material properties at the highest resolution (micro-scale) are assumed to be isotropic. However, the geometry changes are at lower (coarser) resolutions, so orthotropic local material properties are applied to compensate for

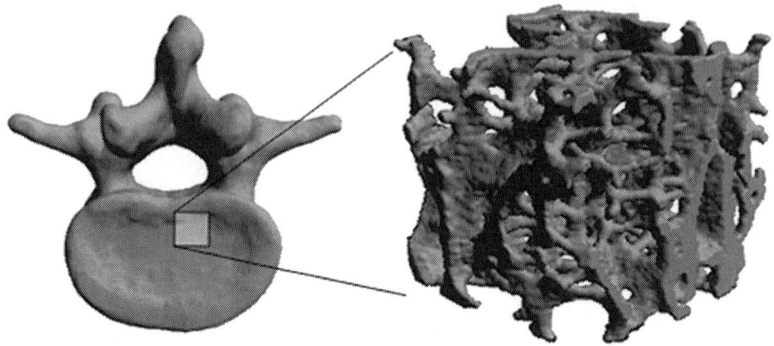

Fig. 9 Vertebra models at different scales: macro-scale on the *left* and micro-scale on the *right*

such structural changes. In effect, when the structure of the geometrical model changes from micro to macro, the geometric anisotropy is replaced by anisotropic material properties.

Stage 4: Verify the computational model by comparing the property of the intermediate models to the known and measurable property of the highest resolution model. For example, models that are used for structural mechanical analysis may be compared in terms of strain energy.

The feasibility of the proposed multiscale domain-based method has been demonstrated on a bio-medical model of L3-vertebra trabecular bone at the microscale structural level. Bones are bio-composite materials characterized by complex multiscale structural geometry and complex behavior, as shown in Fig. 9. Depending on the anatomical site, bone architecture differs significantly at the micro-structural level with respect to shape, thickness, direction and porosity. For example, the rod-like structures of the vertebrae differ significantly from the plate-like structure of the femoral head. Bone architecture can also vary at different locations on the same site due to functionality and locally applied forces characterized by magnitude and direction. Thus, the bone structure of each patient is unique. These characteristics are crucial for diagnosis of bone metabolic disease.

The performance of the multiscale model is demonstrated in Fig. 10. In this example, each model is analyzed with its local material properties, estimated according to the proposed method. The colors represent stresses, as each model was subjected to uniaxial compression. The stress locations are consistent for all levels of resolution. In addition, the "digital magnifying glass" approach can be easily identified. The prominent features of the model are visible at all levels of resolution. The analysis shows that the multiscale method performs according to expectations. The prominent geometric features of the micro-scale models are preserved at all levels of resolution. Moreover, the computational models are reliable and robust for all intermediate levels.

The domain-based approach improves the performance of the hierarchical multi-resolution model and the finite element mechanical analysis. The worst-case

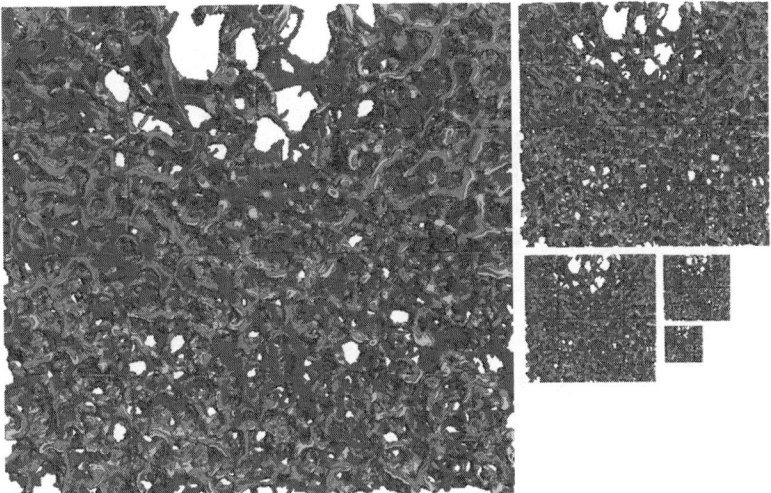

Fig. 10 View-dependent presentation of multiscale mechanical analysis, where colors represent stress intensity

time complexity of the linear octree construction algorithms is evaluated as follows.

Let N be the total number of material voxels and d the number of hierarchy levels of the octree. Encoding a material voxel model requires computational time proportional to the depth d; encoding a region including condensation requires $O(dN)$. An adjacent voxel can be found in $O(d)$. Utilizing the domain-based approach decreases both the number of full voxels and the number of hierarchy levels required to represent them. Assuming that the maximum number of levels in the octree is $d = \log N$ and the number of sub-domains is m, the finite element solver is based on the preconditioned conjugate gradient method (PCG). The dominant operations during iterations are matrix-vector products. In general, matrix-vector multiplication requires $O(k)$ operations, where k is the number of non-zero entries in the matrix. For many problems, including FE analysis, the A matrix is sparse and $kO(N)$. Distributing this task among a number of processors significantly improves algorithm performance and allows solution of models that could not be solved otherwise.

4 Visualization of Bone Micro-Structures

As stated, visualization is not an end by itself but only a means for representing analysis and diagnosis results to physicians or orthopedists. As in geometric modeling, two different representations can be utilized for visualization-surface or volumetric. Since the volumetric model already exists, it seems natural to use a volume for visualization, though there are pros and cons to this approach:

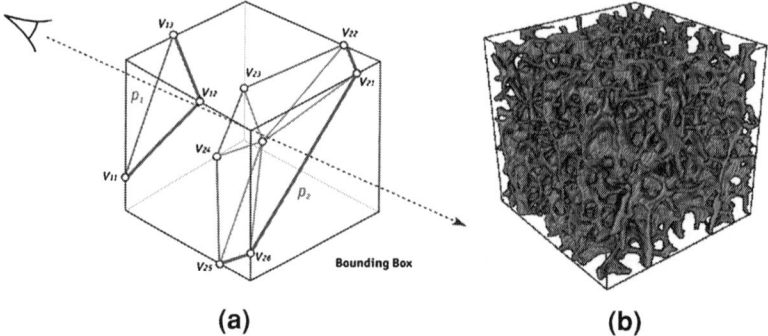

Fig. 11 Visualization of bone micro-structure utilizing direct volume method

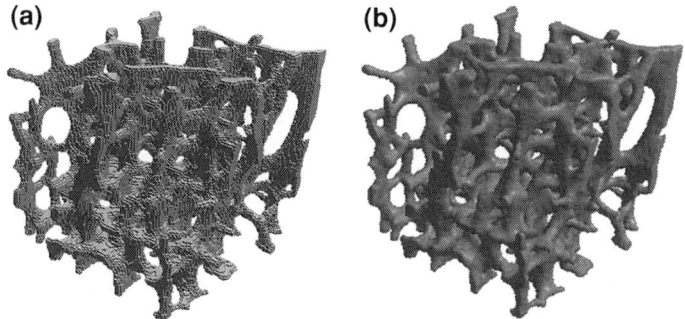

Fig. 12 Visualization of bone micro-structure using only boundary voxels and gradient kernel for calculation of model normal vectors

- Volume visualization methods require special graphic hardware that is not available on commonly used PCs. To overcome such a difficulty, the majority of visualization algorithms utilize texture mapping with a transparency parameter between different layers and blending, as depicted in Fig. 11. Modern graphic cards are designed for rapid processing of textures. Thus, this method is frequently used for volume visualization. It also enables straightforward visualization of multi-layer models, e.g. the human skeleton with different tissues. The main disadvantage of this method is that it is view dependent. Such a process is memory consuming, and its outcome is not always exact and seamless for the viewer.
- For volume representation, hexahedral elements (voxels) are usually utilized. If only surface information is of interest, then just the boundary voxels can be displayed. However, direct voxel visualization will result in jagged surfaces, as depicted in Fig. 12. Calculating model normal vectors using a gradient kernel can overcome this problem. This method permits rapid visualization of cross-sections aligned in the directions of major axes, Fig. 13. But representing a

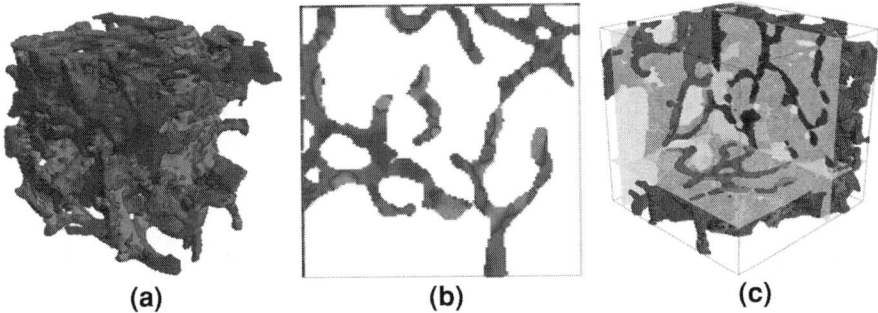

Fig. 13 Visualization possibilities using voxel representation. **a** FE analyzed model. **b** Cross-section of the model. **c** A clipped region of the analyzed model

Fig. 14 Domain-based visualization of bone micro-structures. **a** Color coded sub-domains of the model. **b** The result as presented to the user

cross-section in any other direction is challenging since it requires creating an approximated plane.

The above limitations can be overcome by transforming the volumetric representation to a surface representation, for example utilizing the Marching Cubes algorithm. Surface representation will be smooth and will not require large resources for its visualization. The main disadvantage of this approach is being transformation dependent. Creating a cross-section will require recalculation of the entire surface.

Another difficulty associated with visualizing bone micro-structures relates to the large number of details in these models. Without an efficient visualization algorithm and memory resource management, such a task can be very difficult and sometimes impossible. Therefore, the domain-based approach has become popular in this field as well. Each sub-domain is prepared for visualization on a separate processor or in sequence. The user is unaware of the process since the result is seamless. An example of domain-based visualization is depicted in Fig. 14.

5 Summary and Conclusions

This chapter has reviewed the evolution of the field of patient-specific visualization and diagnosis of bone micro-structure, focusing on medical imaging technology, geometric modeling, structural analysis and visualization. State-of-the-art research focuses on a new multi-scale method that can provide a better understanding of a very complex bio-structure—bone. Moreover, a new 3D multiscale method for mechanical analysis of bone micro-structures was presented. This is one of the very few methods in the field of computational bio-mechanics that allows for the adaptive application of mechanical analysis on large-scale high resolution models. The proposed method is based on domain-based multiresolution hierarchical geometric modeling and multiscale material properties. This method incorporates the following properties:

- The hierarchical geometric model (octree) facilitates continuous bi-directional transition between micro-and macro-scale models by introducing intermediate structural levels.
- The new method for a multiscale material properties model is adapted to porosity changes at different structural levels of the geometrical model.
- The developed complete model provides the synergy between geometric and material models that is required for robust multiscale finite element analysis.

This method is aimed at assisting physicians to better understand and define bone structure strength and bone stability with respect to geometric and material properties. Medical imaging technology and common medical testing will continue to be the major diagnostic tools for initial diagnosis of the above diseases. However, different structures with similar morphological parameters may have significantly different effective material properties. As a result, one structure may withstand the applied load, while another may collapse. Therefore, mechanical testing may better predict structural behavior and improve the quality of diagnosis and prognosis, thus effectively improving patients' quality of life. Moreover, structural simulation can lead to early diagnosis prior to the appearance of symptomatic fractures. The proposed method is a new tool that can be applied to a variety of problems. We propose a number of possibilities for future applications:

- The method can be used as a diagnostic tool for estimating local bone strength in diseases that affect bone micro-structure and material properties, e.g. hypoparathyroidism [62], osteoporosis [21] and metastatic bone cancer [65]. It also can be used for local drug treatment, chemotherapy and calcium enrichment.
- Integrating topology optimization with the proposed method may lead to the design of natural micro-scaffolds [35] which can reduce the healing period.

The proposed multiscale FE method can be integrated as a module into a computerized diagnostic system that, together with other analysis tools, can make a significant impact on diagnostic conclusions. The proposed method offers an alternative to working only with existing macro-and micro-structural levels by introducing intermediate levels. These levels provide continuous transition between macro-and micro-scales by imitating human detail perception. We believe this method can also be extended for modeling and analysis of other materials characterized by irregular and stochastic structure.

References

1. Aboudi, J.: Mechanics of Composite Materials. Elsevier, Amsterdam/New York (1991)
2. Adams, M.F., Bayraktar, H.H., Keaveny, T.M., Papadopoulos, P.: Ultrascalable implicit finite element analyses in solid mechanics with over a half a billion degrees of freedom, proceedings of the 2004 ACM/IEEE conference on supercomputing. IEEE. Comput. Soc., p 34 (2004)
3. Adriano, L.: Improving the robustness and accuracy of the marching cubes algorithm for isosurfacing. IEEE. Trans. Visual. Comput. Graphics **9**, 16–29 (2003)
4. Arbenz, P., van Lenthe, G., Mennel, U., Müller, R., Sala, M.: Multi-level micro-finite element analysis for human bone structures. In: Kågström, B., Elmroth, E., Dongarra, J., Wasniewski, J. (eds.) Applied Parallel Computing, pp 240–250. State of the Art in Scientific Computing. Springer, Berlin/Heidelberg (2010)
5. Azernikov, S., Fischer, A.: A new volume warping method for surface reconstruction. J. Comput. Inf. Sci. Eng. **6**, 355–363 (2006)
6. Azernikov, S., Miropolsky, A., Fischer, A.: Surface reconstruction of freeform objects based on multiresolution volumetric method. J. Comput. Inf. Sci. Eng. **3**, 334–338 (2003)
7. Bajaj, C.L., Coyle, E.J., Lin, K.N.: Arbitrary topology shape reconstruction from planar cross sections. Graph. Models Image Process **58**, 524–543 (1996)
8. Barequet, G., Goodrich, M.T., Levi-Steiner, A., Steiner, D.: Contour interpolation by straight skeletons. Graph. Models **66**, 245–260 (2004)
9. Barequet, G., Sharir, M.: Piecewise-linear interpolation between polygonal slices. Comput. Vis. Image Underst. **63**, 251–272 (1996)
10. Bendsoe, M.P., Kikuchi, N.: Generating optimal topologies in structural design using a homogenization method. Comput. Methods Appl. Mech. Eng. **71**, 197–224 (1988)
11. Bensoussan, A., Lions, J.L., Papanicolaou, G.: Asymptotic Analysis for Periodic Structures. North Holland Pub Co, Amsterdam, New York (1978)
12. Borah, B., Dufresne, T.E., Cockman, M.D., Gross, G.J., Sod, E.W., Myers, W.R., Combs, K.S., Higgins, R.E., Pierce, S.A., Stevens, M.L.: Evaluation of changes in trabecular bone architecture and mechanical properties of minipig vertebrae by three-dimensional magnetic resonance microimaging and finite element modeling. J. Bone Miner. Res. **15**, 1786–1797 (2000)
13. Boyd, S.K., Müller, R.: Smooth surface meshing for automated finite element model generation from 3D image data. J. Biomech. **39**, 1287–1295 (2006)
14. Burstein, A.H., Currey, J.D., Frankel, V.H., Reilly, D.T.: The ultimate properties of bone tissue: the effects of yielding. J. Biomech. **5**, 35–42 , IN31-IN32, 43-44 (1972)
15. Carey, G.F., Jiang, B.N.: Element-by-element linear and nonlinear solution schemes. Commun. Appl. Numer. Methods **2**, 145–153 (1986)
16. Carter, D., Hayes, W.: Bone compressive strength: the influence of density and strain rate. Science **194**, 1174–1176 (1976)
17. Choi, Y.K., Park, K.H.: A heuristic triangulation algorithm for multiple planar contours using an extended double branching procedure. Visual Comput. **10**, 372–387 (1994)
18. Christiansen, H.N., Sederberg, T.W.: Conversion of complex contour line definitions into polygonal element mosaics. SIGGRAPH Comput. Graph. **12**, 187–192 (1978)
19. Cong, G., Parvin, B.: Robust and efficient surface reconstruction from contours. Visual Comput. **17**, 199–208 (2001)

20. Cowin, S.C., Mehrabadi, M.M.: On the identification of material symmetry for anisotropic elastic materials. Q. J. Mech. Appl. Math. **40**, 451–476 (1987)
21. Crawford, R.P., Cann, C.E., Keaveny, T.M.: Finite element models predict in vitro vertebral body compressive strength better than quantitative computed tomography. Bone **33**, 744–750 (2003)
22. Debunne, G., Desbrun, M., Barr, A.H., Cani, M.P.: Interactive multiresolution animation of deformable models, Eurographics Workshop on Computer Animation and Simulation (1999)
23. Feldkamp, L.A., Goldstein, S.A., Parfitt, M.A., Jesion, G., Kleerekoper, M.: The direct examination of three-dimensional bone architecture in vitro by computed tomography. J. Bone Miner. Res. **4**, 3–11 (1989)
24. Frey, P., Sarter, B., Gautherie, M.: Fully automatic mesh generation for 3-D domains based upon voxel sets. Int. J. Numer. Methods Eng. **37**, 2735–2753 (1994)
25. Fritsch, A., Hellmich, C., Dormieux, L.: Ductile sliding between mineral crystals followed by rupture of collagen crosslinks: experimentally supported micromechanical explanation of bone strength. J. Theor. Biol. **260**, 230–252 (2009)
26. Fyhrie, D.P., Hamid, M.S., Kuo, R.F., Lang, S.S.: Direct three-dimensional finite element analysis of human vertebral cancellous bone, 38th meeting of Orthopaedic Research Society, Washington, DC, p 551 (1992)
27. Gomberg, B.R., Saha, P.K., Hee Kwon Song, H., Wehrli, F.W.: Topological analysis of trabecular bone MR images. Anglais **19**, 166–174 (2000)
28. Gonzalez, R.C., Woods, R.E.: Digital Image Processing. Addison-Wesley Longman Publishing Co.Inc., Reading, MA (1992)
29. Goulet, R.W., Goldstein, S.A., Ciarelli, M.J., Kuhn, J.L., Brown, M.B., Feldkamp, L.A.: The relationship between the structural and orthogonal compressive properties of trabecular bone. J. Biomech. **27**, 375–389 (1994)
30. Gupta, H.S., Seto, J., Wagermaier, W., Zaslansky, P., Boesecke, P., Fratzl, P.: Cooperative deformation of mineral and collagen in bone at the nanoscale. Proc. Nat. Acad. Sci. **103**, 17741–17746 (2006)
31. Hellmich, C., Barthélémy, J.F., Dormieux, L.: Mineral-collagen interactions in elasticity of bone ultrastructure—a continuum micromechanics approach. Eur. J. Mech.-A/Solids **23**, 783–810 (2004)
32. Hildebrand, T., Laib, A., Müller, R., Dequeker, J., Rüegsegger, P.: Direct three-dimensional morphometric analysis of human cancellous bone: microstructural data from spine, femur, iliac crest, and calcaneus. J. Bone Miner. Res. **14**, 1167–1174 (1999)
33. Hildebrand, T., Rüegsegger, P.: A new method for the model-independent assessment of thickness in three-dimensional images. J. Microsc. **185**, 67–75 (1997)
34. Hildebrand, T., Rüegsegger, P.: Quantification of bone microarchitecture with the structure model index. Comput. Methods Biomech. Biomed. Eng. **1**, 15–23 (1997)
35. Hollister, S.J.: Porous scaffold design for tissue engineering. Nat. Mater. **4**, 518–524 (2005)
36. Hollister, S.J., Brennan, J.M., Kikuchi, N.: A homogenization sampling procedure for calculating trabecular bone effective stiffness and tissue level stress. J. Biomech. **27**, 433–444 (1994)
37. Hollister, S.J., Fyhrie, D.P., Jepsen, K.J., Goldstein, S.A.: Application of homogenization theory to the study of trabecular bone mechanics. J. Biomech. **24**, 825–839 (1991)
38. Hoppe, H.: Progressive meshes, proceedings of the 23rd annual conference on computer graphics and interactive techniques. ACM, pp 99–108 (1996)
39. Hoppe, H., DeRose, T., Duchamp, T., McDonald, J., Stuetzle, W.: Mesh optimization, proceedings of the 20th annual conference on Computer graphics and interactive techniques. ACM, Anaheim, CA, pp 19–26 (1993)
40. Huiskes, R.: If bone is the answer, then what is the question? J. Anat. **197**, 145–156 (2000)
41. Isaksson, H., Gröngröft, I., Wilson, W., van Donkelaar, C.C., van Rietbergen, B., Tami, A., Huiskes, R., Ito, K.: Remodeling of fracture callus in mice is consistent with mechanical loading and bone remodeling theory. J. Orthop. Res. **27**, 664–672 (2009)
42. Karypis, G.: METIS: a family of multilevel partitioning algorithms, University of Minnesota, USA (1998)

43. Kawagai, M., Sando, A., Takano, N.: Image-based multi-scale modelling strategy for complex and heterogeneous porous microstructures by mesh superposition method. Modell. Simul. Mater. Sci. Eng. **14**, 53–69 (2006)
44. Keppel, E.: Approximating complex surfaces by triangulation of contour lines. IBM J. Res. Dev. **19**, 2–11 (1975)
45. Kharevych, L., Mullen, P., Owhadi, H., Desbrun, M.: Numerical coarsening of inhomogeneous elastic materials. ACM Trans. Graph. **28**, 1–8 (2009)
46. Lorensen, W.E., Cline, H.E.: Marching cubes: a high resolution 3D surface construction algorithm. SIGGRAPH Comput. Graph. **21**, 163–169 (1987)
47. Malladi, R., Sethian, J.A.: Image processing via level set curvature flow. Proc. Natl. Acad. Sci. USA **92**, 7046–7050 (1995)
48. Miropolsky, A., Fischer, A.: Extended geometric filter for reconstruction as a basis for computational inspection. J. Manuf. Sci. Eng. **131**, 1–8 (2009)
49. Morris, D.: Automated Preparation Calibration and Simulation of Deformable Objects. Stanford University Department of Computer Science, Stanford/CA (2007)
50. Newman, T.S., Yi, H.: A survey of the marching cubes algorithm. Comput. Graphics **30**, 854–879 (2006)
51. Nielson, G.M., Hamann, B.: The asymptotic decider: resolving the ambiguity in marching cubes, pProceedings of the 2nd conference on Visualization '91. IEEE Computer Society Press, San Diego, California, pp 83–91 (1991)
52. Nikolov, S., Raabe, D.: Hierarchical modeling of the elastic properties of bone at submicron scales: the role of extrafibrillar mineralization. Biophys. J. **94**, 4220–4232 (2008)
53. Nilsson, O., Breen, D., Museth, K.: Surface reconstruction via contour metamorphosis: an Eulerian approach with lagrangian particle tracking. In Proc. IEEE Visualization, 407–414 (2005)
54. Nowak, M.: Structural optimization system based on trabecular bone surface adaptation. Struct. Multi. Optim. **32**, 241–249 (2006)
55. Odgaard, A., Gundersen, H.J.G.: Quantification of connectivity in cancellous bone, with special emphasis on 3-D reconstructions. Bone **14**, 173–182 (1993)
56. Parfitt, A.M.: Implications of architecture for the pathogenesis and prevention of vertebral fracture. Bone **13**, S41–S47 (1992)
57. Podshivalov, L., Fischer, A., Bar-Yoseph, P.Z.: Multiresolution 2D geometric meshing for multiscale finite element analysis of bone micro-structures. Virtual Phys. Prototyping **5**, 33–43 (2010)
58. Podshivalov, L., Fischer, A., Bar-Yoseph, P.Z.: 3D hierarchical geometric modeling and multiscale FE analysis as a base for individualized medical diagnosis of bone structure. Bone **48**, 693–703 (2011)
59. Podshivalov, L., Holdstein, Y., Fischer, A., Bar-Yoseph, P.Z.: Towards a multi-scale computerized bone diagnostic system: 2D micro-scale finite element analysis. Commun. Numer. Methods Eng. **25**, 733–749 (2009)
60. Rho, J.-Y., Kuhn-Spearing, L., Zioupos, P.: Mechanical properties and the hierarchical structure of bone. Med. Eng. Phys. **20**, 92–102 (1998)
61. Rincón-Kohli, L., Zysset, P.: Multi-axial mechanical properties of human trabecular bone. Biomech. Model. Mechanobiol. **8**, 195–208 (2009)
62. Rubin, M.R., Dempster, D.W., Kohler, T., Stauber, M., Zhou, H., Shane, E., Nickolas, T., Stein, E., Sliney, J., Silverberg, S.J., Bilezikian, J.P., Müller, R.: Three dimensional cancellous bone structure in hypoparathyroidism. Bone **46**, 190–195 (2010)
63. Samet, H.: The Design and Analysis of Spatial Data Structures. Addison-Wesley, Reading, MA (1990)
64. Sheffer, A., Etzion, M., Rappoport, A., Bercovier, M.: Hexahedral mesh generation using the embedded voronoi graph. Eng. Comput. **15**, 248–262 (1999)
65. Tanck, E., Van Aken, J.B., Van der Linden, Y.M., Schreuder, H.W.B., Binkowski, M., Huizenga, H., Verdonschot, N.: Pathological fracture prediction in patients with metastatic lesions can be improved with quantitative computed tomography based computer models. Bone **45**, 777–783 (2009)

66. Ulrich, D., van Rietbergen, B., Weinans, H., Rüegsegger, P.: Finite element analysis of trabecular bone structure: a comparison of image-based meshing techniques. J. Biomech. **31**, 1187–1192 (1998)
67. Van Lenthe, G.H., Voide, R., Boyd, S.K., Muller, R.: Tissue modulus calculated from beam theory is biased by bone size and geometry: implications for the use of three-point bending tests to determine bone tissue modulus. Bone **43**, 717–723 (2008)
68. Van Rietbergen, B.: Micro-FE analyses of bone: state of the art. In: Majumdar, S.B., Brian, K. (eds.) Noninvasive Assessment of Trabecular Bone Architecture and the Competence of Bone. Springer, Berlin/New York (2001)
69. Wehrli, F.W., Saha, P.K., Gomberg, B.R., Song, H.K., Snyder, P.J., Benito, M., Wright, A., Weening, R.: Role of magnetic resonance for assessing structure and function of trabecular bone. Top. Magn. Reson. Imaging **13**, 335–355 (2002)
70. Wehrli, F.W., Song, H.K., Saha, P.K., Wright, A.C.: Quantitative MRI for the assessment of bone structure and function. NMR Biomed. **19**, 731–764 (2006)
71. Weiner, S., Wagner, H.D.: The material bone: structure-mechanical function relations. Annu. Rev. Mat. Sci. **28**, 271–298 (1998)
72. Zaoui, A.: Continuum micromechanics: survey. J. Eng. Mech. **128**, 808–816 (2002)

Patient Specific Modeling of Musculoskeletal Fractures

Eran Peleg

Abstract Choosing and executing an optimal treatment plan for skeletal fractures in clinical practice is a complex procedure. Treatment decisions are often qualitative, based on general guidelines and experience/training of the orthopedic surgeons. Despite its potential to assist in quantifying fracture fixation and thus improve patient outcome, computational patient-specific modeling for selection and planning of fracture treatments is limited at present. During the past 25 years extensive work has been reported regarding patient specific finite element (FE) based modeling. Numerous studies have reported on the development, validation and automation of patient specific FE modeling techniques from quantitative CT data sets. However, a patient specific quantitative process that can be applied in a true clinical environment must cope with profound uncertainties such as; material property assignments, surface geometry and most of all uncertainty of in vivo load amplitudes and gait patterns which are usually only roughly estimated. In this chapter we review common techniques of patient specific modeling of bony structures, and present known limitations and sources of error. Method and experimental validation of a new CT based workflow for patient specific modeling of fracture fixation implementing principal strain ratios is presented.

1 Introduction

Choosing and executing an optimal treatment plan for skeletal fractures in clinical practice is a complex procedure, requiring simultaneous consideration of patient, surgeon, hospital, and fracture-specific factors. Treatment decisions are often

E. Peleg (✉)
Hadassah Medical Center, Jerusalem, Israel
e-mail: eran@hadassah.org.il

qualitative, based on general guidelines and experience/training of the orthopedic surgeons, and therefore, treatment plans for similar medical cases can vary widely among orthopedists despite the use of standard fracture classification and treatment algorithms [2].

Despite its potential to assist in quantifying fracture fixation and thus improve patient outcome, computational patient-specific modeling for selection and planning of fracture treatments is limited at present. Currently, preoperative planning is not routinely used and if performed, usually it consists of digital implant templating on two dimensional X-ray images, and includes angle and distance measurements and position planning. Finite element analysis (FEA) of fracture fixation models derived from computed tomography (CT) scans of individuals holds great potential to provide quantitative patient-specific analysis and evaluation tools but currently is not used in clinical practice.

During the past 25 years extensive work has been reported regarding patient specific finite element (FE) based modeling. Clinical applications have been predominantly aimed at failure prediction of intact bone models under static loading [8, 13, 32, 33, 52] and biomechanical responses to total hip implant insertion [26, 46, 59]. Numerous studies have reported on the development, validation and automation of patient specific FE modeling techniques from quantitative CT data sets [24, 30, 53, 56, 57, 67] however, these investigations primarily focused on strategies for element based application of bony material properties at different densities based on Young's modulus relationships for intact bone models.

A patient specific quantitative process that can be applied in a true clinical environment must cope with profound uncertainties. Previous studies have identified uncertainties related to material property assignments and surface geometry. [29, 56, 65]. Beyond these issues, the most apparent uncertainty is related to the fact that in vivo load amplitudes and directions are not well defined; as such load characteristics including gait patterns are usually roughly estimated. Additional error is expected from uncertainties in interface interactions between implant surfaces. Patient specific FE based studies which involve fracture fixation resembling that of a true clinical workflow are not reported despite the clinical importance of this matter [2, 4, 5, 16], perhaps due in part to the multiple uncertainties inherent in the analyses.

2 CT Based Patient Specific Model Preparation of Bone Structures: Commonly Used Techniques

The basic steps, required for conducting a patient specific quantitative analysis may usually include: segmentation and surface extraction of the desired bone region, creating a volumetric mesh and assigning density based isotropic or ortho-tropic material properties to each element and performing a biomechanical analysis by means of finite element analysis.

2.1 Segmentation and Surface Extraction

Medical image segmentation is the process of labeling each voxel in a medical image data set to indicate its tissue type or anatomical structure. The labels that result from this process have a wide variety of applications in medical research and visualization.

The input to a segmentation procedure is a grayscale digital image data set, for example a CT scan. The desired output of a segmentation process contains labels that classify the input grayscale voxels with the original grayscale dataset used to create the segmentation.

The purpose of segmentation is to provide richer information than that which exists in the original medical images. The collection of labels that is produced through segmentation is frequently used to improve visualization of medical images and to allow quantitative measurements of image structures. Segmentation of medical image data sets allows the creation of a three dimensional view from any angle and more important the creation of surface models using an algorithm such as the Marching Cubes [47]. Though three dimensional models could be created directly from grayscale images using Marching Cubes, the segmentation step is required to provide user defined iso-surfaces. Current methods for segmentation involve the use of algorithms such as; thresholding, snakes, level sets, etc. but eventually manual fine tuning is always required.

2.2 Volumetric Meshing and Finite Element Analysis

Meshing can be defined as the process of breaking up a physical domain into smaller sub-domains (elements). Surface domains may be subdivided into triangle or quadrilateral shapes, while volumes primarily into tetrahedral or hexahedra shapes. The shape and distribution of the elements is usually ideally defined by automatic meshing algorithms. The meshing process is usually needed in order to facilitate the numerical solution of a partial differential equation (mostly for finite element analysis).

The finite element method is a numerical analysis technique for obtaining approximate solutions to a wide variety of engineering problems. Basically, the method envisions the solution region as built up of many small interconnected subregions or elements. A finite element model of a problem gives a piecewise approximation to the governing equations. The basic premise of the finite element method is that a solution region can be analytically modeled or approximated by replacing it with an assemblage of discrete elements. Since these elements can be put together in a variety of ways, they can be used to represent exceedingly complex shapes.

In the field of bone structure analysis, two different methods are generally reported for meshing: (1) voxel based methods [30, 31] and (2) methods based on commercially available automatic mesh generation routines on the other.

The voxel based methods preserve the grid structure as the model elements are cubic elements aligned with the CT grid axis. This procedure allows for a simple identification of the CT grid points which belong to the model element and which are to be considered in the evaluation of the element stiffness. The main disadvantage of the voxel based technique is the unsmooth bone surface which create inaccuracies especially when surface strains and stresses are of interest [61]. When commercial mesh generation routines are used, the mesh is generated starting from the bone surface information (the surface information is usually created by means of segmentation). In this case, CT data set structure is not preserved and irregular shaped elements whose facets are not aligned with the CT grid axis are generated.

2.3 Material Property Assignment to Elements

In addition to geometric requirements of the elements constructing the volumetric grid intended for a FEA, it is a prerequisite that each element contain information that represents real physical properties of the material which is modeled for mechanical response. Specifically, the Young's modulus (elasticity) and Poisson's coefficients are of main concern.

Unlike engineered materials, where elasticity and other material properties are commonly spatially constant and known, the material properties of bone tissue are density based and anisotropic. Various empirical models of the relationship between Young's modulus and apparent bone density can be found in literature [11, 23, 28, 42, 66]. In the particular case of bone tissue, it is possible to extract spatial density of bone by converting the CT Hounsfield Units (HU) to apparent or ash density.

The ability to extract local bone density from a CT data set, and the existence of empirical density based Young's modulus relationships, enables assignment of material properties to all elements of a volumetric grid created from a CT data set.

When dealing with a voxel based mesh, by definition, the HU associated with element is equal to the HU of the voxel hence there is no need for additional calculation of an average HU. The use of a commercial mesh generator, which usually creates tetrahedral shaped elements, require more sophisticated procedure for the assessment of the element material property which in some cases can result in loss of density information. Generally, the average HU in an element is evaluated by averaging the HU of all the voxels and sub-voxels which reside in the element [24, 57, 58].

2.4 Muscle Modeling

The mechanical behavior of muscle has been extensively studied since the earliest reports of Blix and Hill [25]. Since 1938 Hill's model of muscle contraction has dominated the field and a muscle is regarded as an assemblage of sarcomeras.

A number of authors have shown that a muscle is a heterogeneous material structure which induces non-uniform stress and strain distributions [43]. An individual muscle, as a body organ, represents a collection of different fiber types A large range in contractile properties among fiber types enables production of very diverse mechanical responses, from extremely rapid ballistic to slow sustained motions and ultimate postural support.

From a mechanical point of view, a muscle can be considered as a mechanical system, or a structure. Since the constitutive stress–strain relationship of a muscle material is non-linear, the muscle represents a non-linear mechanical system.

Stojanovic et al. [55] developed a numerical procedure for the stress calculation of a multi-fiber Hill's model of muscle as a material model within the finite element method. The procedure is applicable to two- and three-dimensional analyses assuming large strains and large displacements. The work presents an approach applicable to complex (real) muscle structures, with external loading and internal activation (loading). Finite element models simulating the masticatory system biomechanics have also been presented [44].

Although the ability to model muscle structure with internal activation exists, it is un-common whenever the main interest is bone response. In these cases the muscles are simulated as concentrated loads at known insertion locations [9]. At the present time, patient specific models, especially if to be used for clinical trials or in a real clinical workflow, do not model real muscle behavior.

3 A General CT Based Workflow for Patient Specific Modeling of Fracture Fixation

Figure 1 outlines the general features of a CT based clinical workflow that is carried out for patient specific modeling of fracture fixations.

A three dimensional model of the fractured bone is generated from the CT data. Afterwards, the fracture is virtually reduced and the implants are positioned according to the surgeon preferences. The virtual fracture reduction and implant positioning may be carried out by a clinician. For standard cases, the modeling stage may be sufficient for further treatment. If biomechanical analysis is required, the model data is imported into a finite element analysis environment and results are analyzed as part of a decision making process. The modeling and analysis may be carried out pre-operatively, as a decision supporting tool, or as post-operatively for failure prediction assessment and evaluation of recovery.

Different techniques, algorithms and tools may comprise the modeling and analysis parts of the workflow. If such a workflow is to be adapted into a clinical environment, the following conditions are to be met; (1) Generality–ability to provide well-conditioned meshes for all bones, independently from their geometric complexity, (2) Robustness—the ability to generate a mesh of almost every data set [63] and (3) Clinical relevance–the added value to patient treatment must be significant relative to the effort required for completing such a workflow.

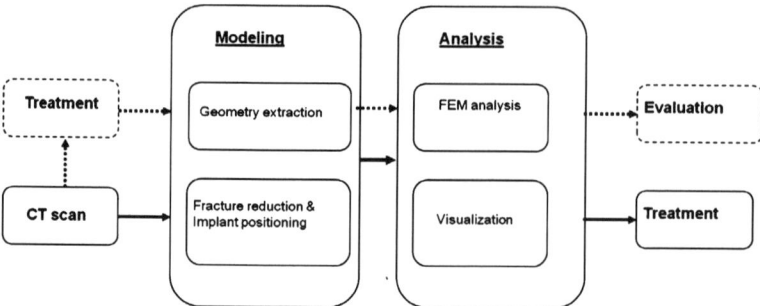

Fig. 1 A general patient specific modeling workflow to be employed in a clinical environment. The patient is CT scanned prior to the operational procedure. The CT data set is used as input for the modeling and the analysis steps. The process can be applied pre-operatively (*solid lines*) or post-operatively (*dotted-lines*)

The above mentioned workflow, which eventually calculates and displays strain and stress patterns, is susceptible to uncertainties of model inputs and one should be critical of the results.

3.1 Sensitivity to Young's Modulus Mathematical Relationship

Information on the mechanical properties of bones is derived from CT-data using a mathematical relationship between the HU values and the mechanical properties of the bone [27, 31, 40, 57, 58]. While the relationship between the HU and physical properties as ash density can be established with direct calibration, the determination of a mathematical relationship between density and mechanical properties is basically empirical and challanging. A large number of mathematical relationships between densitometric measures and mechanical properties (elasticity–density relationship) have been introduced in the literature (Fig. 2). Many of the existing elasticity-density relationships have been adopted in generation of subject specific FE models [10, 11, 15, 28, 36, 37, 66] in most cases with no explicit explanation to the choice of the density-elasticity based relationship. In fact, the influence of the different density-elasticity laws has been demonstrated in a few validation studies [3, 45, 53] but no final conclusion can be derived from the presented results on the best law to adopt, since the predicted accuracy may be affected by a few factors such as; the load case, the chosen anatomy and the region of interest within the anatomy, method of material property assignment, etc. [3, 58, 65].

3.2 Sensitivity to Partial Volume Edge Effects

During the semi-automated segmentation and extraction of surfaces from CT-data sets followed by automatic mesh generation and material property assignment, requires accurate contour extraction. As spatial node coordinates of every element are paired

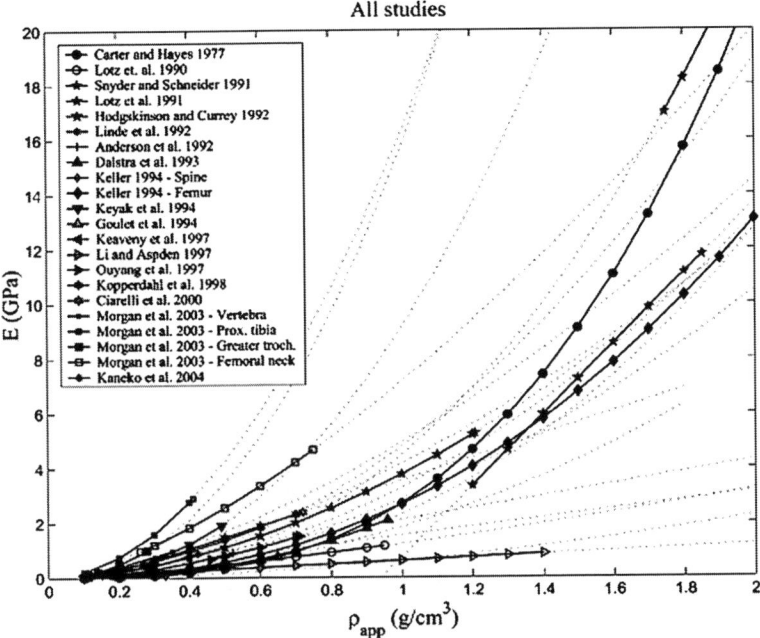

Fig. 2 Young's modulus vs. apparent wet density [23]. In a review study performed by Helgason et al. the various mathematical relationships between bone density and mechanical properties were presented

back to the original CT-data set to calculate the material properties, false location of the surface points will cause errors in both surface geometry and the calculation of elastic modulus values due to inclusion or exclusion of partial voxel volumes (Fig. 3). Such partial volume edge effects (PVEE) can arise from scanning parameters (i.e. insufficient resolution, scan energy, beam hardening) [29, 56], inaccuracy during the segmentation process (evolving from manual errors or weakness of the available automated algorithms) and smoothing (which may result in contraction of the surface and relocation of the surface points). In utilizing semi-automated techniques for the generation of patient specific FE models an additional source of uncertainty can result due to inter- and intra- operator variability. Differing segmentations used to create models from a single data set can result in unique grid topologies with distinct surface strain and stress predictions. Consistent operator-independent generation of surface models is needed as a first step to ensure consistency of FEA results.

3.3 Sensitivity Uncertainty in Simulated Loads

In most in vitro experiments and as well as for numerical analyses, loading conditions represent a major simplification of the actual in vivo loading situation. The simulated loads are usually known only in a general sense. Hence, in a

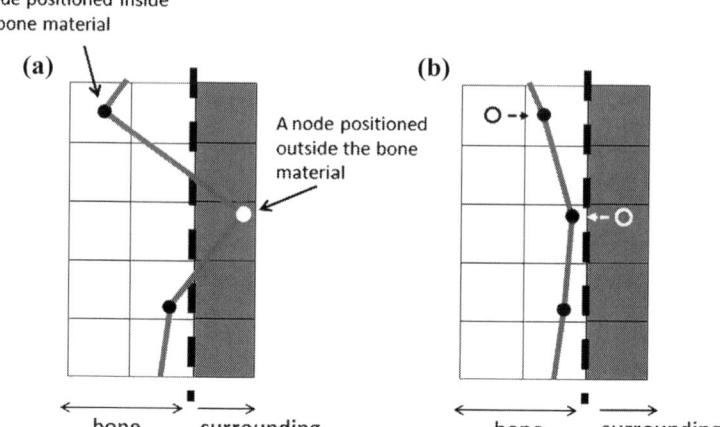

Fig. 3 Partial volume edge effects: Surface nodes originally created outside or inside the cortical bone material can generate partial volume edge errors. Adjusting the location of the surface nodes such that they reside closer to the bone outer boundary may reduce these errors: (**a**) initial segmented surface, (**b**) post-surface adjustment algorithm

computer simulation such uncertainty of loading condition and amplitude could substantially influence the calculated stresses and strains and possibly the amount of calculated stress-shielding and failure prediction in cases were an implant is present.

4 A New CT Based Workflow for Patient Specific Modeling of Fracture Fixation Implementing Principal Strain Ratios

To deal with uncertainties that arise from unknown loading conditions and density-elasticity relationships, it is possible to utilize a principal strain fixation ratio measure (SR). SR is defined as the ratio of principal strains that develop in a fixated bone relative to the principal strains that develop in the same bone in an intact state. By definition, the SR measure is independent of load amplitude as long as a linear response is assumed and modeled. The SR field output variable may provide the ability to view strain shielding (SR < 1 − a) and over strained patterns (SR > 1 + b) at time of fixation and as such may be an indicator for bone adaptation [14, 26, 50, 60, 64] and strain based failure prediction [6, 20, 52] providing an appropriate comparative measure to assist in selection of fixation alternatives.

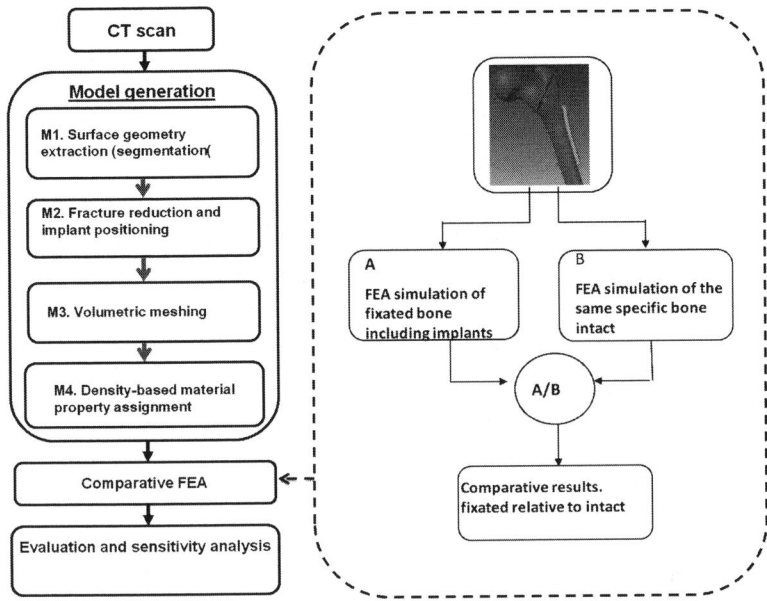

Fig. 4 A CT based workflow implementing principal strain ratios. Two separate FEA simulations are carried out; one for the fixated bone model and another for a virtually intact bone model. For both models, fixated and intact, the external load and boundary conditions are identical. The model preparation phase is identical to the performed in a general CT based workflow

4.1 Workflow Description

The SR based workflow differs from the regular CT based workflow in the FE analysis part. In the proposed workflow (Fig. 4), the analysis stage includes two finite element simulations both creating maximal principal strain values; one of the fixated bone (fractured bone with implants) and one for the same specific virtually intact bone (virtually fused fracture surfaces without implants). This enables to present a comparative biomechanical analysis between the fixated bone relative to its natural state before fracture occurred without the need to construct an intact model from the non-broken opposite side bone which may be asymmetric in anatomy and pathology.

The end result is a three dimensional model presenting stress shielded and over stressed patterns (Fig. 5). The sequence of the workflow is compatible with a clinical environment workflow in which a patient arrives with a fractured bone and there are no healthy control or any bone available for biomechanical analysis.

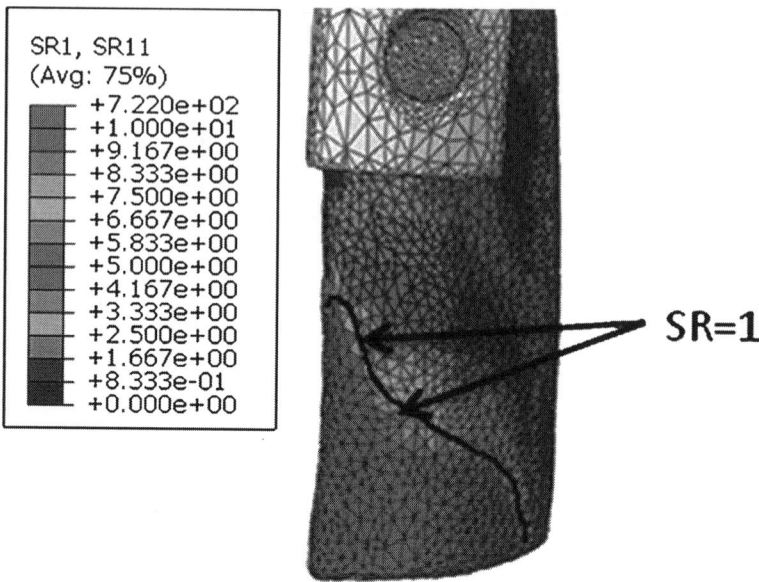

Fig. 5 Three dimensional SR patterns The figure depicts the SR gradients in a fixated femur close to the edge of a extra-medular DHS plate. The SR values change from SR < 1 to SR > 1 within a relatively small region creating a large gradient at the distal end of the femur shaft beneath the implant tip

5 Experimental Validation of the SR Based Workflow

5.1 Experimental Outline

A combined numerical-experimental study was performed comparing FE predicted strains and SR values with strain-gauge measurements obtained on six cadaveric proximal femoral specimens. Each femur was tested twice under the same simplified uniaxial compressive loading condition. The test was first applied to the intact femur (no fracture and no hardware) and subsequently to the femur post fracture following hardware fixation (Fig. 6). Fixation was achieved in all cases with an extra-medullary static hip screw. Post-test CT scanning was performed on each specimen. For each specimen two FE models were created; a fixated model and an intact model. Each model was analyzed utilizing three density-based Young's modulus relationships to yield a total of 6 analyses per specimen. Simulated strains and SR values generated from the FE analyses were compared to the experimental measurements.

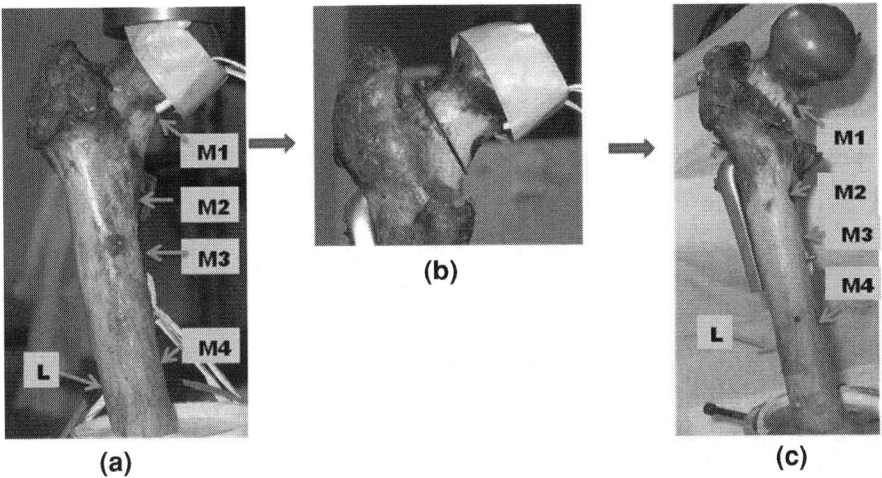

Fig. 6 Experimental setup [45]. **a** An intact specimen prepared for an axial loading test. **b** Insertion of the implant and introduction of a cut to the femoral neck, maintaining the femur geometry is maintained. **c** A fixated specimen

5.2 Methods and Materials

Six fresh frozen cadaveric proximal femurs were obtained from the department of anatomy at the University of Toronto. All the specimens were subjected to volumetric quantitative computed tomography [35] of the proximal region and fell in the range from normal to severe osteoporosis (Table 1). Prior to experimentation each specimen was thawed and the soft tissue removed. The base of each femur was potted in PMMA such that the average angle between the line of action and the femoral shaft lines was 14°. The positioning of the femur at this angle was necessary in order to prevent torque that may cause displacement of the specimen during the experimental procedure. Given the relatively short length of our specimens, the angle was not physiological.

A combination of rosettes and axial strain gauges (FRA-3-11 and FLA-3-11-3LT, Tokyo Sokki Kenkyujo Co, LTD) were placed at the following locations (Fig. 6): femoral neck (M1-rosette), medial shaft opposite to the plate (M2-rosette, M3-rosette), medial-distal region (M4-axial gauge) and lateral- distal region approximately 2 cm beneath the end of the plate (L-axial gauge). Under the loading utilized in this study, the direction of the principal strains were assumed empirically in certain locations (i.e. along the shaft) and confirmed in the FE modeling. As such, and due to a limited number of available channels of our data acquisition unit, axial gauges were used in these locations. The alignment of each axial gauge was assumed to coincide with the highest magnitude principal strain. The axial direction of the gauge was registered in the experimental phase and was accounted for during the simulation phase.

Table 1 Specimen details and diagnostics

Parameter	Value
Sex	Females
Age (years)	56–83
Length (mm)	160–206
BMD[a] (mgr/cm^3)	122–516

[a] BMD was measured by volumetric quantitative CT applied to the trochanteric region

Prior to determining the locations of measurement, preliminary modeling was carried out with a CT-data set of a cadaveric femur with an induced peritrochanteric fracture. SR patterns were examined and the locations which were found to be most affected by the fixation and both eligible in terms of practicality were chosen.

Four fiducials were attached to each specimen and their locations registered to the strain gauges (Microscribe3D dx HCI 2.0). The spatial coordinates of each fiducial were later on extracted from the CT-data set and matched against the respective spatial coordinates as measured during the experimental phase. A transformation matrix was created enabling a rigid transformation of the real world coordinate system to that of the FE model created from the CT-data set.

Loading protocols were consistent for the intact and fixated specimens and were performed in sequence without moving the tested specimen. A uniaxial force was applied (Bionix 858, MTS Systems Corporation) via an actuator. The load amplitude was gradually increased from 100[N] to 800[N] in increments of 100[N]. The load was held at each increment for 15 s while superficial strain data was collected (Daqview DaqBook/2000 V7.13.2) at a frequency of 2[Hz]. Between the load increments the crosshead speed was 0.1 mm/sec. Gradually increasing the load amplitude to the range of a patients' body weight resembled that of a one leg stance [46] and prevented mechanical damage to the specimens [53] or large change in geometry which would have introduced non-linearity. The load amplitude that was used to validate the predicted maximum and minimum principal strain values with the measured values was 400[N]. All the load amplitudes were used to examine the independency of the SR with load amplitude.

Following experimental testing of the intact specimens, fractures were introduced and then fixated with a short extra-medullary plate and a static lag screw. The fracture fixation was performed by the authors following training by an experienced orthopaedic surgeon. All fixations were verified by the surgeon.

The distance between the tip of the lag screw to the femoral head apex was measured for modeling purposes. Only one side screw was used to reduce uncertainties related to the load sharing of a multiple screw system. Based on the OTA classification, the fractures ranged from type pertrochanteric simple (31-A1) to subcapital non-impacted (31-B3).

Following mechanical testing, hardware and strain gauges were removed and each fractured femur was CT scanned (Light-speed VCT- GE Medical Systems) at a slice thickness of 0.625 mm and pixel spacing 0.36 × 0.36 mm^2, together with a

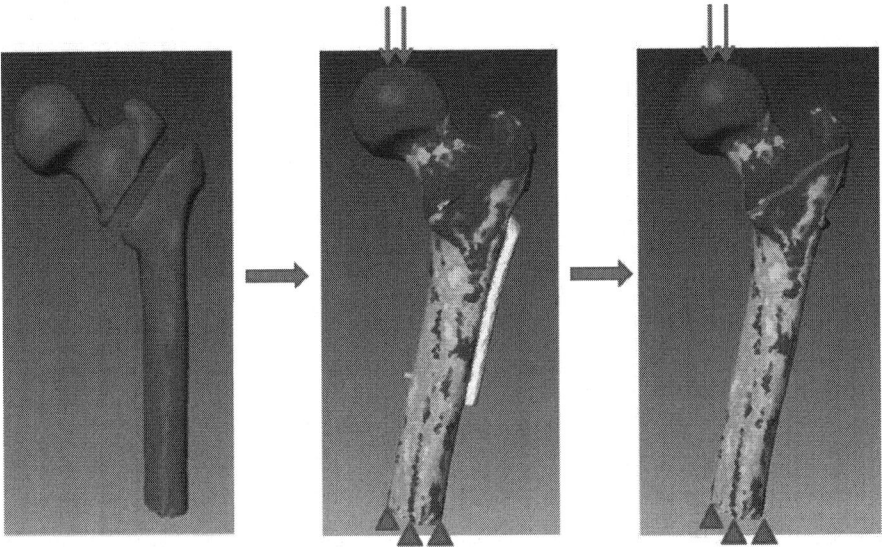

Fig. 7 Modeling and analysis. **a** The fragments of a fixated femur are extracted from a CT data set. **b** A model of the fixated femur after fracture reduction, implant positioning and bone material property assignment. The model was analyzed with boundary conditions resembling the experimental setup. **c** A model of the intact femur in which the implants are removed and the fracture surfaces are bonded. (brighter colors represent higher Young modulus values at the surface of the model)

calibration phantom (Skyscan NV). Each CT data set was calibrated assuming a linear relationship between Hounsfield unit and bone ash density [36]. The apparent density was then calculated assuming a constant ratio $\rho_{ash}/\rho_{app} = 0.6$ [51].

Acquiring the CT of the fractured specimens resembled a clinical scenario in which the diagnostic stage is performed prior to a surgical procedure.

The CT data files and the files containing the geometric data of the side-plate (Richards®) and side screws (prepared in-house) were imported to AmiraDev5.2.2 (Visage Imaging,Inc) for preparation of the models prior to the FE analysis. The following steps were performed: (1) the bone fragments were segmented, (2) volumetric tetrahedral based grids (10-node quadratic elements) were generated using an automatic mesh generator, (3) material properties were assigned to each element as described by Taddei et al. [58], (4) the fractures were reduced and implants positioned (Fig. 7), (5) boolean cutting was conducted between the implants and bone grids and finally, (6) all the model information was exported to Abaqus 6.9-1 (Dassault Systems, USA) for FE analysis.

Three different density-based elasticity relationships were used to describe the bone material properties: Eq. (1) $E = 3,790 \rho_{app}^3$ [11]; Eq. (2) $E = 6,950 \rho_{app}^{1.49}$ [41]; and Eq. (3) $E = 10,500 \rho_{ash}^{2.29}$ [28]. These relationships were chosen because they represent the best approximations for CT based determination of bone material properties as recently compared [53].

Boundary and loading conditions simulated the experimental setup: the distal femur had a zero displacement condition and a spread force was applied to the femoral head in the direction of the distal end (Fig. 7). No contact was applied at the fracture surface between the femoral head and shaft (a constant gap prevailed during the experimental phase). The screw- thread region and the bottom portion of the DHS barrel were bonded to the bone. It was assumed that during the loading phase, force was not exerted between the top portion of the implant and the bone interface.

The contact condition between the screw head and plate seat was modeled as an exponential relationship of pressure to clearance between the contact surfaces ("soft contact" Abaqus 6.9). This type of contact condition has the advantage of fast convergence and is appropriate to use if local tangential forces are not of interest.

A compatible intact model was created for each fixated model. Intact models were created by duplicating the fixated model, removing all implants and bonding the fracture surface between the femoral head and shaft. As such, the intact models were identical in bone grid topology and material property assignments to the fixated models. Based on convergence testing results, all the models were created with an average element size length less than 2 mm [62]. All simulations were carried out assuming linear elastic material properties and small geometrical displacements.

The calculated maximum and minimum principal strains at the surface nodes, corresponding to the sensing area of each strain gauge, were averaged and compared with the experimental measurements. SR values were calculated by dividing the principle strain values in each fixated model by the corresponding values in each intact model. The calculated FE derived SR values were compared to the corresponding experimentally derived SR values. A linear regression between experimental and predicted strains was performed to quantify the prediction accuracy. Root mean square (RMSE) errors and the peak error were calculated.

5.3 Experimental Results

The regression lines between the experimental data and the numerical calculations for the intact specimens exhibited high sensitivity to the different elasticity-density relationships (Fig. 8). The best fit was that related to models created with Eq. (2) (Table 2). The fit for the models created with Eq. (1) and Eq. (3) created regression lines with higher slopes (1.58, 2.2 respectively) and RMSE values (176 µε and 373 µε respectively). This data is in close agreement with that reported by Schileo et al. [53].

The regression lines for the fixated specimens exhibited less sensitivity to density-elasticity relationships than those of the intact models (Fig. 9). The overall range of the slopes was between 1.138 to 1.445. Maximum and average errors were also reduced relative to the intact models (Table 2). As a result of the intense

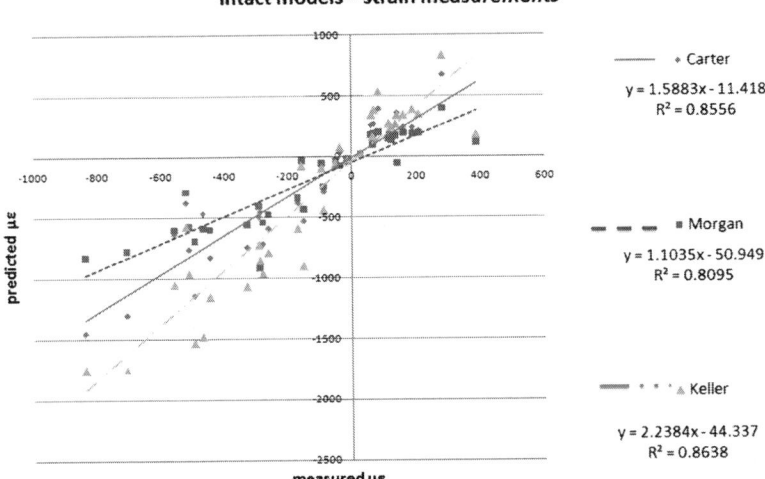

Fig. 8 Predicted vs. measured strains of intact models [45]. The strain measurements were performed prior to the introduction of the fracture and insertion of the implants. The response values were found to be sensitive to different density based elasticity relationship as reported by [53]

Table 2 Goodness of fit parameters for pooled data reported for all model types (intact, fixated strain measurements and strain ratio measurements) under the three different density based elasticity relationships. Eq. (1) $E = 3,790 \rho_{app}^3$ (Carter and Hayes) Eq. (2) $E = 6,950 \rho_{app}^{1.49}$ (Morgan et al.) Eq. (3) $E = 10,500 \rho_{ash}^{2.29}$ (Keller)

Fit parameter	Intact models			Fixated models principal strains			Fixated models strain ratios (SR)		
	Eq. (1)	Eq. (2)	Eq. (3)	Eq. (1)	Eq. (2)	Eq. (3)	Eq. (1)	Eq. (2)	Eq. (3)
R2	0.85	0.81	0.86	0.72	0.77	0.74	0.72	0.73	0.72
Slope	1.59	1.1	2.2	1.145	1.138	1.445	0.63	0.76	0.63
RMSE	176 με	25 με	373 με	21 με	20 με	65 με	0.52	0.12	0.19
Max err%	364	218	536	260	222	325	133	118	79
Average err%	103	69	177	80	71	97	43	32	40

stress shielding (SR < 0.005) at the femoral neck region the strains generated at the M1 location were very small (below the signal to noise ratio) and therefore could not be accounted for as part of the experimental data of the fixated models.

The correlation between the predicted SR values and the experimental values exhibited low sensitivity to the different density-elasticity relationships (Fig. 10). The slope values ranged between 0.63 and 0.76. In terms of R^2 results were not significantly different than those of the fixated models (Table 2), however, RMSE and the average and maximum errors the SR exhibited an improved performance over the intact and fixated strain parameters.

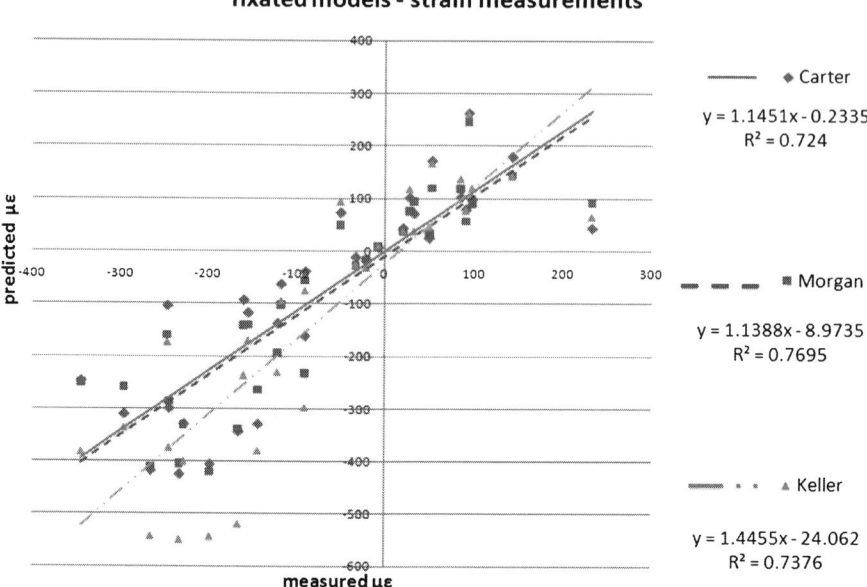

Fig. 9 Predicted vs. measured strains of the fixated models [45]. The strain measurements were performed after the introduction of the fracture and placement of the implants

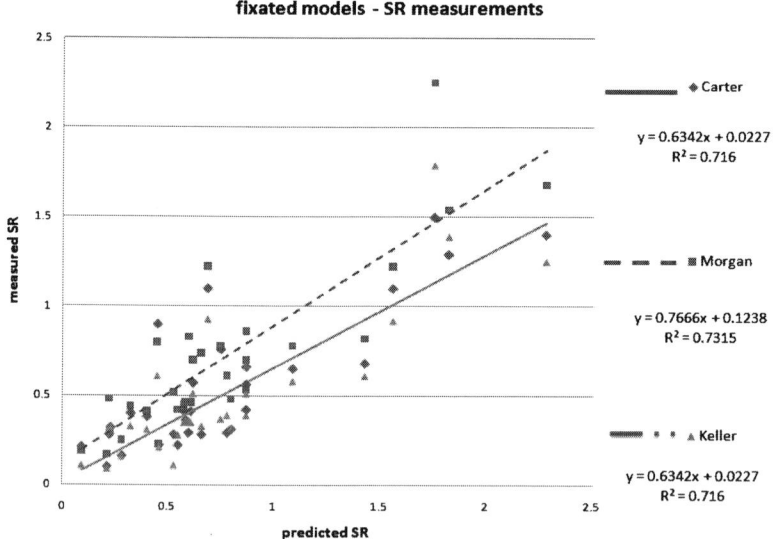

Fig. 10 Predicted vs. measured principal strain ratios (SR) of the fixated models [45]. The experimental values were obtained by dividing the strains measured in the fixated specimen with the results obtained in the intact specimen. The predicted values were obtained by dividing the computed FE strains in the fixated model with that of the intact model

For each specimen, SR values were measured for all the load increments and regressed for linear correlation (confidence of $P < 0.05$), with the respective load amplitudes. The range of slopes of the linear fits was between -0.022 and $+0.029$ (average 0.00099). In all of the specimens the SR values were independent of the load amplitude.

6 Discussion

Many biomechanical studies have been carried out to assess the accuracy of CT based patient specific models for fixated bone constructs [3, 7, 9, 21, 38, 45, 46, 48, 54, 65]. However, despite the proven ability to accurately predict superficial strain values, patient specific modeling and analysis of musculoskeletal fracture fixation is currently limited to in vitro experimental studies. A few drawbacks can be pointed as reasons for the current state

One limitation of the proposed patient specific workflow is its reliance on a CT data set of the fractured anatomy. Due to the high dose of radiation, a CT exam is usually avoided and employed only when fracture geometry and anatomy are complicated [1, 12, 17–19, 22, 34, 39, 49]. Another limitation that still exists is the fact that the proposed workflow is time consuming and the added value to patient care is yet to be proved.

Given the requirements of the clinical environment and the limitations inherent in a patient specific fracture fixation quantitative analysis workflow, it may be more efficient to design a workflow that targets a certain procedures in which there is a choice between two distinct fixative solutions. Another compromise to be considered is reducing model complexity by simplifying loading conditions and sub-modeling. These would enable to substantially reduce model preparation and analysis and would be acceptable if the added value to patient is proved.

To conclude, a patient specific quantitative analysis workflow for fracture fixations may enable comparison between different fixation solutions and with further simplification and automation may be used as a preplanning tool which will assist the surgeons' decisions regarding type of implant and positioning and may also assist in post-operative patient treatment and holds the potential to improve patient treatment and outcome.

References

1. Albrechtsen, J., Hede, J., Jurik, A.G.: Pelvic fractures. Assessment by conventional radiography and CT. Acta Radiologica (Stockholm, Sweden : 1987) **35**, 420–425 (1994)
2. Audige, L., Bhandari, M., Hanson, B., Kellam, J.: A concept for the validation of fracture classifications. J. Orthopaedic Trauma **19**, 401–406 (2005)
3. Barker, D.S., Netherway, D.J., Krishnan, J., Hearn, T.C.: Validation of a finite element model of the human metacarpal. Med. Eng. Phys. **27**, 103–113 (2005)

4. Baumgaertner, M.R., Curtin, S.L., Lindskog, D.M.: Intramedullary versus extramedullary fixation for the treatment of intertrochanteric hip fractures. Clinical Orthopaedics and Related Research **348**, 87–94 (1998)
5. Baumgaertner, M.R., Curtin, S.L., Lindskog, D.M., Keggi, J.M.: The value of the tip-apex distance in predicting failure of fixation of peritrochanteric fractures of the hip. J. Bone Joint Surg. Am. Volume **77**, 1058–1064 (1995)
6. Bayraktar, H.H., Morgan, E.F., Niebur, G.L., Morris, G.E., Wong, E.K., Keaveny, T.M.: Comparison of the elastic and yield properties of human femoral trabecular and cortical bone tissue. J. Biomech. **37**, 27–35 (2004)
7. Behrens, B.A., Wirth, C.J., Windhagen, H., Nolte, I., Meyer-Lindenberg, A., Bouguecha, A.: Numerical investigations of stress shielding in total hip prostheses. Proc. Inst. Mech. Engrs. Part H, J. Eng. Med. **222**, 593–600 (2008)
8. Bessho, M., Ohnishi, I., Matsuyama, J., Matsumoto, T., Imai, K., Nakamura, K.: Prediction of strength and strain of the proximal femur by a CT-based finite element method. J. Biomech. **40**, 1745–1753 (2007)
9. Bitsakos, C., Kerner, J., Fisher, I., Amis, A.A.: The effect of muscle loading on the simulation of bone remodelling in the proximal femur. J. Biomech. **38**, 133–139 (2005)
10. Carter, D.R.: Mechanical loading histories and cortical bone remodeling. Calcified Tissue Int. **36**(Suppl 1), S19–S24 (1984)
11. Carter, D.R., Hayes, W.C.: The compressive behavior of bone as a two-phase porous structure. J. Bone Joint Surg. Am. Volume **59**, 954–962 (1977)
12. Chmelova, J., Sir, M., Jecminek, V.: CT-guided percutaneous fixation of pelvic fractures. Case reports. Biomedical Papers of the Medical Faculty of the University Palacky, Olomouc, Czechoslovakia, vol. 149, pp 177–181 (2005)
13. Cody, D.D., Gross, G.J., Hou, F.J., Spencer, H.J., Goldstein, S.A., Fyhrie, D.P.: Femoral strength is better predicted by finite element models than QCT and DXA. J. Biomech. **32**, 1013–1020 (1999)
14. Cowin, S.C., Hart, R.T., Balser, J.R., Kohn, D.H.: Functional adaptation in long bones: establishing in vivo values for surface remodeling rate coefficients. J. Biomech. **18**, 665–684 (1985)
15. Dalstra, M., Huiskes, R., Odgaard, A., van Erning, L.: Mechanical and textural properties of pelvic trabecular bone. J. Biomech. **26**, 523–535 (1993)
16. Davis, T.R., Sher, J.L., Horsman, A., Simpson, M., Porter, B.B., Checketts, R.G.: Intertrochanteric femoral fractures. Mechanical failure after internal fixation. J. Bone Joint Surg. Br. Volume **72**, 26–31 (1990)
17. Dock, W., Grabenwoger, F., Schratter, M., Farres, M.T., Kwasny, O.: Diagnosis of pelvic fractures: synoptic views of the pelvis versus CT]. RoFo : fortschritte Auf Dem Gebiete Der Rontgenstrahlen Und Der Nuklearmedizin **150**, 280–283 (1989)
18. Duane, T.M., Dechert, T., Wolfe, L.G., Brown, H., Aboutanos, M.B., Malhotra, A.K., Ivatury, R.R.: Clinical examination is superior to plain films to diagnose pelvic fractures compared to CT. The American Surgeon **74**, 476–479 (2008)
19. Falchi, M., Rollandi, G.A.: CT of pelvic fractures. Eur. J. Radiol. **50**, 96–105 (2004)
20. Ford, C.M., Keaveny, T.M.: The dependence of shear failure properties of trabecular bone on apparent density and trabecular orientation. J. Biomech. **29**, 1309–1317 (1996)
21. Gross, S., Abel, E.W.: A finite element analysis of hollow stemmed hip prostheses as a means of reducing stress shielding of the femur. J. Biomech. **34**, 995–1003 (2001)
22. Hansen Jr, S.T.: CT for pelvic fractures. AJR. Am. J. Roentgenology **138**, 592–593 (1982)
23. Helgason, B., Perilli, E., Schileo, E., Taddei, F., Brynjolfsson, S., Viceconti, M.: Mathematical relationships between bone density and mechanical properties: a literature review. Clin. Biomech. (Bristol, Avon) **23**, 135–146 (2008)
24. Helgason, B., Taddei, F., Palsson, H., Schileo, E., Cristofolini, L., Viceconti, M., Brynjolfsson, S.: A modified method for assigning material properties to FE models of bones. Med. Eng. Phys. **30**, 444–453 (2008)
25. Hill, A.V.: The maximum work and mechanical efficiency of human muscles, and their most economical speed. J. Physiol. **56**, 19–41 (1922)

26. Huiskes, R., Weinans, H., Grootenboer, H.J., Dalstra, M., Fudala, B., Slooff, T.J.: Adaptive bone-remodeling theory applied to prosthetic-design analysis. J. Biomech. **20**, 1135–1150 (1987)
27. Keaveny, T.M., Borchers, R.E., Gibson, L.J., Hayes, W.C.: Trabecular bone modulus and strength can depend on specimen geometry. J. Biomech. **26**, 991–1000 (1993)
28. Keller, T.S.: Predicting the compressive mechanical behavior of bone. J. Biomech. **27**, 1159–1168 (1994)
29. Keyak, J.H., Falkinstein, Y.: Comparison of in situ and in vitro CT scan-based finite element model predictions of proximal femoral fracture load. Med. Eng. Phys. **25**, 781–787 (2003)
30. Keyak, J.H., Fourkas, M.G., Meagher, J.M., Skinner, H.B.: Validation of an automated method of three-dimensional finite element modelling of bone. J. Biomed. Eng. **15**, 505–509 (1993)
31. Keyak, J.H., Meagher, J.M., Skinner, H.B., Mote Jr, C.D.: Automated three-dimensional finite element modelling of bone: a new method. J. Biomed. Eng. **12**, 389–397 (1990)
32. Keyak, J.H., Rossi, S.A., Jones, K.A., Les, C.M., Skinner, H.B.: Prediction of fracture location in the proximal femur using finite element models. Med. Eng. Phys. **23**, 657–664 (2001)
33. Keyak, J.H., Rossi, S.A., Jones, K.A., Skinner, H.B.: Prediction of femoral fracture load using automated finite element modeling. J. Biomech. **31**, 125–133 (1998)
34. Killeen, K.L., DeMeo, J.H.: CT detection of serious internal and skeletal injuries in patients with pelvic fractures. Academic Radiol. **6**, 224–228 (1999)
35. Lang, T.F., Keyak, J.H., Heitz, M.W., Augat, P., Lu, Y., Mathur, A., Genant, H.K.: Volumetric quantitative computed tomography of the proximal femur: precision and relation to bone strength. Bone **21**, 101–108 (1997)
36. Les, C.M., Keyak, J.H., Stover, S.M., Taylor, K.T., Kaneps, A.J.: Estimation of material properties in the equine metacarpus with use of quantitative computed tomography. J Orthopaedic Res.: Off. Pub. Orthopaedic Res. Soc. **12**, 822–833 (1994)
37. Lotz, J.C., Gerhart, T.N., Hayes, W.C.: Mechanical properties of trabecular bone from the proximal femur: a quantitative CT study. J. Comput. Assist. Tomogr. **14**, 107–114 (1990)
38. Luria, S., Hoch, S., Liebergall, M., Mosheiff, R., Peleg, E.: Optimal fixation of acute scaphoid fractures: finite element analysis. J. Hand Surg. **35**, 1246–1250 (2010)
39. Martos, J., Bohar, L., Fekete, G.: Three-dimensional CT studies of pelvic fractures]. Magyar Traumatologia, Orthopaedia Es Helyreallito Sebeszet **35**, 116–119 (1992)
40. Merz, B., Niederer, P., Muller, R., Ruegsegger, P.: Automated finite element analysis of excised human femora based on precision -QCT. J. Biomech. Eng. **118**, 387–390 (1996)
41. Morgan, E.F., Bayraktar, H.H., Keaveny, T.M.: Trabecular bone modulus-density relationships depend on anatomic site. J. Biomech. **36**, 897–904 (2003)
42. Morgan, E.F., Keaveny, T.M.: Dependence of yield strain of human trabecular bone on anatomic site. J. Biomech. **34**, 569–577 (2001)
43. Pappas, G.P., Asakawa, D.S., Delp, S.L., Zajac, F.E., Drace, J.E.: Nonuniform shortening in the biceps brachii during elbow flexion. J. Appl. Physiol. (Bethesda, Md.: 1985) **92**, 2381–2389 (2002)
44. Peck, C.C., Hannam, A.G.: Human jaw and muscle modelling. Arch. Oral Biol. **52**, 300–304 (2007)
45. Peleg, E., Beek, M., Joskowicz, L., Liebergall, M., Mosheiff, R., Whyne, C.: Patient specific quantitative analysis of fracture fixation in the proximal femur implementing principal strain ratios. Method and experimental validation. J. Biomech. **43**, 2684–2688 (2010)
46. Pettersen, S.H., Wik, T.S., Skallerud, B.: Subject specific finite element analysis of stress shielding around a cementless femoral stem. Clin. Biomech. (Bristol, Avon) **24**, 196–202 (2009)
47. Rajon, D.A., Bolch, W.E.: Marching cube algorithm: review and trilinear interpolation adaptation for image-based dosimetric models. Computerized Med. Imaging Graphics : Off. J. Computerized Med. Imaging Soc. **27**, 411–435 (2003)
48. Reggiani, B., Cristofolini, L., Varini, E., Viceconti, M.: Predicting the subject-specific primary stability of cementless implants during pre-operative planning: preliminary validation of subject-specific finite-element models. J. Biomech. **40**, 2552–2558 (2007)

49. Robertson, D.D., Sutherland, C.J., Chan, B.W., Hodge, J.C., Scott, W.W., Fishman, E.K.: Depiction of pelvic fractures using 3D volumetric holography: comparison of plain X-ray and CT. J. Comput. Assist. Tomogr. **19**, 967–974 (1995)
50. Ruimerman, R., Hilbers, P., van Rietbergen, B., Huiskes, R.: A theoretical framework for strain-related trabecular bone maintenance and adaptation. J. Biomech. **38**, 931–941 (2005)
51. Schileo, E., Dall'ara, E., Taddei, F., Malandrino, A., Schotkamp, T., Baleani, M., Viceconti, M.: An accurate estimation of bone density improves the accuracy of subject-specific finite element models. J. Biomech. **41**, 2483–2491 (2008)
52. Schileo, E., Taddei, F., Cristofolini, L., Viceconti, M.: Subject-specific finite element models implementing a maximum principal strain criterion are able to estimate failure risk and fracture location on human femurs tested in vitro. J. Biomech. **41**, 356–367 (2008)
53. Schileo, E., Taddei, F., Malandrino, A., Cristofolini, L., Viceconti, M.: Subject-specific finite element models can accurately predict strain levels in long bones. J. Biomech. **40**, 2982–2989 (2007)
54. Seral, B., Garcia, J.M., Cegonino, J., Doblare, M., Seral, F.: Finite element study of intramedullary osteosynthesis in the treatment of trochanteric fractures of the hip: Gamma and PFN. Injury **35**, 130–135 (2004)
55. Stojanovic, B., Kojic, M., Rosic, M., Tsui, C.P., Tang, C.Y.: An extension of Hill's three-component model to include different fiber types in finite element modelling of muscle. Int. J. Num. Meth. Eng. **71**, 801–817 (2007)
56. Taddei, F., Martelli, S., Reggiani, B., Cristofolini, L., Viceconti, M.: Finite-element modeling of bones from CT data: sensitivity to geometry and material uncertainties. IEEE Trans. Bio-Med. Eng. **53**, 2194–2200 (2006)
57. Taddei, F., Pancanti, A., Viceconti, M.: An improved method for the automatic mapping of computed tomography numbers onto finite element models. Med. Eng. Phys. **26**, 61–69 (2004)
58. Taddei, F., Schileo, E., Helgason, B., Cristofolini, L., Viceconti, M.: The material mapping strategy influences the accuracy of CT-based finite element models of bones: an evaluation against experimental measurements. Med. Eng. Phys. **29**, 973–979 (2007)
59. van Rietbergen, B., Huiskes, R.: Load transfer and stress shielding of the hydroxyapatite-ABG hip: a study of stem length and proximal fixation. J. Arthroplasty **16**, 55–63 (2001)
60. Van Rietbergen, B., Huiskes, R., Weinans, H., Sumner, D.R., Turner, T.M., Galante, J.O.: ESB Research Award 1992. The mechanism of bone remodeling and resorption around press-fitted THA stems. J. Biomech. **26**, 369–382 (1993)
61. Viceconti, M., Bellingeri, L., Cristofolini, L., Toni, A.: A comparative study on different methods of automatic mesh generation of human femurs. Med. Eng. Phys. **20**, 1–10 (1998)
62. Viceconti, M., Davinelli, M., Taddei, F., Cappello, A.: Automatic generation of accurate subject-specific bone finite element models to be used in clinical studies. J. Biomech. **37**, 1597–1605 (2004)
63. Viceconti, M., Pancanti, A., Dotti, M., Traina, F., Cristofolini, L.: Effect of the initial implant fitting on the predicted secondary stability of a cement less stem. Med. Biol. Eng. Comput. **42**, 222–229 (2004)
64. Weinans, H., Huiskes, R., van Rietbergen, B., Sumner, D.R., Turner, T.M., Galante, J.O.: Adaptive bone remodeling around bonded noncemented total hip arthroplasty: a comparison between animal experiments and computer simulation. J. Orthopaedic Res.: Official Pub. Orthopaedic Res. Soc. **11**, 500–513 (1993)
65. Weinans, H., Sumner, D.R., Igloria, R., Natarajan, R.N.: Sensitivity of periprosthetic stress-shielding to load and the bone density-modulus relationship in subject-specific finite element models. J. Biomech. **33**, 809–817 (2000)
66. Wirtz, D.C., Schiffers, N., Pandorf, T., Radermacher, K., Weichert, D., Forst, R.: Critical evaluation of known bone material properties to realize anisotropic FE-simulation of the proximal femur. J. Biomech. **33**, 1325–1330 (2000)
67. Zannoni, C., Mantovani, R., Viceconti, M.: Material properties assignment to finite element models of bone structures: a new method. Med. Eng. Phys. **20**, 735–740 (1998)

Part II
Spine

Advancements in Spine FE Mesh Development: Toward Patient-Specific Models

Nicole A. Kallemeyn, Kiran H. Shivanna, Nicole A. DeVries, Swathi Kode, Anup A. Gandhi, Douglas C. Fredericks, Joseph D. Smucker and Nicole M. Grosland

Abstract Laboratory-driven experimental studies are capable of delineating the biomechanical characteristics of the spine. They are limited, however, to external responses; that is, internal stresses and strains throughout the structures are not readily attained. Mathematical simulations provide a unique opportunity to serve as an adjunct to experimental studies to predict the external responses, while complementing the experiments by providing such internal responses. Musculoskeletal finite element (FE) analyses have emerged as an invaluable tool in orthopaedic-related research. While it has provided significant insight into the biomechanics of the spine, the demands associated with modeling the geometrically complex structures often limit its utility. Individualized models are important for future development of this field, as they offer a means of correlating mechanical predictions with clinical outcomes. However, relatively few FE studies to date have employed specimen- or patient-specific models. Spine modeling is by no means an exception. In this chapter we describe multiblock methods for generating subject-specific spine meshes to alleviate the current limitations of spine meshing. In addition, we

N. A. Kallemeyn · K. H. Shivanna · N. A. DeVries · S. Kode ·
A. A. Gandhi · N. M. Grosland (✉)
Department of Biomedical Engineering, The University of Iowa,
Iowa City, IA, USA
e-mail: grosland@engineering.uiowa.edu

N. A. Kallemeyn · K. H. Shivanna · N. A. DeVries · S. Kode ·
A. A. Gandhi · N. M. Grosland
Center for Computer Aided Design, The University of Iowa,
Iowa City, IA, USA

D. C. Fredericks · J. D. Smucker · N. M. Grosland
Department of Orthopaedics and Rehabilitation,
The University of Iowa, Iowa City, IA, USA

demonstrate additional computational tools to perform "virtual surgery," and show examples of how the techniques have been applied to date.

Abbreviations

FE	Finite element
ACDF	Anterior cervical discectomy and fusion
FSU	Functional spinal unit
CT	Computed tomography
MR	Magnetic resonance
IVD	Intervertebral disc
ALL	Anterior longitudinal ligament
PLL	Posterior longitudinal ligament
CL	Capsular ligament
LF	Ligamentum flavum
IS	Interspinous ligament
ODL	Open door laminoplasty
DDL	Double door laminoplasty

1 Introduction

The spine consists of discrete bony elements (vertebrae) joined by passive ligamentous restraints, separated by intervertebral discs and articulating joints, and dynamically controlled by muscular activation. This complex mechanical structure serves to protect the spinal cord and nerve roots while providing flexibility via motion to the neck and trunk. It transmits the weight of the upper body to the pelvis and is subjected to internal forces exceeding many times body weight. The spine is broadly divided into five regions: the cervical spine, the thoracic spine, the lumbar spine, the sacrum, and the coccyx. Each region has its own unique set of kinematic functions, pathologies, and treatments. The cervical spine can be further subdivided into the upper cervical spine segment (C0–C1–C2) and the lower cervical spine (C3–C7), each having distinct anatomic and functional features.

Likely secondary to the large amount of motion, the cervical and lumbar spine are the most prevalent regions affected by degenerative disorders. Disruption of any of these structures via degeneration, injury, or surgical procedures may alter the physical properties of the motion segment either by their direct biomechanical effects or the accentuation of a pre-existing instability [34]. These alterations may ultimately lead to pain, physical dysfunction, and medical disability. Low back pain alone is the most expensive disease process in the 30–50 year age range [89]. The direct costs of care for individuals affected by low back pain have been estimated at $20 billion and total costs exceeding $100 billion [49].

Surgical intervention in the cervical spine has been generally limited to treatment of nerve root disorders, radiculopathy, spinal cord compression, and myelopathy using conservative measures. Anterior cervical discectomy and fusion (ACDF) has a long-standing track record in this regard [3]. However, fusion-based interventions for neck pain have not been seen as beneficial in the long term. Despite the surgical indication, ACDF maintains a known incidence of adjacent functional spinal unit (FSU) degeneration of 2.9% per annum resulting in clinically relevant symptoms [38]. Segments adjacent to a fused segment are prone to recurrent neurologic symptoms [32, 33], have increased range of motion [20], and have increased intradiscal pressures [16]. Prevention of adjacent FSU degeneration has been postulated to be an important secondary measure of success of new technologies such as cervical total disc arthroplasty. Despite the desire to intervene in a way that is "motion sparing," the understanding and modeling of this complex environment is lacking.

Preventive measures and treatment modalities for correcting spinal disorders benefit significantly from advancements aimed at understanding the biomechanics of the human spine in the normal as well as altered states [31]. Experimental investigations alone, however, may fail to completely delineate the biomechanics of such disorders. The complexities that precede experimental measurements, the difficulties in experimental simulation of pathology at various stages of the disorder, and the effect of unrecognized parameters on measurements in an experimental investigation have been cited as experimental inadequacies [70]. Such limitations are sometimes confounded by the drive to create interventions prior to defining the pathological process and expenses ensued during in vivo investigations. These ambient factors have provided an impetus to develop mathematical models (e.g., finite element models) of the spine to complement the in vivo investigations.

1.1 *Rationale for the Use of the Finite Element Method in Spine Biomechanics Research*

The finite element (FE) method constitutes one of numerous analytical techniques used in conjunction with experimental procedures to solve problems in spinal biomechanics. A number of mathematical tools available for such analyses, however, are not suitable for the highly irregular structural geometry of the spinal column. The unique capability of the FE method to evaluate stresses in structures of complex geometry, loading, and material behavior makes the technique advantageous over many other mathematical alternatives [28]. Utilization of this method requires that the model be defined initially as a geometric entity. The model is then subdivided into discrete, small divisions referred to as *elements*. These elements are assumed to be joined solely at the corners (*nodes*), establishing a geometrical grid of the entire structure. Consequently, the nodes grossly define the geometric shape of an element. A wide variety of elements with different geometric form (i.e., beams, plates, blocks, etc.) are available; thus making it possible to model the spine with varying degrees of complexity. As a result, preliminary estimates may be attained with a simplistic model rather quickly, en route to developing a more realistic/detailed model. It should be noted that the solution obtained via the FE method is approximate in the sense that it converges to the exact solution for the model when the mesh density approximates infinity. Accordingly, the literature reveals that FE models of varying complexities have been developed to study the spine [26, 28, 63, 87, 94]. Although FE modeling is in principle suited for structures of arbitrary complexity (i.e., biological structures), it must not be relied upon as a tool to generate solutions without having carefully formulated the problem. Modeling techniques often require a number of simplifications in geometry, material properties, and loading that may adversely affect the validity of the results. Its application to three-dimensional and/or nonlinear structures remains tedious, expensive, and prone to error. The expense of a FE analysis of the spine is frequently underestimated (i.e., software-licensing fees, computational time, personnel time for model development, in addition to running and analyzing the results, etc.). One of the primary bottlenecks currently remains the time devoted to model development.

Nevertheless, the past two decades have shown a growing interest in computational modeling of the spine. In 1987, Yoganandan et al. [95] reviewed mathematical modeling of the spine, including continuum models, discrete and lumped parameter models, in addition to models based on the advanced finite element method. In 1995, Goel and Gilbertson [28] summarized applications of the finite element method to the thoracolumbar spine. That same year, Gilbertson et al. [26] reviewed ongoing developments in spine biomechanics research through the finite element method. Shortly thereafter, Yoganandan et al. [94] and Voo and Pintar [87] followed up with a critical review of cervical spine finite element modeling applications. More recently, Natarajan et al. [63] provided a review of the most recent advances in the development of poroelastic analytical models. These key review articles have concentrated on developments in model construction, constitutive law (material properties) identification, loading and boundary condition details, and efforts toward validating the models. The advancements in imaging techniques and the vast improvements in computational speed have permitted continued increases in the complexity of the models.

1.2 Representative FE Models of the Spine

The compliance of the human spine needed to perform physiologic tasks is provided solely by its functional spinal units (FSUs) [70]. The mechanical response of a single motion segment relies on its substructures (i.e., the intervertebral disc, vertebrae, ligaments, and facet joints). Further subdivisions may be implemented by dividing each substructure into a number of constituent elements. For example, the intervertebral disc can be partitioned into the cartilaginous endplates, the nucleus, and the annular region. Additionally, the latter may be subdivided into the collagenous fiber layers and the ground substance. Such subdivisions may be performed until the desired degree of refinement is attained.

The highly complex nature of the human spine as a whole renders the foregoing substructuring technique valuable to researchers attempting to study the biomechanics of the individual motion segments. There exist advantages as well as limitations, however, to such investigations. Although appropriately applied loading and boundary conditions enable the spinal unit to experience realistic states of stress and deformation as if it were acting as a part of the whole, the response of the spinal column cannot necessarily be computed via single motion segment studies. The most primitive analyses [2] related to the intervertebral motion segment were restricted to the isolated disc-body unit, placing emphasis on the disc. Such analyses [2, 48, 66, 72, 75–77] varied in degree of complexity, from the development of simplistic models to the formulation of FE models of various degrees of refinement.

1.3 Simplistic FE Models

'Simple' FE models [67, 71] of the spine typically represent each motion segment (vertebra-disc-vertebra unit) with minimal elements, ranging in number from one to three, thereby reducing computational time and costs. Such models may be augmented by the addition of finite elements representing the ribcage to simulate the trunk. In addition, the effects of muscles, soft tissues, internal organs, as well as spinal instrumentation (i.e., Harrington rods, orthoses, etc.) may be incorporated as well. This approach has been popular for the study of lengthy regions of the spine—necessary for the study of vibration, thoracic impact biomechanics, scoliosis, spinal bracing of thoracolumbar injuries, and spinal stability. In general, simplistic FE models of the spine are used to obtain motion segment displacements/loads in response to various external loading and boundary conditions.

1.4 Detailed FE Models

Detailed FE models of the spine utilize the aforementioned substructuring technique to model the individual spinal components (i.e., disc, ligaments, facet joints, etc.) with various element types and mesh densities. Due to the excessive number of nodes and elements typically required to accurately represent the complex geometries in a detailed FE model, computational storage and time restraints have, until recently, limited detailed FE analyses of the spine to one- or two- motion segments. The progressive evolution of FE models of the spine has influenced the development of detailed multimotion-segment models (i.e., models consisting of two or more motion segments). Multimotion-segment FE models are capable of providing information with respect to the biomechanics of the spine that a single motion segment model is incapable of predicting. For example, clinical reports suggest that spinal fusions alter the biomechanics and kinematics of contiguous spinal segments. Consequently, the extent to which a stabilization procedure or injury affects the biomechanics of the adjacent levels may be delineated via multimotion-segment models.

Today, the majority of spine models share the following characteristics. The geometry is based on image datasets, typically axial slice contours from computed tomography (CT) image datasets from a cadaveric specimen (age: late 60s or older). The models are multi-segment in nature, consisting of a given number of vertebrae, each separated by an intervertebral disc, bilateral articulating facet joints, and numerical representations of the ligamentous structures at each segmental level. Nonlinear cable or truss elements or springs are used to define the ligaments, while contact interactions at the facets are modeled via surface interactions or more often using gap elements with no resistance in tension and a stiffening resistance in compression. The intervertebral disc is often represented by an isotropic material reinforced by crisscrossing fibers, while the nucleus is represented as an incompressible fluid filled cavity. The intervertebral disc height is often based on values reported in the literature. Furthermore, the assumption of sagittal symmetry is oftentimes adopted to minimize the time devoted to mesh development as well as computational run time. This assumption also aids model validation in the sense that the number of loading conditions considered may be reduced. Commonly accepted material property assignments are taken from the literature.

Despite these similarities, the geometric fidelity of these models differs considerably. Oftentimes, models that claim to be "accurate 3D representations" simply resemble the anatomic structure under consideration, yet remain rudimentary. Again, this is most likely attributed to the lack of available software and meshing techniques for addressing the geometric complexities of the spine, and hence the vast amount of time required to generate such models.

1.5 Experimental Validation

Validation is one of the most challenging aspects of modeling musculoskeletal mechanics, as it requires accurate experimental measurements of quantities that are difficult to measure. As a result, more often than not, the validation of musculoskeletal finite element models has not been given the attention that it merits. The long-term success of FE modeling, however, hinges on the proper validation of computational models [90]. There are numerous ways to validate a spine model. Traditionally, a validated spine model is one that compares favorably with previously reported experimental behaviors. The most important model parameters to validate are those that the model will be used to address. For example, a model to be used to study ligament strains must be validated against both kinematics and ligament strains. If, however, the model will be used to predict spinal kinematics, validation of the bone strains is impractical.

Spinal kinematics are generally nonlinear, making validation even more complex. Spine FE models validated against averaged experimental kinematics are often compared to only a single data point, i.e., the rotation as a result of a single load. However, knowledge about the entire loading curve can be valuable when analyzing the effects of spinal interventions, especially motion sparing technologies. It is important to be able to reproduce the motion path and determine the individual effects each of the nonlinear soft tissues has on the resulting motion [6, 43].

Subject-specific validation experiments are challenging; consequently, the majority of models are validated against published experimental data from a range of subjects [5]. Intersubject variability lends itself to high standard error measurements associated with such studies, thereby making validation of the kinematics difficult. Furthermore, experimental testing conditions vary throughout the literature. For example, the number of segments tested, specimen orientation, the loading and boundary conditions, type of testing system, and combinations thereof, differ considerably. Nevertheless, a validated model of the normal spine with biological fidelity is a necessary precursor to accurate simulation of pathological or post-surgical spinal biomechanics.

In addition to validation, mesh convergence is of importance so that the FE modeler is assured that an appropriate discretization of the domain has been performed, and that the mesh refinement will produce accurate numerical results [85]. A typical convergence study involves generating multiple meshes of varying refinement and applying the same loading and boundary conditions to each mesh. A relevant resultant quantity is then plotted with respect to the number of elements or nodes and convergence is defined where there is little change in the measured result with an increase in mesh refinement. The measured quantity should be carefully chosen based on what the model will ultimately be used to measure. In the case of the spine, it is usually segmental rotation/displacement or stress.

Convergence studies are not typically performed in spine finite element analyses due to the lack of remeshing capabilities for most mesh development methods; in fact, it is rarely mentioned in most spine FE studies [17]. In the studies which do report convergence, details are often not included. Villarraga et al. [86] developed a canine cervical FE model and performed a displacement convergence test on one-half of a cervical body using nine interior and exterior points to monitor the model convergence. Others have given generic statements on mesh refinement such as "owing to the geometrical complexity of the cervical spine, the FE mesh of C46 was fairly fine" [82]. Lu et al. [59] discussed that the mesh density of their model was selected based on a convergence study but did not publish the methods employed to obtain the results. The mesh was determined to be "at least as fine as other models." In another lumbar FE study [15], a 'coarse' mesh and 'refined' mesh were created and moment-rotation (flexion) curves were generated for both models. Often the limiting factor in performing in depth convergence studies is the time and effort required to develop multiple meshes of varying densities.

2 Mesh Generation Overview

Mesh generation techniques are broadly employed in the various engineering fields that make use of physical models based on partial differential equations. Numerical simulations of such models are routinely used for design dimensioning and validation purposes. The finite element method, via a mesh, obtains a solution to an intractable complex continuum problem by replacing it with a discrete one for computational purposes, thus affording an approximate solution to the original problem. An important step for the numerical simulation of physical phenomena is the transformation of the underlying differential equation into a finite dimensional space. In the domain of interest the resulting partial differential equation is approximated using numerical methods, on finite intervals. Determining the finite intervals requires a discretization of the domain. The exact type of discretization is determined by the numerical solution method. Furthermore, the nature of the discretization influences the quality of the numerical solution (i.e., by the shape of the elements and the accurate approximation of the domain).

While the effects of element shape on FE solutions are still being investigated, there are two broad types of mesh generation schemes—routines for structured and unstructured meshes [4, 25, 62]. The techniques for generating structured grids are based on rules for geometrical grid-subdivisions and mapping techniques. Structured grids, as the name implies, have a clear structure and can be recognized by the fact that all interior nodes of the mesh have an equal number of adjacent elements. The techniques used to generate them produce triangular or quadrilateral elements in two-dimensional analyses, and tetrahedral and hexahedral elements in three-dimensions. Structured meshing algorithms generally involve complex iterative smoothing techniques that attempt to align elements with boundaries or physical domains. Where non-trivial boundaries are required, "block-structured" techniques can be employed which allow the user to break the domain up into topological blocks. Structured grid generators are most commonly used when strict elemental alignment is mandated by the analysis code or it is necessary to capture physical phenomenon. Unstructured mesh generation, on the other hand, relaxes the node valence requirement, allowing any number of elements to meet at a single node. Triangle and tetrahedral meshes are most commonly thought of when referring to unstructured meshing. While some overlap exists between structured and unstructured mesh generation technologies, the main feature that distinguishes the two approaches is the unique iterative smoothing algorithm(s) employed by structured grid generators. While free-form meshing schemes using tetrahedral elements are widely employed, it is well known that hexahedral elements would, in many cases, be more effective for analysis [9, 57, 84, 88]. Unfortunately, an effective general unstructured mesh generation scheme of hexahedral elements for complex geometries does not exist.

2.1 Automated Meshing Algorithms

Three-dimensional models of human bones are usually derived from image data sets, whether they be CT or magnetic resonance (MR), using various approaches, most of which are scarcely automated. The manual generation of a three-dimensional mesh, although often a highly accurate method, requires significant time and operator effort to complete a single mesh. Consequently, most of the analyses reported in the literature refer to a single, or 'average,' bony geometry, although in many cases the anthropometric variability of bone size and shape should not be neglected. Furthermore, mesh refinements and convergence checks are rarely reported for this type of mesh. As a result, additional compromises may include sub-optimal mesh refinement, homogenously modeled regions of (heterogeneous) bone [64] or simplifying assumptions of symmetry [14, 26, 27, 29, 30, 39, 78]. These limitations are ever more prevalent when detailed models of the spine are considered due to the complexities involved.

In an effort to unencumber the mesh development process, several automated meshing algorithms have been implemented, thus making the notion of patient-specific models a reality. Automated mesh generators typically rely on tetrahedral meshing capabilities. Geometrically complex structures, such as the musculoskeletal anatomy, can readily be represented by a collection of tetrahedral elements, as these elements tend to be more geometrically versatile; whereas hexahedral definitions prove more challenging. Tetrahedral elements, however, are constant-strain elements, and tend to be outperformed by hexahedral elements. Robust in nature, hexahedral elements are highly favored. They yield a mesh of better quality and require a lower element count to fill a physical domain, thereby reducing the computational costs.

All hex mesh definitions emerged in the form of a voxel mesh, which builds cubic 8-noded elements directly from CT data sets. The fact that the voxel meshing technique does not rely on any pre-processing of the dataset made it a favorable choice and, in some cases, perhaps the only viable option.

The premise of voxel-based meshing is the transformation of CT image data directly into elements for FE models using eight-noded hexahedral elements [50]. Because there exists a linear relationship between the digital CT Hounsfield number and the apparent density of bone [61], not only is the geometry preserved, but the material properties are also directly assigned throughout the volume. The surfaces of direct voxel-based models, however, are characterized by abrupt right angles—a stair-step effect. These abrupt transitions have remained largely in the realm of annoyance, in that they render the computed solutions suspect only near the boundaries [41, 42, 51]. However, for applications where surface quantities are the focus of interest (e.g., joint contact analyses, bone surface remodeling, surface initiation of fractures, and models validated by surface strain data), a means for surface smoothing is critical to overcome the geometric irregularities inherent to the voxel-based models.

Automated hexahedral meshing methods such as plastering [7], whisker weaving [81], and octree-based [40, 96] practices have been reported. Although such techniques yield meshes of high quality, element size and orientation remains a challenge. Moreover, they have proven to be less than robust.

Teo et al. [84] have developed a patient-specific model of the human spine based on the Virtual Human dataset. They proposed a mesh development technique based on a grid plane, used to divide the anatomic boundaries and its inclusions into discrete meshes. A grid frame was then used to connect the grid planes between any two adjacent planes. The resulting mesh consisted predominantly of hexahedral elements, with pentahedral and tetrahedral elements residing at the boundaries. A limitation of their model was the use of an index parameter governing the degree of external smoothing. For example, if the parameter is set to 0, the algorithm does not smooth the steps between the contours, thereby yielding a stairstep appearance similar to the voxel-based mesh definitions. In contrast, if the parameter is set to 1, the algorithm best fits the elements to smooth the surface, thereby resulting in a tradeoff between mesh quality and geometric conformance. Furthermore, the different element types may result in a mathematical stiffness mismatch, which may yield discontinuities. Lastly, the model has yet to be fully validated. Not to mention that it is based on the Visible Human dataset; which precludes the model from being validated against the physical cadaveric specimen, but rather relies on reports in the literature.

A tetrahedronization scheme developed by Wang et al. [88] was used to create marching-cubes surface smoothened finite element models of the spine. The lumbar (L3) spine model conforms to the spinal anatomy and is more accurate than traditional voxel models because of the smoothed surfaces. However, the meshes produced with this technique consist of tetrahedrons at the outer surface and hexahedral elements internally, which is not ideal. Another limitation of this method is that it is voxel-based, resulting in a higher number of elements (computationally expensive).

Kaminsky et al. [46] developed a spine meshing tool which uses a 3D region growing function to accelerate the image segmentation process. A rotation transformation and warped dissection plane are applied to separate vertebral bodies once they have been segmented. A fully automated routine has been developed which ultimately results in a mesh consisting of mostly brick (hexahedral) elements, with small parts of the mesh replaced by tetrahedral elements to avoid calculation errors. This method is an improvement over voxel meshing because smoothing inaccuracies are not present.

Non-rigid deformations have also been used for patient-specific finite element mesh generation. This method is based on a surface matching algorithm through which a least squares minimization of the distance between two surfaces is performed [79]. In these cases, a reference or template mesh can be matched to the geometry of a particular subject. The key benefits of this technique are that all hexahedral subject-specific meshes can be created from one template mesh (assuming the template mesh is all hex), and that mesh generation time is generally reduced compared to manual methods. However, the main drawback is that a high quality template mesh must be produced initially. In addition, if the surfaces of interest are relatively different in size and/or shape, the resulting mesh quality may be less than ideal.

Couteau et al. [10] developed a mesh matching algorithm which generates custom 3D FE meshes from an existing mesh using an elastic registration method. The registration method reshapes an object to match another object using a hierarchical and adaptive 3D displacement grid to represent the transform between the object and reference mesh, and uses gradient features to match the object. The technique was evaluated using the femur. CT images of a femur were segmented using a threshold technique (edge detection algorithm); cubic splines were fit to the contour lines and stacked to represent the 3D bony surface. Patran (MSC Software, Santa Ana, CA) was used to create the reference mesh with hexahedral and wedge elements; this mesh was validated using a vibration technique as well as comparison with experimental strain measurements. The surface points of ten other femora were obtained; the reference and femora surface points were aligned at their centroids and the reference mesh was matched to each femur. The quality of the meshes was evaluated and they were considered satisfactory. When compared to the voxel technique, the advantage of this method is the smooth surface representation. The disadvantages of this method are the need to create the reference mesh, and the elastic registration method does not accept shapes for matching which are very different. Excessive distortion of elements can occur if the template mesh is much larger than the structure to be meshed [84]. In this case, the template mesh was not comprised of all hexahedral elements.

Our research group has developed an automated hexahedral meshing technique for anatomical structures using a mapped meshing method [60, 36]. The objective was to map a predefined template mesh of high quality onto a different subject-specific bony surface definition. The process involves an iterative closest point registration to define an affine transformation to bring the template mesh into rough correspondence with the subject surface, thereby accounting for initial size, position, and orientation differences between the objects. A deformable registration was then applied to bring the template mesh into correspondence with the subject surface. The technique successfully generated meshes of the phalanx bones.

The techniques described above, among others, can be used for subject-specific spine meshing. However, they either do not consist entirely of hexahedral elements or require a high quality hexahedral template mesh. In the case of the spine, the template mesh must often be generated manually. For mapped meshing, even if a high quality template mesh can be made, there may still be element quality issues due to differences in geometry between the template and the subject meshes. In this instance, a different template mesh could provide better results; therein requiring the ability to remesh easily and quickly. Needless to say, an automatic purely hexahedral mesh generator does not currently exist. Moreover, semiautomatic hexahedral mesh generators account for a large portion (50–90%) of the budget for a large nonlinear model [5].

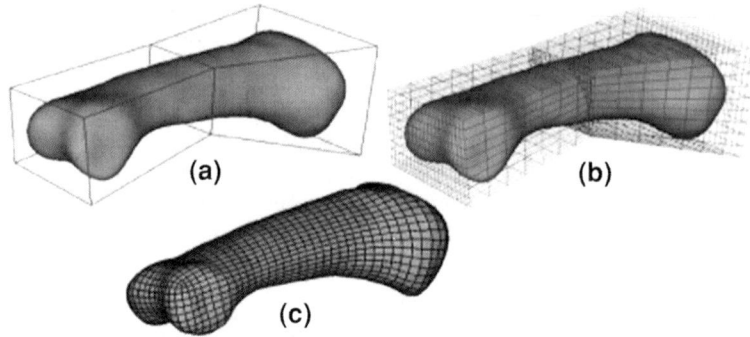

Fig. 1 a Phalanx bone with building block structure. **b** The blocks are assigned mesh seeds which are then projected onto the surface. **c** The resulting mesh

2.2 Multiblock Mesh Generation

The semi-automated multiblock method allows the user to break up a complex domain of interest into topological blocks. These multiblock grids are a powerful extension of the structured mesh. Structured meshing techniques are applied to a series of interconnected sub-grids or 'blocks.' While the individual blocks remain structured, the blocks fit together in an unstructured manner. As a result, the advantages of structured and unstructured meshes are harnessed. The multiblock technique affords geometric flexibility while retaining computational efficiency [37, 65]. Our research group has developed a multiblock approach to finite element mesh development in the form of a standalone software package called IA-FEMesh [37]. The focus of the package is the development of tools for meshing the complex geometries associated with anatomic structures. The goal was to create an intuitive, easy to use interface that nearly guides the user through the model development process. The modeling techniques initiate with a triangulated isosurface of the bone(s) of interest, generated directly from a segmented CT or MR image data set. An interactive technique is used to create one or more building blocks around the bone surface (Fig. 1a); mesh seeds are defined for each block (Fig. 1b) which are then projected onto the surface to be surface nodes for the mesh. The interior nodal positions are calculated using elliptical grid generation, and the volume is then filled with hexahedral elements through connectivity of the nodes (Fig. 1c). Laplacian smoothing can be performed to improve the quality of the elements at the mesh surface. Thereafter, the material properties, loading conditions, and boundary conditions are assigned to the mesh and the model is exported in ABAQUS (DS Simulia, Providence, RI) format.

As an initial evaluation of the software [13], phalanx bone meshes were created and compared to meshes created with conventional methods using Patran

(MSC Software, Santa Ana, CA) and an automatic tetrahedral mesh generator called NetGen [68]. The time to generate hexahedral meshes of similar mesh density was 6 min using IA-FEMesh as compared to 330 min using Patran. NetGen produced a tetrahedral mesh in 4 min. Overall, the stress distributions between the meshes were similar yet the tetrahedral meshes required more elements to obtain smooth distributions. Meshes created with Patran resulted in flat, inaccurate articulating surfaces because the mesh was created from stacked traces from CT segmentations. Because IA-FEMesh uses smoothed surface representations of the bone for mesh generation, it was able to capture the geometry of the articulating surfaces, making phalanx meshes that were suitable for contact analysis.

In addition to having the ability to capture the geometry of articulating surfaces, the multiblock method allows for rapid remeshing. Once the block structure has been created, the vertices can be easily modified to change where the nodes are projected. The mesh seeds can be edited to increase or decrease mesh refinement, making it very practical for convergence studies. In addition to global mesh refinement, localized refinement can be achieved by modifying the mesh seeds of a smaller set of blocks (or a single block) as opposed to the entire structure.

3 Multiblock Spine Meshing

The aforementioned multiblock meshing techniques were developed with the intention of meshing anatomic structures represented by single closed surface representations. These tools have been used to successfully mesh a number of structures, ranging from the phalanx bones of the hand to a porcine skull. The spinal anatomy, however, presents additional meshing challenges. As a result we developed novel meshing tools (currently under development in-house) to be used in conjunction with IA-FEMesh [45]. For example, the intervertebral disc is often modeled to include the different regions of the annulus fibrosis and nucleus pulposus. To further model the criss-crossing layers of fibers within the annulus, it is desirable to control the number of concentric "rings" which make up the annulus. These meshing features as well as the mesh seeds must propagate through the adjacent vertebral bodies to satisfy mesh connectivity in those regions. The modeler must consider these issues when building the mesh.

Spine multiblock meshing initiates with the vertebral surfaces of interest segmented from CT images. Each vertebra is meshed in two stages: first the vertebral body, followed by the posterior region. These two regions are then merged to create a full vertebral mesh. Next, the intervertebral discs are interpolated between the corresponding vertebral body meshes. Multiple spinal levels are created by merging the intervertebral discs with the adjacent vertebral bodies. Finally, ligaments are defined between the adjacent vertebrae.

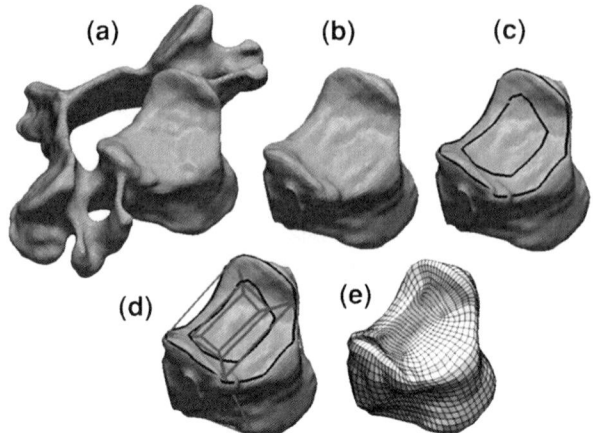

Fig. 2 a The full vertebral surface is **b** clipped at the pedicles. **c** The superior and inferior endplates are traced and **d** a building block structure is used to generate the **e** mesh

3.1 The Vertebral Body

An imposed constraint of the spine model is the mesh pattern defining the vertebral body. This pattern was chosen based on experience, namely modeling the connectivity of the intervertebral disc (i.e., distinguishing the annulus fibrosis and the nucleus). Most notably, the pattern simulates the concentric layers of the annulus and accommodates annular fiber definitions in each element. Again, the vertebral mesh initiates with the surface defined from the image dataset. The STL surface definition of the entire vertebra is initially clipped using two cutting planes at the pedicles enabling the posterior region to be removed (Fig. 2a, b). Delaunay triangulation is used to patch the holes introduced at the pedicles, thereby yielding a closed continuous surface.

By coupling a novel interactive tracing technique with our projection method, the vertebral centrum can be meshed with relative ease. The first step consists of delineating the superior and inferior endplates of the vertebral body, as well as the periphery of the adjacent nucleus pulposus, by interactively tracing these respective limits [73] (Fig. 2c). These traces set the bounds for a "butterfly" building block structure. This block assemblage is defined automatically around the vertebral body surface at the request of the user and can then be edited (Fig. 2d). Thereafter, mesh seeds are assigned throughout the block structure and the nodes are projected onto the vertebral body surface using closest point projection (Fig. 2e). Solid elliptical interpolation is used to generate the interior nodes [74].

3.2 The Vertebral Arch

One of the challenges associated with meshing the irregular geometry of the vertebrae is the connectivity between the posterior elements and the vertebral body. One option to alleviate this challenge could have been to use additional

Fig. 3 a,b Points on the vertebral mesh are picked to generate initial building blocks for the posterior region at the pedicles. **c** The block structure for the entire posterior region is then generated and **d** the vertebral mesh is created

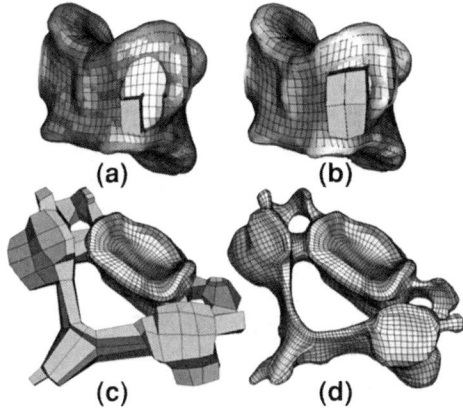

building blocks aligned with the pedicles, further subdividing the vertebral body block structure. Blocks could then be created for the posterior region to create a full vertebral mesh. This option was abandoned, however, for a novel meshing scheme which allows the posterior region to be meshed separately and merged with the existing vertebral body mesh. Using the new meshing technique, new blocks are extruded directly from the nodes of a predefined vertebral body mesh at the pedicle regions (Fig. 3a, b). These blocks become the starting point for the user to create a new building block structure surrounding the posterior surface. They also provide a mutual platform for later merging the body and posterior meshes.

After the posterior building blocks are created (Fig. 3c), mesh seeds are assigned to the block structure based on a global average element edge length. Within IA-FEMesh, the user has the ability to change the mesh seeding for individual blocks if local mesh refinements are desired. Mesh seed changes for individual blocks propagate throughout the block structure via shared block edges.

To create the full vertebral mesh, the vertebral body mesh is automatically modified at the pedicle regions to match the posterior region mesh on the planes where the cuts were made. The seeds for the posterior building blocks at the pedicles are automatically governed by the seeding of the corresponding vertebral body mesh to ensure mesh connectivity in that region. The posterior mesh seeds are then projected onto the posterior vertebral surface using closest point projection, and elliptic interpolation [74] is used to calculate the interior node locations of the hexahedral mesh. Laplacian smoothing [19] and mesh untangling [8] are subsequently performed to improve the mesh quality. The separate vertebral body and posterior region meshes can then merged to create the entire vertebral mesh (Fig. 3d).

Once a vertebral mesh has been generated, the ability to map it to a surface representing another vertebral level and/or patient would be desirable. As described previously we have developed techniques for mapping such meshes [36, 60]. Moreover, we have taken advantage of the multiblock method by extending our mapped meshing routine to the block structures themselves [44]. Block structures are considerably less refined than meshes and may be readily

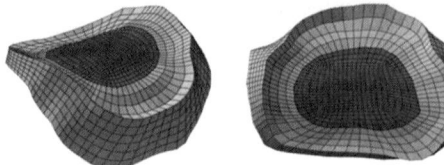

Fig. 4 Intervertebral disc mesh: the outer lighter colored elements represent the annulus and the inner darker colored elements represent the nucleus

edited (via mesh seeding or adjusting the vertices) after mapping to maximize element quality, thereby yielding a patient-specific mesh of high quality in less time than mapping the mesh itself. For large patient-specific spine datasets, this can be particularly beneficial.

Multilevel spine models require not only adjacent vertebral meshes but the supporting soft tissues as well. Namely, the intervertebral disc(s), ligaments and the cartilaginous articulations at the facets. The following sections outline meshing strategies for each of these structures.

3.3 The Intervertebral Disc

The intervertebral disc (IVD) mesh is automatically generated by interpolating between adjacent vertebral body endplate mesh definitions (Fig. 4). Creation of the IVD is straightforward. As described earlier, the vertebral bodies are assigned the same mesh seeding and pattern; this pattern is then carried through the cross-section of the IVD. The user has the ability to choose the number of elements that are created in the superior-inferior direction of the disc.

3.4 The Ligaments

Ligaments are modeled using two-noded truss or spring elements and are defined to act nonlinearly in tension only. The origin and insertion locations are based on anatomical descriptions [69, 93]. Cross-sectional areas are assigned based on experimental data from the literature. For the cervical spine, five cervical spine ligaments are usually modeled: the anterior longitudinal ligament (ALL), posterior longitudinal ligament (PLL), capsular ligaments (CL), ligamentum flavum (LF), and the interspinous ligament (IS) (Fig. 5a).

3.5 The Facets

There are a number of different methods that can be used to model facet cartilage in a spine finite element model. The facet joint is commonly modeled by defining contact interactions either via surface interactions or gap elements. Gap elements

Fig. 5 a Typical ligaments included in cervical spine modeling. **b** C2–C7 multilevel finite element model

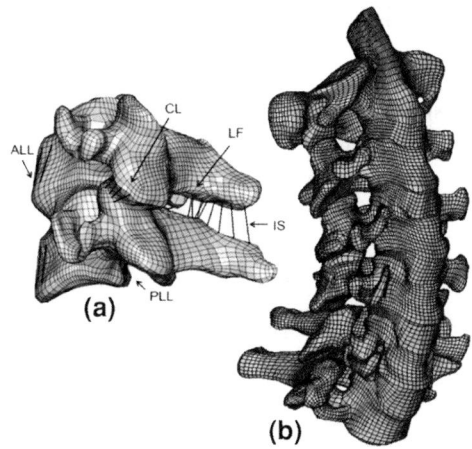

are two-noded elements which define a direction of contact. The facet joints can also be treated as a finite-sliding surface interaction problem, where cartilage layers on each facet surface can be represented by an exponential pressure-over-closure relationship. This relationship can be defined so that the as the distance between the two surfaces decreases, the contact pressure increases exponentially until eventually reaching the modulus of posterior bone.

Kumaresan et al. [55] examined other methods for modeling the facet joint. In addition to gap elements and contact interactions, they specifically modeled synovial fluid with hydrostatic incompressible fluid elements and hyperelastic solid elements. An isolated facet joint capsule model was created and the force–displacement response of the joint was examined with respect to each method for modeling the facet joint. Inclusion of the fluid element definition for the synovial fluid most accurately represented the anatomy of the human facet joint and provided uniform stresses across the joint. However, the downside of this study was that an idealized, isolated facet joint was used. Implementing these techniques into a full spine model presents many challenges with respect to meshing on a patient-specific basis.

A more detailed finite element study has been done on a C3–C7 spine model, which analyzed the effects of different facet cartilage representations including constant thickness, flat surface, and two anatomy-based thickness distribution models [91]. The anatomy-based distribution models were established from averaged experimental data using serial sections of fresh-frozen cervical vertebrae [92]. The results of the finite element study were that the cartilage thickness representation does not impact the overall range of motion or global contact force predictions, but does affect facet contact pressure, area, center of pressure, and joint congruence.

Because it is important to accurately represent facet cartilage in a finite element model, more work is needed in this area to improve patient-specific spine modeling. Ideally, detailed MR scans of the facet joints could be performed to obtain subject-specific information about the facet cartilage (i.e., coverage, thickness, etc.) to mesh the cartilage using hexahedral elements. This information could then be spatially registered with the CT images to obtain subject-specific cartilage within a spine model with contact interactions defined between them at each level.

Using the multiblock techniques described, we have developed a C2–C7 multilevel subject-specific model of the cervical spine with all of the relevant anatomic structures (Fig. 5b). We have validated this model using subject-specific flexibility data [43].

4 Applications

After an intact spine model has been created and validated, it can be used for an infinite number of applications. Spinal disorders and degeneration often result in the need for surgical intervention. Common surgical procedures include anterior cervical discetomy and fusion, laminectomy, and laminoplasty. In recent years, total disc arthroplasty devices have been developed with the underlying goal of preventing adjacent level degeneration that can result after fusion [1]. Most of these methods require insertion of devices into the spine such as cages, plates, rods, and/or screws. Finite element modeling is an invaluable tool that can be used to study the complex biomechanical situation created by these devices.

Several spine models have been used to study the effects of fusion and artificial disc replacements [14, 18, 21–24, 35, 47, 54, 56, 58, 83, 97]. In most cases, an intact mesh was analyzed and then modified to simulate the surgical procedure. Fusion is usually modeled via removal of the intervertebral disc and insertion of bone grafts, plates, or using posterior fixation. Artificial disc replacements are often modeled based on the geometry of an existing commercial implant, or using simplified geometry. The interface between the endplates and bone is often assumed to be tied to simulate complete union.

Modeling the surgical interventions of the spine can be a challenge, especially with regard to meshing. Often the modeler desires to make a direct comparison between the intact and altered states. There are generally two methods for modeling surgical interventions; altering the existing intact mesh or making modifications at the surface level and then remeshing. The most common method is to modify an existing mesh; however, the original mesh does not always readily accommodate a device due to the pattern and/or geometry of the initial elements. As a result, simplifications of the implant and/or contact interfaces between the bone and implant are often made.

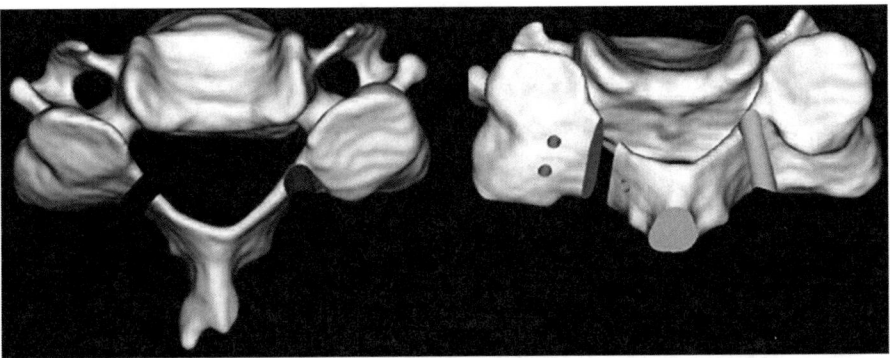

Fig. 6 Using the virtual tools, a series of cuts and holes were introduced to prepare the bony surface for a laminoplasty procedure

4.1 Surgical Simulations

We have developed a suite of tools (virtual instruments and guides) which enable surgical procedures to be readily simulated directly on the bony surfaces generated from CT scans [80]. These tools may be used to cut bone via a planar or box cut, perform Boolean operations on surfaces (e.g., introduce an implant into a host bone), transform/rotate surfaces in space, as well as other relevant operations. From these resulting surfaces, IA-FEMesh can be used to generate all-hexahedral subject-specific meshes. We have used these tools to study the effects of laminoplasty on a single vertebral level by performing "virtual surgery" on a cervical vertebra [80] (Fig. 6). For example, a box cut was used to create a bicortical defect on one side of the lamina, while a Boolean operation between the bony surface definition and a cylindrical surface was used to create a unicortical hinge defect on the contralateral side, simulating a high speed burr. The planar cut function was used to resect the spinous process. The resulting surface was then meshed in IA-FEMesh and a load to failure analysis was performed.

4.2 Laminoplasty and Laminectomy Procedures

If element alignment permits, it is often more convenient to modify the existing intact mesh. In an effort to study the effects of multilevel laminoplasty and laminectomy procedures on the cervical spine, we developed techniques to modify the intact mesh, and when appropriate introduce stabilizing implants [52, 53]. Our intact cervical (C2-T1) mesh [53] was modified to address three scenarios; open door laminoplasty (ODL), double door laminoplasty (DDL), and laminectomy. The affected levels were C3–C6.

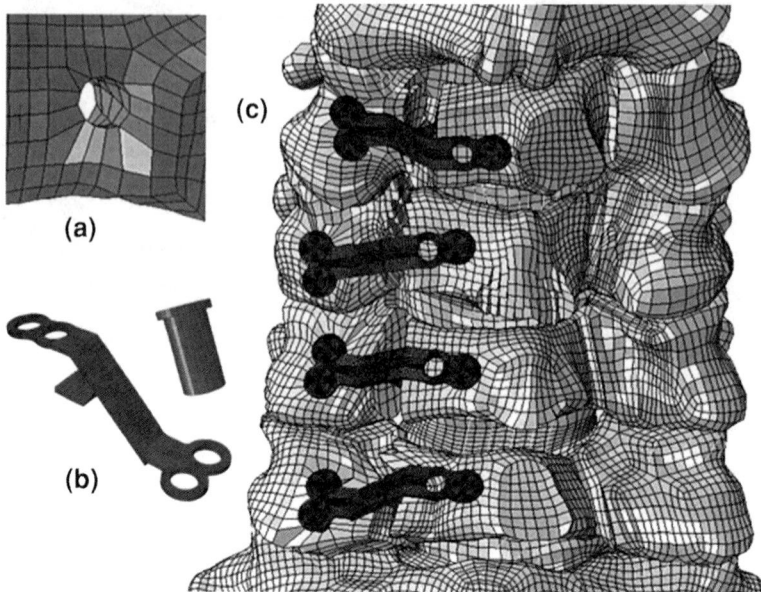

Fig. 7 The ODL procedure. **a** A screw hole introduced into an intact mesh. **b** Screw and OD plate meshes used for laminoplasty study. **c** Plates and screws introduced in the multilevel cervical spine FE model

A cut (i.e., door) was introduced at the junction of the lamina and the lateral mass by completely removing the bicortical layer of elements. A hinge of approximately 3–4 mm was created along the contralateral junction of the lamina and lateral mass by removing elements representing the unicortical layer. Using in-house code, the hinge region was smoothed to the shape of a cylinder to simulate the use of a high speed burr.

To accommodate screw holes in the lamina and lateral mass for the ODL, in-house tools were developed to remove elements in the posterior region of the vertebrae. The screw surfaces were placed in the desired locations using the Surgical Suite and elements of the intact mesh that fell within them were automatically removed from the mesh. The locations of the nodes of the screw hole were modified to match the cylindrical screw shape via projection onto the shaft (Fig. 7a). The bone/screw and screw/plate surfaces were assumed to be tied during the analysis, but contact interactions were defined at the plate/bone interfaces. To aid in the laminoplasty procedure, the interspinous ligaments of the involved levels were resected along with the unilateral ligamentum flavum at the unaltered levels. Additionally, the spinous processes of the involved levels were resected. The lamina of each C3 through C6 vertebra was opened towards the hinge by applying a uniform load. The stresses developed in the vertebrae during each of the laminar openings were incorporated into the model as initial conditions once the plate and screws were inserted into the model to stabilize the laminar opening.

Fig. 8 The DDL procedure. **a** The spinous process was split and opened. **b** Example of a graft mesh used to stabilize the spinous process opening. **c** Grafts introduced into the multilevel cervical spine FE model

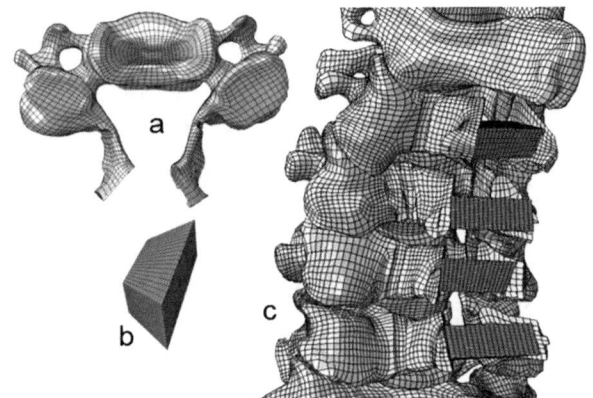

The bone graft, screw surfaces, and implants were generated in ProE (PTC, Needham, MA) and meshed with hexahedral elements using IA-FEMesh (Fig. 7b). Figure 7c shows the final mesh with the OD plates inserted at levels C3–C6.

The multilevel DDL procedure required two bilateral hinges at the junction of the lateral mass and the lamina for levels C3 through C6. The hinges were created using the same process as described above. The spinous process was sagitally split until the desired laminar opening was obtained (Fig. 8a). A trapezoidal shaped graft was placed in the open ends of the spinous process (Fig. 8b, c). The interspinous ligaments at the involved levels were removed along with partial removal of the ligamentum flavum at the midline from C2 to C7 to allow for laminar opening. The stresses induced during the opening of the laminae were incorporated as initial conditions for further analyses, and contact between the graft and bone was tied after the spacer was inserted into the model. Laminectomy was simulated by removing the spinous process and lamina on both sides of each vertebra for levels C3–C6.

Efforts are underway to extend the cervical spine model to address additional motion preserving technologies such as total disc replacements; namely the behavior at and adjacent to the operative level.

4.3 Additional Spine Models

The aforementioned multiblock meshing techniques are not limited to the cervical spine or human models. The methods are versatile enough to account for not only the differences in geometry between the cervical and lumbar regions (Fig. 9a), but between species. Animal models, for example, have proven valuable for investigating the performance of various spinal devices and procedures. In particular, the disc space and vertebral body widths of the sheep spine are relatively similar to that of the human; thereby making the sheep model ideal for studying the spine. It is important to understand both the similarities and differences between the human and sheep spine when constructing a valuable study and interpreting the results.

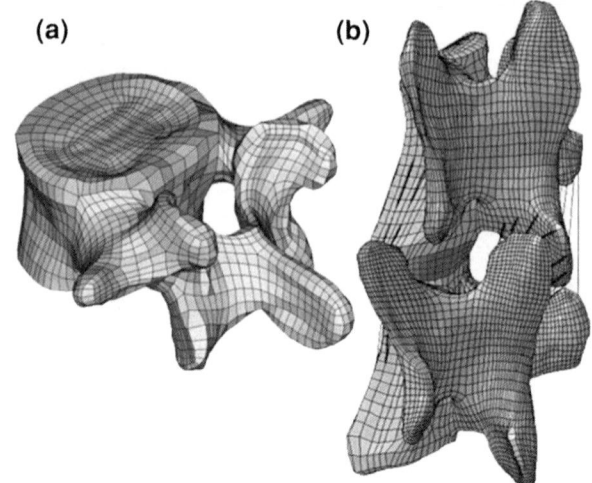

Fig. 9 Examples of other vertebrae meshed using the multiblock methods. **a** Human lumbar spine mesh. **b** Levels C3–C4 of a C2–C7 multilevel sheep cervical spine model

Although experimental testing can provide relative ranges of motion and stiffness parameters, it is difficult if not impossible to gather biomechanical data such as facet joint contact area and pressure, disc pressure, and bone stresses. Finite element modeling has often been used for this purpose. We developed a multilevel C2–C7 finite element model of the sheep spine using the same multiblock approach used for the human spine (Fig. 9b). The sheep cervical spine model [12] will be validated against intact multilevel experimental data [11] and will be used to complement in vitro studies involving the sheep cervical spine.

5 Summary

We believe that the ability to readily develop reliable models (i.e., finite element) on a patient/subject specific basis may facilitate a greater understanding of spinal biomechanics, and may be used in a variety of applications in medical treatment and safety. The long term goals for the resulting models are to aid in providing assessments of (1) the healthy spine, (2) the spine as altered by disease, degeneration, aging, trauma or surgery, (3) the spine with spinal instrumentation and (4) to assist in the design and development of spinal technologies, such as implants and the necessary surgical instrumentation.

References

1. Acosta, F.L.: Cervical disc arthroplasty: general introduction. Neurosurg. Clin. N. Am. **16**(4), 603–607 (2005)
2. Belytschko, T., Kulak, R.F., Schultz, A.: A finite element stress analysis of an intervertebral disc. J. Biomech. **7**, 277–285 (1972)

3. Bohlman, H.H., Emery, S.E., Goodfellow, D.B., Jones, P.K.: Robinson anterior cervical discectomy and arthrodesis for cervical radiculopathy. Long-term follow-up of one hundred and twenty-two patients. J. Bone Jt. Surg. **75**(9), 1298–1307 (1993)
4. Bornemann, F., Erdmann, B., Kornhuber, R.: Adaptive multilevel methods in three space dimensions. Int. J. Numer. Methods Eng. **36**, 3187–3203 (1993)
5. Bowden, A.: Finite element modeling of the spine. In: Kurtz, S.M., Edidin, A.A. (eds.) Spine Technology Handbook. Elsevier Inc., USA (2006)
6. Bowden, A.E., Guerin, H.L., Villarraga, M.L., Patwardhan, A.G., Ochoa, J.A.: Quality of motion considerations in numerical analysis of motion restoring implants of the spine. Clin. Biomech. **23**, 536–544 (2008)
7. Blacker, T.D., Meyers, R.J.: Seams and wedges in plastering: a 3-D hexahedral mesh generation algorithm. Eng. Comput. **9**(2), 83–93 (1993)
8. Brewer, M., Diachin, L.F., Knupp, P., Leurent, T., Melander, D.: The Mesquite mesh quality improvement toolkit. Proceedings of 12th International Meshing Roundtable, pp. 239–250 (2003)
9. Cifuentes, A.O., Kalbag, A.: A performance study of tetrahedral and hexahedral elements in 3-D finite element structural analysis. Finite Ele. Anal. Des. **12**(3–4), 313–318 (1992)
10. Couteau, B., Payan, Y., Lavallée, S.: The mesh-matching algorithm: an automatic 3D mesh generator for finite element structures. J. Biomech. **33**(8), 1005–1009 (2000)
11. De Vries, N.A., Gandhi, A.A., Fredericks, D.C., Smucker, J.D., Grosland, N.M.: In vitro study of the C2–C7 sheep cervical spine. ASME Summer Bioengineering Conference, Farmington, PA (2011)
12. DeVries, N.A., Kallemeyn, N.A., Shivanna, K.H., Grosland, N.M.: Finite element analysis of the C2–C7 sheep spine. ASME Summer Bioengineering Conference, Naples, FL (2010)
13. De Vries, N.A., Shivanna, K.H., Tadepalli, S.C., Magnotta, V.A., Grosland, N.M.: IA-FEMESH: anatomic FE models—a check of mesh accuracy and validity. Iowa Orthopaedic J. **29**, 48 (2009)
14. Dooris, A.P., Goel, V.K., Grosland, N.M., Gilbertson, L.G., Wilder, D.G.: Load-sharing between anterior and posterior elements in a lumbar motion segment implanted with an artificial disc. Spine **26**(6E), 122–129 (2001)
15. Eberlein, R., Holzapfel, G.A., Fröhlich, M.: Multi-segment FEA of the human lumbar spine including the heterogeneity of the annulus fibrosus. Comput. Mech. **34**(2), 147–163 (2004)
16. Eck, J.C., Humphreys, S.C., Lim, T.-H., Jeong, S.T., Kim, J.G., Hodges, S.D., An, H.S.: Biomechanical study on the effect of cervical spine fusion on adjacent-level intradiscal pressure and segmental motion. Spine **27**, 2431–2434 (2002)
17. Fagan, M.J., Julian, S., Mohsen, A.M.: Finite element analysis in spine research. Proc. Inst. Mech. Eng. Part H. J. Eng. Med. **216**(5), 281–298 (2002)
18. Faizan, A., Goel, V.K., Garfin, S.R., Bono, C.M., Serhan, H., Biyani, A., Elgafy, H., Krishna, M., Friesem, T.: Do design variations in the artificial disc influence cervical spine biomechanics A finite element investigation. Eur. Spine J. (2009) (Epub ahead of print)
19. Field, D.A.: Laplacian smoothing and delaunay triangulations. Commun. Appl. Numer. Methods **4**(6), 709–712 (1988)
20. Fuller, D.A., Kirkpatrick, J.S., Emery, S.E., Wilber, R.G., Davy, D.T.: A kinematic study of the cervical spine before and after segmental arthrodesis. Spine **23**(15), 1649–1656 (1998)
21. Galbusera, F., Anasetti, F., Bellini, C.M., Costa, F., Fornari, M.: The influence of the axial, antero-posterior and lateral positions of the center of rotation of a ball-and-socket disc prosthesis on the cervical spine biomechanics. Clin. Biomech. **25**(5), 397–401 (2010)
22. Galbusera, F., Bellini, C.M., Costa, F., Assietti, R., Fornari, M.: Anterior cervical fusion: a biomechanical comparison of 4 techniques. J. Neurosurg. Spine **9**, 444–449 (2008)
23. Galbusera, F., Bellini, C.M., Raimondi, M.T., Fornari, M., Assietti, R.: Cervical spine biomechanics following implantation of a disc prosthesis. Med. Eng. Phy. **30**(9), 1127–1133 (2008)

24. Galbusera, F., Fantigrossi, A., Raimondi, M.T., Assietti, R., Sassi, M., Fornari, M.: Biomechanics of the C5–C6 spinal unit before and after placement of a disc prosthesis. Biomech. Model. Mechanobiol. **5**(4), 253–261 (2006)
25. George, P.L.: Automatic Mesh Generation. Wiley, London (1993)
26. Gilbertson, L.G., Goel, V.K., Kong, W.Z., Clausen, J.D.: Finite element methods in spine biomechanics research. Crit. Rev. Biomed. Eng. **23**(5–6), 411–473 (1995)
27. Goel, V.K., Clausen, J.D.: Prediction of load sharing among spinal components of a C5–C6 motion segment using the finite element approach. Spine **23**(6), 684–691 (1998)
28. Goel, V.K., Gilbertson, L.G.: Applications of the finite element method to thoracolumbar spine research—past, present, and future. Spine **20**, 1719–1727 (1995)
29. Goel, V.K., Kong, W., Han, J.S., Weinstein, J.N., Gilbertson, L.G.: A combined finite element and optimization investigation of lumbar spine mechanics with and without muscles. Spine **18**(11), 1531–1541 (1993)
30. Goel, V.K., Ramirez, S.A., Kong, W., Gilbertson, L.G.: Cancellous bone Young's modulus variation within the vertebral body of a ligamentous lumbar spine—applications of bone adaptive remodeling concepts. J. Biomech. Eng. **117**(3), 226–271 (1995)
31. Goel, V.K., Weinstein, J.N.: Biomechanics of the spine: Clinical and surgical perspective. CRC Press, Boca Raton (1990)
32. Goffin, J., Geusens, E., Vantomme, N., Quintens, E., Waerzeggers, Y., Depreitere, B., Van Calenbergh, F., van Loon, J.: Long-term follow-up after interbody fusion of the cervical spine. J. Spinal Disord. **17**(2), 79–85 (2004)
33. Goffin, J., van Loon, J., van Calenbergh, F., Plets, C.: Long-term results after anterior cervical fusion and osteosynthetic stabilization for fractures and/or dislocations of the cervical spine. J. Spinal Disord. **8**(6), 500–508 (1995)
34. Grobler, L.J., Frymoyer, J.W., Robertson, P.A., Novotny, J.E.: Biomechanics of lumbar spinal surgery. Semin. Spine Surg. **5**(1), 59–72 (1993)
35. Grosland, N., Goel, V., Grobler, L.: Comparative biomechanical investigation of multiple interbody fusion cages: a finite element analysis. In: 45th Annual Meeting, Orthopaedic Research Society, Anaheim, CA, p 1002, 1999
36. Grosland, N.M., Bafna, R., Magnotta, V.A.: Automated hexahedral meshing of human anatomical structures using deformable registration. Comput. Methods Biomech. Biomed. Eng. **12**(1), 35–43 (2009a)
37. Grosland, N.M., Shivanna, K.H., Magnotta, V.A., Kallemeyn, N.A., DeVries, N.A., Tadepalli, S.C., Lisle, C.: IA-FEMesh: an open-source, interactive, multiblock approach to anatomic finite element model development. Comput. Methods Programs Biomed. **94**(1), 96–107 (2009b)
38. Hilibrand, A.S., Carlson, G.D., Palumbo, M.A., Jones, P.K., Bohlman, H.H.: Radiculopathy and myelopathy at segments adjacent to the site of a previous anterior cervical arthrodesis. J. Bone Jt. Surg. Am. **81**(4), 519–528 (1999)
39. Huiskes, R., Heck, J.V.: Stresses in the femoral head-neck region after surface replacement. A three-dimensional finite element analysis. Trans. Orthop. Res. Soc. **6**, 174 (1981)
40. Ito, Y., Shih, A.M., Soni, B.K.: Octree-based reasonable quality hexahedral mesh generation using a new set of refinement templates. Int. J. Numer. Methods Eng. **77**(13), 1809–1833 (2009)
41. Jacobs, C.R., Davis, B.R., Rieger, C.J., Francis, M., Saad, M., Fyhrie, D.P.: The impact of boundary conditions and mesh size on the accuracy of cancellous bone tissue modulus determination using large-scale finite element modeling. J. Biomech. **32**, 1159–1164 (1999)
42. Jacobs, C.R., Mandell, J.A., Beaupre, G.S: A comparative study of automatic finite element mesh generation techniques in orthopaedic biomechanics. In: Bioengineering Conference of ASME, pp. 512–514 (1993)
43. Kallemeyn, N.A., Gandhi, A.A., Kode, S., Shivanna, K.H., Smucker, J.D., Grosland, N.M.: Validation of a C2–C7 cervical spine finite element model using specimen-specific flexibility data. Med. Eng. Phy. **32**(5), 482–489 (2010)

44. Kallemeyn, N.A., Natarajan, A., Magnotta, V.A., Grosland, N.M.: Hexahedral meshing of subject-specific anatomic structures using mapped building blocks. Under Review (2011)
45. Kallemeyn, N.A., Tadepalli, S.C., Shivanna, K.H., Grosland, N.M.: An interactive multiblock approach to meshing the spine. Comput. Methods Programs Biomed. **95**(3), 227–235 (2009)
46. Kaminsky, J., Rodt, T., Gharabaghi, A., Forster, J., Brand, G., Samii, M.: A universal algorithm for an improved finite element mesh generation: Mesh quality assessment in comparison to former automated mesh-generators and an analytic model. Med. Eng. Phys. **27**(5), 383–394 (2005)
47. Kang, H., Park, P., La Marca, F., Hollister, S.J., Lin, C.Y.: Analysis of load sharing on uncovertebral and facet joints at the C5-6 level with implantation of the Bryan, Prestige LP, or ProDisc-C cervical disc prosthesis: an in vivo image-based finite element study. Neurosurg. Focus **28**(6), 9 (2010)
48. Kasra, M., Shirazi-Adl, A., Drouin, G.: Dynamics of human lumbar intervertebral joints: Experimental and finite-element investigations. Spine **17**(1), 93–101 (1992)
49. Katz, J.N.: Lumbar disc disorders and low-back pain: socioeconomic factors and consequences. J. Bone Jt. Surg. **88**(Suppl 2), 21–24 (2006)
50. Keyak, J.H., Meagher, J.M., Skinner, H.B., Mote, C.D.J.: Automated three-dimensional finite element modeling of bone: a new method. J. Biomed. Eng. **12**, 389–397 (1990)
51. Keyak, J.H., Skinner, H.B.: Three-dimensional finite element modeling of bone; effects of element size. J. Biomed. Eng. **14**, 483–489 (1992)
52. Kode, S., Kallemeyn, N.A., Smucker, J.D., Fredericks, D.C., Grosland, N.M.: Biomechanical effects of laminoplasty and laminectomy on the stability of cervical spine. In: ASME Summer Bioengineering Conference, Farmington, PA (2011)
53. Kode, S., Kallemeyn, N.A., Smucker, J.D., Grosland, N.M.: Finite element model of multi level cervical laminoplasty. In: ASME Summer Bioengineering Conference, Naples, FL (2010)
54. Kumaresan, S., Yoganandan, N., Pintar, F.A.: Finite element analysis of anterior cervical spine interbody fusion. Biomed. Mater. Eng. **4**, 221–230 (1997)
55. Kumaresan, S., Yoganandan, N., Pintar, F.A.: Finite element modeling approaches of human cervical spine facet joint capsule. J. Biomech. **31**(4), 371 (1998)
56. Le Huec, J.C., Lafage, V., Bonnet, X., Lavaste, F., Josse, L., Liu, M., Skalli, W.: Validated finite element analysis of the Maverick total disc prosthesis. J. Spinal Disord. Tech. **23**(4), 249 (2010)
57. Lepi, S.M.: Practical guide to finite elements: a solid mechanics approach. Mechanical engineering: A Series of Textbooks and Reference Books. Marcel Dekker Inc., New York (1988)
58. Lopez-Espina, C.G., Amirouche, F., Havalad, V.: Multilevel cervical fusion and its effect on disc degeneration and osteophyte formation. Spine **31**(9), 972 (2006)
59. Lu, Y.M., Hutton, W.C., Gharpuray, V.M.: Do bending, twisting, and diurnal fluid changes in the disc affect the propensity to prolapse? A viscoelastic finite element model. Spine **21**(22), 2570–2579 (1996)
60. Magnotta, V.A., Li, W., Grosland, N.M.: Comparison of displacement-based and force-based mapped meshing. Insight Journal 2008 MICCAI Workshop, Computational Biomechanics for Medicine, 1–10 (2008)
61. McBroom, R.J., Hayes, W.C., Edwards, W.T., Goldberg, R.P., White, A.A.: Prediction of vertebral body compressive fracture using quantitative computed tomography. J. Bone Jt. Surg. **67A**, 1206–1214 (1985)
62. Mitchell, S.A., Vavasis, S.A.: Quality mesh generation in three dimensions. In: Proc. 8th ACM Conf. on Comp. Geometry pp. 212–221 (1992)
63. Natarajan, R.N., Williams, J.R., Andersson, G.B.: Recent advances in analytical modeling of lumbar disc degeneration. Spine **29**(23), 2733–2741 (2004)
64. Oonishi, H., Isha, H., Hasegawa, T.: Mechanical analyis of the human pelvis and its application to the artificial hip—by means of the three dimensional finite element. J. Biomech. **16**, 427–444 (1983)

65. Owen, S.J.: A survey of unstructured mesh generation technology. In: Proceedings of the 7th International Meshing Roundtable vol. 3(6), (1998)
66. Rao, A.A., Dumas, G.A.: Influence of material properties on the mechanical behavior of the L5-S1 intervertebral disc in compression: a nonlinear finite element study. J. Biomed. Eng. 13, 139–151 (1991)
67. Rohlmann, A., Calisse, J., Bergmann, G., Weber, U.: Internal spinal fixator stiffness has only a minor influence on stresses in the adjacent discs. Spine 24(12), 1192–1195 (discussion 1195–1196) (1999)
68. Schoberl, J.: NETGEN-An advancing front 2D/3D-mesh generator based on abstract rules. Comput. Visual. Sci. 1, 41–52 (1997)
69. Sherk, H.H., Parke, W.W.: Anatomy. In: Committee, C.S.R.S.E. (ed.) The Cervical Spine. Lippincott, Philadelphia (1989)
70. Shirazi-Adl, A.: Models of the functional spinal unit. Semin. Spine Surg. 5(1), 23–28 (1993)
71. Shirazi-Adl, A.: Nonlinear stress analysis of the whole lumbar spine in torsion–mechanics of facet articulation. J. Biomech. 27(3), 289–299 (1994)
72. Shirazi-Adl, S.A., Shrivatsava, S.C., Ahmed, A.M.: Stress analysis of the lumbar disc-body unit in compression. Spine 9(2), 120 (1984)
73. Shivanna, K.H., Tadepalli, S.C., Grosland, N.M.: Feature-based multiblock finite element mesh generation. Comput. Aided Des. 42, 1108–1116 (2010)
74. Spekreijse, S.P.: Elliptic grid generation based on Laplace equations and algebraic transformations. J. Comput. Phy. 118(1), 38–61 (1995)
75. Spilker, R.L.: Mechanical behavior of a simple model of an intervertebral disc under compressive loading. J. Biomech. 13, 895–901 (1980)
76. Spilker, R.L., Daugirda, D.M., Schultz, A.B.: Mechanical response of a simple finite element model of the intervertebral disc under complex loading. J. Biomech. 17(2), 103–112 (1984)
77. Spilker, R.L., Jakobs, D.M., Schultz, A.B.: Material constants for a finite element model of the intervertebral disk with a fiber composite annulus. J. Biomech. Eng. 108, 1–11 (1986)
78. Stokes, I.A., Laible, J.P.: Three-dimensional osseo-ligamentous model of the thorax representing initiation of scoliosis by asymmetric growth. J. Biomech. 23(6), 589–595 (1990)
79. Szeliski, R., Lavallée, S.: Matching 3-D anatomical surfaces with non-rigid deformations using octree-splines. Int. J. Comput. Vis. 18(2), 171–186 (1996)
80. Tadepalli, S.C., Shivanna, K.H., Magnotta, V.A., Kallemeyn, N.A., Grosland, N.M.: Toward the development of virtual surgical tools to aid orthopaedic FE analyses. EURASIP J. Adv. Sig. Process. (2010)
81. Tautges, T.J., Blacker, T., Mitchell, S.A.: The whisker weaving algorithm: A connectivity-based method for constructing all-hexahedral finite element meshes. Int. J. Numer. Methods Eng. 39(19), 3327–3349 (1996)
82. Teo, E.C., Ng, H.W.: Evaluation of the role of ligaments, facets and disc nucleus in lower cervical spine under compression and sagittal moments using finite element method. Med. Eng. Phy. 23(3), 155–164 (2001)
83. Teo, E.C., Yang, K., Fuss, F.K., Lee, K.K., Qiu, T.X., Ng, H.W.: Effects of cervical cages on load distribution of cancellous core: a finite element analysis. J. Spinal Disord. Tech. 17(3), 226 (2004)
84. Teo, J.C.M., Chui, C.K., Ong, S.H., Yan, C.H., Wang, S.C., Wong, H.K., Teoh, S.H.: Heterogeneous meshing and biomechanical modeling of human spine. Med. Eng. Phy. 29, 277–290 (2007)
85. Viceconti, M., Olsen, S., Nolte, L.P., Burton, K.: Extracting clinically relevant data from finite element simulations. Clin. Biomech. 20(5), 451–454 (2005)
86. Villarraga, M.L., Anderson, R.C., Hart, R.T., Dinh, D.H.: Contact analysis of a posterior cervical spine plate using a three-dimensional canine finite element model. J. Biomech. Eng. 121(2), 206–214 (1999)
87. Voo, L., Pintar, F.A.: Finite element applications in human cervical spine modeling. Spine 21, 1824–1834 (1996)

88. Wang, Z.L., Teo, J.C.M., Chui, C.K., Ong, S.H., Yan, C.H., Wang, S.C., Wong, H.K., Teoh, S.H.: Computational biomechanical modelling of the lumbar spine using marching-cubes surface smoothened finite element voxel meshing. Comput. Methods Programs Biomed. **80**(1), 25–35 (2005)
89. Wassell, J.T., Gardner, L.I., Landsittel, D.P., Johnston, J.J., Johnston, J.M.: A prospective study of back belts for prevention of back pain and injury. J. Am. Med. Assoc. **284**, 2727–2732 (2000)
90. Weiss, J.A., Gardiner, J.C., Ellis, B.J., Lujan, T.J., Phatak, N.S.: Three-dimensional finite element modeling of ligaments: Technical aspects. Med. Eng. Phy. **27**, 845–861 (2005)
91. Womack, W., Ayturk, U.M., Puttlitz, C.M.: Cartilage thickness distribution affects computational model predictions of cervical spine facet contact parameters. J. Biomech. Eng. **133**, 1–10 (2011)
92. Womack, W., Woldtvedt, D., Puttlitz, C.M.: Lower cervical spine facet cartilage thickness mapping. Osteoarthr. Cartil. **16**, 1018–1023 (2008)
93. Yoganandan, N., Kumaresan, S., Pintar, F.A.: Geometric and mechanical properties of human cervical spine ligaments. J. Biomech. Eng. **122**, 623 (2000)
94. Yoganandan, N., Kumaresan, S., Voo, L., Pintar, F.A.: Finite element applications in human cervical spine modeling. Spine **21**(15), 1824–1834 (1996)
95. Yoganandan, N., Myklebust, J.B., Ray, G., Sances Jr., A.: Mathematical and finite element analysis of spine injuries. Crit. Rev. Biomed. Eng. **15**(1), 29–93 (1987)
96. Zhang, Y., Hughes, T.J.R., Bajaj, C.L.: An automatic 3D mesh generation method for domains with multiple materials. Comput. Methods Appl. Mech. Eng. **199**, 405–415 (2010)
97. Zhong, Z.C., Chen, S.H., Hung, C.H.: Load-and displacement-controlled finite element analyses on fusion and non-fusion spinal implants. Proc. Inst. Mech. Eng., Part H: J. Eng. Med. **223**(2), 143–157 (2009)

Patient-Specific Modeling of Scoliosis

J. Paige Little and Clayton J. Adam

Abstract Current complication rates for adolescent spinal deformity surgery are unacceptably high and in order to improve patient outcomes, the development of a simulation tool which enables the surgical strategy for an individual patient to be optimized is necessary. In this chapter we will present our work to date in developing and validating patient-specific modeling techniques to simulate and predict patient outcomes for surgery to correct adolescent scoliosis deformity. While these simulation tools are currently being developed to simulate adolescent idiopathic scoliosis patients, they will have broader application in simulating spinal disorders and optimizing surgical planning for other types of spine surgery. Our studies to date have highlighted the need for not only patient-specific anatomical data, but also patient-specific tissue parameters and biomechanical loading data, in order to accurately predict the physiological behaviour of the spine. Even so, patient-specific computational models are the state-of-the-art in computational biomechanics and offer much potential as a pre-operative surgical planning tool.

1 Introduction

Scoliosis is a spinal deformity involving a side-to-side curvature of the vertebrae in the frontal plane (Fig. 1a) and axial rotation of the vertebrae in the transverse plane (Fig. 1e). As such, scoliosis is a three-dimensional deformity with patients often demonstrating an overly prominent ribcage and shoulder blade due to the abnormal

---

J. P. Little · C. J. Adam (✉)
Paediatric Spine Research Group, Queensland University of Technology,
Brisbane, Australia
e-mail: c.adam@qut.edu.au

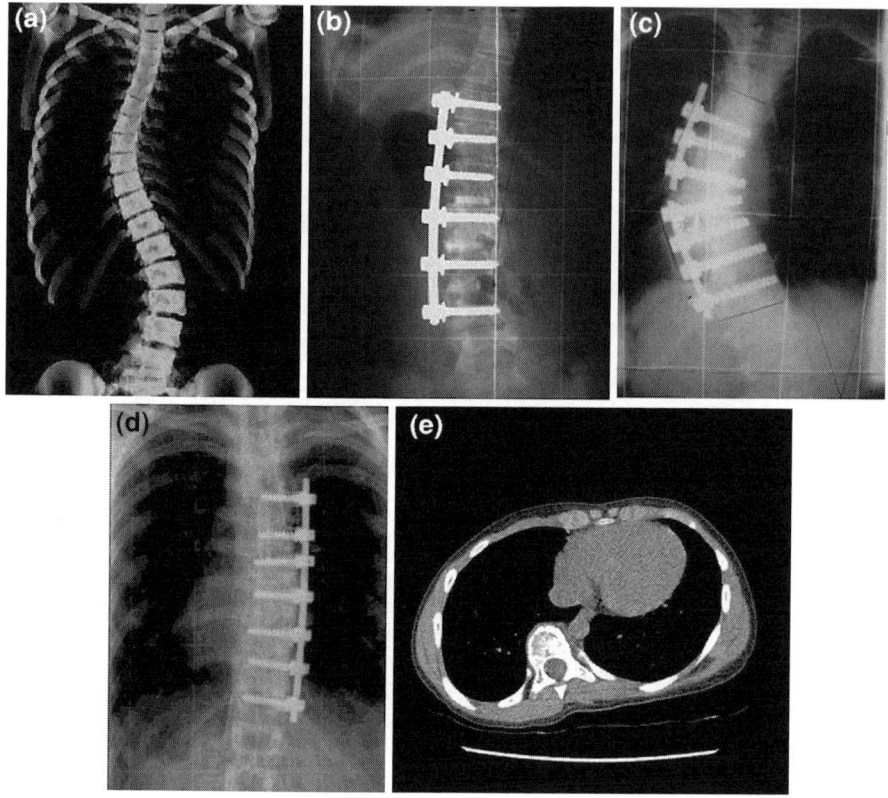

Fig. 1 a Scoliosis spinal deformity. **b** Anterior scoliosis implant. **c** Implant complication—rod fracture. **d** Implant complication—*top* screw pullout (*circled*). **e** Transverse slice showing axial rotation of vertebra, posterior rib hump and anterior ribcage prominence

axial rotation of the vertebrae as well as hip and shoulder asymmetry (Fig. 1e). Adolescent Idiopathic Scoliosis (AIS) is the most common type of spinal deformity, affecting 2–4% of the population (American National Scoliosis Foundation) and there is neither a known cause nor a known cure for the condition. As such, treatment can only attempt to prevent curve progression and/or reduce the spinal deformity.

In conservative treatment for AIS, orthotic braces are used in an attempt to prevent curve progression, but in cases where this is not successful in arresting progression, or in patients with a severe deformity, corrective surgery is the only option. This involves attachment of rods to the patient's spine using screws inserted into the spinal vertebrae (Fig. 1b). The implant construct shown in Fig. 1b is an anterior construct, however posterior constructs (not shown), in which long pairs of rods are attached to the back of the spine using screws, hooks and wires, are also commonly used in scoliosis correction surgery.

In either approach, bone graft is applied at the time of surgery to encourage fusion of adjacent vertebrae, so that in a successful procedure the metal implant is required to provide stiffness and stability to the spine for the 3–6 months after surgery so that bony fusion can occur. After this the spinal loading is borne by the fused vertebrae and while the implant often remains in situ, it has no significant load bearing role.

Despite significant stated advantages of the anterior single rod procedure over posterior techniques, including reduced blood loss, reduced muscle dissection and improved cosmesis [22, 29, 41], the anterior fusion procedure still brings with it high (15–20%) implant-related complication rates [4, 31]. Such complications include screw pullout (Fig. 1d) and rod fracture (Fig. 1c). The most significant outcomes for the patient from the surgical screw and rod failures are the potential onset of back pain and loss of deformity correction [4].

Currently, surgical planning decisions are based on factors such as scoliosis curve severity and curve type (assessed using various radiographic classification systems), pre-operative patient assessment (e.g. flexibility tests) and surgeon judgement. These factors introduce a level of subjectivity into the surgical planning and as such, the surgical procedure chosen for a particular patient may differ considerably between surgeons [3]. The ability of the surgeon to correct spinal deformity while avoiding complications requires a balance between the applied corrective forces (excessive forces will cause implant breakage or tissue damage) and the degree of deformity correction attempted (insufficient correction will leave an unbalanced spine after surgery which is prone to further deformity progression even after skeletal maturity). Attaining this balance is a complex biomechanical challenge. To address this challenge our research group has developed spinal simulation techniques, enabling patient-specific finite element (FE) models of individual patient's spinal anatomy to be created. Using these FE models, the forces and deformations in the implant and spinal tissues are predicted for intra-operative and physiological spinal loads. Spinal deformity correction is essentially a structural problem, so the use of patient-specific finite element models during pre-operative planning of corrective surgery for scoliosis can potentially provide surgeons with a powerful support tool in making more informed decisions regarding the most appropriate procedure for an individual patient. While our research to date has focused on anterior, single rod scoliosis correction constructs, the patient-specific spinal modeling techniques we describe here are broadly applicable for simulating spinal disorders and optimizing surgical planning for other types of spinal surgery.

Other researchers in the field of spine biomechanics and simulation have made substantial contributions to addressing the need for a better understanding of scoliosis biomechanics (for example, [35]) and moreover, the need for patient-specific thoracolumbar spine models to inform surgeons in their clinical decision making (for example, [2, 16, 32]). However, the models developed to date often use highly idealized representations for the spinal anatomy, with either rigid bodies [2] or elastic beam elements [16] representing the spinal vertebrae; and lumped parameter (multi degree-of-freedom spring) representations to simulate the collective behaviour of the spinal ligaments, annulus fibrosus and nucleus pulposus in the intervertebral disc [2, 16]. Physiologically, the spinal vertebrae at each joint are connected by seven separate ligaments, with lines of action between specific bony landmarks on the adjacent vertebral bones. The intervertebral disc alone is a highly complex structure whose mechanics are governed by a complex interplay between collagen fibres embedded within adjacent lamellae of the annulus fibrosus and the fluid-filled, pressurized nucleus pulposus which the annulus surrounds [20]. Furthermore, in previous models the contribution of the ribcage is either not included or represented using an idealized spring stiffness for adjoining spinal segments [32]. However, the ribcage and its articulation with the spinal column (costovertebral joints) have been shown to be of considerable importance in governing the biomechanics of the spine, with as much as 40% decrease in spinal stiffness observed in cadaveric human spines following ribcage removal [27, 40]. All these structures play an important role in the overall biomechanics of both the intact and surgically altered spine.

Modern imaging techniques, visualization algorithms and image post-processing methods permit patient-specific anatomical representations for the osseous structures to be simulated. However, of equal importance in developing patient-specific spine models which are clinically useful, is the ability to derive not only patient specific osseous anatomy but patient specific material properties for the spinal tissues. Of particular importance are the soft tissue structures in the spine (ligaments, muscles and intervertebral discs) as these facilitate the articulation between adjacent vertebrae during physiological motion and primarily govern patient flexibility.

The focus for our spinal modeling has been to develop techniques for generating patient-specific models of scoliosis patients, which are sufficiently detailed (in terms of the spinal anatomy and individual spinal structures represented) to allow accurate simulation of the effects of surgery on the thoracolumbar spine. To this end, our patient-specific spine model includes detailed representations for:

- the vertebrae (both vertebral body and posterior elements);
- the intervertebral discs (annulus fibrosus lamellae, collagen fibres and nucleus pulposus);
- the seven spinal ligaments spanning each joint;
- the zygapophyseal (facet) joints;
- the vertebrae-rib articulations (costo-vertebral and costo-transverse joints); and
- the ribcage (ribs, sternum and costo-sternal joints).

The following sections will detail our methods for simulating both pre-operative flexibility tests and intra-operative deformity corrections; and our investigations to determine patient-specific soft tissue parameters using physiological load cases on the intact and surgically altered scoliotic spine. This chapter concludes with a discussion of current challenges and future work with regard to developing clinically useful and physiologically accurate computer simulations of an individual patient's spine.

2 Patient-Specific Modeling and Analysis

2.1 Model Development

The three-dimensional, low dose (2.0–3.7 m Sv radiation dose) computed tomography (CT) dataset (supine positioning, 512 × 512 pixels per slice, 1.25 mm axial slice spacing, 2.5 mm slice thickness) for an individual patient is obtained prior to surgery using a helical CT protocol (GE Lightspeed Plus, General Electric Medical Systems, Milwaukee, WI, USA). CT scans are clinically indicated prior to keyhole anterior scoliosis surgery as CT scanning allows safer screw sizing and positioning [13]. The CT dataset is then imported into custom image processing software (Matlab R2007b, The Mathworks, Natick, MA) where user-selected landmarks on the osseous anatomy (vertebral bodies, superior and inferior endplates, posterior elements, articulating surfaces on the zygapophyseal joints and sternum) are selected and exported as three-dimensional co-ordinate data (Fig. 2). These co-ordinate data define key landmarks at each vertebral level in the thoracolumbar spine.

These data points are imported into custom preprocessing software (Python 2.5 programming language, Python Software Foundation) where they are used to reconstruct the three-dimensional, patient-specific geometry using parametric descriptions for the bony structures of the spine and ribcage. Using the custom pre-processor, the osseous anatomy is meshed for analysis using Abaqus 6.7 (Simulia, Providence, RI) (Fig. 3) and it is possible to control the mesh density in each of the model components, to optimize the solution time and model accuracy.

The parametric description for the outer profile of the vertebral endplates uses a series of elliptical and cubic equations with C^1 continuity [21] (Fig. 4d). The axial profile for the vertebral bodies is extrapolated from the endplates, using a second order polynomial derived with user-selected points on the vertebral cortex that describe the concavity in the frontal and sagittal planes. Similarly, the intervertebral discs are interpolated between the transverse profiles of the adjacent vertebral endplates, thus defining the outer profile of the annulus fibrosus. The outer boundary of the nucleus pulposus is defined by scaling the outer anulus profile about its centroid. The curved transverse profile of the zygapophyseal joint articulating surfaces is described using sinusoidal curves derived from user-selected points on the joint surfaces at each vertebral level. The representation of the three-dimensional anatomy for the stiff posterior regions of the vertebrae (spinous processes, transverse processes, pedicles and lamellae) are simplified, using quasi-rigid, linear beams between user-selected points on the posterior elements (Fig. 4d).

The upper and lower edges of the ribs were separately represented using fifth order polynomials to describe the variation of the rib profile with distance along the edge from the costo-vertebral joint. Three separate polynomial equations described the profile of the rib edge in each plane and the polynomial constants were derived from user-selected landmarks along the rib edge. Similarly, user-selected landmarks

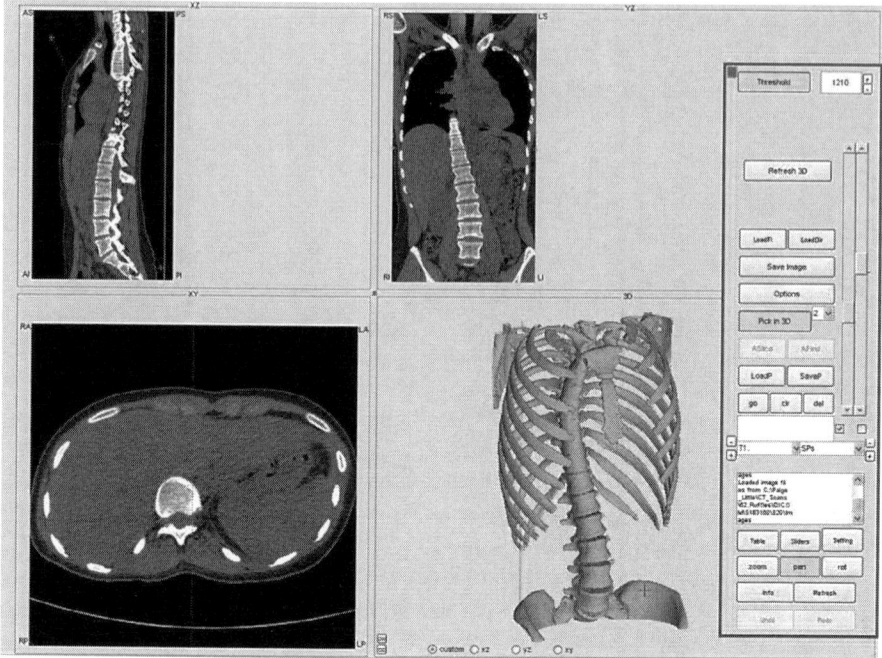

Fig. 2 Custom developed image processing interface for defining user-selected landmarks on pre-operative CT datasets

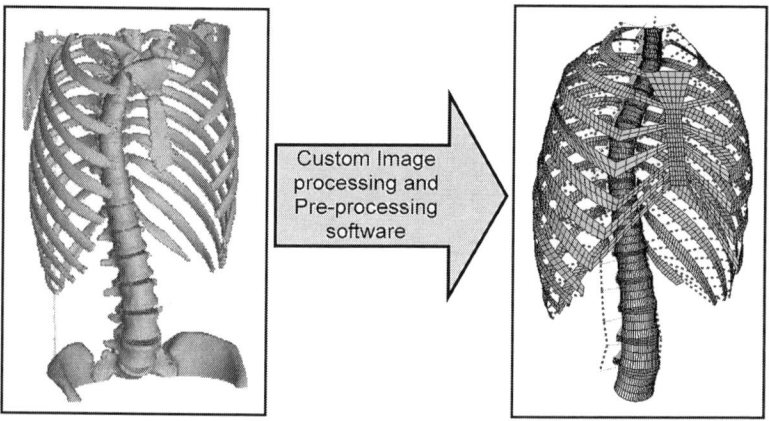

Fig. 3 Three-dimensional osseous reconstruction converted to a finite element mesh for analysis using Abaqus 6.7

defined the manubrium and sternum, with these structures represented as planar surfaces. The costal cartilage was modeled using a linear interpolation between the sternum/manubrium and the medial rib ends, while the costo-sternal connections were modeled as beam connectors for thoracic levels T1–T7.

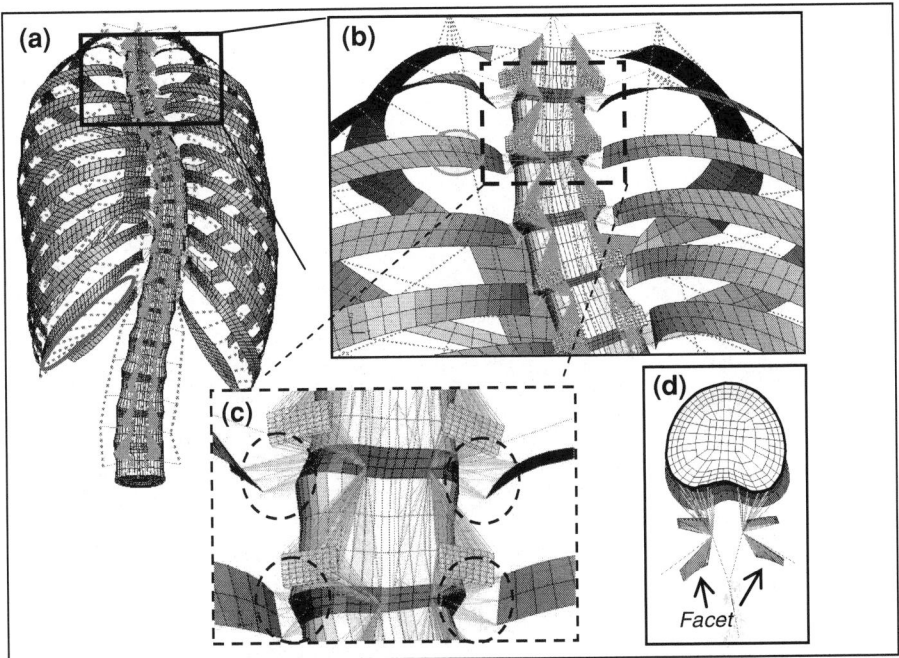

Fig. 4 a A posterior view of the meshed thoracolumbar spine and ribcage, showing the intervertebral discs in purple, vertebra in *grey* and *blue broken lines* denoting the inter-costal soft tissue connections (*circled in red* for spine level T11–T12) between ribs of adjacent thoracic levels. **b** Posterior view of the upper thoracic spine, *blue broken lines* denote the costo-transverse connections (*circled* in *green* for spine level T3) between the transverse process and the adjacent rib surface. **c** Posterior view of T2, showing the costo-vertebral beam connections as clusters of *light-blue* beams (*circled in blue* for the T1–T2 and T2–T3 connections). **d** T12 vertebra, showing the transverse profile of the vertebral endplate highlighted in purple, the posterior bony elements represented as beam elements (shown in *green*) and the articulating surfaces of the zygapophyseal (*facet*) joints

Anatomically, the costo-vertebral joint articulates with both the superior and inferior postero-lateral cortices of the adjacent vertebra as well as the lateral surface of the intermediate intervertebral disc. As mentioned, the costo-vertebral joints have been shown to be of key importance in governing the biomechanics of the spine [27, 40] and in light of this, our thoracolumbar spine model represents these joints in detail [19]. The costo-vertebral joints at each spinal level are represented using 'clusters' of beam elements, connecting the medial rib with the adjacent vertebral cortices and intervertebral disc surface (Fig. 4b, c). The beam properties were defined to provide a representative cross-sectional joint area of 46 mm^2 which was based on the approximate articulating region between the rib and the postero-lateral vertebrae/disc surfaces. Our computational representation for the costo-vertebral joints is relatively complex in an attempt to capture the biomechanical behaviour of this joint. As such, we have conducted a separate finite element study on a single thoracic motion segment to ensure this representation correctly reproduces the physiological behaviour of a human joint (see below). The left and right costo-transverse joints were represented as rigid, kinematic constraints between the lateral aspect of the transverse processes and the posterior most surface of the adjacent rib (Fig. 4b).

The anterior longitudinal ligament (ALL), posterior longitudinal ligament (PLL), inter/supra-spinous ligament (SL), inter-transverse ligament (ITL), ligamentum flavum (LF) and capsular ligaments (CL) were represented as linear connectors between the bony attachment points for each ligament. No ligament wrapping has been represented. Similarly, the inter-costal soft tissue connections were represented as obliquely oriented linear connectors between the edges of adjacent ribs (Fig. 4a—circled in red for level T11–T12). Anatomically, these soft tissues connect adjacent ribs and vary in orientation along the rib and between ribs, so connector elements were inclined at 50–60° to the edge of the inferior rib.

The element types used to simulate each of the structures represented in the thoracolumbar spine model are detailed in Table 1.

There is a paucity of experimental data characterising the mechanical behaviour of soft tissues from adolescent, scoliotic spines; with the only methods to derive such parameters being either direct, intra-operative measurement (which is highly challenging due to difficulties in physically accessing and accurately testing tissues) or inverse determination based on biomechanical modeling of pre-operative patient flexibility tests. Few researchers attempt to incorporate patient-specific tissue properties in thoracolumbar spine models for investigating scoliosis biomechanics [16].

In the absence of material parameters describing the mechanical behaviour of the adolescent osseous and soft tissue structures, materials in our models were initially derived from data for adult spinal tissues (Table 1). Bending and torsional stiffness parameters prescribed for the beams representing the costo-vertebral joints were derived from Lemosse et al. [17], who quantified the mechanical behaviour of human cadaveric costo-vertebral joints in the three axes of motion. The material parameters listed in Table 1 were defined as a benchmark set of material data.

The following sections will detail our methods for defining physiological loading conditions on both the intact and surgically altered spine and our recent findings relating to determination of individualized patient tissue properties.

2.2 Validation of the Single Motion Segment Representation

We have previously conducted FE investigations of a thoracic single motion segment, in order to validate the modeling assumptions employed in our full thoraculumbar spine model [19] (Fig. 5a). This investigation reproduced the loading and boundary conditions employed by Oda et al. [27] and since anatomical data for the motion segments tested in this previous study were not available,

Table 1 Element types, material types and material parameters defined for each structure in the thoracolumbar spine model

Structure	Element type	Material type	Material parameters	Reference
Cortical bone	4-node shell	Linear elastic	$E = 11{,}300$ MPa $v = 0.2$	[23]
Cancellous bone	First order brick	Linear elastic	$E = 140$ MPa $v = 0.2$	[23]
Posterior elements (pedicles, lamellae, transverse processes, spinous processes)	Linear beam	Linear elastic	Quasi-rigid	
Zygapophyseal joint bone	4-node shell	Linear elastic	As for cortical	[25]
Anulus fibrosus—ground substance	First order brick	Hyperelastic, Mooney-Rivlin	$C_{10} = 0.7$ $C_{01} = 0.2$	[25]
Anulus fibrosus—collagen fibres	Tension-only, embedded rebar	Linear elastic	$E = 500$ MPa $v = 0.3$	[15, 38]
Nucleus pulposus	3D, 4-node fluid	Hydrostatic fluid	Incompressible	[20, 24]
Costal rib bone/Sternum	4-node shell	Linear elastic	$E = 9{,}860$ MPa $v = 0.3$	[14]
Costal cartilage	4-node shell	Linear elastic	$E = 49$ MPa $v = 0.4$	[14]
Costo-vertebral joint	Linear beam	Linear elastic	$E_{compr} = 245$ N mm^{-1} Torsion stiffness, $k = 4{,}167$ N mm rad^{-1} Bending stiffness, $k = 6706$ N mm rad^{-1} (average antero-posterior and cranio-caudal flexion stiffness)	[1, 17]
Ligaments—ALL and PLL	Spring	Nonlinear elastic	Piecewise, nonlinear	[26]
Ligaments—SL, ITL, LF,CL	Axial connectors	Nonlinear elastic	Piecewise, nonlinear	CL, SSP–[34]; ITL–[5]; LF–[26]
Inter-costal connections	Axial connectors	Linear elastic	$E = 25$ MPa	[36]

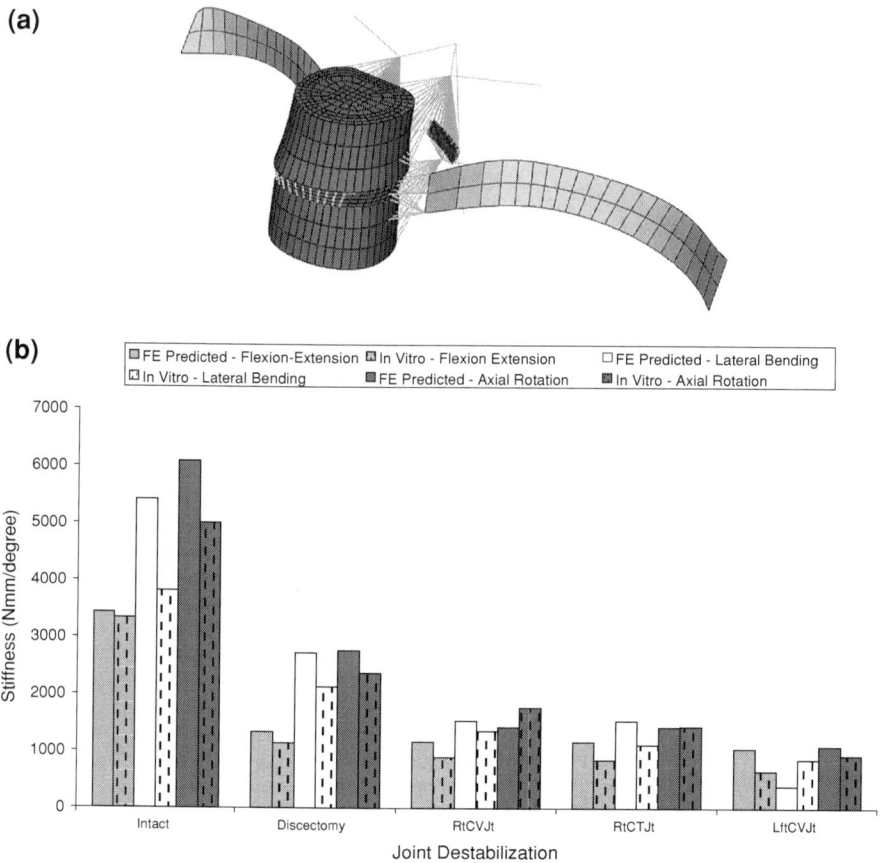

Fig. 5 a FE model representation of thoracic single motion segment. **b** Comparison of in vitro and FE predicted joint stiffness for an applied moment of 2,000 N mm, for the intact case [28] and four successive stages of joint destabilization [27] (Discectomy—complete removal of the intervertebral disc, RtCVJt—removal of the *right* costo-vertebral joint, RtCTJt—removal of the *right* costo-transverse joint, LftCVJt—removal of the *left* costo-vertebral joint)

anatomical data from the Visible Woman (The Visible Human Project, US National Library of Medicine) was simulated. Oda et al. [27] conducted an experimental study of a thoracic motion segment, to determine the change in segmental stiffness with successive removal of joint structures (intervertebral disc, right cost-vertebral joint, right costo-transverse joint and left costo-vertebral joint). For a 2,000 N mm applied moment, the average error between the simulated rotations about the three axes of motion in the intact joint and the in vitro data of Panjabi et al. [28] was 17% (Fig. 5b). The computational results for the simulated joint with four stages of destabilisation showed similar reductions in segmental stiffness to those observed experimentally by Oda et al. [27] (Fig. 4b). Given the uncertainty in anatomy mentioned above, these results provided significant

Table 2 Patient details and clinical Cobb angle measurements (degrees)

Patient ID	Age	Gender	Pre-operative Cobb angle	Fulcrum bending Cobb angle	Surgically corrected Cobb angle
1	14	M	53	34	30
2	14	F	44	26	14
3	13	F	40	15	1.6
4	10	F	58	15	5
5	22	F	42	8	7

confidence in our modeling methodology for representing both the spinal motion segments (including rib attachments) and by extension the full thoracolumbar spine, since the full model is a series of inter-connected motion segments.

3 Simulating Load Cases in the Intact and Surgically Altered Spine

To date, our patient-specific modeling research has focused on two load cases; (1) a pre-operative clinical assessment known as the fulcrum bending radiograph, and (2) the actual intra-operative deformity correction performed by the surgeon to attach a single rod anterior implant and correct the scoliotic spinal deformity. These load cases are both well suited to initial model development and validation because they are biomechanically well-defined (muscle activation in both cases is minimal), and we have access to detailed clinical and radiographic data for assessment of model predictions. Given the previously mentioned lack of material property data for paediatric spinal tissues, we sought to investigate the importance of patient-specific soft tissue parameters in accurately predicting both pre-operative patient flexibility and intra-operative deformity correction.

In an initial series of patients, we used CT datasets for five AIS patients to generate patient-specific, full thoracolumbar spine models (Table 2). Each of these models was analysed under the two load cases just mentioned and these are described in more detail below. The Cobb angle is the standard measure of scoliosis severity, defined as the maximum angle between the most tilted upper and lower vertebrae in a scoliosis curve, and is shown graphically in Fig. 7 below. Cobb angle is measured clinically using two dimensional plane radiographs, but can also be readily calculated from the deformed configuration of the computer models.

3.1 Fulcrum Bending Radiograph: Clinical Load Case 1

The first of these load cases, the fulcrum bending radiograph, is taken pre-operatively to clinically assess patient flexibility prior to surgery [6]. The patient lays laterally over a padded cylindrical bolster (Fig. 6a), with the convex side of

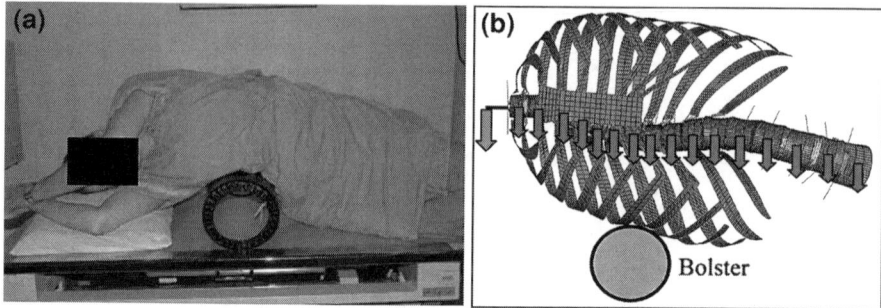

Fig. 6 **a** Fulcrum bending—patient positioned over the bolster. **b** Patient-specific spine model positioned over the bolster, showing segmental torso loads as *green arrows* and point load representing the head/neck weight as *red arrow* (NB. Arrows are not to scale)

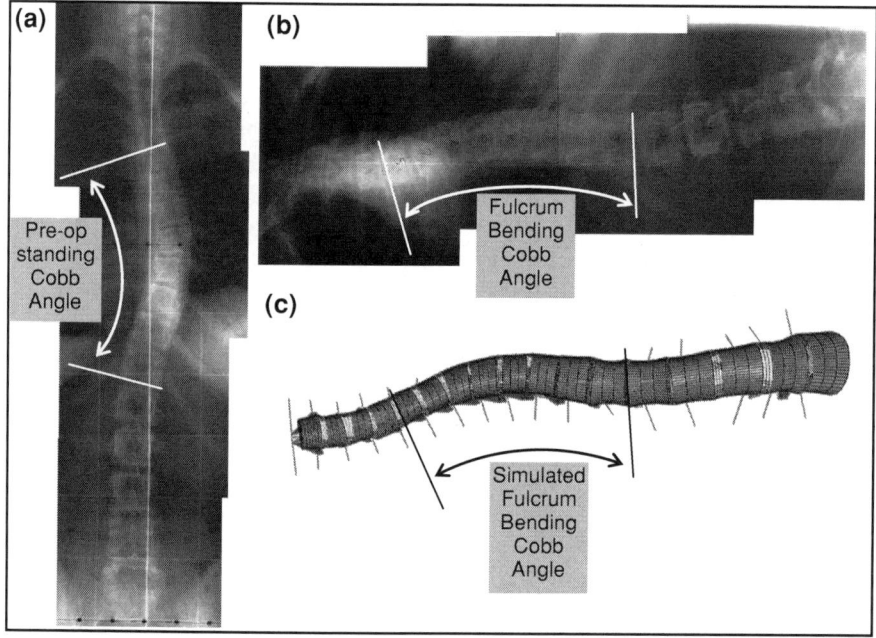

Fig. 7 Comparison of lateral deformity, measured using Cobb angle, between **a** the standing radiograph, **b** the fulcrum bending radiograph and **c** the deformed patient-specific spine model for the fulcrum bending load case

their curve toward the bolster surface. When in the final position, the patient's shoulder should be just suspended above the table with their pelvis and lowermost arm being the only point of contact between their upper-body and the table (Fig. 6a). In lying over the bolster, the patient's lateral deformity will reduce under the action of gravitational forces due to body weight—note that this is a passive deformity correction since there is no voluntary muscle activation during the activity.

The magnitude of the deformity reduction over the fulcrum is a clinically useful parameter because it indicates how much deformity correction the surgeon can expect to achieve clinically. Due to its passive nature (minimal muscle activation), this loading case is well suited for analysis with a passive osseo-ligamentous FE model and was simulated in the *intact* (surgically un-altered) model for each patient.

Patient-specific, CT-derived segmental torso weights for each vertebral level were determined using custom-developed software (Matlab R2007b). These weights were calculated using an average tissue density of 1.04×10^{-3} g mm^{-3} [10] and were applied at the CT-derived centroid of each vertebral level in the transverse plane (Fig. 6b). A load vector representing the full weight of the uppermost arm and half the weight of the lowermost arm was applied at the centroid of the T1 vertebrae in each model. The magnitude of this vector equated to 9% body weight [10]. Additionally, a load vector equivalent to the weight of the head and neck (8% body weight) [10] was simulated as a point load, applied superior to the uppermost vertebrae in each model. The bolster was simulated as a rigid body and rigidly constrained. A frictionless, tangential contact relationship was defined between the bolster surface and each rib and an exponential, normal contact relationship was used to represent the foam layer covering the bolster surface. The spine was free to rotate about a point simulating the contact between the pelvis and the table on which the patient rests.

Clinically, the reduction in Cobb angle while the patient lies over the bolster is compared with the Cobb measured from a standing radiograph, in order to provide a clinical measure of patient flexibility. This measure is called the Fulcrum Flexibility [6], FF, and is calculated according to Eq. 1. Lateral deformity is measured using Cobb Angle [7] (Fig. 7a, b).

$$\text{Fulcrum Flexibility, FF} = \frac{(\text{Standing Cobb Angle} - \text{Fulcrum Cobb Angle})}{\text{Standing Cobb Angle}} \times 100\%$$
(1)

Following analysis of the model under the fulcrum bending load case, the simulated FF was calculated for each of the five patient spine models and compared with the clinically measured value. The simulated FF was calculated using the predicted Cobb angle measured from the initial, undeformed spine geometry and the Cobb angle once the spine was positioned over the bolster (Fig. 7c). However, since CT scans on which the initial geometry of the FE models are based are performed while the patient is supine, the Cobb angle measured from the initial, unloaded model spine geometry is always less than the same patient's Cobb angle measured clinically on a standing radiograph. This difference between scoliosis severity in supine and standing positions has been previously measured, and therefore can be corrected for. On average, the difference between the Cobb angle measured from supine CT images and standing radiographs is 9° [37]. So that the predicted Cobb angle for the undeformed spine geometry could be compared to the clinically measured value (measured from standing radiographs),

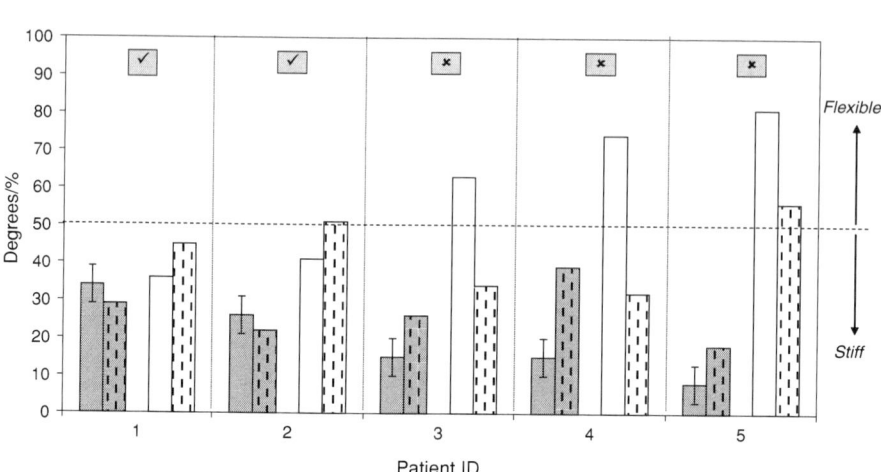

Fig. 8 Comparison of clinical and predicted values for the fulcrum bending Cobb angle (degrees) and FF (%). *Tick* symbol denotes simulation results within the clinically accepted error in Cobb angle measurement of ±5°, *cross* denotes simulation results outside the clinically accepted error range

the predicted model value was therefore increased by 9° before the simulated FF was calculated.

Furthermore, an acceptable error range of ±5° was defined for comparisons of the predicted and clinically measured Cobb angles, since this is the accepted variability associated with clinical measurements of plane radiographs [39]. The model predicted results for the fulcrum bending Cobb angle agreed with the clinically measured fulcrum flexibility (FF) for only two of the five patients and these two patients were those with a clinical FF of less than 50% (Fig. 8). In the context of the fulcrum bending radiograph, patients with a clinical FF of less than 50% are defined as 'stiff' and with a clinical FF greater than 50% are defined as 'flexible'.

Figure 9 shows a qualitative comparison of the predicted and the clinically observed deformed shapes for the spinal column—to obtain this figure the outline of the model-predicted, deformed spinal column was overlaid on the clinically obtained fulcrum bending radiograph for two patients (patients two and four). (We note that other researchers [11, 33] have conducted a quantitative investigation of spinal curvature measured from radiographs and we intend to adopt similar quantitative techniques in future investigations comparing the clinical and FE-predicted deformity.) Clinically, patient two was classified as a 'stiff' curve and the predicted results were within the acceptable error range for the deformed Cobb angle and FF (Fig. 8). Qualitatively, the predicted curvature of this patient's spinal column once over the bolster was similar to the clinically observed curvature for the vertebral levels T5–T12—which was the deformed region of the thoracic spine over which the Cobb angle was measured (Fig. 9a). Conversely, the

Patient-Specific Modeling of Scoliosis

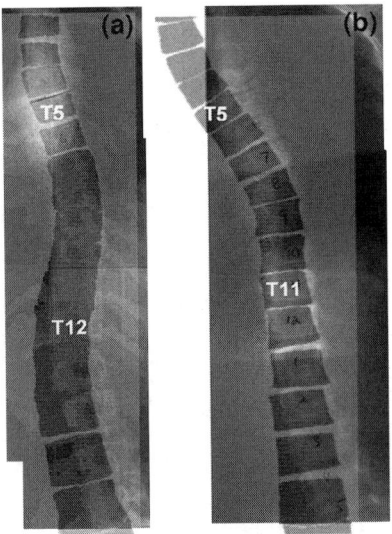

Fig. 9 Overlay of the binary outline from the deformed spinal column and the coronal fulcrum bending radiograph. **a** Patient 2, stiff patient; **b** Patient 4, flexible patient

predicted results for patient four were outside the acceptable error range for deformed Cobb angle and FF and the qualitative comparison of predicted and clinically observed spinal column curvature showed dissimilarities in deformed spine shape for vertebral levels T1–T8 (Fig. 9b).

The mismatch between simulated and clinical spine curve shapes displayed in Fig. 9b was similar for all three patients with a predicted FF outside the acceptable clinical error range (Fig. 8, patients 3–5). Both the qualitative and quantitative comparison of the model predicted and clinical FF results suggested that the patient-specific spine models for these patients (using adult spine tissue properties from Table 1) were overly stiff, with the predicted segmental rotations in the coronal plane too small for the given segmental torso weights. For these patients, the use of personalized tissue parameters which characterize the patient-specific tissue response would be expected to result in improved predictions of the spine behaviour. However, the predicted results for patients with a clinical FF less than 50% (clinically 'stiff'), showed good agreement (i.e. less than $\pm 5°$ error) with the clinical data, suggesting that for a subset of patients the benchmark tissue material properties derived from adult data are an appropriate representation for the adolescent tissues.

3.2 Towards Patient-Specific Soft Tissue Properties: What Have We Learned from the Fulcrum Bending Load Case?

Having compared the clinical and predicted FF for the benchmark set of (adult) tissue parameters with a pilot group of five patients, the next challenge is to determine an appropriate method to adjust these tissue parameters to achieve accurate patient-specific tissue properties for those patients where the benchmark (adult) tissue properties from existing literature do not provide a correct prediction of the physiological spine biomechanics. Spinal flexibility is influenced by several soft tissue structures, including the spinal ligaments, the intervertebral disc (annulus fibrosus and nucleus pulposus), the zygapophyseal joints and the cost-vertebral connections. Moreover, the influence of these structures may vary depending on the particular loading condition experienced by the spinal joints. Since spinal flexibility as measured by the fulcrum bending radiograph is a single numeric parameter (FF), the problem of determining a unique set of patient-specific tissue parameters which characterize this flexibility is difficult. However, our previous studies investigating patient-specific tissue parameters have concluded that only a few of these structures are of key importance in governing spinal flexibility [9, 19] and these structures will now be discussed.

In a previous FE modeling study, we conducted a detailed FE investigation of a single motion segment (SMS) to better understand the influence of the individual spinal ligaments and disc structures on the segmental stiffness (as measured by vertebral rotation) [9]. The SMS model was developed using the same modeling parameters and methods as the full thoracolumbar spine model. The loading condition simulated the physiological limits of rotational displacement in the three planes of motion (flexion, extension, left/right lateral bending, left/right axial rotation) [30], with rotations applied about a centre of rotation which simulated the physiological joint. The results from these analyses showed firstly, that by individually removing the spinal ligaments (setting stiffness parameters equal to zero), the SMS stiffness could be reduced (Fig. 10) and secondly, that only a few soft tissue structures have an appreciable effect on the SMS joint biomechanics. Specifically, this study demonstrated the importance of the intervertebral disc in governing spinal flexibility, showing that removal of the disc annular collagen fibres significantly reduced the SMS stiffness. The study also concluded that correct representation of the capsular ligament mechanical properties was important in simulating the biomechanics of the SMS.

Following this, as a first step toward determining patient-specific soft tissue properties, we used the fulcrum bending load case to determine which soft tissue structures were of most importance in governing spinal flexibility [18]. In this study, the disc collagen fibre elastic modulus and the ligament nonlinear elastic stiffnesses were reduced by up to 40% in separate analyses and the effect on predicted vertebral rotations assessed. Note that in this study, stiffnesses for all ligaments were reduced at the same time. Results showed an average increase in coronal vertebral rotation of 3.1 and 13.4% with a 40% reduction in the ligament and collagen fibre stiffnesses, respectively. In a separate analysis, the change in FF due to complete removal of the intervertebral discs was investigated (this was a limiting case since this represented a 100% reduction in the stiffness of all disc materials). The results of this disc removal simulation demonstrated a three-fold increase in the predicted FF compared to the results for the fulcrum bending simulation using the intact model and benchmark tissue stiffnesses (FF = 39% for intact spine, increasing to FF = 120% after

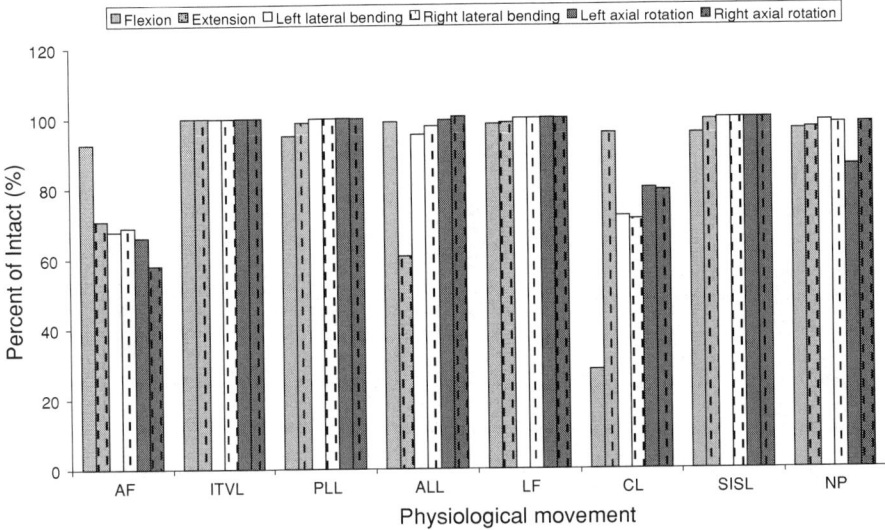

Fig. 10 Predicted changes in single motion segment (*SMS*) joint stiffness following removal of individual soft tissue structures, expressed as a percentage of the Intact value, following removal of SMS joint structures. *AF* annulus fibrosus, *ITVL* inter-transverse ligament, *PLL* posterior longitudinal ligament, *ALL* anterior longitudinal ligament, *LF* ligamentum flavum, *CL* capsular ligament, *SISL* inter-/supra-spinous ligament, *NP* nucleus pulposus

intervertebral disc removal). Note that just reducing the disc collagen fibre stiffness had a minimal effect on FF, only increasing it from FF = 39% (intact spine) to FF = 40% (60% reduction in disc collagen fibre stiffness). This study confirmed the importance of the intervertebral disc in influencing spinal flexibility (as measured by vertebral rotation) and showed that a lumped approach to altering the ligament stiffnesses does not necessarily capture the complex interaction in ligament load sharing which occurs across the spinal joints.

The costo-vertebral joint (CVJt) was the only soft tissue structure which was not investigated in this prior study [18] and is the subject of a current material property sensitivity study (Little and Adam, unpublished data), exploring the effect of changes in costo-vertebral joint stiffness on FF for a group of 10 AIS patients. As already mentioned, these rib-vertebra joints are of key structural importance in governing spinal stiffness [27, 40] and are represented in detail in our full thoracolumbar spine model. In this material sensitivity study, for patients where the benchmark (adult) tissue parameters (Table 1) did not provide a predicted FF within the accepted clinical measurement error (predicted Fulcrum Cobb angle within ±5° of the clinically measured Cobb angle) [39], the costo-vertebral joint (CVJt) stiffness parameters were either increased or decreased accordingly. Specifically, if the predicted FF differed from the clinical value by less than -10% (i.e. the simulated spine was too flexible) the CVJt stiffness was increased by 100% (doubled) and if the predicted FF differed by more than 10% (i.e. the simulated spine was too stiff) the CVJt stiffness was decreased by 99%. Reducing

Table 3 Clinical and predicted Fulcrum flexibility values (%), showing predicted results for both the benchmark tissue properties (Table 1) and altered CVJt stiffness

Patient §	1	2	3	4	5	6	7	8	9	10
Age	14	14	22	14	14	17	16	27	13	10
Gender	M	F	F	F	F	F	F	F	F	F
Pre-operative clinical Cobb angle	53	44	52	53	60	52	42	45	40	48
Clinical FF	36	41	42	43	50	52	55	58	63	74
Predicted FF using Benchmark Tissue Properties (Table 1)	45*	51	37*	60	25	45*	47*	47	34	32
Changed CVJt Stiffness	No	Yes ↑	No	Yes ↑	Yes ↓	No	No	Yes ↓	Yes ↓	Yes ↓
Predicted FF using altered CVJt Properties		23*		23	44			18*	23	37

(↑ = increase CVJt stiffness by 100%, ↓ = decrease CVJt stiffness by 99%, * = predicted FF results are within the acceptable error range) (§ Note that the patient identifier numbers in Table 3 do not correspond to those stated in Table 2 as this table refers to a separate study)

the CVJt stiffness by 99% was intended to approximate complete removal of the joint stiffness and to represent the maximum possible effect of reducing the stiffness of these joints. These coarse changes in CVJt stiffness were intended to provide initial sensitivity data which can then inform subsequent, more finely grained analyses. The results of this series of simulations are summarized in Table 3.

For the benchmark tissue parameters, the predicted FF was outside the acceptable error range for six of the 10 patients (Table 3). Large alterations in CVJt stiffness improved the FF results for all these six patients, but only brought the predicted FF results for two of them within the acceptable error range (Table 3). For the overall group of 10 patients, the mean error between model predicted and clinically measured fulcrum Cobb angle and FF was reduced when altered CVJt stiffness parameters were used (Fig. 11) and six of the 10 patients demonstrated a predicted FF within the acceptable clinical error after the changes in CVJt stiffness. For the remaining four patients even the large alterations in CVJt stiffness (100% increase or 99% decrease) did not sufficiently alter the flexibility of the simulated spine to provide agreement with the clinical values.

We conclude that changes to the CVJt stiffness can have an appreciable effect on spinal flexibility and this material sensitivity study went some way towards determining which of the spinal soft tissue structures are of most importance in governing spinal flexibility. Taken together with our previous studies, it is apparent that there is neither an individual soft tissue structure which completely governs spinal flexibility, nor is a lumped approach (varying all tissue properties at the same time) necessarily appropriate for varying soft tissue parameters to achieve patient-specific spine behaviour.

Patient-Specific Modeling of Scoliosis

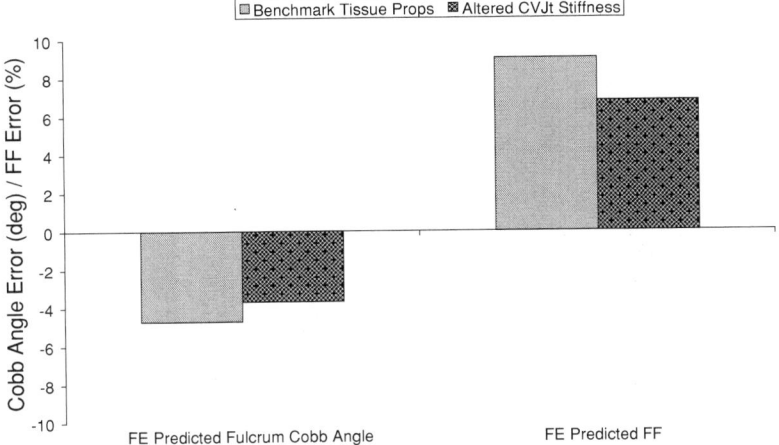

Fig. 11 Average error in FE predicted Fulcrum Cobb angle (degrees) and FF (%) after analyzing the 10 patient-specific spine models under the fulcrum bending load case. Results are presented for both the benchmark tissue parameters and after altering the CVJt stiffness

Rather, when patients are observed to be clinically 'stiff' (FF < 50%) or 'flexible' (FF < 50%) following assessment using the fulcrum bending test, this spinal behaviour is likely governed by a group of soft tissue structures acting in partnership. This suggests that, more detailed pre-operative flexibility assessment data than the currently available single scalar parameter (i.e. FF) is necessary to isolate and characterize patient-specific tissue properties for these structures. Ideally, clinical data characterizing the forces applied to the patient's spine by the bolster while undergoing the fulcrum bending test would be measured for comparison with FE predicted forces, to assess both force and spinal deformation (and thus, vertebral segmental stiffnesses) in the fulcrum bending radiograph. Possibly, the distribution of segmental torso weights differs between clinically 'stiff' and clinically 'flexible' patients, resulting in variations in segmental weights which are currently not captured in the patient-specific segmental weights we derive from patient CT datasets. Since the fulcrum bending test has not yet been biomechanically characterized, the forces applied to the patient's spine in contact with the bolster as well as the forces acting at regions of contact with the X-ray bed are not known. These measurements are a required step toward deriving patient-specific spinal tissue properties.

3.3 Intra-Operative Surgical Corrective Forces: Clinical Load Case II

During anterior scoliosis surgery, the key biomechanical step to reduce the spinal deformity occurs when the surgeon attaches a titanium alloy rod to the patient's spine using screws which are inserted into the vertebrae within the extents of the structural curve.[1] The surgeon applies stepwise compressive forces between pairs of screw heads in adjacent vertebrae (starting with the distal two screws and moving toward the proximal two screws in the construct) and once the desired level-wise correction has been achieved, each screw is locked onto the rod. This successive, level-wise deformity correction results in a reduction in the overall deformity of the structural curve. Prior to compression of the screw heads, the intervertebral discs between these vertebrae are partially removed (approximately the lateral half) and bone graft material is inserted into this disc space to encourage the vertebrae to fuse together in the corrected position.

This surgical process is the focus of our intra-operative surgical modeling, in which we attempt to develop a load case which simulates forces applied by the surgeon to the patient's spine while undergoing the anterior, single rod, scoliosis correction procedure. As already mentioned, our research to date has focused on the anterior, single rod, scoliosis correction procedure since we have access to a detailed set of clinical data including clinically indicated pre-operative CT scans for AIS patients who undergo this surgery thorascopically (via a minimally invasive or key-hole approach). As such, this patient dataset has provided detailed clinical data for model validation of our patient-specific simulation techniques.

In order to simulate the intra-operative load case, it was necessary to first model the surgically altered spinal anatomy and implant components. All five patients chosen for intra-operative simulation underwent a single rod, anterior corrective procedure (Fig. 1b). Clinical data for the surgical procedure (including which vertebral levels were fused, number of screws implanted, rod diameter and alloy) were used to simulate the individual surgical procedure for each patient-specific model. Using our custom pre-processing software it was possible to re-generate each patient's spine model (both geometry and FE mesh) with surgically altered spinal geometry (Fig. 12a, b), incorporating details for:

- Screw placement/orientation—The vertebral screws were represented as embedded continuum elements in the vertebral bodies, assuming an idealized, perfectly bonded contact relationship between the surface of the screw shaft and underlying cancellous bone elements in the vertebral body (Fig. 12b). Note that this representation does not take into consideration the screw threads in contact with the underlying bone;
- Discectomy levels—During surgery, the intervertebral discs between the vertebral bodies attached to the rod are partially removed (half the disc) and bone graft material is inserted into the disc space to promote bony fusion in the 3–6 months after surgery. These discectomy levels were simulated in the surgically altered spinal geometry by removing half the continuum elements representing the annulus fibrosus mesh, and by removing the entire fluid filled cavity representing the nucleus pulposus (Fig. 12b). Since this was an

---

[1] The region of the spine with an intrinsic lateral curvature rather than a deformity related to a functional imbalance of the spinal soft tissues. The structural curve includes vertebrae which are within the extents of the spinal deformity measured by the Cobb angle.

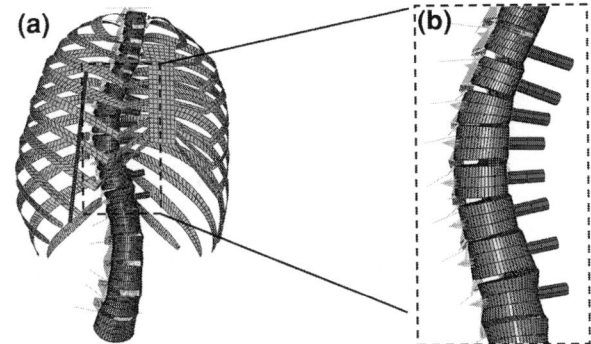

Fig. 12 a Full patient-specific spine model (undeformed mesh) showing the single anterior rod and screws prior to insertion. **b** Detail of thoracic spinal levels showing partial thoracic discectomies and screws (*green*) embedded within vertebral bodies. Note that the screw heads have been enlarged for visualization purposes

intra-operative load case, no fusion material was simulated as the bone graft offers no appreciable stiffness to the spine at the time of surgery. Interfacial contact between the exposed endplate surfaces of the adjacent vertebrae was simulated using an exponential, softened contact algorithm (normal contact) and Coulomb friction, $\mu = 0.3$ (tangential sliding).

- Rod geometry—The rod was represented with a user-defined length and diameter to match clinical data for each patient. The rod and screws are manufactured from Titanium alloy, which was represented as a linear elastic material (E = 108,000 MPa, $\upsilon = 0.3$).

This surgical load case simulated the intra-operative compressive forces applied by the surgeon in order to correct the spinal deformity. Following removal of the intervertebral discs and insertion of the screws, the screw-to-screw compressive forces used to reduce the scoliosis deformity are simulated between successive vertebral levels within the structural curve. A compressive force is applied between the screw heads at adjacent vertebrae using connector elements, and once the required segmental correction has been simulated, locking of the screw onto the rod is simulated using bonded contact between the two FE meshes, thus simulating the cumulative, level-wise intra-operative correction of the spinal deformity.

Using a novel clinical measurement technique, our group (Cunningham et al. [8]) have determined the in vivo compressive forces applied intra-operatively for a series of patients undergoing the single rod, anterior corrective procedure. Using a strain-gauged surgical compression tool and data logging system, the study recorded a continual force tracing during the correction of each spinal joint, up until the point at which the screw was locked onto the rod. Using this dataset, Cunningham et al. [8] defined the average level-wise compressive force applied by the surgeons intra-operatively at each spinal level, relative to the apex of the scoliotic curve.[2]

---

[2] The apex of the scoliotic curve is the most laterally deviated vertebra on a frontal radiograph.

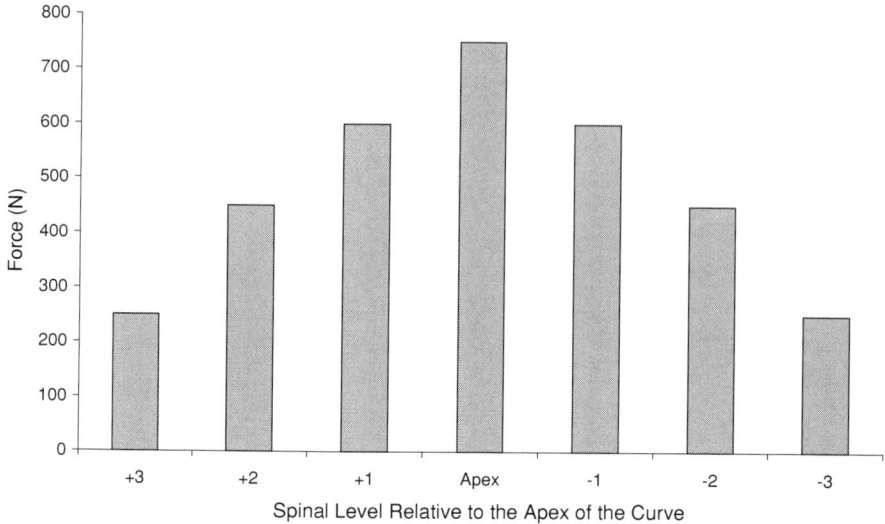

Fig. 13 Intra-operative compressive forces applied to the model at vertebral level, based on the mean forces measured by Cunningham et al. [8]. Vertebral level is normalized relative to the apex of the structural curve, with positive levels indicating vertebrae cephalic to the apex and negative indicating vertebrae inferior to the caudal to the apex

The compressive force data from this experimental study were utilized to simulate the intra-operative load case for each of the five patient-specific spine models (Fig. 13). The incremental compression steps carried out surgically were simulated with the spine model and the L5 vertebra was constrained from motion during all load steps. To assess the validity of the patient-specific spine models in predicting surgical deformity correction, the corrected Cobb angle (at the end of the surgical deformity correction) predicted by each patient's FE model was compared with the clinically measured (standing) Cobb angle obtained one week post-operatively. In addition to this, for patient three a supine, coronal radiograph taken immediately post-operatively (while the patient was still in Intensive Care, ICU) was also available. As with the FF load case, an acceptable tolerance in Cobb angle of ±5° was used [39].

Predicted results for the surgically corrected Cobb angle were within the acceptable error range for two of the five patient-specific models when comparing the predicted results to the clinical measurements obtained from radiographs taken one-week post-operatively (Fig. 14). However, for patient three, when comparing the predicted Cobb angle with the clinical value measured immediately post-operatively (ICU radiograph), the predicted result was within the acceptable error range. The ICU radiograph is obtained while the patient is supine, while the one-week post-operative radiographs are obtained while the patient is standing. While previous studies have shown that on average, Cobb angle increases by 9° when comparing supine to standing [37], little is known about the comparative change in Cobb angle for an instrumented spine. These data suggest there may be

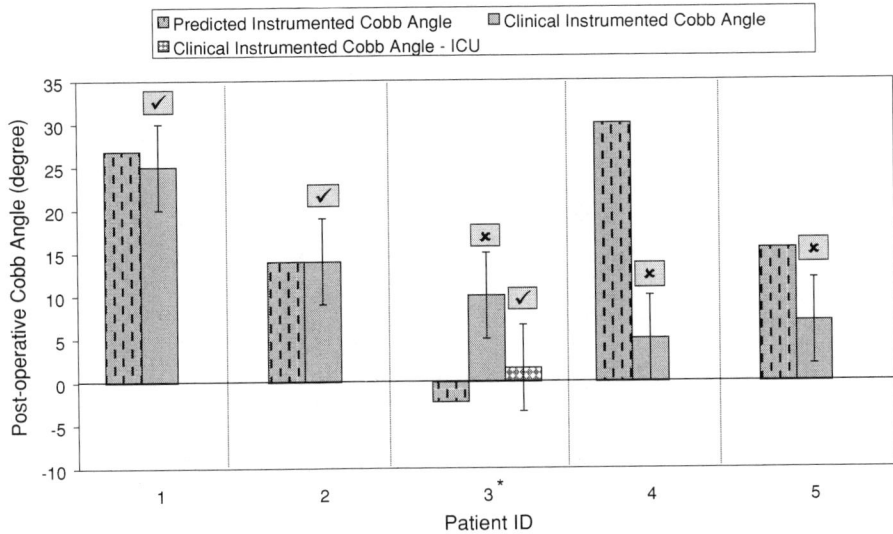

Fig. 14 Comparison of the clinical and model-predicted Cobb angle after surgery. The clinical Cobb angle was measured from radiographs taken one week post-operatively. *For patient three, the clinical Cobb angle was also measured from a radiograph taken immediately post-operatively (patient supine and in ICU). For patient 3, the negative Cobb angle indicates the simulated surgical procedure 'over-corrected' the spinal deformity, creating a 2.3° lateral deformity on the contra-lateral side of the spine to the initial deformity

an increase in Cobb angle of a similar magnitude for the instrumented spine, however, supine radiographs are not routinely obtained immediately post-operatively, so this data was available for only one of the patients in this series.

Patients four and five demonstrated a predicted Cobb angle after surgery of 30 and 15.5°, respectively, which were substantially larger than the actual clinical values of 5 and 7°, respectively. These two patients also demonstrated a predicted FF outside the acceptable error range (Fig. 8). Figure 15 shows the predicted shape of the surgically altered spine for patient two. By comparing the undeformed shape of the spinal column shown in Fig. 12b with the deformed shape shown in Fig. 15b, the reduction in the intervertebral disc space resulting from compressive forces applied between the screw heads can be seen. In some instances this compressive force was sufficient to result in contact between adjacent vertebral endplates.

Qualitative comparison of the predicted and clinically observed surgical deformity correction showed that for patients with a predicted corrected Cobb within the acceptable error range, both the shape of the instrumented curve and the shape of the spinal levels above and below the instrumented curve were similar to the coronal radiograph. Conversely, for the three patients whose simulated surgical correction was outside the acceptable error range compared to the clinical result, the predicted shape of both the instrumented region as well as the spinal curves above and/or below this region tended to deviate from the post-operative radiograph (Fig. 16).

Fig. 15 a Full thoracic spine after intra-operative compressive forces have been applied (deformed mesh, Patient 2). **b** Detail showing deformed rod and decreased height of disc space following successive compression of vertebral levels

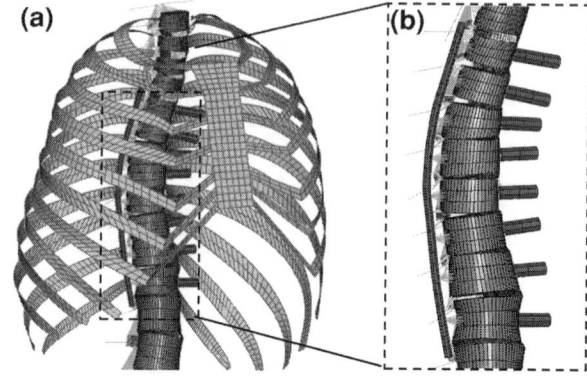

Fig. 16 Overlay of the binary outline from the deformed spinal column and the 1 week post-operative coronal radiograph. This overlay shows the orientation of vertebra both within (**a** Patient 3) or above (**b** Patient 4) the instrumented curve deviated from the orientation on the radiographic image

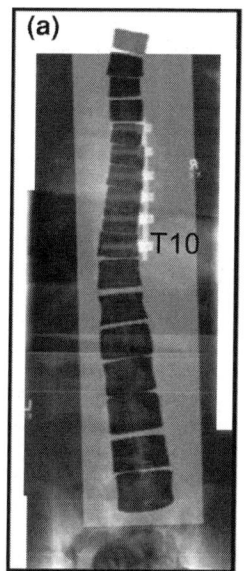
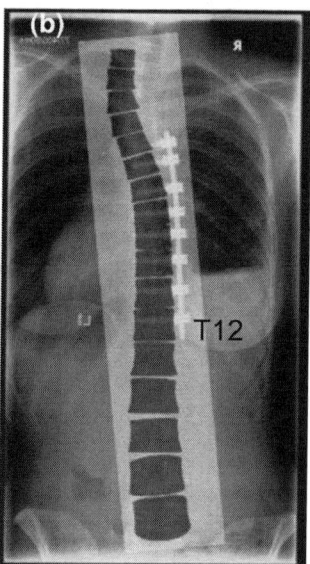

3.4 Towards Patient-Specific Soft Tissue Properties: Insights from the Surgical Load Case

Scoliosis is a complex 3D deformity which results in abnormal curvature of the spine in all three anatomical planes. Generally, the spinal deformity will be observed as both a primary structural curve—which is the region where the lateral, side-to-side curvature is a manifestation of abnormal growth—and compensatory curves—which develop above and/or below the primary curve and are non-structural, occurring in order for the spine to maintain balance (a compensatory lumbar curve is visible below vertebral level T12 in Fig. 9a). In addition to the correction of the primary thoracic deformity adjacent to the implant, previous clinical studies of single rod anterior scoliosis surgery have shown that surgical correction of the primary structural deformity also results in a reduction in the compensatory deformity observed distal to the primary thoracic curve [12]. Qualitative comparisons between the model and clinical results in the patient group just described, showed that the FE simulations predicted changes in both the primary (thoracic) and compensatory (lumbar) spinal curves following analysis of the intra-operative load case simulation. This suggests that the current passive osseo-ligamentous spine models are capable of reproducing at least the passive osseo-ligamentous aspect of this compensatory curve behaviour in the adolescent scoliotic spine.

The results of the intra-operative load case suggest that for a subset of patients, the use of a set of benchmark (adult) tissue properties based on existing literature provides a reasonable approximation to adolescent spinal tissue properties. This is in keeping with the conclusions drawn from the results of the Fulcrum Bending load case. However, for the remaining patients, the simulated deformity correction was considerably low in comparison to the clinical value, implying that the simulated spine was overly stiff in comparison to the actual tissues. In order to improve the agreement between the clinical and predicted results, the use of patient-specific paediatric soft tissue properties is essential.

The intra-operative deformity correction load case was applied using data for compressive forces measured in vivo [8] and such data is novel in the field of scoliosis biomechanics, with few previous studies having included measured force data in computational simulations of scoliosis surgery. However, the loading profile (Fig. 13) applied to each of the models in our pilot analyses to date, is averaged data from force measurements carried out on six patients. Preliminary FE investigations currently underway in our group suggest that the use of individual, patient-specific loading profiles (based on the in vivo data from Cunningham et al. [8]) will yield different predicted deformity corrections than the use of an 'average' loading profile for a particular type of surgical procedure. Thus as well as the need for patient-specific anatomy and patient specific tissue mechanical properties, patient-specific intra-operative surgical forces may also be required for realistic simulation of scoliosis surgery.

4 Conclusions and Future Directions

Patient-specific spine modeling presents a complex and multi-factorial challenge to not only accurately simulate the anatomy of the spine but also to appropriately replicate the physiological behaviour of the spinal soft tissues and to individualize the loading conditions applied to the spinal structures during surgical procedures. Our computational studies to date have demonstrated there is no one single soft tissue structure which governs the flexibility of the spine, but rather that there is a complex interaction between the spinal soft tissues which regulates the overall biomechanics of the spine.

In an attempt to better understand and ultimately derive patient-specific soft tissue properties it is appealing to take advantage of existing clinically performed patient flexibility assessments. Patient flexibility as assessed using the Fulcrum bending radiograph is strongly dependent on the relative stiffness of the various spinal soft tissues and derivation of tissue stiffnesses using these assessments provides in vivo data without the necessity for highly invasive testing of patient tissues. However, in order for such patient assessments to be useful in deriving patient-specific tissue properties, it is necessary for biomechanical characterization of these assessments to be carried out, through measurements of the forces applied to the patient's spine.

Our studies to date demonstrate that the use of patient-specific anatomy alone is not sufficient to accurately simulate the biomechanical behaviour of the adolescent spine. The use of a set of benchmark material properties based on adult data can adequately reproduce the clinically observed behaviour of AIS patients for only a subset of patients. Patient-specific tissue properties are of key importance in creating accurate simulations of the spine and ribcage and in reproducing the clinically observed flexibility and motion of the spine.

In our work to date the focus has been on model development and validation by comparison with readily available clinical data such as FF and post-operative Cobb angle. However, once appropriately validated, the patient-specific models potentially provide a wealth of biomechanical information which can be used to assess the likely success of a proposed deformity correction procedure. For example, predicted rod strain is related to the likelihood of rod fracture under cyclic loads, predicted relative motion between adjacent vertebrae is related to the likelihood of successful bony fusion (as opposed to pseudarthrosis) following surgery, and predicted changes in tissue strain after a surgical intervention are valuable indicators of the degree of tissue remodeling which would be expected to occur in response to the altered spinal loading conditions.

The integration of patient-specific finite element simulations in a pre-operative surgical planning tool shows much potential. By utilizing a spine model which is truly personalized to the patient; in terms of anatomy, tissue properties and intra-operative corrective forces; accurate predictions for the outcomes of a particular surgical procedure should be possible prior to surgery. The overall aim of this approach to surgery simulations is to assist surgeons to plan the optimal biomechanical correction for a particular patient, while avoiding the risk of implant or tissue damage. This concept represents a paradigm shift in scoliosis surgery planning—by complementing a surgeon's expertise with biomechanical simulation tools, it will be possible to accurately predict treatment outcomes for individual patients and ultimately, improve overall patient health post-operatively. There are significant challenges in realizing the clinical potential of patient specific biomechanical simulations such as these. Imaging modalities must allow pre-operative 3D imaging with sufficient resolution to accurately define patient-specific spinal anatomy, but without excessive radiation dose. Secondly, reliable biomechanical characterization techniques (based on clinical assessments such as the fulcrum bending radiograph for example) must be developed to ensure that patient-specific tissue mechanical properties can be prescribed which will correctly simulate spinal stiffness. Thirdly, prediction of the biomechanics of the surgically altered spine must be extended beyond the operating theatre to simulate physiologically relevant load cases after surgery (i.e. during gait and other post-operative activities involving muscle activation). Finally, issues of model convergence and solution time must be addressed to ensure that results can be obtained in a clinically relevant timeframe. These challenges can only be overcome through sustained effort by the biomechanical modeling community, but the goal of improved outcomes for complex spinal surgery patients make this research effort highly worthwhile.

References

1. Andriacchi, T., Schultz, A., Belytschko, T., Galante, J.: A model for studies of mechanical interactions between the human spine and rib cage. J. Biomech. **7**(6), 497–507 (1974)
2. Aubin, C.E., Labelle, H., Chevrefils, C., Desroches, G., Clin, J., Eng, A.B.: Preoperative planning simulator for spinal deformity surgeries. Spine **33**(20), 2143–2152 (2008)
3. Aubin, C.E., Labelle, H., Ciolofan, O.C.: Variability of spinal instrumentation configurations in adolescent idiopathic scoliosis. Eur. Spine J. **16**(1), 57–64 (2007)
4. Betz, R.R., Harms, J., Clements, D.H., 3rd, Lenke, L.G., Lowe, T.G., Shufflebarger, H.L., Jeszenszky, D., Beele, B.: Comparison of anterior and posterior instrumentation for correction of adolescent thoracic idiopathic scoliosis. Spine **24**(3), 225-239 (1999)
5. Chazal, J., Tanguy, A., Bourges, M., Gaurel, G., Escande, G., Guillot, M., Vanneuville, G.: Biomechanical properties of spinal ligaments and a histological study of the supraspinal ligament in traction. J. Biomech. **18**(3), 167–176 (1985)
6. Cheung, K.M., Luk, K.D.: Prediction of correction of scoliosis with use of the fulcrum bending radiograph. J. Bone Joint Surg. **79**(8), 1144–1150 (1997)
7. Cobb, R.J.: Outline for study of scoliosis. In: American Academy of Orthopaedic Surgeons, Instructional Course Lectures. CV Mosby, St Louis, pp. 261–275 (1948)
8. Cunningham, H., Little, J.P., Adam, C.J.: The measurement of applied forces during anterior single rod correction of adolescent idiopathic scoliosis. Paper presented at the ACSR annual meeting, Adelaide, August 2009
9. Cunningham, H., Little, J.P., Pearcy, M.J., Adam, C.J.: The effect of soft tissue properties on overall biomechanical response of a human lumbar motion segment: A preliminary finite element study. In: Brebbia, C.A. (ed.) Modelling in Medicine and Biology VII, WIT Transactions on Biomedicine and Health. WIT Press, Southampton, pp. 93–102 (2007)
10. Erdmann, W.S.: Geometric and inertial data of the trunk in adult males. J. Biomech. **30**(7), 679–688 (1997)
11. Hasler, C.C., Hefti, F., Buchler, P.: Coronal plane segmental flexibility in thoracic adolescent idiopathic scoliosis assessed by fulcrum-bending radiographs. Eur. Spine J. **19**(5), 732–738 (2010)
12. Hay, D., Izatt, M.T., Adam, C.J., Labrom, R.D., Askin, G.N.: Radiographic outcomes over time after endoscopic anterior scoliosis correction: a prospective series of 106 patients. Spine **34**(11), 1176–1184 (2009)
13. Kamimura, M., Kinoshita, T., Itoh, H., Yuzawa, Y., Takahashi, J., Hirabayashi, H., Nakamura, I.: Preoperative CT examination for accurate and safe anterior spinal instrumentation surgery with endoscopic approach. J. Spinal Disord. Tech. **15**(1), 47–51; discussion 51–42 (2002)
14. Kimpara, H., Lee, J.B., Yang, K.H., King, A.I., Iwamoto, M., Watanabe, I., Miki, K.: development of a three-dimensional finite element chest model for the 5(th) percentile female. Stapp Car Crash J. **49**, 251–269 (2005)

15. Kumaresan, S., Yoganandan, N., Pintar, F.A.: Finite element analysis of the cervical spine: a material property sensitivity study. Clin. Biomech. (Bristol, Avon) **14**(1), 41–53 (1999)
16. Lafage, V., Dubousset, J., Lavaste, F., Skalli, W.: 3D finite element simulation of cotrel-dubousset correction. Comput. Aided Surg. **9** (1–2), 17–25 (2004)
17. Lemosse, D., Le Rue, O., Diop, A., Skalli, W., Marec, P., Lavaste, F.: Characterization of the mechanical behaviour parameters of the costo-vertebral joint. Eur. Spine J. **7**(1), 16–23 (1998)
18. Little, J.P., Adam, C.J.: The effect of soft tissue properties on spinal flexibility in scoliosis: biomechanical simulation of fulcrum bending. Spine **34**(2), E76–E82 (2009)
19. Little, J.P., Adam, C.J.: Effects of surgical joint destabilization on load sharing between ligamentous structures in the thoracic spine: a finite element investigation. Clin. Biomech. (Bristol, Avon) (in press)
20. Little, J.P., Adam, C.J., Evans, J.H., Pettet, G.J., Pearcy, M.J.: Nonlinear finite element analysis of anular lesions in the L4/5 intervertebral disc. J. Biomech. **40**(12), 2744–2751 (2007a)
21. Little, J.P., Pearcy, M.J., Pettet, G.J.: Parametric equations to represent the profile of the human intervertebral disc in the transverse plane. Med. Biol. Eng. Comput. **45**(10), 939–945 (2007b)
22. Lonner, B.S., Kondrachov, D., Siddiqi, F., Hayes, V., Scharf, C.: Thoracoscopic spinal fusion compared with posterior spinal fusion for the treatment of thoracic adolescent idiopathic scoliosis. J Bone Joint Surg. **88**(5), 1022–1034
23. Lu, Y.M., Hutton, W.C., Gharpuray, V.M.: Do bending, twisting, and diurnal fluid changes in the disc affect the propensity to prolapse? A viscoelastic finite element model. Spine **21**(22), 2570–2579 (1996)
24. Nachemson, A.: Lumbar Intradiscal Pressure: experimental studies on post-mortem material. Acta Orthopaedica Scandinavica **43**, (1960)
25. Natali, A., Meroi, E. Nonlinear analysis of intervertebral disc under dynamic load. Trans. ASME J. **112**, 358–362 (1990)
26. Nolte, L.P., Panjabi, M., Oxland, T.: Biomechanical properties of lumbar spinal ligaments. In: Heimke, G., Soltesz, U., Lee, A.J.C. (eds.) Clinical Implant Materials. Elsevier, Amsterdam, pp. 663–668 (1990)
27. Oda, I., Abumi, K., Cunningham, B.W., Kaneda, K., McAfee, P.C.: An in vitro human cadaveric study investigating the biomechanical properties of the thoracic spine. Spine **27**(3), E64–E70 (2002)
28. Panjabi, M.M., Brand, R.A., Jr., White, A.A., 3rd., Mechanical properties of the human thoracic spine as shown by three-dimensional load-displacement curves. J. Bone Joint Surg. **58**(5), 642–652
29. Papin, P., Arlet, V., Marchesi, D., Laberge, J.M., Aebi, M.: Treatment of scoliosis in the adolescent by anterior release and vertebral arthrodesis under thoracoscopy. Rev. Chir. Orthop. Reparatrice Appar. Mot. **84**(3), 231–238 (1998)
30. Pearcy, M.J.: Stereo radiography of lumbar spine motion. Acta Orthopaedica Scandinavica Supplementum **56**(212), 1–45 (1985)
31. Reddi, V., Clarke, D.V., Jr., Arlet, V.: Anterior thoracoscopic instrumentation in adolescent idiopathic scoliosis: a systematic review. Spine **33**(18), 1986–1994 (2008)
32. Rohlmann, A., Richter, M., Zander, T., Klockner, C., Bergmann, G.: Effect of different surgical strategies on screw forces after correction of scoliosis with a VDS implant. Eur. Spine J. **15**(4), 457–464 (2006)
33. Sangole, A.P., Aubin, C.E., Labelle, H., Stokes, I.A., Lenke, L.G., Jackson, R., Newton, P.: Three-dimensional classification of thoracic scoliotic curves. Spine **34**(1), 91–99 (2008)
34. Shirazi-Adl, A., Ahmed, A.M., Shrivastava, S.C.: Mechanical response of a lumbar motion segment in axial torque alone and combined with compression. Spine **11**(9), 914–927 (1986)
35. Stokes, I.A., Gardner-Morse, M.: Muscle activation strategies and symmetry of spinal loading in the lumbar spine with scoliosis. Spine **29**(19), 2103–2107 (2004)
36. Stokes, I.A., Laible, J.P.: Three-dimensional osseo-ligamentous model of the thorax representing initiation of scoliosis by asymmetric growth. J. Biomech. **23**(6), 589–595 (1990)

37. Torell, G., Nachemson, A., Haderspeck-Grib, K., Schultz, A.: Standing and supine Cobb measures in girls with idiopathic scoliosis. Spine **10**(5), 425–427 (1985)
38. Ueno, K., Liu, Y.K.: A three-dimensional nonlinear finite element model of lumbar intervertebral joint in torsion. J. Biomech. Eng. **109**(3), 200–209 (1987)
39. Vrtovec, T., Pernus, F., Likar, B.: A review of methods for quantitative evaluation of spinal curvature. Eur. Spine J. **18**(5), 593–607 (2009)
40. Watkins, R., 4th., Watkins, R., 3rd., Williams, L., Ahlbrand, S., Garcia, R., Karamanian, A., Sharp, L., Vo, C., Hedman, T.: Stability provided by the sternum and rib cage in the thoracic spine. Spine **30**(11), 1283–1286 (2005)
41. Wong, H.K., Hee, H.T., Yu, Z., Wong, D.: Results of thoracoscopic instrumented fusion versus conventional posterior instrumented fusion in adolescent idiopathic scoliosis undergoing selective thoracic fusion. Spine **29**(18), 2031–2038 (2004)

Subject-Specific Models of the Spine for the Analysis of Vertebroplasty

Alison C. Jones, Vithanage N. Wijayathunga, Sarrawat Rehman and Ruth K. Wilcox

Abstract Treatments for the spine still lag behind other orthopaedic interventions in terms of their technical maturity and success rates. The use of computational models for pre-clinical testing of new spinal devices is attractive but, as yet, these models have not reached their full potential for the evaluation of treatments across a patient population. This chapter examines the development of subject-specific models of the spine with a particular focus on the analysis of fracture fixation using vertebroplasty. The development of specimen-specific models allows direct comparison to be made between the model predictions and the behaviour of the same specimens in laboratory tests, thus allowing the features and sensitivity of the models to be thoroughly examined. The progress made in this field is examined and key results presented. Current work and future challenges are also discussed in the context of extending the technique to simulate the performance of new treatments across a population.

1 Introduction

It is estimated that eight out of ten adults experience an episode of low back pain during their lifetime [1] and the symptoms of a significant number will continue to progress despite conservative treatment. The global spine market has seen substantial growth in the last decade but despite huge investment, the success of spinal treatments remains relatively low, with higher complication rates than orthopaedic

A. C. Jones · V. N. Wijayathunga · S. Rehman · R. K. Wilcox (✉)
Institute of Medical and Biological Engineering,
School of Mechanical Engineering,
University of Leeds, Leeds, UK
e-mail: r.k.wilcox@leeds.ac.uk

procedures in the hip or knee. One reason for the poor clinical outcomes of spinal interventions is likely to be the variation in the anatomy and tissue properties between patients, and this was recently identified as being a major factor in the high failure rates of an interspinous device [2, 3]. Both mechanical and morphological differences between patients may also lead to differences in the structural effect of less invasive treatments such as vertebroplasty and kyphoplasty, where clinical outcomes have been highly variable [4, 5].

In order to evaluate spinal treatments effectively, and to optimise device design before introduction into clinical practice, it is therefore necessary to capture these variations in the population within the pre-clinical testing process. Traditionally, this evaluation has taken the form of experimental testing on isolated device components or in vitro cadaveric specimens. In the latter case, the cost and availability of specimens tend to lead to low sample sizes; in addition, the age and level of degeneration in the specimens do not necessarily represent the target patient group. In more recent decades, computational finite element (FE) models of the spine have been developed and used by researchers to evaluate a range of spinal treatments [6]. However, since these models have usually been based on generic geometries and material properties, validation lacks robustness because of the wide range of experimental and in vivo data with which the model predictions can be compared. Whilst limited parametric variations in material properties have been undertaken, models have yet to capture the morphological and mechanical differences that occur over the patient cohort. These limitations have curbed the extent to which computational models are currently used as part of the pre-clinical testing process for spinal devices.

Advances in imaging and computational techniques have now enabled the possibility of developing models based on the geometry and mechanical properties of individual spinal specimens. This approach paves the way for patient-specific analysis and the potential for the evaluation of treatments across a population. It also enables more robust validation through direct comparison between models and corresponding specimens tested in vitro.

This chapter outlines the progress that has been made in the development of subject-specific spinal models. The emphasis is on the vertebra, which has received greater attention than the soft tissues of the spine, and the treatment of vertebral fractures through cement augmentation, known as vertebroplasty. The development stages and current state of art are described and the future challenges in this field are discussed.

2 Specimen-Specific Vertebral Modelling Methods

2.1 The Need for Specificity

If generic FE models are not sufficient to capture the variance in behaviour of spinal vertebrae across a population, then it is necessary to understand which aspects of the models need to be assigned in a subject-specific manner in order to

Table 1 Concordance correlation coefficients for the predicted stiffness of porcine vertebrae. The values are shown for three different types of FE models and a purely numeric method, each compared to corresponding experimental data. Taken from Wilcox [7]

Method	Concordance coefficients	
	Stiffness	Strength
Numerical: using average BV/TV and cross sectional area	0.638	0.609
FE: geometry-generic, material property-specific	0.058	0.125
FE: geometry-specific, material property-generic	0.724	0.429
FE: geometry-specific, material property-specific	0.881	0.752

achieve sufficient accuracy. A study of porcine vertebrae was undertaken to address this question by generating FE models of specimens tested in the laboratory [7]. Each combination of generic and subject-specific, geometry and material properties were assigned to the models and the results compared (i.e. generic geometry with subject-specific material properties, etc.). It was shown that individually assigning both the material properties and the geometry in a specimen-specific way substantially increased the accuracy of stiffness and strength predictions (Table 1). The stiffness and strength are key indicators of vertebral integrity and health. As such these are the aspects which procedures, such as vertebroplasty, seek to restore back to pre-fracture levels.

2.2 Model Generation

In order to generate vertebral models where both the geometry and the material property distribution are specific to an individual vertebra, some form of imaging is usually necessary to provide the input data. One such methodology is described in the following section.

The method was developed to rapidly and accurately generate vertebral models and assign subject-specific geometry and material properties from high resolution image data [8, 9]. The aim in this case was to generate subject-specific finite element models of individual vertebral bones which did not require high-performance computers, but would run on a powerful desktop machine. The model development process is presented in (Fig. 1); this method has subsequently been adopted and modified by a number of other studies.

The model generation process is dependent on images from micro computed tomography (μCT) scans of each specimen. These images are used to capture the geometric features and the inhomogeneous material properties of the vertebra, which are then incorporated into the matching finite element model. Using μCT imaging provides 3D, high-resolution images of each vertebral specimen. As well as capturing the shape of the external bone surface, the resolution (typically 0.074 mm for the studies presented here) captures the trabecular architecture

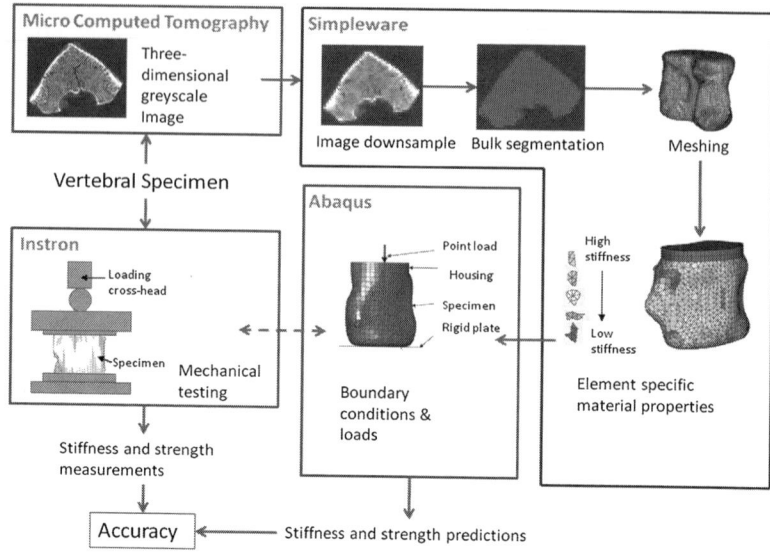

Fig. 1 Flow chart of specimen-specific vertebral finite element model generation and validation against equivalent experimental data

within the bone. The resulting images are converted from the Hounsfield units generated by the μCT software to a standard greyscale image format (containing a greyscale level range of 0–255). The finite element model is then generated from the greyscale images. In the studies presented here, this step was achieved using a semi-automated image processing and meshing software package (ScanIP/FE, Simpleware Ltd, Exeter, UK).

The process of separating the whole vertebra from the image background is referred to here as "bulk segmentation". This segmentation is performed using various standard image processing tools within the image processing software and aims to capture a single vertebral object with no internal gaps. Any marrow, blood or other fluid within the trabecular structure is therefore included in this object.

Since μCT imaging uses X-ray absorption to map out the 3D picture of the bone, it is reasonable to assume a relationship between image brightness and local material density. It is therefore possible to establish a conversion formula which relates image greyscale, through density, to the material Young's modulus. In the work described here, a linear conversion was assumed and a single factor was used to relate image greyscale directly to material Young's modulus, but other methods have been investigated and are discussed in Sect. 3. Each element within the finite element mesh is assigned a Young's modulus based on the greyscale values of the image in that area, creating an inhomogeneous material property distribution throughout the vertebra.

In order to generate finite element models which require relatively low computational power, the elements in the mesh are necessarily much larger than the voxels in the underlying image. Each element therefore covers an area which will include both trabecular bone tissue and trabecular space. The greyscale value used to calculate the Young's modulus for an individual element represents the average of the greyscale values of all of the image voxels contained within that element.

Once the vertebral object had been identified through the bulk segmentation process, a mesh can then be generated within this region and individual material properties assigned based on underlying image greyscale values.

The resulting models are then exported for finite element analysis. In the studies presented here, the processing and post processing has been undertaken using a proprietary finite element modelling package (Abaqus, Dassault Systèmes Simulia Corp., Providence, RI, USA).

3 Validation and Sensitivity Testing

The specimen-specific models, described in this section, contain the vertebral bone in isolation and do not attempt to simulate any soft tissue that would surround the bone in vivo. Developing isolated vertebral bone models allows for one-to-one matches between computational models and laboratory specimens, making direct validation of each case possible. The general approach has been to generate models of the lowest complexity necessary to provide a sufficiently accurate representation of the tissue behaviour. This makes it easier to understand which aspects introduce the greatest inaccuracy.

This section outlines the validation of these models and the sensitivity tests that have been undertaken to evaluate the levels of complexity necessary for different model inputs.

3.1 Validation

The method outlined above enables direct comparisons to be made between subject-specific FE models predictions and corresponding experimental data. This subject-specific vertebral model generation method, was originally developed using porcine vertebral bodies [8] and was later adapted for use with human vertebra [9]. In these studies, the boundary conditions and loading regime were set up to match those of the corresponding experimental tests. To ensure the boundary conditions were applied to the correct region, the cement end caps used to hold the experimental specimens were also included in the FE models in one study [9]. In both cases, the lower surface was prevented from moving or rotating in any direction, and a point load was applied to a rigid plate on the superior surface. Both the computational model and the corresponding experimental specimen were subjected to axial compression. Measurements of stiffness and strength were taken in both cases and compared in order to check the accuracy of the model generation method.

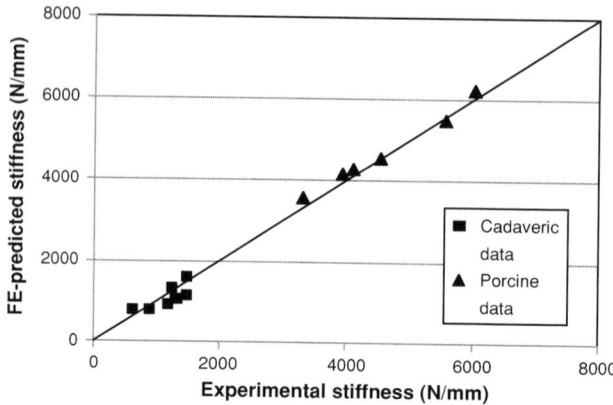

Fig. 2 Agreement between FE predicted stiffness and experimentally measured stiffness for porcine [8] and human [9] vertebrae

For the human vertebrae [9], a training set of six vertebrae was used to derive a suitable factor for the conversion of greyscale levels to material Young's modulus values. This conversion factor was then used on an independent set of vertebrae and compared with corresponding experimental test data. Good agreement was found between the computational predictions and the experimental measurement (Fig. 2) and the method was therefore considered to be valid for human vertebrae.

3.2 Finite Element Mesh Resolution

The initial porcine study [8] looked in detail at the interaction between bone geometry and material properties when both were derived from μCT images. For complex geometries, such as living tissue, a refinement of the mesh will also cause a change in the accuracy with which that geometry is captured. In this method changing the mesh size involves also changing the resolution of the underlying image. The coarser the mesh is, the more neighbouring greyscale values are merged together. These greyscale values provide an approximate representation of the trabecular and cortical bone architecture. The mesh resolution therefore affects not only the external geometry accuracy but also that of the internal architecture [8]. Convergence testing showed that an element size of $1 \times 1 \times 1$ mm represented a converged solution (Fig. 3). These element sizes were later applied for models of individual human vertebrae [9].

A method of separating the mesh resolution effect from that of the inhomogeneous material properties has been implemented for cubic bone specimens [10]. The inhomogeneous material property values and distribution for a low resolution mesh were maintained as the mesh was refined, allowing the generation of a series

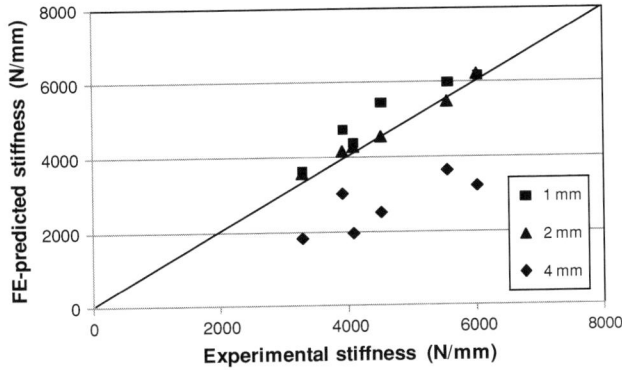

Fig. 3 Finite element mesh resolution tests for porcine vertebral models with specimen-specific geometry and material properties and 1, 2 and 4 mm element sizes [8]

of models of different resolutions containing identical material values. These models showed excellent convergence for an element size below 1.5 mm. This was in contrast to the results of the equivalent models where the element size and material property distribution were linked. In those cases the stiffness prediction continued to change below an element size of 1.5 mm.

3.3 Boundary Conditions and Loading

3.3.1 Simulation of In Vitro Loading

Since the computational models discussed in the previous section matched in vitro specimens, their boundary conditions reflect those of the specimen mounted in the materials testing machine. Several sensitivity tests have been performed, on porcine [8] and human [9] specimens, to establish the sensitivity of the stiffness prediction to the position of the applied load. As little as 2 mm of change in the position of the point load was found to cause up to 20% difference in the stiffness prediction (Fig. 4). These results were taken into account in subsequent experimental tests where extra steps were taken to control the load positions and record them more accurately for translation to the equivalent computer models.

3.3.2 Simulation of In Vivo Loading

The loads acting on the vertebra in vivo are far more complex than the experimental testing set-up because they are distributed between the zygapophysial joints and across the vertebral body. To apply the appropriate boundary conditions and

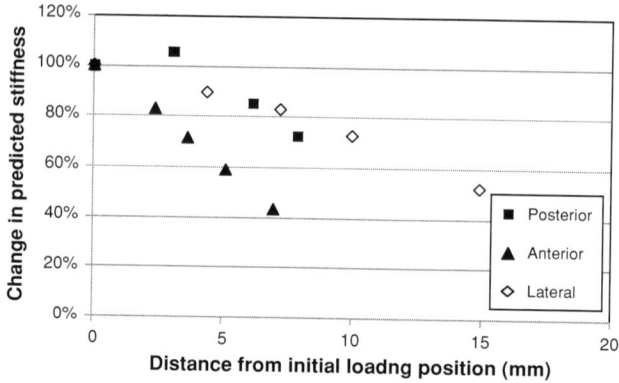

Fig. 4 Sensitivity of vertebral stiffness predictions to the position of the point load on the superior plate [8]

loads to the vertebral body, the method of load distribution across the body of a single vertebra has been investigated. Distributed loads applied to the vertebral body via a cement end cap conforming to the superior vertebra geometry were found to be sensitive to the point of loading as described for the in vitro specimens above, the orientation of the original specimen and the surface area captured by the conforming cement end cap. To eliminate the effect of specimen orientation and the variable contact area between the cement end cap and the superior vertebral body, a pressure applied to a surface defined on the vertebral body was considered by applying a force per unit area that acted in-plane with the global coordinate system. The maximum variability in the predicted vertebral stiffness was found to be 8% for models generated from repeated scans of the same specimen that were aligned in the same orientation. For in vivo boundary conditions it is therefore important to consider both the surface area and direction of the applied load.

4 Material Properties

As discussed in Sect. 2, specimen-specific models require specimen-specific material properties to accurately capture the variation both from one specimen to another and also within each individual vertebra.

4.1 Number of Material Properties

Using the method described in Sect. 2, each element in the FE model is assigned material properties based on its underlying greyscale value derived from the

Table 2 Percentage change in predicted stiffness when the number of material properties assigned to the vertebral model are banded and compared against the non-banded results

	Five property bands (% change from non-banded)	Two property bands (% change from non-banded)
Vertebra 1	5.2	6.1
Vertebra 2	4.7	5.4
Vertebra 3	−0.1	5.2
Vertebra 4	10.5	12.3

scanned image. In principle, since the models are based on 'tiff' images, this could result in 255 Young's Modulus values, representing the vertebral bone. In actual practice, the range of greyscale values observed for a typical vertebral μCT scan is far less than 255, and usually in the range of 60–100.

A study was conducted in order to understand the influence of the number of Young's Modulus values required to represent the vertebral bone. Four cadaveric vertebrae were imaged using a microCT scanner, and finite element models were developed based on the image data, using the method described previously ('non-banded' models).

The greyscale values were then grouped together so that for each vertebra, FE models were also generated with two and five greyscale bands. For each band, the average greyscale value was obtained, and the conversion to Young's modulus was based on this average greyscale. This resulted in a collection of elements with same material properties representing each band in the finite element model.

Subsequently, all of the finite element models for each vertebrae were subjected to a simulated axial compression and the vertebral axial stiffness computed. The results from these simulations, shown in Table 2, indicate that the overall stiffness (in this case, compressive stiffness) tends to increase with the grouping of elements, prior to material property assignment. However, the errors appear to be within a reasonable margin. In other words, the methodology appears to be iterating towards an average homogeneous modulus value for the whole vertebra.

4.2 Linear Versus Non-Linear Relationships

In addition to the number of material properties assigned, the method of conversion from the image data to the material properties the method by which they are derived from the μCT image data is also an important consideration in the model generation process.

A study was conducted to find out the most appropriate conversion criteria. Four cadaveric vertebrae were imaged using a microCT scanner, and finite element models were developed based on the image data, using the process shown in Fig. 1. Thereafter the underlying greyscale data for each element in the model was converted to material properties, employing the following three relationships

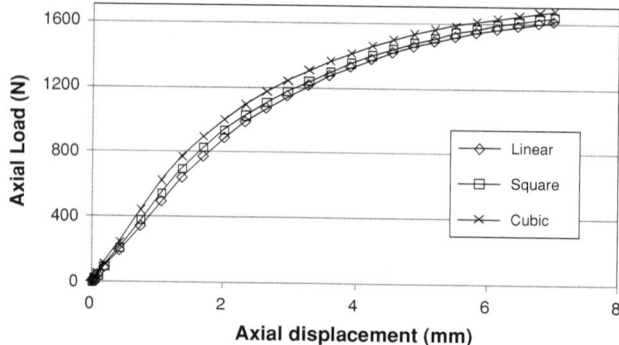

Fig. 5 Load–displacement curves for the vertebral model generated using different relationships [9]

between the greyscale value (GS), the elastic modulus (E), and the yield stress (σ_y):

$$\text{Linear}: E = GS \times k_1, \sigma_y = GS \times k_2$$

$$\text{Square}: E = GS^2 \times k_3, \sigma_y = GS^2 \times k_4$$

$$\text{Cubic}: E = GS^3 \times k_5, \sigma_y = GS^3 \times k_6$$

Subsequently all the models were subjected an axial compression simulation, and the load–displacement behaviour representing each criteria was recorded. The results are shown in Fig. 5 and it was observed that for these vertebral models under simple axial compression, the difference between the curves was slight and was not considered to be significant compared with model prediction errors due to other factors.

4.2.1 Density Based Properties Verses Bone Volume Based Properties

Broadly, there are two methods that are used to derive the properties of the vertebral bone from the μCT image data. In the studies described in the previous sections, the properties are based on the underlying image greyscale, the method will be referred to as the 'greyscale approach'. There is another method in which the properties are based on information derived from the segmented images, described here as the 'segmentation approach'. In the greyscale approach, the average greyscale of all of the voxels within each finite element is calculated and, since this will be related to the density of the bone, the material properties are calculated from this value. In the segementation approach, the μCT images are first segmented to capture only the trabecular bone tissue itself and not the trabecular space or any interstitial fluids. The material properties of each finite element are

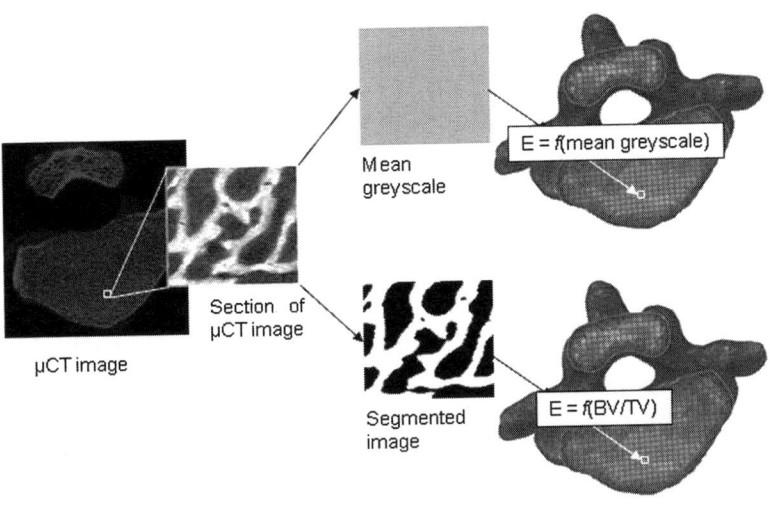

Fig. 6 Example of two different approaches for the derivation of mechanical properties from the µCT image data. For each element within the model, the elastic modulus is calculated based on either the mean grayscale (grayscale approach) or the BV/TV (segmentation approach) of that region

then based on the bone volume fraction (BV/TV) or other trabecular structural characteristics for that region in the segmented images. Examples of the two approaches are shown in Fig. 6.

Each method has advantages and disadvantages. The greyscale approach takes into account the spatial variation in bone tissue properties [11], since regions of higher mineralisation will generate a brighter signal than regions of lower mineralisation. However, since the averaging process also includes all of the voxels in the trabecular space, this method is only applicable if the specimens were imaged in a standard state (e.g. fully hydrated with marrow intact), and any changes in the specimen state would necessitate the derivation of new functions to determine the mechanical properties. In addition, the imaging system itself requires calibration to ensure there is consistency in the greyscale levels between specimens, although this is possible with the use of hydroxyapetite phantoms.

With the segmentation approach, the state of the specimens is less important, since providing the tissue can be segmented from the trabecular space, it does not matter if the trabecular space is filled with marrow or air. The disadvantage is that the segmentation process means that the greyscale information on the trabecular tissue level itself is lost, and effectively it has to be assumed that the tissue modulus within each specimen and between specimens is the same.

The two methods have been compared on the same image datasets, and the level of error was found to be similar, in both cases being less than 10% (Table 3). Clearly it would also be possible to combine both approaches, but this would require additional steps as well as the maintenance of the specimens in the same state, so a combined approach has yet to be fully investigated.

Table 3 Average absolute error in the stiffness predictions for porcine [24] and human [9] vertebra using the two different model generation methods compared to corresponding experimental data

Species	Average absolute error in stiffness predictions	
	Greyscale approach	Segmentation approach
Pig (n = 6)	5.7%	7.5%
Human (n = 4)	12.2%	5.0%

Table 4 Absolute errors between experimental stiffness values and specimen-specific finite element models based on BV/TV, comparing species-specific conversion factors to a global conversion factor, in all cases using a linear relationship (E = BV/TV × k). In this example, different threshold values were used to segment the images from the two different species and in all cases, the coefficient k was optimised to produce the lowest possible error

Method	Species		
	Porcine	Human	Combined
Optimised coefficient k (GPa)	0.38	0.25	0.33
Average absolute error	7.5%	5.0%	21%

4.2.2 Derivation of Material Properties for Different Species

The approaches discussed above have also been applied to bone from several different species of animal. Whilst good agreement has been found between specimen-specific finite element model predictions and experimental test results for a number of different species, the conversion factors used to determine the elastic modulus from the image greyscale have been different.

To compare models between species, a study was undertaken in which specimens from cadaveric and porcine spines were converted to models. All the specimens were imaged in the same calibrated μCT system with the same settings. The images were converted to models using the process outlined in Sect. 2. In this case, the segmentation approach was used to derive the element elastic modulii following image segmentation. For each species, the BV/TV to elastic modulus conversion factor was optimised to produce the best fit to the experimental data and a third conversion factor was also applied and optimised to best fit to the whole specimen set. When species-specific BV/TV to elastic modulus conversion values were used, the errors were low, but when a single conversion factor was applied across the whole set, the errors were much higher, as shown in Table 4. Inspection of the images indicates that the porcine bone is far less mineralised than the human bone, with a much lower greyscale distribution within the trabecular voxels (Fig. 7); it is therefore likely to have a lower tissue elastic modulus. This highlights the potential weakness of the BV/TV method, in that it is based on the assumption that the tissue level elastic modulus is uniform.

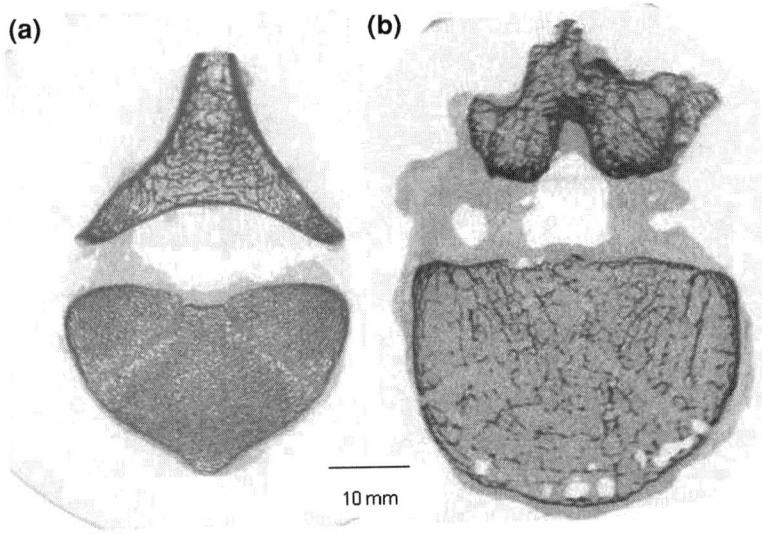

Fig. 7 Example images of **a** porcine and **b** human μCT data both imaged at the same resolution (74 × 74 × 74 um) and X-ray beam settings, showing the difference in trabecular bone structure and density. For clarity, the images have been reversed so denser bone is shown as a *darker grey*

4.3 The Effect of Tissue Preservation

With the limited availability of fresh human tissue it is desirable to recruit dry-bone and embalmed specimens from medical dissections, anatomical and archaeological collections to generate finite elements from a larger population set.

During the embalming process, tissue is preserved by chemical fixation to prevent autolysis and bacterial attack, i.e. breakdown or decay of the tissue, for its intended long term storage and handling. This process involves the removal of water and use of aldehydes based fixative. Aldehydes are known to stabilise soluble proteins which mainly exist in soft tissue by forming cross-links [12]; however, the mineral composition of bone has not been found to alter [13]. The process of preparing dry-bone specimens for anatomical collections also requires the physical removal of soft tissue by the means of de-fleshing, controlled decomposition and chemically de-fatting i.e. maceration, and the complete removal of water preserving the mineralised tissue only.

The mechanical properties of biological tissue are largely dependent on the content of water. In the case of embalmed and dry-bone specimens the removal of water and structural changes to the soft tissue, are known to alter the mechanical properties of the bone. Formalin fixation has been reported to reduce the compressive strength of bone [14], [15]. Whereas dry bone specimens compared to wet have shown an increase in the ultimate tensile strength and Young's Modulus [16]. However, given that there are no indications of change in the mineralised bone,

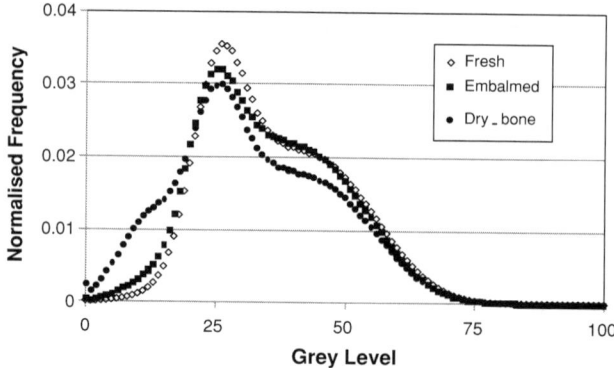

Fig. 8 Distribution of normalised mean greyscale values for ovine specimens in their fresh state and after they have been treated in their embalmed and dry-bone states

it is possible that the volume of vertebral bone from these preserved tissues can be used in the segmentation approach. The linear conversion factor validated for fresh vertebral specimens can then be used to generate finite element models from these preserved tissues.

There is some question over selecting the right threshold level to distinguish bone from the background in images of preserved tissue. The removal of water and soft tissues will alter the range of intensity values in μCT images of bone. Thus, a shift in the distribution of greyscales values of embalmed or dry-bone compared to fresh tissue can be anticipated as marrow and fluids occupying the trabecular space and the surrounding soft tissues are no longer present. However, experimental studies have shown that this shift is not constant (Fig. 8) and the threshold levels determined for fresh, embalmed and dry-bone specimens are very similar.

Finite element models generated from a central volume of interest within the vertebral body show that the difference in stiffness between the corresponding fresh and preserved tissue is almost within the repeatability of measurements. Also this difference is not greater than that due to other factors such as the boundary conditions and loading mentioned above.

5 Models of the Vertebroplasty Procedure

Over the past three decades, there has been a rapid increase in the use of vertebroplasty and kyphoplasty to treat vertebral compression fractures. Although originally developed for the augmentation of metastatic vertebrae, the rapid growth in the technique was due mainly to its use in the treatment of osteoporotic fractures. However, the results of recent randomized clinical trials have been contradictory and there remains considerable doubt over the mechanical consequences [17] or clinical effectiveness [4, 5] of the procedure.

Fig. 9 A μCT scan showing the cement ingression into the trabeculae

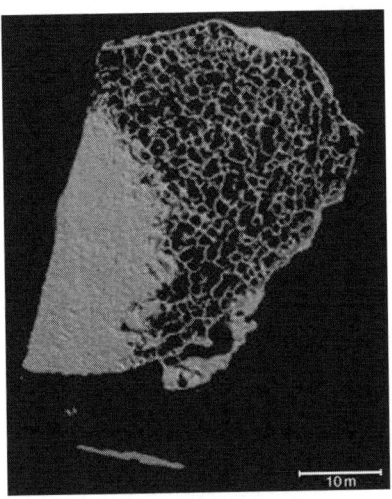

Vertebroplasty involves the injection of a small quantity of liquid bone cement into the fractured vertebrae. Once set, the cement is thought to stabilize the fracture and reduce pain. Concerns about its effectiveness and the longer term effects on the adjacent vertebrae have led to a number of FE studies to evaluate the mechanical effects of the technique. This section examines the use of specimen-specific models for this application and the additional challenges of modeling the cement-bone composite.

5.1 Representation of Cement in a Vertebral Body

During image processing of computer tomography scans, it is possible to visualise and identify the cement augmented region within the vertebral body, using greyscale thresholding techniques (Fig. 9). The finite element mesh for this region could then be associated with a different set of material properties rather than subjecting it to the greyscale based property conversion used for bone. In the majority of currently published work, the cemented region within the vertebral body has been represented using a uniform material, usually with the same properties as pure cement [18, 19] or in some cases, with relatively lower elastic properties than pure cement [20].

In order to evaluate the appropriateness of using the properties of pure cement to represent the augmented region within the vertebral body, Wijayathunga et.al. (2008) conducted a cadaveric study using the same approach to generate specimen-specific vertebral FE models as was described previously. The results obtained for prophylactic augmentation of cadaveric vertebrae show that associating the cement-augmented region with the material properties of pure cement does not provide good agreement in comparison to experimental observations, as illustrated in Fig. 10.

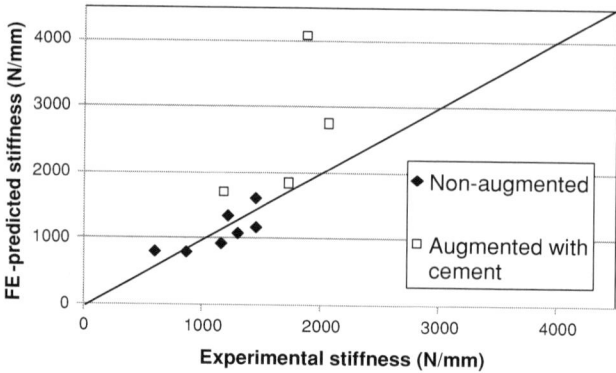

Fig. 10 Predicted versus experimentally measured stiffness of all of the specimens from Wijayathunga et al.[9]

5.2 Representation of Cement in Laboratory Specimens

In reality it is clear that when cement is injected to the vertebral body, it flows within the trabecular architecture, displacing the bone marrow and other fluids. Hence, after its curing, the cement will form an interface with the trabecular strands and any remaining fluids, effectively giving rise to a composite for the augmented region within the vertebral body.

Several studies ([21, 22], Buckley et al. 2003) have indicated that the average mechanical properties of this cement bone composite are different from the properties of pure cement as well as cancellous bone. The Race et al. [22] experiment used plugs extracted from vertebral bodies that were completely compacted with cement and in the other two studies, the bone plugs or blocks were completely filled with cement after extraction. In both cases, they represent a significant deviation from the actual vertebroplasty clinical scenario, and there are still doubts whether such homogeneous property values for the cement-bone composite could be used to computationally simulate a vertebroplasty procedure.

Similar tests have also been undertaken using cores of synthetic bone augmented with cement and compared to finite element models [23]. Here three sets of specimens were tested without augmentation, after full immersion in cement and partially filled with cement injected into the central region, simulating the vertebroplasty procedure. In all cases, the specimens were imaged with μCT and specimen-specific finite element models generated using a similar approach to that shown in Fig. 1. The first two sets were used to derive greyscale-based conversion factors for the elastic modulus of the synthetic bone and bone-cement composite and these values applied to the partially injected specimen set. Again, the predicted stiffness values overestimated the results obtained from the laboratory tests, as shown in Fig. 11. It was found that a reasonable fit to the experimental data could be achieved by using a much lower modulus for the cement region (Fig. 11), indicating that the cement-bone

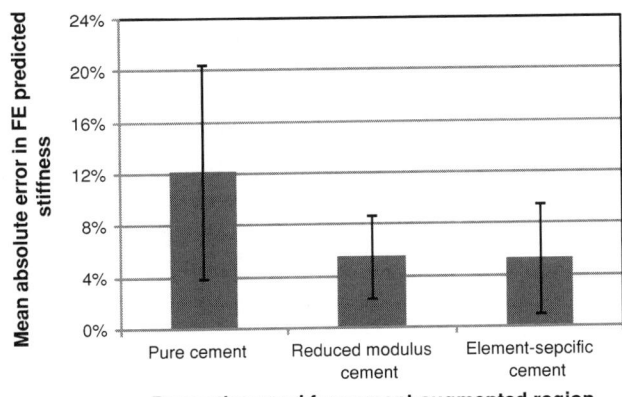

Fig. 11 Absolute error in FE predictions of stiffness compared to experimental results for cores of synthetic bone augmented with cement. The cement-augmented region was assigned properties of either pure cement (E = 2,500 MPa), a reduced modulus (E = 345 MPa) or an element-specific modulus based on the grayscale. Results taken from Zhao et al. [23]

composite behaviour is more complex than had previously been assumed in finite element simulations of vertebroplasty. This may be due to the cement not fully penetrating the 'cement region' under injection, leaving pockets of air (in this case) or fluid (in the case of real bone). However 'rule of mixture' calculations indicate that if this were the only reason, the percentage air/fluid would have to be far higher than is likely [10]. It is therefore likely that other factors such as local effects at the cement-bone interfaces, also play an important role.

5.3 The Reduced Modulus Approach

By using the reduced modulus for the cement region derived by Zhao et al. [23] and applying it to the models of the augmented cadaveric vertebrae described in Sect. 4.1, the level of error was considerably reduced, indicating that this 'reduced modulus' method may be sufficient to evaluate the gross effect of cement augmentation on vertebral stiffness. However work is still required to further understand the mechanics of the cement-bone interface and the effect of interstitial fluids in order to develop FE models of cement augmentation to the next level.

6 Spinal Segment Models

Research on separate vertebral segments provides a wealth of information on, for example, bone disease conditions and fracture risk. However understanding the details of the articulation and the biomechanical behaviour of the spine, which are

Fig. 12 A three-vertebral segment FE model

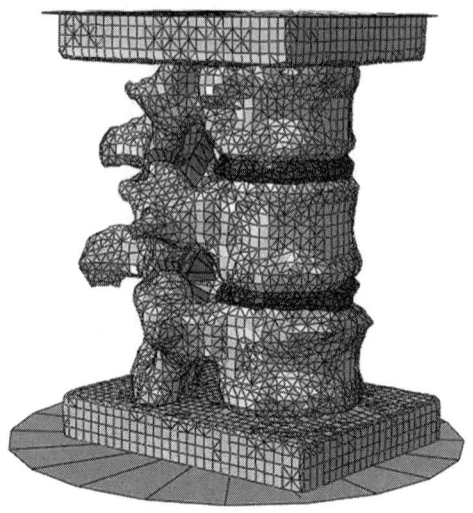

largely influenced by the soft tissue components such as the intervertebral discs and ligaments, is not possible without considering a length of the column that consists of several vertebral segments. Usually, lengths with more than two vertebral segments are used for this purpose, so that there will be at least one vertebra in the middle, that is, reasonably far from the specimen fixation at the ends. Furthermore the existence of multiple vertebral segments allows the incorporation of the intervertebral discs, spinal ligaments and the facet joint capsule.

Having the specimen specific geometry is undoubtedly useful in these multiple segment models, for a number of reasons. The intervertbral discs, facet joints and the spinal ligaments play a key role with regard to load transfer, stability and the articulation of the spine. The intervertebral disc shape directly depends on the end plate geometry and the curvature. With degeneration, the vertebral endplate undergoes geometrical, structural and composition changes. The facet surface orientation, curvature and angle, impinge on the facet contact mechanics. Accurate representations of the ligament anchor locations and orientations also depend on the specimen geometry and morphology.

The development of specimen-specific FE models at a similar level of detail to the vertebral models is still in relative infancy. It is possible to image larger sections of the spine using CT or, increasingly, μCT and such methods have been used to generate specimen-specific models (Fig. 12). This approach has been used to evaluate the effect of vertebroplasty on the adjacent structures and has indicated a shift in load transfer towards the annulus during an axial compression when the intervertebral discs are degenerated (Fig. 13).

The major challenge in the development of specimen-specific spinal segment models lies in the derivation of information on the soft tissues from image data. Whilst CT or μCT can capture the bone sufficiently to derive both geometrical

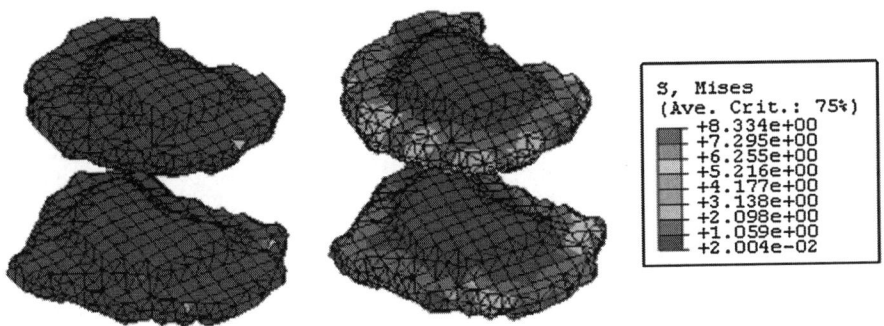

Fig. 13 The intervertebral discs of a three-vertebral segment model under axial load demonstrating the variation in stress distribution for healthy (*left*) and degenerated (*right*) conditions

information and material properties, the same modalities have not yet been successfully used with soft tissue. Magnetic resonance imaging (MRI) provides the most feasible alternative, but there is still considerable progress to be made before similar levels of repeatability and agreement can be achieved for specimen-specific models of the discs and other soft tissue as have been seen for the vertebrae.

7 Discussion

This chapter has outlined the progress that has been made in the development of specimen-specific FE models of the spine, with a focus on models of the vertebrae and the analysis of vertebroplasty. During the model development phase, the major advantage of specimen-specific modelling is that it allows direct comparisons to be made between the model predictions and corresponding experimental specimens. In terms of FE models of single vertebrae, a relatively high level of robustness has been achieved, with errors typically of the order of 10% or less. The breadth of sensitivity analysis of these models has also enabled a good understanding of the necessary level of detail required for the different input parameters.

In terms of clinical application, this methodology could potentially be used to generate patient-specific models for fracture risk screening. However, since relatively high resolution imaging is required, the translation to clinic will not be immediate. The driver for the development of the models was not so much for clinical practice applications but more for the development of pre-clinical testing tools, particularly for the analysis of vertebroplasty. The results of the cement-augmentation FE models indicates that the simulation of the bone-cement composite is not straightforward and whilst progress has been made in terms of representing the gross behaviour of the augmented vertebrae, further work is still necessary in this area.

One salient lesson from these studies has been the importance of validation. After the relatively high levels of agreement that were achieved in modeling the non-augmented vertebrae, it might have seemed logical that these could then be used with confidence to evaluate a range of clinical procedures. The results shown in Fig. 10 indicate that the inclusion of any new assumptions into the modelling process, such as how the cement-bone composite behaves, must be checked against experimental data to ensure the model robustness is maintained.

The extension of single vertebra models to spinal segment models was outlined in Sect. 5. It is clear that there is much development work still to be undertaken. The soft tissues, and in particular the intervertebral disc, have considerably higher levels of non-linearity and time-dependence than the vertebrae and it is likely that a greater number of parameters will be necessary to define them. Whilst developing specimen-specific models of these tissues is a major undertaking, the types of sensitivity study outlined in Sect. 3 can help to direct efforts to where there will be greatest gain in accuracy. From the vertebra, for example, it was found that higher order relationships between the image grayscale and the bone elastic modulus were unnecessary, and it may be similarly true with the soft tissue that a smaller number of lower order parameters can be identified which dominate the overall response. These processes of sensitivity analysis and validation are necessary if both specimen-specific and, in the longer term, patent-specific models of spinal segments are to gain the necessary credibility to be used in pre-clinical testing or patient evaluation.

With the single vertebra models, the next step in the model development is to apply the technique to larger image datasets. One of the major barriers to this is the collation of sufficient number of scans, particularly from the younger population. At present, the μCT images required to build the models can only be acquired from cadaveric specimens. The work outlined in Sect. 4 on preserved tissue and the use of the segmentation approach paves the way for models to be generated from μCT images of embalmed and dry bone specimens from anatomy and museum collections. This will allow much larger datasets to be developed than if the models could only be built from the much more limited supply of fresh cadaveric tissue. A large bank of models would enable new treatments to be virtually evaluated, following the necessary validation, over a patient population rather than in a much smaller set of cadaveric tissue within the laboratory. Another challenge in the use of these models is the application of appropriate loads and boundary conditions to ensure that the FE outcomes are comparable across the population set. The studies presented in Sect. 3 highlighted how factors such as the orientation of the specimens, the area of the applied load and boundary conditions can have a major effect on the predicted vertebral stiffness. The work presented on loading and the orientation of the specimen (Sect. 3.3.2) indicates that the direction of loading is critical, and therefore registration of all the models to one alignment is likely to be a necessary step in the model generation process.

In addition to using image data to build banks of separate FE models, the data could also be used to build statistical models of different patient cohorts. Whilst geometric shape modeling has reached a level where this would undoubtedly be possible, the challenge remains in generating realistic underlying bone properties. Progress in this area is continuing however and such an approach is likely to be achieved in the foreseeable future. This could eventually lead full circle, back to the 'generic' model, but now one that could, at the push of a button, be morphed to represent any percentile within the patient population.

In conclusion, considerable progress has been made in the development of subject-specific FE models of the spinal vertebrae. These models have enabled direct validation against experimental data and good levels of agreement have been achieved. The models are now being extended to larger spinal segments and to represent patient populations. The use of the models to simulate the effects of vertebroplasty has illustrated their potential as a pre-clinical testing tool, but also the importance of the validation step in each new model application.

References

1. Nachemson, A., Waddell, G., Norlund, A.: Epidemiology of neck and low back pain. In: Nachemson, A., Jonsson, E. (eds.) Neck and Back Pain: The Scientific Evidence of Causes, Diagnosis and Treatment. Lippincott, Williams and Wilkins, Philadelphia (2000)
2. Barbagallo, G., Olindo, G., Corbino, L., Albanese, V.: Analysis of complications in patients treated with the X-Stop interspinous process decompression system: Proposal of a novel anatomical scoring system for patient selection and review of the literature. Neurosurgery **65**(1), 111–120 (2009)
3. Bowers, C., Amini, A., Dailey, A.T., Schmidt, M.H.: Dynamic interspinous process stabilization: Review of complications associated with the X-Stop device. Neurosurg. Focus **28**(6), E8 (2010)
4. Kallmes, D.F., Comstock, B.A., Heagerty, P.J., Turner, J.A., Wilson, D.J., Diamond, T.H., Edwards, R., Gray, L.A., Stout, L., Owen, S., Hollingworth, W., Ghdoke, B., Annesley-Williams, D.J., Ralston, S.H., Jarvik, J.G.: A randomised trial of vertebroplasty for osteoporotic spinal fractures. N. Engl. J. Med. **361**, 569–579 (2009)
5. Wardlaw, D., Cummings, S.R., Meirhaeghe, J.V., Bastian, L., Tillman, J.B., Ranstam, J., Eastell, R., Shabe, P., Talmadge, K., Boonen, S.: Efficacy and safety of balloon kyphoplasty compared with non-surgical care for vertebral compression fracture (FREE): A randomised controlled trial. Lancet **373**, 1016–1024 (2009)
6. Jones, A.C., Wilcox, R.K.: Finite element analysis of the spine: Towards a framework of verification, validation and sensitivity analysis. Med. Eng. Phys. **30**, 1287–1304 (2008)
7. Wilcox, R.K.: The influence of material property and morphological parameters on specimen-specific finite element models of porcine vertebral bodies. J. Biomech. **40**, 669–673 (2007)
8. Jones, A.C., Wilcox, R.K.: Assessment of factors influencing finite element vertebral model predictions. J. Biomech. Eng. **129**, 898–903 (2007)
9. Wijayathunga, V.N., Jones, A.C., Oakland, R.J., Furtado, N.R., Hall, R.M., Wilcox, R.K.: Development of specimen-specific finite element models of human vertebrae for the analysis of vertebroplasty. Proc. Inst. Mech. Eng. [H] J. Eng. Med. **222**, 221–228 (2008)
10. Zhao, Y.: Finite element modelling of cement augmentation and fixation for orthopaedic applications. PhD Thesis, School of Mechanical Engineering, University of Leeds (2010)
11. Zysset, P.K., Guo, X.E., Hoffler, C.E., Moore, K.E., Goldstein, S.A.: Elastic modulus and hardness of cortical and trabecular bone lamellae measured by nanoindentation in the human femur. J. Biomech. **32**, 1005–1012 (1999)

12. Bancroft, J.D., Stevens A.: Theory and Practice of Histological Techniques, 4th edn. Churchill Livingstone (1996)
13. Pleshko, L.P., Boskey, A.L., Mendelson, R.: An FT-IR microscopic investigation of the effects of tissue preservation of bone. Calcif. Tissue Int. **51**(1), 72–77 (1992)
14. Currey, J.D., Brear, K., Zioupos, P., Reilly, G.C.: Effect of formaldehyde fixation on some mechanical properties of bovine bone. Biomaterials **16**(16), 1267–1271 (1995)
15. McElhaney, J., Fogle, J., Byars, E., Weaver, G.: Effect of embalming on the mechanical properties of beef bone. J. Appl. Physiol. **19**, 1234–1236 (1964)
16. Lievers, W.B., Poljsak, A.S., Waldmana, S.D., Pilkeya, A.K.: Effects of dehydration-induced structural and material changes on the apparent modulus of cancellous bone. Med. Eng. Phys. **32**(8), 921–925 (2010)
17. Wilcox, R.K.: The biomechanical effect of vertebroplasty on the adjacent vertebral body: A finite element study. Proc. Inst. Mech. Eng. [H] J. Eng. Med. **220**(4), 565–572 (2006)
18. Kosmopoulos, V., Keller, T.S.: Damage-based finite-element vertebroplasty simulations. Eur. Spine J. **13**, 617–625 (2004)
19. Liebschner, M.A.K., Rosenberg, S.W., Keaveny, T.M.: Effects of bone cement volume and distribution on vertebral stiffness after vertebroplasty. Spine **26**, 1547–1554 (2001)
20. Baroud, G., Nemes, J., Heini, P., Steffen, T.: Load shift of the intervertebral disc after a vertebroplasty: a finite-element study. Eur Spine J **12**, 421–426 (2003)
21. Baroud, G., Nemes, J., Ferguson, S.J., Steffen, T.: Material changes in osteoporotic human cancellous bone following infiltration with acrylic bone cement for a vertebral cement augmentation. Comp Methods Biomech Biomed Eng **6**, 133–139 (2003)
22. Race, A., Mann, K.A., Edidin, A.A.: Mechanics of bone/PMMA composite structures: An in vitro study of human vertebrae. J. Biomech. **40**(5), 1002–1010 (2007)
23. Zhao, Y., Jin, Z.M., Wilcox, R.K.: Modelling cement augmentation: A comparative experimental and finite element study at the continuum level. Proc. Inst. Mech. Eng. [H] J. Eng. Med. **224**, 903–911 (2010)
24. Tarsuslugil, S.: PhD Thesis, School of Mechanical Engineering, University of Leeds (2011)
25. Buckley, P.J., Orr J.F.: The compressive strength of cancellous bone/cement composite materials. 12th conference of the European society of biomechanics, p. 325. Dublin (2000)
26. Edmondston, S.J., Singer, K.P., Day, R.E., Breidahl, P.D., Price, R.I.: Formalin fixation effects on vertebral bone density and failure mechanics: an in vitro study of human and sheep vertebrae. Clin Biomech **9**(3), 175–179 (1994)

Part III
Heart and Circulation

Morphological and Functional Modeling of the Heart Valves and Chambers

Razvan Ioan Ionasec, Dime Vitanovski and Dorin Comaniciu

Abstract Personalized cardiac models have become a crucial component of the clinical workflow, especially in the context of complex cardiovascular disorders, such as valvular heart disease. In this chapter we present a comprehensive framework for the patient-specific modeling of the valvular apparatus and heart chambers from multi-modal cardiac images. An integrated model of the four heart valves and chambers is introduced, which captures a large spectrum of morphologic, dynamic and pathologic variations. The patient-specific model parameters are estimated from four-dimensional cardiac images using robust learning-based techniques. These include object localization, rigid and non-rigid motion estimation, and surface boundary estimation from dense 4D data (TEE, CT) as well as regression-based techniques for surface reconstruction from sparse 4D data (MRI). Clinical applications based on the patient-specific modeling approach are proposed for decision support in Transcatheter Aortic Valve Implantation and Percutaneous Pulmonary Valve Implantation while performance evaluation is conducted on a population of 476 patients.

1 Introduction

The unprecedented increase in life expectancy over the current and past century propelled Cardiovascular Disease (CVD) to become the deadliest plague, which today causes approximately 30% of fatalities worldwide and nearly 40% in high-income regions. Valvular Heart Disease (VHD) is a representative class of CVD, which affects 2.5% of the global population and requires yearly over

R. Ioan Ionasec (✉) · D. Vitanovski · D. Comaniciu
Image Analytics and Informatics, Siemens Corporate Research, Princeton, NJ, USA
e-mail: razvan.ionasec@siemens.com

100,000 surgeries in the United States alone. Yet, heart valve operations are the most expensive and riskiest cardiac procedures, with an average cost of $141,120 and 4.9% in-hospital death rate [29].

Due to the strong anatomical, functional and hemodynamic inter-dependency of the heart valves, VHDs do not affect only a single valve, but rather several valves are impaired. Recent studies demonstrate strong influence of pulmonary artery systolic pressure on the tricuspid regurgitation severity [31]. In Lansac et al. [28], Timek et al. [44] the simultaneous evaluation of aortic and mitral valves is encouraged, given the fibrous aortic-mitral continuity, which anchors the left side valves and facilitates the reciprocal opening and closing motion during the cardiac cycle. Moreover, in patients with mitral and tricuspid valve regurgitation, joint surgery is recommended to minimize the risk for successive heart failure or reduced functional capacity. Morphological and functional assessment of the complete heart valve apparatus is crucial for clinical decision making during diagnosis and severity assessment as well as treatment selection and planning.

Decisions in valvular disease management increasingly rely on non-invasive imaging. Techniques like Transesophageal Echocardiography (TEE), cardiac Computed Tomography (CT) and Cardiovascular Magnetic Resonance (CMR) imaging, enable dynamic four dimensional scanning of the beating heart over the whole cardiac cycle. Precise morphological and functional knowledge about the valvular apparatus is highly esteemed and it is considered a prerequisite for the entire clinical workflow including diagnosis, therapy-planning, surgery or percutaneous intervention as well as patient monitoring and follow-up. Nevertheless, most non-invasive investigations to date are based on two-dimensional images, user-dependent processing and manually performed, potentially inaccurate measurements [4].

The progress in medical imaging is matched by important advances in surgical techniques, bioprosthetic valves, robotic surgery and percutaneous interventions, which have led to a twofold increase in the number of valve procedures performed in the United States since 1985 [26]. There has been a major trend in cardiac therapy towards minimally invasive transcatheter procedures to reduce the side effects of classical surgical techniques. Without direct access and view to the affected structures those interventions are usually performed in so-called Hybrid ORs, equipped with advanced imaging technology. Thus, procedures such as the Transcatheter Aortic Valve Implantation (TAVI) are permanently guided via real-time intra-operative images provided by X-ray Fluoroscopy and Transesophageal Echocardiography systems [1]. Powerful computer-aided tools for extensive non-invasive assessment, planning and guidance are mandatory to continuously decrease the level of invasiveness and maximize effectiveness of valve therapy.

Except for the past decade, cardiac modeling was almost exclusively focused on the left ventricle (LV) [33, 42]. Rueckert and Burger [37], Fritz et al. [14] achieved a combined model of the two ventricles, LV and right ventricle (RV). Few methods in the literature also consider the left and right atria [12, 19, 30, 56, 58, 59] but none explicitly handle the heart valves. The majority of existent valve models presented in the literature are generic and rough approximations of the true valvular anatomy. Their primary application is the analysis of the blood-tissue interaction during the cardiac cycle as well as mechanical and functional behavior of the valvular apparatus. The first cardiac model to include the heart valves was proposed by Peskin and McQueen [35]. De Hart et al. [11] introduced a refined computational model of the aortic valve while Soncini et al. [41] presented a realistic finite element model of the physiological aortic root from medical imaging data. Kunzelman et al. [27] introduced the first three-dimensional finite element model of the mitral valve. Votta et al. [51] presented an extended mitral valve model based on *in vivo* data. Watanabe et al. [53] introduced a geometrical model of the mitral valve, obtained from real-time three-dimensional TEE. The study by Veronesi et al. [48] also considers the aortic valve to investigate the functional dependency between the two left-side valves. Schievano et al. [39] proposed an analysis protocol of the pulmonary trunk based on rapid prototyping systems. Recently introduced models of the aortic valve [22, 52] the mitral valve [10, 40] aortic-mitral coupling [23, 24, 47] address important aspect of data-driven valve models, yet do not offer a unified approach for the patient-specific modeling of the entire valvular apparatus.

In this chapter we present a complete patient-specific model of the valvular apparatus and heart chambers estimated from multi-modal cardiac images. Section 2 describes a morphological and functional representation of the aortic, mitral, tricuspid and pulmonary valves as well as the four heart chambers. The patient-specific parameters of the cardiac models are estimated from four-dimensional cardiac images using learning-based methods. Section 3 describes robust algorithms for object localization and estimation, non-rigid motion estimation, and surface boundary estimation from dense 4D data (Computed Tomography and Echocardiography) as well as regression-based techniques for surface reconstruction from sparse 4D data (Magnetic Resonance Imaging). Based on the patient-specific modeling techniques, in Sect. 5 we introduce two clinical applications that support analysis and planning in Transcatheter Aortic Valve Implantation (TAVI) and Percutaneous Pulmonary Valve Implantation (PPVI). Part of this work has been reported in our previous publications [16, 23–25, 49, 50].

2 Modeling of the Heart Valves and Chambers

This section introduces an explicit mathematical representation of the cardiac valves and chambers that parameterizes relevant clinical aspects observable through non-invasive imaging modalities (see Fig. 1). The proposed model of the aortic, mitral, tricuspid and pulmonary valves and the four chambers precisely captures morphological, dynamical and pathological variations. To handle the inherent complexity, the representation is structured on three abstraction layers: global location and rigid motion, non-rigid landmark motion model, and comprehensive surface model. Each model abstraction naturally links to anatomical and dynamical aspects at a specific level of detail, while the hierarchical interconnection of the individual parameterizations is driven by the physiology of the valves.

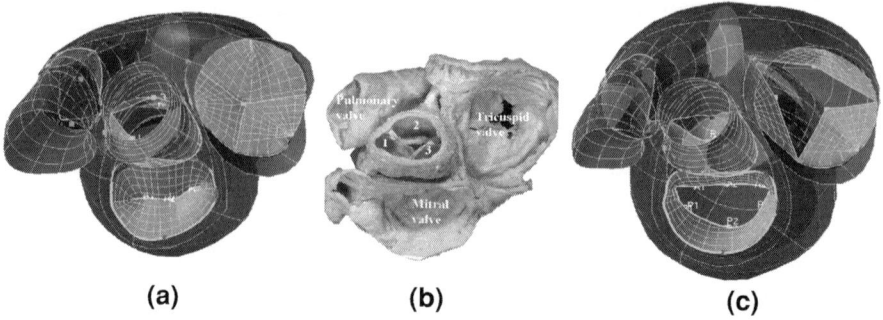

Fig. 1 The proposed model of the heart valves and chambers in systole (**a**), and diastole (**c**). **b** Explanted specimen of the heart valves-reproduced with permission of the author and the European association for cardio-thorac. surgery from: Anderson RH. The surgical anatomy of the aortic root. Multimedia man cardiothorac surg doi:10.1510/mmcts.2006.002527

2.1 Parametrization

Global Location and Rigid Motion: The global location of the valves is parameterized through a similarity transformation in the Euclidean three-dimensional space, which includes nine parameters. A time variable t is augmenting the representation to capture the temporal variation during the cardiac cycle and model the rigid valve motion:

$$\theta(t) = \{(c_x, c_y, c_z), (\alpha_x, \alpha_y, \alpha_z), (s_x, s_y, s_z), t\} \quad (1)$$

where (c_x, c_y, c_z), $(\alpha_x, \alpha_y, \alpha_z)$, (s_x, s_y, s_z) are the position, orientation and scale parameters. The location and rigid motion of each valve is modeled independently through its individual set of parameters $\theta(t)$, which results into a total of $36 \times T$ parameters for a given volume sequence $I(t)$ of length T.

Non-Rigid Landmark Motion Model: The second abstraction level models anatomical landmarks that are robustly identifiable by doctors, possess a particular visual pattern, and serve as anchor points for qualitative and quantitative clinical assessment. Normalized by the time-dependent similarity transformation, the motion of each anatomical landmark j can be parameterized by its corresponding trajectory $\mathbf{L}^j$ over a full cardiac cycle. For a given volume sequence $I(t)$, one trajectory $\mathbf{L}^j$ is composed by the concatenation of the spatial coordinates:

$$\mathbf{L}^j(\theta_i) = [\mathbf{L}^j(0), \mathbf{L}^j(1), \cdots, \mathbf{L}^j(t), \cdots, \mathbf{L}^j(T-1)] \quad (2)$$

where $\mathbf{L}^j$ are spatial coordinates with $\mathbf{L}^j(t) \in \mathscr{R}^3$ and t an equidistant discrete time variable $t = 0, \cdots, T-1$.

Comprehensive Surface Model: The full geometry of the valves and chambers is modeled using surface meshes constructed along rectangular grids of vertices. For each anatomic structure A_k, the underlying grid is spanned along two

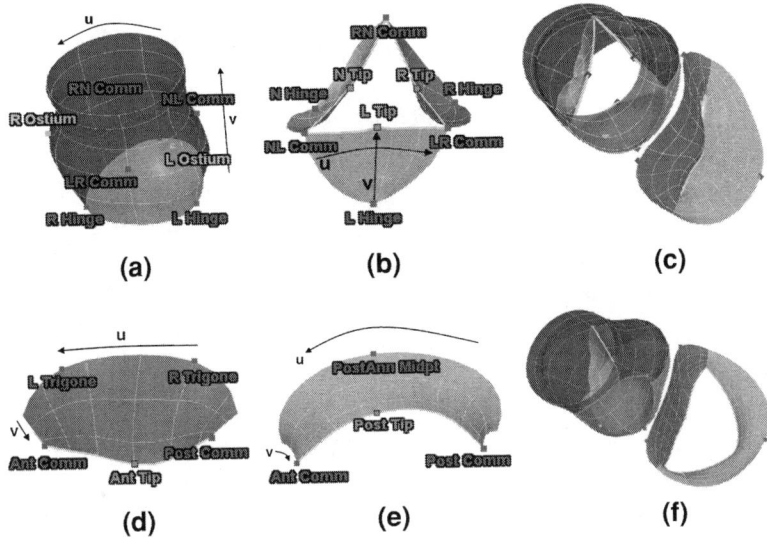

Fig. 2 Isolated surface components of the aortic and mitral models with parametric directions and spatial relations to anatomical landmarks: **a** aortic root, **b** aortic leaflets, **c** aortic-mitral in end-systole, **d** anterior mitral leaflet, **e** posterior mitral leaflet, and **f** aortic-mitral in end-diastole

physiologically aligned parametric directions, **u** and **v**. Each vertex $\mathbf{p}^{A_k} \in \mathcal{R}^3$ has four neighbors, except the edge and corner points with three and two neighbors, respectively. Therefore, a rectangular grid with $n \times m$ vertices is represented by $(n-1) \times (m-1) \times 2$ triangular faces. The model M of a certain component at a particular time step t is uniquely defined by vertex collections of the anatomic structures A_k. The time parameter t extends the representation to capture dynamics:

$$M(\mathbf{L}^j, \theta_i) = \left[\overbrace{\{\mathbf{p}_0^{A_1}, \cdots, \mathbf{p}_{N_1}^{A_1}\}}^{\text{first anatomy}}, \cdots, \overbrace{\{\mathbf{p}_0^{A_n}, \cdots, \mathbf{p}_{N_n}^{A_n}\}}^{n\text{-th anatomy}}, t \right] \quad (3)$$

where n is the number of represented anatomies, and $N_1 \ldots N_n$ are the numbers of vertices for a particular anatomy given in the following sections.

2.2 Left-Heart Valves

Aortic Valve: The aortic valve (AV) connects the left ventricular outflow tract to the ascending aorta and includes the aortic root and three leaflets/cusps [left (L) aortic leaflet, right (R) aortic leaflet and none (N) aortic leaflet]. The root extends from the basal ring to the sinotubular junction and builds the supporting structure for the leaflets. These are fixed to the root on a crown-like attachment and can be thought of as semi-lunar pockets (see Fig. 2).

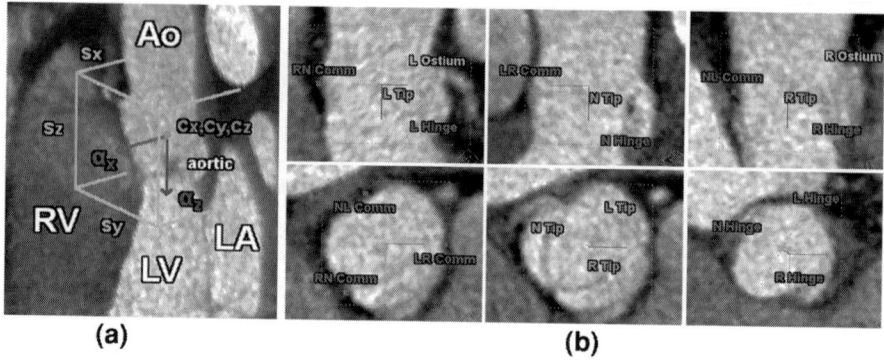

Fig. 3 Anatomical landmark model of the aortic valve. **a** Perspective view including the bounding box. **b** Landmarks relative to the anatomical location illustrated in long and short axis from an example CT study

Three aortic commissure points, LR-Comm, NL-Comm and RN-Comm, describe the interconnection locations of the aortic leaflets, while three hinges, L-Hinge, R-Hinge, and N-Hinge, are their lowest attachment points to the root. For each leaflet of the valve, the center of the corresponding free-edge is marked by the leaflet tip point: L/R/N-Tip. The interface between the aorta and coronary arteries is symbolized using the L/R-Ostium, the two coronary ostia (see Fig. 3b).

The anatomical landmarks are also used to describe the global location and rigid motion as follows (see Fig. 3a): $(c_x, c_y, c_z)_{aortic}$ equals to the gravity center of the aortic landmarks, except aortic leaflet tips. α_z is the normal vector to the LR-Comm, NL-Comm, RN-Comm plane, α_x is the unit vector orthogonal to α_z which points from $(c_x, c_y, c_z)_{aortic}$ to LR-Comm, α_y is the cross-product of α_x and α_z. $(s_x, s_y, s_z)_{aortic}$ is given by the maximal distance between the center $(c_x, c_y, c_z)_{aortic}$ and the aortic landmarks, along each of the axes $(\alpha_x, \alpha_y, \alpha_z)$.

Four surface structures represent the aortic valve: aortic root, left coronary leaflet, right coronary leaflet and non coronary leaflet. The aortic root connects the ascending aorta to the left ventricle outflow tract and is represented through a tubular grid (see Fig. 2a). This is aligned with the aortic circumferential **u** and ascending directions **v** and includes 36×20 vertices and 1368 faces. The root is constrained by six anatomical landmarks, i.e. three commissures and three hinges, with a fixed correspondence on the grid. The three aortic leaflets, the L-, R- and N-leaflet, are modeled as paraboloids on a grid of 11×7 vertices and 120 faces (see Fig. 2b). They are stitched to the root on a crown like attachment ring, which defines the parametric **u** direction at the borders. The vertex correspondence between the root and leaflets along the merging curve is symmetric and kept fixed. The leaflets are constrained by the corresponding hinges, commissures and tip landmarks, where the v direction is the ascending vector from the hinge to the tip (see Fig. 2c).

Mitral Valve: Located in between the left atrium (LA) and the left ventricle (LV), the mitral valve (MV) includes the posterior leaflet, anterior leaflet, annulus

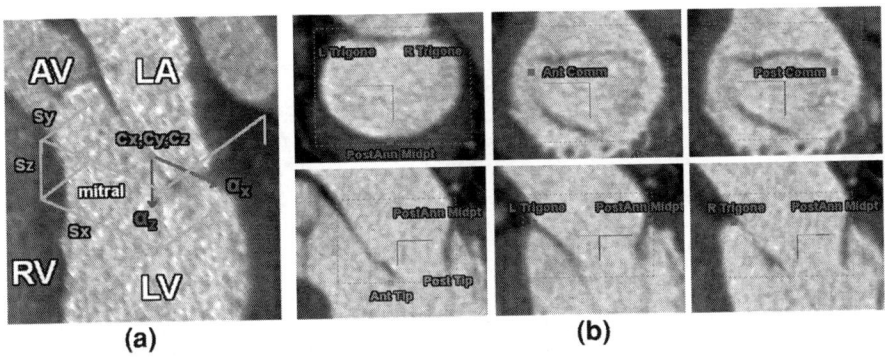

Fig. 4 Anatomical landmark model of the mitral valve. **a** Perspective view including the bounding box. **b** Landmarks relative to the anatomical location illustrated in long and short axis from an example CT study

and the subvalvular apparatus. The latter consists of the chordae tendiane and papillary muscles, which are not explicitly treated in this work (see Fig. 2).

The two interconnection points of the mitral leaflets at their free edges are defined by the mitral anterior and posterior commissures, while the mitral annulus is fixed by the L/R-Trigone and posteroannular midpoint (PostAnn MidPoint). The center of the two mitral leaflets' free-edges is marked by the leaflet tip points, the Ant and Post-Tip (anterior/posterior) leaflet tips (see Fig. 4b).

The barycentric position $(c_x, c_y, c_z)_{mitral}$ is computed from the mitral landmarks, except mitral leaflet tips (see Fig. 4a). α_z is the normal vector to the L/R-Trigone, PostAnn MidPoint plane, α_x is orthogonal to α_z and points from $(c_x, c_y, c_z)_{mitral}$ towards the PostAnn MidPoint. The scale parameters $(s_x, s_y, s_z)_{mitral}$ are defined as for the aortic valve, to comprise the entire mitral anatomy.

The mitral leaflets separate the (LA) and (LV) hemodynamically and are connected to the endocardial wall by the saddle shaped mitral annulus. Both are modeled as paraboloids and their upper margins implicitly define the annulus. Their grids are aligned with the circumferential annulus direction **u** and the orthogonal direction **v** pointing from the annulus towards leaflet tips and commissures (see Fig. 2d, e). The anterior leaflet is constructed from 18×9 vertices and 272 faces while the posterior leaflet is represented with 24×9 vertices and 368 faces. Both leaflets are fixed by the mitral commissures and their corresponding leaflet tips. The left / right trigones and the postero-annular midpoint further confine the anterior and posterior leaflets, respectively (see Fig. 2f).

2.3 Right-Heart Valves

Pulmonary Valve: The pulmonary trunk emerges out of the (RV) and branches into the left and right pulmonary arteries, which connect to the corresponding lung. It supports a semilunar valve, geometrically and topologically similar to the aortic valve.

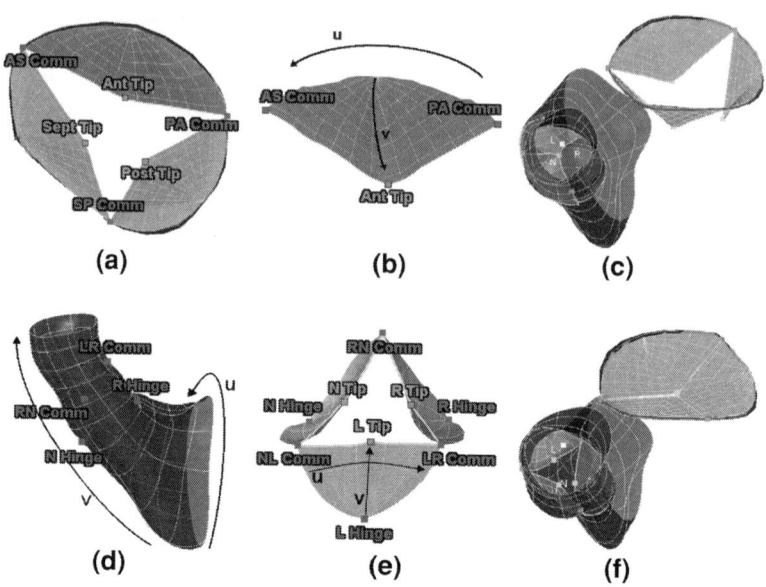

Fig. 5 Isolated surface components of the tricuspid and pulmonary models with parametric directions and spatial relations to anatomical landmarks: **a** tricuspid leaflet, **b** tricuspid annulus and leaflets, **c** tricuspid-pulmonary in end-diastole, **d** pulmonary trunk, **e** pulmonary leaflets, and **f** tricuspid-pulmonary in end-systole

The three leaflets of the pulmonary valve (PV) are named according to their relationship with respect to the (AV) as left and right facing leaflet, and none facing leaflet (see Fig. 5).

Identical as for the aortic valve, three commissure points, LR-Comm, NL-Comm and RN-Comm, describe the interconnection locations of the pulmonary leaflets, while three hinges, L-Hinge, R-Hinge, and N-Hinge, are their lowest attachment points to the pulmonary trunk. For each leaflet of the valve, the center of the corresponding free-edge is marked by the leaflet tip point: L/R/N-Tip. The pulmonary trunk is bounded by two landmarks, the RVOT on the right ventricle side and Bifurcation distal to the valve location. The model is completed by right ventricle Trigone landmark (see Fig. 6b).

The location $(c_x, c_y, c_z)_{pulmonary}$ is equal to the gravity center of the commissures and hinges landmarks (see Fig. 6b). α_z is the normal vector to the LR-Comm, NL-Comm, RN-Comm plane, α_x is the unit vector orthogonal to α_z which points from $(c_x, c_y, c_z)_{aortic}$ to LR-Comm, α_y is the cross-product of α_x and α_z. $(s_x, s_y, s_z)_{aortic}$ is given by the maximal distance between the center $(c_x, c_y, c_z)_{aortic}$ and the aortic landmarks, along each of the axes $(\alpha_x, \alpha_y, \alpha_z)$.

The representation of the pulmonary valve is composed out of four structures: pulmonary trunk, left facing leaflet, none facing leaflet and right facing leaflet. The pulmonary trunk emerges out of the right ventricular outflow tract, supports

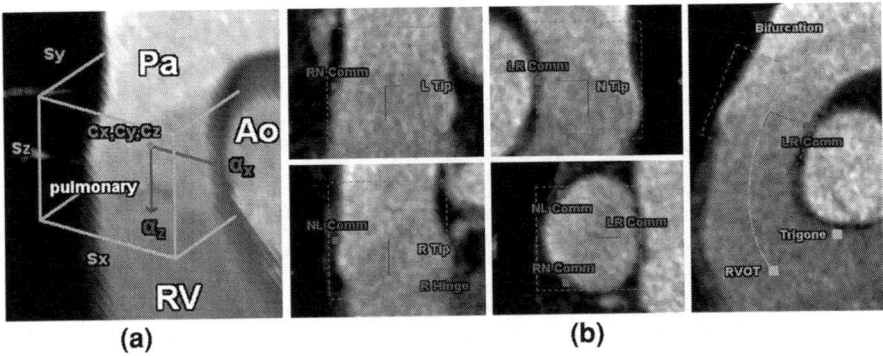

Fig. 6 Anatomical landmark model of the pulmonary valve. **a** Perspective view including the bounding box. **b** Landmarks relative to the anatomical location illustrated in long and short axis from an example CT study

the pulmonary valves and its three leaflets, and ends at the level of the pulmonary artery bifurcation. The grid, which spans the pulmonary trunk surface, is aligned with the circumferential **u** and longitudinal direction v of the valve (see Fig. 5d). It includes 50×40 vertices and 3822 faces confined through the pulmonary commissures, hinges and the (RV) trigone. Additionally, the RVOT and Bifurcation landmarks determine its longitudinal span. The attached L-, R- and N- leaflets, are modeled as paraboloids along the annulus circumferential direction **u** and vector **v** pointing from the corresponding hinge to the leaflet tip (see Fig. 5e). Each includes 11×7 vertices and 120 faces bounded by the associated two commissures, hinge and tip (see Fig. 5f).

Tricuspid Valve: The tricuspid valve (TV), also called the right atrioventricular valve, separates the (RA) from the (RV). It mainly consists of the annulus and subvalvular apparatus as well as three leaflets: septal, inferior and anterosuperior leaflet (see Fig. 5).

The three leaflets of the tricuspid valve, septal-, anterior- and posterior leaflet interconnect in three points marked by three tricuspid commissures, namely the AS Comm, the PA Comm and the SP Comm landmarks. The tricuspid landmark model is completed by the Sept Tip, Ant Tip and Post Tip landmarks, which mark the center of the leaflets' free edge (see Fig. 7b).

The barycentric position $(c_x, c_y, c_z)_{tricuspid}$ is computed from the tricuspid commissures, AS Comm, PA Comm and SP Comm (see Fig. 7a). α_z is the normal vector to the commissural plane, α_x is orthogonal to α_z and points towards the AS Comm. The scale parameters $(s_x, s_y, s_z)_{tricuspid}$ are defined to comprise the entire tricuspid valve anatomy.

The function of the tricuspid valve is to regulate the blood flow from the (RA) to the (RV), staying closed during systole and open during diastole. The model is constrained by three surfaces: septal-, anterior- and posterior leaflet (see Fig. 5a). The tricuspid leaflets are modeled as hyperbolic paraboloids and implicitly

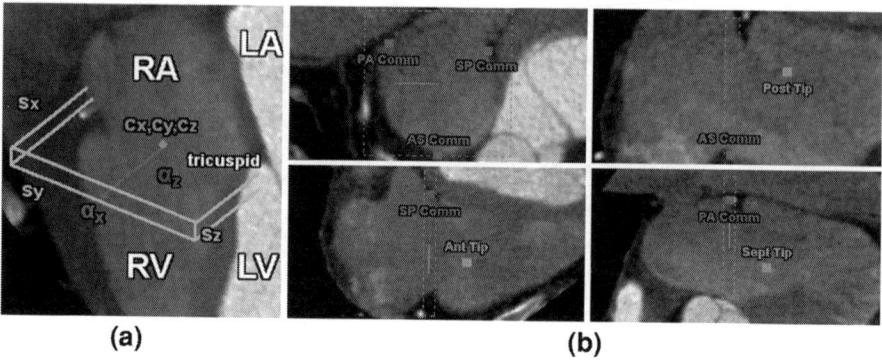

Fig. 7 Anatomical landmark model of the tricuspid valve. **a** Perspective view including the bounding box. **b** Landmarks relative to the anatomical location illustrated in long and short axis from an example CT study

describe the tricuspid annulus. Their grids are spanning along the annulus circumferential direction **u** and the perpendicular vector v pointing for the annulus towards the corresponding leaflet tip, and consist out of 22×14 vertices and 546 faces (see Fig. 5b). Each leaflet is constrained by the corresponding two zcommissures and one leaflet tip (see Fig. 5c).

2.4 Heart Chambers

Left Ventricle and Atrium: The left ventricle is constructed from 78 landmarks (16 mitral lateral, 15 mitral septum, 16 left ventricle output tract and 32 aortic valve control points) and four surface geometries (LV epicardium, LV endocardium and LV output tract). The left atrial surface is connected to the left ventricle via the aortic valve control points (Fig. 8a) Zheng et al. [57].

Right Ventricle and Atrium: The right ventricle is composed of 74 landmarks (16 tricuspid lateral, 15 tricuspid septum, 28 tricuspid valve and 18 pulmonary valve control points) and four surface geometries (RV apex, RV output tract and RV inflow tract). The right atrial surface is constrained by 28 tricuspid valve control points and links to the right ventricle (Fig. 8b) Zheng et al. [56].

3 Parameter Estimation from Cardiac Images

3.1 Object Localization and Rigid Motion Estimation

The goal is to determine the location of each specific anatomy from multimodal cardiac images. Thus, the location and motion parameters θ of each valve, defined in Sect. 2.1, are estimated from a sequence of volumes I:

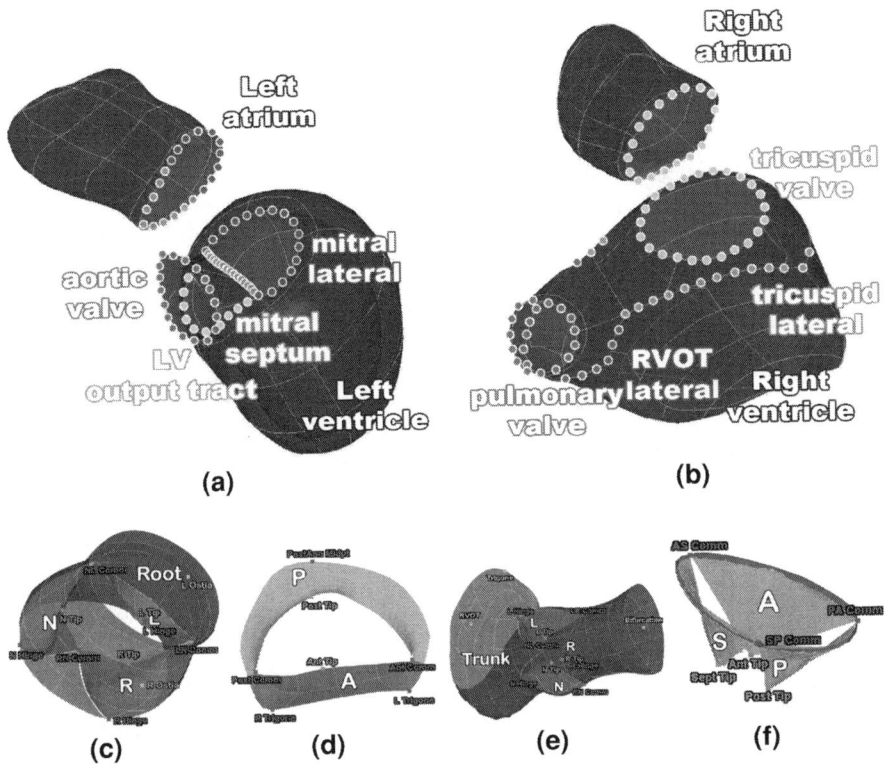

Fig. 8 Cardiac model components: **a** the *left* heart (*left* ventricle and *left* atrium), **b** *right* heart (*right* ventricle and *right* atrium), **c** aortic valve, **d** mitral valve, **e** pulmonary valve, and **f** tricuspid valve

$$\arg\max{}_\theta p(\theta|I) = \arg\max{}_\theta p(\theta(0), \cdots, \theta(n-1)|I(0), \cdots, I(n-1)) \quad (4)$$

Location Estimation: To solve Eq. 4, we formulate the object localization as a classification problem and estimate $\theta(t)$ for each time step t. The probability $p(\theta(t)|I(t))$ can be modeled by a learned detector D, which evaluates and scores a large number of hypotheses for $\theta(t)$. To avoid exhaustive search along a nine-dimensional space we apply the Marginal Space Learning (MSL) framework [56] and break the original parameter space into a subset of increasing marginal spaces:

$$\Sigma_1 \subset \Sigma_2 \subset \cdots \subset \Sigma_n = \Sigma$$

The nine-dimensional space described by the similarity transform in a three-dimensional Euclidean space is decomposed as follows:

$$\Sigma_1 = (c_x, c_y, c_z)$$
$$\Sigma_2 = (c_x, c_y, c_z, \alpha_x, \alpha_y, \alpha_z)$$
$$\Sigma_3 = (c_x, c_y, c_z, \alpha_x, \alpha_y, \alpha_z, s_x, s_y, s_z)$$

where Σ_1 represents the position marginal space, Σ_2 the position + orientation marginal space and Σ_3 the position + orientation + scale marginal space, which coincides with the original domain. In practice, the optimal arrangement for MSL sorts the marginal spaces in a descending order based on their variance. In our case, due to the CT, MRI and TEE acquisition protocols and physiological variations of the heart, the highest variance comes from translation followed by orientation and scale.

From the marginalization of the search domain, the target posterior probability can be expressed as:

$$p(\theta(t)|I(t)) = p(c_x, c_y, c_z|I(t))$$
$$p(\alpha_x, \alpha_y, \alpha_z|c_x, c_y, c_z, I(t))$$
$$p(s_x, s_y, s_z|\alpha_x, \alpha_y, \alpha_z, c_x, c_y, c_z, I(t))$$

Instead of using a single detector D, we train detectors for each marginal spaces D_1, D_2 and D_3, and estimate $\theta(t)$ by gradually increasing the dimensionality. Detectors are trained using the Probabilistic Boosting Tree [46] with Haar and Steerable Features [56]. After each stage only a limited number of high-probability candidates are kept to significantly reduce the search space: 100 highest score candidates are retained in Σ_1, 50 in Σ_2 and 25 in Σ_3.

Robust Motion Aggregation: $[\theta_0(0)...\theta_{25}(0)]...[\theta_0(n-1)...\theta_{25}(n-1)]$ are the candidates with the highest score estimated at each time step t, $t = 0,...,n-1$. To obtain a temporally consistent global location and motion $\theta(t)$, a RANSAC estimator is employed. We assume a constant model for the cardiac motion, which drives the global movement of the entire valvular apparatus. From randomly sampled candidates, the one yielding the maximum number of inliers is picked as the final motion. Inliers are considered within a distance of $\sigma = 7$ mm from the current candidate and extracted at each time step t. The distance measure $d(\theta(t)_1, \theta(t)_2)$ is given by the maximum L1 norm of the standard unit axis deformed by the parameters $\theta(t)_1$ and $\theta(t)_2$, respectively:

$$L1(\mathbf{a_1}, \mathbf{a_2}) = max\{|x1 - x2|, |y1 - y2|, |z1 - z2|\}$$
$$d(\theta(t)_1, \theta(t)_2) = \frac{1}{4}(L1(\mathbf{c_1}, \mathbf{c_2}) + L1(\mathbf{X}_1 s_{x1}, \mathbf{X}_2 s_{x2}) + L1(\mathbf{Y}_1 s_{y1}, \mathbf{Y}_2 s_{y2})$$
$$+ L1(\mathbf{Z}_1 s_{z1}, \mathbf{Z}_2 s_{z2})) \tag{5}$$

where X, Y and Z are the unit axes obtained from the Euler angles $(\alpha_x, \alpha_y, \alpha_z)$, $\mathbf{c}$ the position vectors, and s_x, s_y, s_z scale parameters.

3.2 Trajectory Spectrum Learning for Non-Rigid Motion Estimation

Based on the determined global location and rigid motion, in this section we introduce a trajectory spectrum learning algorithm to estimate non-linear landmark movements from volumetric sequences [20]. Considering the representation in Sect. 2.1 Eq. 2, the objective is to find for each landmark j its trajectory $\mathbf{L}^j$, with the maximum posterior probability from a series of volumes $I(t)$, given the rigid motion $\theta(t)$:

$$\arg\max_{\mathbf{L}^j} p(\mathbf{L}^j|I,\theta) = \arg\max_{\mathbf{L}^j} p(\mathbf{L}^j(0),\ldots,\mathbf{L}^j(n-1)|I(0),\ldots, \\ I(n-1),\theta(0),\ldots,\theta(n-1)) \quad (6)$$

It is difficult to solve Eq. 6 directly, thus various assumptions, such as the Markovian property of the motion [55], have been proposed to the posterior distribution over $\mathbf{L}^j(\mathbf{t})$ given images up to time t. However, results are often not guaranteed to be smooth and may diverge over time, due to error accumulation. These fundamental issues can be addressed effectively if both, temporal and spatial appearance information is considered over the whole sequence at once. A trajectory can be uniquely represented by the concatenation of its Discrete Fourier Transform (DFT) coefficients,

$$\mathbf{s}^j = [\mathbf{s}^j(0), \mathbf{s}^j(1), \cdots, \mathbf{s}^j(n-1)] \quad (7)$$

where $\mathbf{s}^j(f) \in \mathscr{C}^3$ is the frequency spectrum of the x, y, and z components of the trajectory $\mathbf{L}^j(t)$, and $f = 0, 1, \cdots, n-1$. The magnitude of $\mathbf{s}^j(f)$ is used to describe the shift-invariant motion according to the shift theorem of DFT, while the phase information is used to handle temporal misalignment. Equation 6 can be reformulated as finding the DFT spectrum $\mathbf{s}^j$, with the maximal posterior probability:

$$\arg\max_{\mathbf{s}^j} p(\mathbf{s}^j|I,\theta) = \arg\max_{\mathbf{s}^j} p(\mathbf{s}^j(0),\ldots,\mathbf{s}^j(n-1)|I(0),\ldots, \\ I(n-1),\theta(0),\ldots,\theta(n-1)) \quad (8)$$

Instead of estimating the motion trajectory directly, we apply discriminative learning to detect the spectrum $\mathbf{s}^j$ in the frequency domain by optimizing Eq. 8.

Search Space Marginalization: Inspired by the MSL, we efficiently perform trajectory spectrum learning and detection in DFT subspaces with gradually increased dimensionality. The intuition is to perform a spectral coarse-to-fine motion estimation, where the detection of coarse level motion (low frequency) is incrementally refined with high frequency components representing fine deformations.

As described earlier, the motion trajectory is parameterized by the DFT spectrum components $\mathbf{s}^j(f), f = 0, \ldots, n-1$. We differentiate between two types of subspaces, individual component subspaces $\Sigma^{(k)}$ and marginalized subspaces Σ_k defined as:

$$\Sigma^{(k)} = \{\mathbf{s}(k)\} \tag{9}$$

$$\Sigma_k = \Sigma_{k-1} \times \Sigma^{(k)} \tag{10}$$

$$\Sigma_0 \subset \Sigma_1 \subset \ldots \subset \Sigma_{r-1}, r = |\zeta| \tag{11}$$

The subspaces $\Sigma^{(k)}$ are efficiently represented by a set of corresponding hypotheses $\mathcal{H}^{(k)}$ obtained from the training set. The pruned search space enables efficient learning and optimization:

$$\Sigma_{r-1} = \mathcal{H}^{(0)} \times \mathcal{H}^{(1)} \times \ldots \times \mathcal{H}^{(r-1)}, r = |\zeta|$$

Learning in Marginal Trajectory Spaces: The algorithm starts by learning the posterior probability distribution in the DC marginal space Σ_0. Subsequently, the learned detector D_0 is applied to identify high probable candidates $\mathscr{C}_0$ from the hypotheses $\mathcal{H}^{(0)}$. In the following step, the dimensionality of the space is increased by adding the next spectrum component (in this case the fundamental frequency, $\Sigma^{(1)}$). Learning is performed in the restricted space defined by the extracted high probability regions and hypotheses set $\mathscr{C}_0 \times \mathcal{H}^{(1)}$. The same operation is repeated until reaching the genuine search space Σ_{r-1}.

For each marginal space Σ_k, corresponding discriminative classifiers D_k are trained on sets of positives Pos_k and negatives Neg_k. We analyze samples constructed from high probability candidates $\mathscr{C}_{k-1}$ and hypotheses $\mathcal{H}^{(k)}$. The sample set $\mathscr{C}_{k-1} \times \mathcal{H}^{(k)}$ is separated into positive and negative examples by comparing the corresponding trajectories to the ground truth in the spatial domain using the following distance measure:

$$d(\mathbf{L}^j{}_1, \mathbf{L}^j{}_2) = \max_t \|\mathbf{L}^j{}_1(t) - \mathbf{L}^j{}_2(t)\| \tag{12}$$

where $\mathbf{L}^j{}_1$ and $\mathbf{L}^j{}_2$ denote two trajectories for the j-th landmark. Positives are in a certain distance $dist_{pos}$ (e.g. 1.5 mm) to the ground-truth over the whole trajectories. Negatives, however, are selected individually for each time step, if the tested position in space and time is larger than $dist_{neg}$ (e.g. 3.5 mm). The probabilistic boosting tree PBT is applied to train a strong classifier D_k.

Motion Trajectory Estimation: The local non-rigid motion is parameterized by both magnitude and phase of the trajectory spectrum $\mathbf{s}^j(f)$. The parameter estimation is conducted in the marginalized search spaces $\Sigma_0, \ldots, \Sigma_{r-1}$ using the trained spectrum detectors $D_0, \ldots, D_{r-1}$ as illustrated in Fig. 9. Starting from an initial zero-spectrum, we incrementally estimate the magnitude and phase of each frequency component $\mathbf{s}(k)$. At the stage k, the corresponding robust classifier D_k is exhaustively scanned over the potential candidates $\mathscr{C}_{k-1} \times \mathcal{H}^{(k)}$. The probability of a candidate $\mathbf{C_k} \in \mathscr{C}_{k-1} \times \mathcal{H}^{(k)}$ is computed by the following objective function:

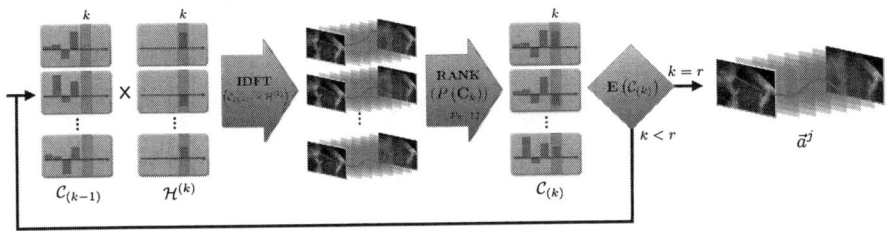

Fig. 9 Diagram depicting the estimation of non-rigid landmark motion using trajectory spectrum learning

$$p(\mathbf{C_k}) = \prod_{t=0}^{n-1} D_k(IDFT(\mathbf{C_k}), I, t) \qquad (13)$$

where $t = 0, \ldots, n-1$ is the time instance (frame index). After each step k, the top 50 trajectory candidates $\mathscr{C}_k$ with high probability values are preserved for the next step $k+1$. The procedure is repeated until a final set of trajectory candidates $\mathscr{C}_{r-1}$, defined in the full space Σ_{r-1}, is computed.

3.3 Comprehensive Model Estimation

The final stage in our hierarchical model estimation algorithm is the delineation of the full morphology and dynamics of the anatomies:

$$\arg\max_M p(M|I, \theta, \mathbf{L}) = \arg\max_M p(M(0), \ldots, M(n-1)|I(0), \ldots,$$
$$I(n-1), \theta(0), \ldots, \theta(n-1), \mathbf{L}(0), \ldots, \mathbf{L}(n-1)) \qquad (14)$$

The shape model is first estimated in the End-Diastole (ED) and End-Systole (ES) phases of the cardiac cycle and then the non-rigid deformation is propagated to the remaining phases using a learned motion prior.

Estimation in Cardiac Key Phases: Using the previously estimated model parameters, a pre-computed mean shape of the comprehensive valvular model is placed into the volumes $I(t_{ED})$ and $I(t_{ES})$ through a TPS transform Bookstein [5]. The initial estimate is then deformed to fit the true valvular anatomy using learned object boundary detectors, regularized by statistical shape models (see Fig. 10).

Learning based methods provide robust results Zheng et al. [56], Yang et al. [55] by utilizing both gradients and image intensities at different image resolutions and by incorporating the local context. Hence, the non-rigid deformation is guided by a boundary detector D_b learned using the probabilistic boosting-tree and steerable features [56]. After initialization, D_b evaluates hypotheses for each discrete boundary point along its corresponding normal direction. The new boundary points are set to the hypotheses with maximal probability. To guarantee physiologically

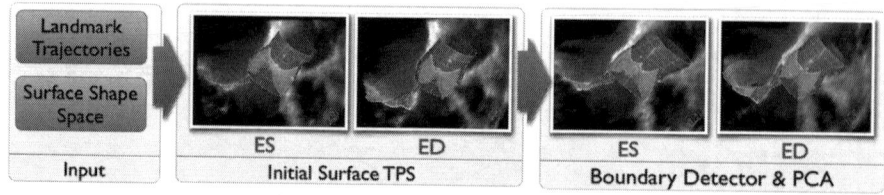

Fig. 10 Diagram depicting the estimation of the comprehensive valve model. Estimation in cardiac key phases, end-diastole and end-systole

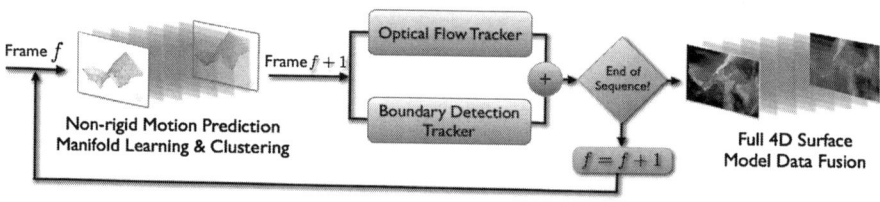

Fig. 11 Diagram depicting the estimation of the comprehensive valve model. Estimation in the full cardiac cycle

compliant results, the final model is obtained after projecting the estimated points to the statistical shape space [24].

Motion Estimation: Starting from the detection results in the ED and ES phases, the model deformations are propagated in both forward and backward directions using learned motion priors similar as in Yang et al. [55] (see Fig. 11). The motion prior is estimated at the training stage using motion manifold learning and hierarchical K-means clustering, from a pre-annotated database of sequences containing one cardiac cycle each. First the temporal deformations are aligned by 4D generalized procrustes analysis. Next a low-dimensional embedding is computed from the aligned training sequences using the ISOMAP algorithm Tenenbaum et al. [43], to represent the highly nonlinear motion of the heart valves. Finally, in order to extract the modes of motion $\overline{X}_m$, the motion sequences are clustered with hierarchical K-means based on the Euclidean distance in the lower dimensional manifold.

One-step forward prediction is used to select the correct motion mode for predicting time step t. Therefore the previous shapes $M(t)$ from time steps $t = 1 \cdots T - 1$ and the corresponding time steps in each of the motion modes $\overline{X}_m$ are sub-sampled by a constant factor k and the TPS transform T_{TPS} computed. The mean error between the warped shape and the corresponding shape on each motion mode is computed, and the motion mode with minimum distance is selected for prediction:

$$E_{TPS}(\overline{\mathbf{X}}_m(t), M(t)) = \frac{k}{N} \sum_{j=1}^{N/k} \|\overline{\mathbf{X}}_m^j(t) - T_{TPS}(M^j(t))\| \quad (15)$$

$$\overline{\mathbf{X}}(T) = \arg\min_{m} \frac{1}{T-1} \sum_{t=1}^{T-1} E_{TPS}(\overline{\mathbf{X}}_m(t), M(t)) \quad (16)$$

where N denotes the number of points in $M(t)$, $\overline{\mathbf{X}}_m^j$ and M^j are shape vertices, and $\overline{\mathbf{X}}(T)$ the selected motion mode. The shape prediction $M(T)'$ for the following frame T is then computed by inverse TPS mapping $M(T)' = T_{TPS}^{-1}(\overline{\mathbf{X}}(T))$ and the boundary detector D_b deforms the initialization to make it fit the data in the update step. To ensure temporal consistency and smooth motion and to avoid drifting and outliers, two collaborative trackers, an optical flow tracker and a boundary detection tracker D_b, are used in our method. The results are then fused into a single estimate by averaging the computed deformations and the procedure is repeated until the full 4D model is estimated for the complete sequence.

3.4 Regression-based Surface Reconstruction

Sections 3.1–3.3 describe the framework for parameter estimation from dense 4D images. In this section we present a regression-based method for estimating model parameters from sparse 4D data, usually produced during Cardiac Magnetic Resonance (CMR) exams.

In *regression* a solution to the following optimization problem is normally sought Zhou et al. [57]:

$$\hat{\mathcal{R}}(\mathbf{x}) = argmin_{\mathcal{R} \in \mathfrak{S}} \sum_{n=1}^{N} L(y(\mathbf{x}_n), \mathcal{R}(\mathbf{x}_n))/N \quad (17)$$

where $\mathfrak{S}$ is the set of possible regression functions, $L(\circ, \circ)$ is a loss function that penalizes the deviation of the regressor output $\mathcal{R}(\mathbf{x}_n)$ from the true output, and N is the number of available training examples. In our case the reconstruction task is defined as a regression problem between the full surface model of a heart valve and the respective sparse data acquired using the proposed CMR protocol:

$$\mathbf{M} = \hat{\mathcal{R}}(\mathbf{x}_{sparse}) + \varepsilon \quad (18)$$

In our regression problem both for input and output data we focus on shape information and ignore respective volume data. Thus, the output M is always a set of m 3D points as defined in Sect. 2.1

Invariant Shape Descriptors: The input, $\mathbf{x}_{sparse}$, are shape descriptors (SD) describing the cloud of points belonging to surface in the sparse CMR data. The simplest but reliable solution is to use the coordinates of known points as input Vitanovski et al. [50]. A different solution, which we exploit here, is to use angles, distances and areas between random sampled points as point cloud descriptors Osada et al. [32]:

- A3: Measures the Angle between three random points;
- D2: Measure the distance between two random points;
- D3: Measures the square root of the area of the triangle between three random points;

For the different shape descriptors proposed by Osada et al. [32] we measured feature importance by analysing the features selected by additive boosting. We have identified (A3, D2 and D3) to be most informative in our context with the average probability of occurrence 0.11, 0.07, and 0.13, correspondingly. In addition, all three types are translation, rotation and scale invariant descriptors which overcome the necessity of point correspondences. Finally, histogram bins and the four first normalized central moments describing the histogram distribution are computed from the descriptors and incorporate in the regression model as input.

Ensembles of Additive Boosting Regressors: Each component m regression problem $\hat{\mathscr{R}}^m$ is solved by learning using additive boosting regression (ABR) Friedman [13]. In ABR, the weak regressors ρ_t are sequentially fit to the residuals, starting from the mean $\overline{M}$ and proceeding with the residuals of the availabe set of weak regressors themselves. In ABR, the output function is assumed to take a linear form as follows Friedman [13]:

$$\hat{\mathscr{R}}(\mathbf{x}) = \sum_{t=1}^{T} \alpha_t \rho_t(\mathbf{x}); \rho_t(\mathbf{x}) \in \Im \quad (19)$$

where $\rho_t(\mathbf{x})$ is a base (weak) learner and T is the number of boosting iterations.

We use very simple weak regressors as the base learners: **simple 1D linear regression** (SLR), **logistic stumps** (LS) and **decision stumps** (DS). For SLR, in each boosting iteration a feature which results in the smallest squared loss with linear regression is added to the pool of already selected features. Each weak learner is thus a simple linear regressor of the form $(y = \beta_1 x + \beta_0)$ where x is the selected shape descriptor and y is a scalar output coordinate. LS is a simple logistic function on one shape descriptor x:

$$y = \frac{1}{1+e^{-z}}, z = \beta_1 x + \beta_0 \quad (20)$$

Finally, DS is a piecewise linear threshold function where threshold θ is selected so that the variance in the subsets of instances produced is minimized. Generalization performance improvement of the underlying regression models and avoidance of overfitting is achieved by injecting randomization in the input data and random features sub spacing (BRFS), similar to Webb [54]. In particular,

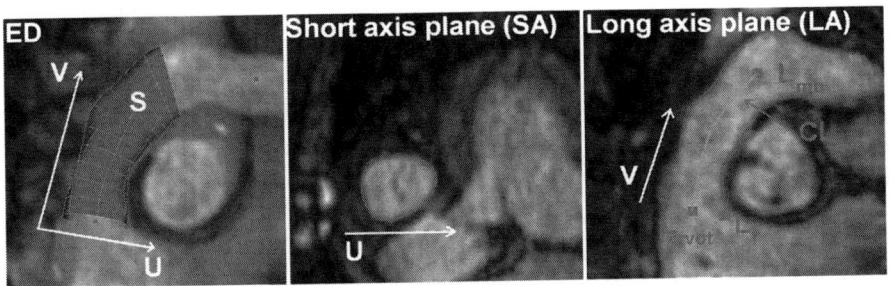

Fig. 12 *Left:* 3D MRI scan of the whole heart in the ED phase. *Middle:* short axis imaging plane. *Right:* long axis imaging plane

instead of providing a single model $\mathcal{R}$ for the training set X, we generate a set of models $\mathcal{R}_i^j$, each obtained using the same additive regression procedure but on a *random sample* of the data, with instances S_i obtained using random sampling with replacement, and a subset of features F_j including 50% features randomly sampled without replacement from the original set. The final solution is then simply the mean surface for the surfaces obtained with the regressors generated from the random samples:

$$\mathcal{R} = mean_{i,m}(\mathcal{R}_i^m) \qquad (21)$$

Regression-based Pulmonary Valve Model Estimation from 2D+t CMR: In clinical settings the long acquisition time characteristic to CMR is reduced by scanning the Pulmonary Valve (PV) with a sparse 4D CMR protocol. Thus, one end-diastolic 3D volume I_{ED} and two orthogonal 2D+t cine images, short axis (SA) and long axis (LA) (see Fig. 12) instead of a dense 4D image. We apply the regression-based surface reconstruction to estimate a 4D patient-specific model of the pulmonary valve $M_{pulmonary}$ from the sparse data:

$$M_{pulmonary} = \mathcal{R}(SD[M(t_{ED}), (C_{LA}, C_{SA})_{1...T}]) \qquad (22)$$

where $M(t_{ED})$ is the pulmonary valve model in I_{ED} and $(C_{LA}, C_{SA})_{1...T}$ the pulmonary valve contours in the LA and SA images over the entire sequence of length T. $M(t_{ED}), (C_{LA}, C_{SA})_{1...T}$ are estimated using the pipeline from Sects. 3.1–3.3. The method is similarly applied to estimate other anatomical structures from sparse images.

4 Experimental Results

In this section we demonstrate the performance of the proposed patient-specific parameter estimation framework from multi-modal images. Experiments are performed on a large and heterogeneous data set acquired using CT, TEE and MRI

Table 1 Accuracy of the global location and rigid motion estimation, quantified from the box corners and reported using the mean error and standard deviation distribution over each valve and employed modality

Mean / STD (mm)	Aortic valve	Mitral valve	Pulmonary valve	Tricuspid valve
TEE	4.78 ± 3.26	5.00 ± 2.02	–	–
Cardiac CT	4.40 ± 1.98	6.94 ± 2.19	7.72 ± 10.03	–
CMR	–	–	7.19 ± 3.50	–

Table 2 Accuracy of the non-rigid landmark motion estimation, quantified by the Euclidean distance and reported using the mean error and standard deviation distribution over each valve and employed modality

Mean / STD (mm)	Aortic valve	Mitral valve	Pulmonary valve	Tricuspid valve
TEE	2.79 ± 1.26	3.60 ± 1.56	–	–
Cardiac CT	2.72 ± 1.52	2.79 ± 1.20	3.50 ± 2.70	–
CMR	–	–	4.30 ± 3.00	–

scanners from 476 patients affected by a large spectrum of cardiovascular and valvular heart diseases: regurgitation, stenosis, prolapse, aortic root dilation, bicuspid aortic valve and Tetralogy of Fallot. The imaging data set includes 1,330 cardiac CT, 5,061 TEE and 83 CMR volumes, which were collected from medical centers around the world.

Each volume in our data set is associated with an annotation obtained through an expert-guided process that includes the manual placing of anatomical landmarks and delineation of anatomical surfaces. The obtained models are consider as ground truth and were used for training and testing of the proposed algorithms. Three-fold cross validation was performed for all experiments and reported results reflect performance on unseen, test data.

Performance of the Object Localization and Rigid Motion Estimation: The performance of the global location and rigid motion estimation, θ, described in Sect. 3.1 is quantified at the box corners of the detected time-depenedent similarity transformation. The average Euclidean distance between the eight bounding box points, defined by the similarity transform parameters $\{(c_x, c_y, c_z), (\alpha_x, \alpha_y, \alpha_z), (s_x, s_y, s_z), t\}$ and the ground-truth box is reported. Table 1 illustrates the mean errors and corresponding standard deviations distributed over the four valves and employed image modalities. The average accuracy of the individual detection stages is 3.09 ± 3.02 mm for position, 9.72 ± 5.98 for orientation, and 6.50 ± 4.19 for scale.

Performance of the Non-Rigid-Landmark Motion Estimation: The accuracy of the Trajectory Spectrum Learning algorithm (see Sect. 3.2), which estimate the non-rigid landmark motion model, L, is measured using the Euclidean distance between detected and corresponding ground truth landmark trajectories. Table 2 demonstrates the precision expressed in mean errors and standard deviations, distributed over the four valves and three data sources. Note that reported values are obtained by averaging the performance of individual landmarks with respect to the corresponding valve.

Table 3 Accuracy of the comprehensive valve model estimation, quantified by the Point-to-Mesh distance and reported using the mean error and standard deviation distribution over each valve and employed modality

Mean / STD (mm)	Aortic valve	Mitral valve	Pulmonary valve	Tricuspid valve
TEE	1.35 ± 0.54	2.29 ± 0.64	–	–
Cardiac CT	1.22 ± 0.38	2.02 ± 0.57	1.60 ± 0.20	–
CMR	–	–	1.90 ± 0.20	–

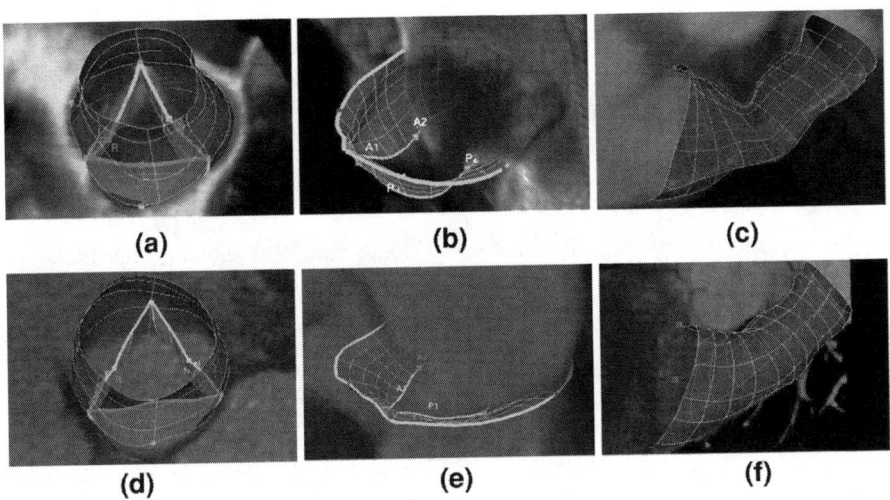

Fig. 13 Examples of comprehensive valves model estimation: **a** aortic valve in TEE, **b** mitral valve in TEE, **c** pulmonary valve in CMR, **d** aortic valve in cardiac CT, **e** mitral valve in cardiac CT, and **f** pulmonary valve in cardiac CT

Performance of the Comprehensive Valve Model Estimation: The accuracy of the comprehensive valvular model estimation, M, (see Sect. 3.3) is evaluated by utilizing the point-to-mesh distance. For each point on a surface $\mathbf{p}$, we search for the closest point (not necessarily one of the vertices) on the other surface to calculate the Euclidean distance. To guarantee a symmetric measurement, the point-to-mesh distance is calculated in two directions, from detected to ground truth surfaces and vice versa. Table 3 contains the mean error and standard deviation distributed over the four valves and image types. Examples of estimation results are given in Fig. 13.

Overall, the estimation accuracy of the patient-specific valvular parameters from multi-modal images is 1.73 mm. On a standard PC with a quad-core 3.2 GHz processor and 2.0 GB memory, the total computation time for the all three estimation stages is 4.8 s per volume (approx 120 s for average length volume sequences), from which the global location and rigid motion estimation requires 15% of the computation time (approx 0.7 s), non-rigid landmark motion 54% (approx 2.6 s), and comprehensive valvular estimation 31% (approx 1.5 s).

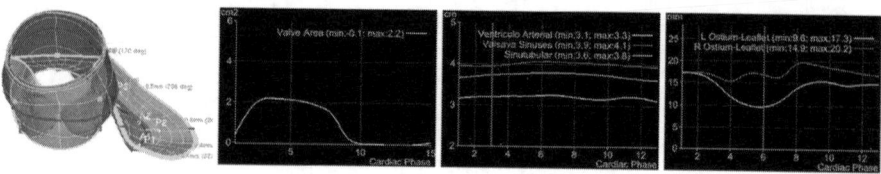

Fig. 14 Examples of aortic-mitral morphological and functional measurements. From *left* to *right*: aortic valve model with measurement traces, aortic valve area, aortic root diameters and ostia to leaflets distances

5 Clinical Applications

5.1 Valve Analysis for Transcatheter Aortic Valve Implantation

It is axiomatic that precise quantification of the anatomy and function is fundamental in the medical management of valvular heart disease. Emerging percutaneous and minimally invasive valve interventions, such as the Transcatheter Aortic Valve Implantation (TAVI), require extensive non-invasive assessment, as clinicians have restricted direct access to the sensitive anatomies [2, 36]. Measurements of the aortic annulus are mandatory for the correct sizing the prosthetic implant, while precise knowledge of the coronary ostia position prevents hazardous ischemic complications by avoiding the potential misplacement of aortic valve implants. Data about the integral three-dimensional configuration of critical structures (ostia, commissures, hinges, etc.) and their relative location over the entire cardiac cycle becomes increasingly relevant.

We proposed a paradigm shift in the clinical evaluation of the valvular apparatus, which replaces manual analysis based on 2D images with automated model-based quantification from 4D data. At the center of the proposed approach is the dynamic valvular model introduced in Sect. 2, which captures comprehensive patient-specific information of the morphology and functions from multi-modal images following the methods described in Sect. 3 The explicit mathematical model is exploited to express a wide-ranging collection of quantitative parameters that support the overall clinical decision making process (see Fig. 14). In comparison with the gold standard, which processes 2D images and performs manual measurements, the key benefits of the proposed model-based analysis are: increased precision and reproducibility, decreased processing time, and integrated and comprehensive analysis. In the following we present a series of clinical validation experiments performed jointly with various clinical collaborators [7, 9, 15, 21].

Aortic Valve Opening Area Analysis from cardiac CT: The usefulness of cardiac CT to assess the aortic valve opening area (AVA) has been exhaustively documented Halpern [17]. However, manual valve planimetry is cumbersome and time-consuming. In the following experiment we evaluated the accuracy and time-effectiveness of automated model-based AVA computation compared to manual planimetry. Retrospectively ECG-gated cardiac CT data of 32 patients scanned with dual-source CT (n = 21) or 64-slice CT (n = 11) were included. Data were reconstructed at 10% increments across the cardiac cycle with 1.5 mm section thickness and 1mm increment. Two independent observers performed manual planimetric measurements by tracing the maximal systolic orifice on double oblique short axis multiplanar reconstructions. The same data were then analyzed using an automatic model-based method. The leaflets' geometries during maximal opening define the course of the free margins. The encompassed AVA can be computed as a surface integral.

Data was analyzed using linear regression and Bland Altman plots. Interobserver and intermethod variances were calculated. Analysis times for both methods were recorded. Mean AVA by CT planimetry was $3.62 \pm 1.21\,cm^2$. Mean AVA derived from the model was $3.74 \pm 1.34\,cm^2$. Excellent correlation was found between planimetric and automated quantification (r = 0.963, P < 0.0001). Bland Altman plots revealed a systematic bias of $0.12 \pm 0.38\,cm^2$. Intermethod variance did not differ significantly from interobserver variance (0.28 vs $0.25\,cm^2$), placing 82% of model measurements between user measurements. Mean analysis time was significantly reduced for model-based measurements (mean 125 s), compared with manual planimetry (mean 230 s). The proposed model-based method allows automated, patient specific morphologic and dynamic quantification of AVA. Measurement results are within the interobserver variance of manual planimetry. Quantification of AVA derived from an aortic valve model enables fast, accurate assessment in excellent agreement with manual planimetry and has the potential to improve cardiac imaging workflow.

Aortic Valve and Root Analysis from Cardiac CT and 3D TEE: Accurate anatomical and functional assessment of AV and aortic root is crucial for understanding the pathophysiology of abnormalities and for managemening decision-making in patient with aortic valve disease and aortic aneurysm. The aim of this study was to evaluate the feasibility of the modal-based method to assess the aortic valve and aortic root from volumetric 3-D Echo compared to CT. Volume-rendered 3-D TEE data were obtained using V5M transducer, Siemens Sequoia. Volumetric CT images were acquired using 64-Slice CT(Avanto, Siemens). We dynamically measured the AVA(cm2), diameter of sinotubular junction(d-STJ, mm), sinus of Valsalva(d-SV, mm) and basal ring(d-BR, mm). 364 CT volumes from 41 patients and 23, 3-D TEE volumes from 15 patients with normal to mild AR were acquired. 3-D TEE data about AV and root showed strong correlation with CT data, as illustrated in Table 4. This novel automated model-based approach provides accurate dimensions of the AV and the aortic root and may aid in valve and root repair procedures (see Fig. 15, Table 4).

Aortic Valve and Root in Aortic Regurgitation from 3D TEE: In this study we applied the model-based analysis approach to automatically quantify the aortic valve and the root from 3-D TEE data in patients with aortic regurgitation (AR). Volumetric 3-D TEE of the AV and proximal root from 15 patients with AR was analyzed. The conventional measures were compared to 2-D, and the

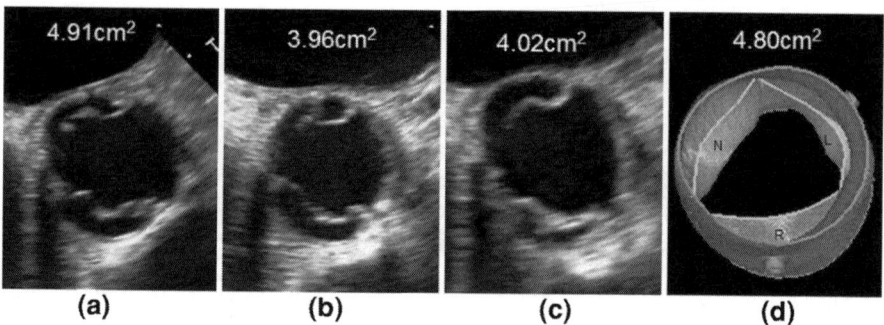

Fig. 15 Aortic valve area measured manually from 2D images versus the automatic 3D model-based quantification: **a, b, c** show the manual measurement from three different 2D TEE images of the same patient, which demonstrate the sensitivity of the current gold standard approach to the position of the 2D-dimensional section, and **d** shows the proposed precise and reproducible 3D model-based measurement

Table 4 Comparison of AVA and aortic diameter between cardiac CT and 3D TEE

	3D TEE	Cardiac CT	r-value	p-value
AVA (cm^2)	3.09 ± 0.85	4.33 ± 1.36	0.707	0.013
Max. Ventriculoarterial junct. ø (mm)	2.42 ± 0.27	2.74 ± 0.36	0.982	0.018
Max. Valsava sinuses ø (mm)	3.16 ± 0.32	3.92 ± 0.46	0.993	0.007
Max. Sinotubular junct. ø (mm)	2.69 ± 0.26	3.19 ± 0.21	0.775	0.042

nonconventional measures were compared to known normal database. Conventional measures- 2-D and the model-based measures of AV area (r = 0.98), STJ diameter (r = 0.73) and SV diameter (r = 0.79) showed good correlation; annular diameter was discordant (r = 0.58) consistent with its complex geometry in AR. Nonconventional measures (abnormal vs. normal, mm) by the model-based method—Inter-commissural distance (mm) was increased (Left: 25.9 + 3 Vs. 25, Right: 27.1 + 3 Vs. 25.9 and Non: 27.2 + 3 Vs. 25.5), Annulus to coronary ostia distance (mm) was increased (Right: 19.3 + 3 Vs. 17.2 + 3 and Left 16.9 + 3 Vs. 14.4 + 3); also, leaflet tip to ostia minimum distance was 5 + 1.6 (R) and 8 + 1.2 (L). The directly measured 3-D ERO in mild AR was 10–20 mm² and moderate AR was 30 mm². Automated quantification of the aortic and the root yields vital and incremental measures which may be valuable to guide surgical and percutaneous interventions to improve outcomes (Fig. 16).

5.2 Patient Selection for Percutaneous Pulmonary Valve Implantation

Until recently, pulmonary valve replacement has been exclusively performed trough open heart surgery Boudjemline MD et al. [6], with all associated risks:

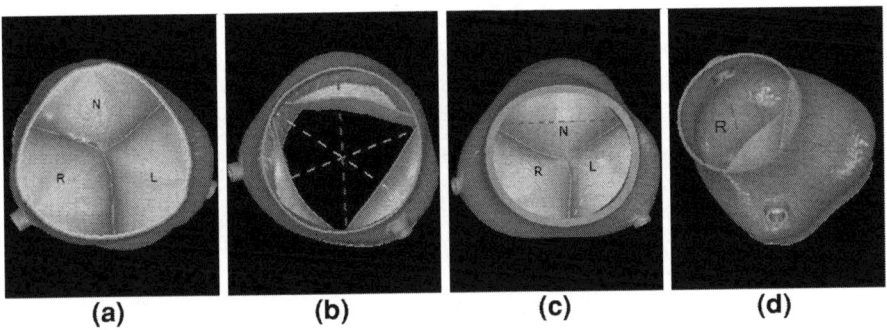

Fig. 16 Examples of model-based aortic valve measurements: **a** aortic annular diameter (ventriculoarterial junction), **b** sinus width, **c** inter-commissural distances, and **d** coronary ostia to leaflet tip distances

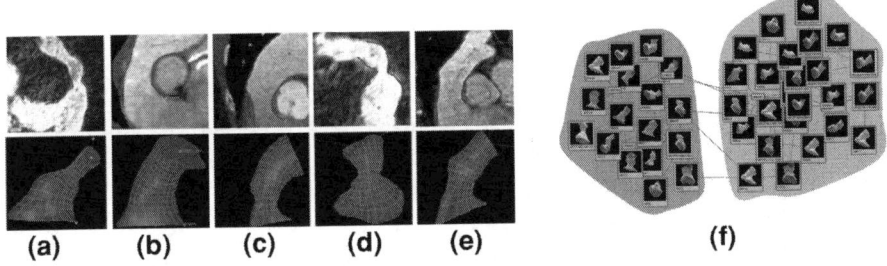

Fig. 17 Types of pulmonary trunk morphologies: **a** pyramidal shape, **b** constant diameter, **c** inverted pyramidal shape, **d** narrowed centrally but wide proximally and distally, **e** wide centrally but narrowed proximally and distally Schievano et al. [38], and **f** relative neighborhood graph with unsuitable (*blue* cluster) and suitable (*red* cluster) individuals

high morbidity, incidence of neurological damage, stroke and repeated valve replacement Parr et al. [34]. Novel Percutaneous Pulmonary Valve Implantation (PPVI) techniques Schievano et al. [39], offer a less traumatic and safer treatment of the pulmonary valve Carnaghan [8].

The selection of patient for PPVI treatment is largely based on the morphology of the pulmonary trunk Bonhoeffer et al. [3]. Intervention in unsuitable anatomies exposes patients to unnecessary invasive catherization, for which the implanted device has a high probability of proximal device dislodgment. In an effort to standardize the selection process, Schievano et al. [38] proposed a classification of various morphologies, based on geometric measures and appearance of the right-ventricular outflow tract and the pulmonary trunk, into five groups: pyramidal shape (type I), constant diameter (type II), inverted pyramidal shape (type III), wide centrally but narrowed proximally and distally (type IV), and narrowed centrally but wide proximally and distally (type V) (see Fig. 17). Patients from type I are considered to be unsuitable for PPVI due to the narrow artery and high probability of device migration. Therefore the main challenge lies in discriminating anatomies of type I from other four classes.

We propose a discriminative distance function learned using Random Forest in the product space to provide automated patient selection, namely classification of the subjects into two classes: PPVI suitable and PPVI unsuitable (see Fig. 17f). We formulate the problem as follows:

$$\hat{y} = \underbrace{argmax}_{y \in \{-1,+1\}} (p(y|C)) \quad C = (\mathbf{p}_0, \ldots, \mathbf{p}_N, F_0, \ldots, F_Q) \tag{23}$$

where $y \in \{-1,+1\}$ are the application specific labels and each model instance is represented by a parameter vector C, composed out of N surface vertices $\mathbf{p}_i \in M_{pulmonary}$ (see Sect. 2) and Q application dependent features F_i derived from the model. Instead of learning directly the posterior probability, a distance learning followed by the actual classification is performed, where each step requires search in a less complex functional space than in the immediate learning Tsymbal et al. [45]. Thus, learning is performed from weak representations described through equivalence constraints (C^1, C^2, y), where C^1 and C^2 are feature vectors (see Eq. 23) and $y \in \{+1,-1\}$ is a label indicating whether the two instances are similar or dissimilar Hertz [18]. Using Random Forest (RF) for the learning algorithm, for a given forest f the similarity between two instances C^1 and C^2 is calculated as follows: 1) the instances are propagated down all K trees within f and their terminal positions z in each of the trees ($z_1 = (z_{11}, \cdots, z_{1K})$ for C^1, similarly z_2 for C^2) are recorded, and 2) the similarity between the two instances then equals to (I is the indicator function):

$$S(C^1, C^2) = \frac{1}{K} \sum_{i=1}^{K} I(z_{1i} = z_{2i}) \tag{24}$$

The proposed automated PPVI patient selection method was evaluated on a population of 102 patients that include all five types of pulmonary trunk geometry (Fig. 17) and cumulated into 50 patient of type I (i.e. unsuitable for PPVI) and 52 patients with suitable geometries. The accuracy of methods is validated by leave-one-out cross-validation for various classification approaches including k-Nearest Neighbors (kNN), AdaBoost (AB) and Random Forests (RF) in the canonical space, as well as AdaBoost and Random Forests [23] in the product and difference spaces (AB-pr, RF-pr, AB-di and RF-di) and intrinsic RF distance (RF-dist). AdaBoost in the product space showed highest performance for the PPVI suitability selection with 91% correct classification. The proposed approach has the potential to significantly improve accuracy and reproducibility of patient selection for PPVI and pre-procedural PPVI planning.

6 Conclusion

The main subject addressed in this chapter is the patient-specific estimation of physiological heart valvular and chamber models from multi-modal cardiac images. We proposed a physiological model of the heart valves to precisely capture their anatomical, dynamical and pathological variations. Our model is hierarchically defined and comprehensively represents the location and rigid motion, anatomical landmarks and the comprehensive shape and dynamics of all four cardiac valves: aortic, mitral, tricuspid and pulmonary valves. We presented discriminative learning-based framework that permits the estimation of patient-specific model parameters from cardiac images. In the first step Marginal Space Learning and RANdom SAmple Consensus are applied for the time-coherent detection of the valvular location and motion from an arbitrary four-dimensional cardiac scan. Subsequently, Trajectory Spectrum Learning robustly estimates the parameters of the anatomical landmarks from four-dimensional image sequences. The last method performs object delineation of dynamic models, using boundary detectors and motion manifold learning techniques. The performance of the proposed estimation framework is demonstrated through extensive experiments on 476 patients, which results into an average precision of 1.73 mm and speed of 4.8 s per volume. Two clinical applications based on the modeling and estimation techniques are described: a novel paradigm for the clinical analysis of the valvular apparatus with application to Transcatheter Aortic Valve Implantation, and automated patient selection method for Percutaneous Pulmonary Valve Implantation based on learning-based distance functions. The technology described in this chapter can potentially advance the management of patients affected by valvular heart disease by reducing clinical investigation costs, risks for complications during procedures, and ultimately by improving the overall outcome of valvular treatment.

References

1. Agarwal, A., Triggs, B.: Tracking articulated motion using a mixture of autoregressive models. In: Proceedings of the European Conference Computer Vision, pp. III 54–65 (2004)
2. Akhtar, M., Tuzcu, E.M., Kapadia, S.R., Svensson, L.G., Greenberg, R.K., Roselli, E.E., Halliburton, S., Kurra, V., Schoenhagen, P., Sola, S.: Aortic root morphology in patients undergoing percutaneous aortic valve replacement: Evidence of aortic root remodeling. J. Thorac. Cardiovasc. Surg. **137**, 950–956 (2009)
3. Bonhoeffer, P., Boudjemline, S.A., Qureshi, Y., Bidois, J.L., Iserin, L., Acar, P., Merckx, J., Kachaner, J., Sidi, D.: Percutaneous insertion of the pulmonary valve. J. Am. Coll. Cardiol. **39**, 1664–1669 (2002)
4. Bonow, R.O., Carabello, B.A., Chatterjee, K., de Leon, A.C.J., Faxon, D.P., Freed, M.D., Gaasch, W.H., Lytle, B.W., Nishimura, R.A., O'Gara, P.T., O'Rourke, R.A., Otto, C.M., Shah, P.M., Shanewise, J.S.: Acc/aha 2006 guidelines for the management of patients with valvular heart disease: a report of the american college of cardiology/american heart association task force on practice guidelines (writing committee to develop guidelines for the management of patients with valvular heart disease). Circulation **114**, 84–231 (2006)

5. Bookstein, F.L.: Principal warps: thin-plate splines and the decomposition of deformations. IEEE Trans. Pattern Anal. Mach. Intell. **11**, 567–585 (1989)
6. Boudjemline, Y., Agnoletti, G., Bonnet, D., Sidi, D., Bonhoeffer, P.: Percutaneous pulmonary valve replacement in a large right ventricular outflow tract: An experimental study. Am. Coll. Cardiol. **43**, 1082–1087 (2004)
7. Calleja, A., Razvan, I., Houle, H., Liu, S., Dickerson, J., Thavendiranathan, P., Sai-Sudhakar, C., Crestanello, J., Ryan, T., Vannan, M.: Automated quantitative modeling of the aortic valve and root in aortic regurgitation using volume 3-d transesophageal echocardiography. In: American College of Cardiology Annual Meeting-ACC 2010, Atlanta, USA (2010)
8. Carnaghan, H.: Percutaneous pulmonary valve implantation and the future of replacement. Sci. Technol. **20**, 319–322 (2006)
9. Choi, J.H., Georgescu, B., Ionasec, R.I., Raman, S., Hong, G.R., Liu, S., Houle, H., Vannan, M.A.: Novel semi-automatic quantitative assessment of the aortic valve and aortic root from volumetric 3d echocardiography: comparison to volumetric cardiac computed tomography (ct). In: AHA, New Orleans, USA (2008)
10. Conti, C., Stevanella, M., Maffessanti, F., Trunfio, S., Votta, E., Roghi, A., Parodi, O., Caiani, E., Redaelli, A.: Mitral valve modelling in ischemic patients: finite element analysis from cardiac magnetic resonance imaginge. In: Computing in Cardiology (2010)
11. De Hart, J., Peters, G., Schreurs, P., Baaijens, F.: A three-dimensional computational analysis of fluid-structure interaction in the aortic valve. J. Biomech. **36**, 103–110 (2002)
12. Ecabert, O., Peters, J., Schramm, H., Lorenz, C., von Berg, J., Walker, M.J., Vembar, M., Olszewski, M.E., Subramanyan, K., Lavi, G., Weese, J.: Automatic model-based segmentation of the heart in CT images. IEEE Trans. Med. Imaging **27**, 1189–1201 (2008)
13. Friedman, J.H.: Greedy function approximation: a gradient boosting machine. Ann. Stat. **29**, 1189–1232 (2000)
14. Fritz, D., Rinck, D., Dillmann, R., Scheuring, M.: Segmentation of the left and right cardiac ventricle using a combined bi-temporal statistical model. SPIE Med. Imaging **6141**, 605–614 (2006)
15. Gassner, E., Ionasec, R.I., Georgescu, B., Vogt, S., Schoepf, U., Comaniciu, D.: Performance of a dynamic aortic valve model for quantification of the opening area at cardiac mdct . comparison to manual planimetry. In: Radiological Society of North American (RSNA), Chicago, USA (2008)
16. Grbić, S., Ionasec, R., Vitanovski, D., Voigt, I., Wang, Y., Georgescu, B., Navab, N., Comaniciu, D.: Complete valvular heart apparatus model from 4D cardiac CT. Medical image computing and computer-assisted intervention: MICCAI. Int. Conf. Med. Image Comput. Computer-Assisted Intervention **13**, 218–226 (2010)
17. Halpern, E.: Clinical Cardiac CT: Anatomy and Function. Thieme Medical Publishers, New York, USA (2008)
18. Hertz, T.: Learning distance functions: algorithms and applications. Ph.D. thesis. The Hebrew University of Jerusalem (2006)
19. Huang, J., Huang, X., Metaxas, D., Axel, L.: Dynamic texture based heart localization and segmentation in 4-d cardiac images. Biomedical Imaging From Nano to Macro 2007 ISBI 2007 4th IEEE International Symposium, pp. 852–855 (2007)
20. Ionasec, R.I., et al.: Robust motion estimation using trajectory spectrum learning: application to aortic and mitral valve modeling from 4d tee. In: Proceedings of the Int'l Conference Computer Vision (2009)
21. Ionasec, R.I., Georgescu, B., Comaniciu, D., Vogt, S., Schoepf, U., Gassner, E.: Patient specific 4d aortic root models derived from volumetric image data sets. Radiological Society of North American (RSNA), Chicago, USA (2008a)
22. Ionasec, R.I., Georgescu, B., Gassner, E., Vogt, S., Kutter, O., Scheuering, M., Navab, N., Comaniciu, D.: Dynamic model-driven quantification and visual evaluation of the aortic valve from 4d ct. MICCAI **1**, 686–694 (2008b)

23. Ionasec, R.I., Voigt, I., Georgescu, B., Houle, H., Hornegger, J., Navab, N., Comaniciu, D.: Personalized modeling and assessment of the aortic-mitral coupling from 4D TEE and CT. In: MICCAI, Heidelberg, pp. 767–775 (2009)
24. Ionasec, R.I., Voigt, I., Georgescu, B., Wang, Y., Houle, H., Vega-Higuera, F., Navab, N., Comaniciu, D.: Patient-specific modeling and quantification of the aortic and mitral valves from 4-D cardiac CT and TEE. IEEE Trans. Med. Imaging **29**, 1636–1651 (2010)
25. Ionasec, R.I., Wang, Y., Georgescu, B., Voigt, I., Navab, N., Comaniciu, D.: Robust motion estimation using trajectory spectrum learning: application to aortic and mitral valve modeling from 4d tee. In: Proceedings of the 12th International Conference on Computer Vision (ICCV), IEEE, Kyoto, Japan (2009)
26. Jablokow, A.: National center for health statistics: national hospital discharge survey: annual summaries with detailed diagnosis and procedure data. Journal Data on Health Resources Utilization 13 (2009)
27. Kunzelman, K., Einstein, D., Cochran, R.: Fluid-structure interaction models of the mitral valve: function in normal and pathological states. Philos. Trans. R. Soc. Lond., B, Biol. Sci. **362**, 1393–1406 (2007)
28. Lansac, E., Lim, H., Shomura, Y., Lim, K., Rice, N., Goetz, W., Acar, C., Duran, C.: A four-dimensional study of the aortic root dynamics. Eur. J. Cardio-Thorac. Surg: Official J. Eur. Assoc. Cardio-Thorac. Surg. **22**, 497–503 (2002)
29. Lloyd-Jones, D., Adams, R., Carnethon, M., De Simone, G., Ferguson, T.B., Flegal, K., Ford, E., Furie, K., Go, A., Greenlund, K., Haase, N., Hailpern, S., Ho, M., Howard, V., Kissela, B., Kittner, S., Lackland, D., Lisabeth, L., Marelli, A., McDermott, M., Meigs, J., Mozaffarian, D., Nichol, G., O'Donnell, C., Roger, V., Rosamond, W., Sacco, R., Sorlie, P., Stafford, R., Steinberger, J., Thom, T., Wasserthiel-Smoller, S., Wong, N., Wylie-Rosett, J., Hong, Y.: American heart association statistics committee and stroke statistics subcommittee, heart disease and stroke statistics–2009 update: a report from the american heart association statistics committee and stroke statistics subcommittee. Circulation **119**, e21–e181 (2009)
30. Lorenz, C., von Berg, J.: A comprehensive shape model of the heart. Med. Image Anal. **10**, 657–670 (2006)
31. Mutlak, D., Aronson, D., Lessick, J., Reisner, S., Dabbah, S., Agmon, Y.: Functional tricuspid regurgitation in patients with pulmonary hypertension. CHEST **135**, 115–121 (2009)
32. Osada, R., Funkhouser, T., Chazelle, B., Dobkin, D.: Shape distributions. ACM Trans. Graph. **21**, 807–832 (2002)
33. Park, J., Metaxas, D., Young, A., Axel, L.: Deformable models with parameter functions for cardiac motion analysis from tagged mri data. IEEE Trans. Med. Imaging **15**, 278–289 (1996)
34. Parr, J., Kirklin, J., Blackstone, E.: The early risk of re-replacement of aortic valves. Ann. Thorac. Surg. **23**, 319–322 (1977)
35. Peskin, C.S., McQueen, D.M.: Case Studies in Mathematical Modeling: Ecology, Physiology, and Cell Biology. Prentice-Hall, Englewood Cliffs, NJ, USA (1996)
36. Piazza, N., de Jaegere, P., Schultz, C., Becker, A., Serruys, P., Anderson, R.: Anatomy of the aortic valvar complex and its implications for transcatheter implantation of the aortic valve. Circ. Cardiovasc. Interventions **1**, 74–81 (2008)
37. Rueckert, D., Burger, P.: Geometrically deformable templates for shape-based segmentation and tracking in cardiac mr images. In: EMMCVPR '97: Proceedings of the First International Workshop on Energy Minimization Methods in Computer Vision and Pattern Recognition, Springer-Verlag, London, UK, pp. 83–98 (1997)
38. Schievano, S., Coats, L., Migliavacca, F., Norman, W., Frigiola, A., Deanfield, J., Bonhoeffer, P., Taylor, A.: Variations in right ventricular outflow tract morphology following repair of congenital heart disease: implications for percutaneous pulmonary valve implantation. J. Cardiovasc. Magn. Reson. **9**, 687–695 (2007)
39. Schievano, S., Migliavacca, F., Coats, S., Khambadkone, L., Carminati, M., Wilson, N., Deanfield, J., Bonhoeffer, P., Taylor, A.: Percutaneous pulmonary valve implantation based

on rapid prototyping of right ventricular outflow tract and pulmonary trunk from mr data. Radiology **242**, 490–49 (2007)
40. Schneider, R.J., Perrin, D.P., Vasilyev, N.V., Marx, G.R., Del Nido, P.J., Howe, R.D.: Mitral annulus segmentation from 3d ultrasound using graph cuts. IEEE Trans. Med. Imaging **29**, 1676–1687 (2010)
41. Soncini, M., Votta, E., Zinicchino, S., Burrone, V., Mangini, A., Lemma, M., Antona, C., Redaelli, A.: Aortic root performance after valve sparing procedure: a comparative finite element analysis. Med. Eng. Phys. **31**, 234–243 (2009)
42. Staib, L.H., Duncan, J.S.: Model-based deformable surface finding for medical images. IEEE Trans. Med. Imaging **15**, 720–731 (1996)
43. Tenenbaum, J.B., de Silva, V., Langford, J.C.: A global geometric framework for nonlinear dimensionality reduction. Science **290**, 2319–2323 (2000)
44. Timek, T., Green, G., Tibayan, F., Lai, F., Rodriguez, F., Liang, D., Daughters, G., Ingels, N., Miller, D.: Aorto-mitral annular dynamics. Ann. Thorac. Surg. **76**, 1944–1950 (2003)
45. Tsymbal, A., Huber, M., Zhou, S.K.: Discriminative distance functions and the patient neighborhood graph for clinical decision support. Springer. Chapter Advances in Computational Biology, pp. 515–522 (2010)
46. Tu, Z.: Probabilistic boosting-tree: learning discriminative methods for classification, recognition, and clustering. ICCV **2**, 1589–1596 (2005)
47. Veronesi, F., Corsi, C., Sugeng, L., Mor-Avi, V., Caiani, E., Weinert, L., Lamberti, C., Lang, R.: A study of functional anatomy of aortic-mitral valve coupling using 3D matrix transesophageal echocardiography. Circ. Cardiovasc. Imaging **2**, 24–31 (2009)
48. Veronesi, F., Corsi, C., Sugeng, L., Mor-Avi, V., Caiani, E., Weinert, L., Lamberti, C., Lang, R.M.: A study of functional anatomy of aortic-mitral valve coupling using 3D matrix transesophageal echocardiography. Circ. Cardiovasc. Imaging **2**, 24–31 (2009)
49. Vitanovski, D., Ionasec, R.I., Georgescu, B., Huber, M., Taylor, A., Hornegger, J., Comaniciu, D.: Personalized pulmonary trunk modeling for intervention planning and valve assessment estimated from ct data. In: International Conference on Medical Image Computing and Computer-Assisted Intervention (MICCAI), London, USA, pp. 17–25 (2009)
50. Vitanovski, D., Tsymbal, A., Ionasec, R., Georgescu, B., Huber, M., Hornegger, J., Comaniciu, D.: Cross-modality assessment and planning for pulmonary trunk treatment using ct and mri imaging. In: International Conference on Medical Image Computing and Computer-Assisted Intervention (MICCAI), Beijing, China (2010)
51. Votta, E., Caiani, E., Veronesi, F., Soncini, M., Montevecchi, F., Redaelli, A.: Mitral valve finite-element modelling from ultrasound data: a pilot study for a new approach to understand mitral function and clinical scenarios. Philos. Transact. A Math. Phys. Eng. Sci. **366**, 3411–3434 (2008)
52. Waechter, I., et al.: Patient specific models for planning and guidance of minimally invasive aortic valve implantation. In: MICCAI 2010. Springer Berlin/Heidelberg. Volume 6361 of Lecture Notes in Computer Science, pp. 526–533 (2010)
53. Watanabe, N., Ogasawara, Y., Yamaura, Y., Kawamoto, T., Toyota, E., Akasaka, T., Yoshida, K.: Quantitation of mitral valve tenting in ischemic mitral regurgitation by transthoracic real-time three-dimensional echocardiography. J. Am. Coll. Cardiol. **45**, 763–769 (2005)
54. Webb, G.I.: Multiboosting: a technique for combining boosting and wagging. Mach. Learn. **40**, 159–196 (2000)
55. Yang, L., Georgescu, B., Zheng, Y., Meer, P., Comaniciu, D.: 3d ultrasound tracking of the left ventricle using one-step forward prediction and data fusion of collaborative trackers. In: IEEE Conference on Computer Vision and Pattern Recognition (2008)
56. Zheng, Y., Barbu, A., Georgescu, B., Scheuering, M., Comaniciu, D.: Four-chamber heart modeling and automatic segmentation for 3d cardiac ct volumes using marginal space learning and steerable features. IEEE TMI **27**, 1668–1681 (2008)
57. Zhou, S.K., Georgescu, B., Zhou, X.S., Comaniciu, D.: Image based regression using boosting method. ICCV **1**, 541–548 (2005)

58. Zhuang, X., Rhode, K.S., Razavi, R.S., Hawkes, D.J., Ourselin, S.: A registration-based propagation framework for automatic whole heart segmentation of cardiac mri. IEEE Trans. Med. Imaging **29**, 1612–1625 (2010)
59. Zhuang, X., Yao, C., Ma, Y.L., Hawkes, D., Penney, G., Ourselin, S.: Registration-based propagation for whole heart segmentation from compounded 3D echocardiography. IEEE pp. 1093–1096 (2010)

Patient-Specific Analysis of Blood Flow and Mass Transport in Small and Large Arteries

X. Y. Xu, N. Sun, D. Liu and N. B. Wood

Abstract Recent advances in medical image-based computational modelling have made it possible for patient-specific analysis of blood flow and mass transport in the circulation. In this chapter, we describe mathematical models for (1) flow and transport of macromolecules in large arteries, and (2) flow and oxygen delivery in an arteriolar network. Detailed examples are given of mass transport in specific cases, including applications to atherosclerotic arteries and microvascular network. The first example shows transport of a large molecule, low-density lipoprotein, in a human right coronary artery with a mild stenosis, whilst the second is about a small dissolved molecule, oxygen, in an abdominal aortic aneurysm illustrating the effect of intraluminal thrombus. Finally, the additional effects of non-Newtonian properties of blood and haemoglobin-bound oxygen on oxygen transport in the retinal arteriolar network of normal and hypertensive subjects are described. Patient-specific geometries obtained by means of a variety of imaging techniques are used in these examples.

1 Introduction

Patient-specific haemodynamic analysis has become increasingly common in recent years, usually aimed at testing hypotheses related to the role of haemodynamics in atherogenesis or in development and progression of arterial diseases (e.g. [19, 44, 54, 59, 84, 89, 94]), determining the effect of cardiovascular devices

---

X. Y. Xu (✉) · N. Sun · D. Liu · N. B. Wood
Department of Chemical Engineering,
Imperial College, London, UK
e-mail: Yun.xu@imperial.ac.uk

(e.g. [11, 46]), predicting and comparing outcomes of new and existing vascular interventional procedures [21, 76, 78, 85], and evaluating the effects of therapeutic drugs on arterial anatomy and the associated changes on flow and wall shear stress (WSS) [2]. In the former the influence of WSS, the frictional force exerted on the endothelium, was initially hypothesised by Caro et al. [10] and supported subsequently by in vitro or animal experimental data, as discussed by Malek et al. [37] and Slager et al. [57, 58], as well as by the understanding derived from molecular biology as reviewed by Davies [16] and Lehoux et al. [29].

The interaction of haemodynamics with the endothelium is not limited to its connection with atherogenesis. Its importance is much wider than that; their symbiosis involves health as well as disease. The range of agonists stimulated in the endothelium by low and normal or high WSS is large [37], but we could mention as an example the importance of just one substance, nitric oxide (NO). NO expression is promoted or depressed in reaction to high or low WSS respectively and, in the normal range is one of the main control mechanisms for haemodynamic resistance via arterial tone, being available through a range of reactions through the release of eNOS (endothelial nitric oxide synthase) from caveolae just inside the endothelial cell wall. The NO itself is transported sub-luminally to the intima, where smooth muscle cells respond by relaxing in the case of high WSS, allowing for increased demand for flow. It is also atheroprotective, whereas low and oscillating WSS in regions of disturbed flow is atherogenic, setting up a chain of events involved in atherogenesis, including ingress of low-density lipoprotein (LDL), lipid oxidation, recruitment of monocytes and leucocytes, leading to inflammation. Thus the accurate determination of WSS is important for understanding both atherogenesis and cardiovascular physiology in general; normal levels of WSS, amongst other actions, inhibit inflammatory activation and increases expression of antioxidant enzymes and eNOS [57]. WSS can also influence expression of nNOS (neural NOS) [32].

Initial trials investigating the feasibility of combining computational fluid dynamics (CFD) with medical imaging, originally using magnetic resonance imaging (MRI) led to demonstration of its successful implementation on healthy subjects (e.g. [35, 44, 88, 90]). The combination of medical imaging with CFD has become known as image-based modelling. It was also realised that image-based modeling would provide a practical means for determining WSS and, in the opinion of the authors, is superior to methods based on extrapolation to the wall of phase contrast MRI data, particularly in regions where velocity gradients at and near the wall are both large (high WSS) and variable, e.g. along flow dividers at bifurcations [38, 49]. Moreover, the detection and measurement of turbulent flow, which may occur in part of the circulation, is not yet routine in MRI [48], although there has been some success with Doppler ultrasound [26, 77]. However, with suitable models, physiologically relevant low Reynolds number turbulent flows can be simulated using RANS-based CFD (e.g. [69]), and large eddy simulation [70, 79]. The latter is expected to become more prevalent for simulating turbulent flow and WSS in complex geometries in the near future, having already achieved spectacular success in complex engineering flows.

Subsequently, image-based modeling using MRI for patient-specific geometry reconstruction was extended to include other imaging modalities, such as ultrasound [28, 56, 61] and computed tomography (CT) (e.g. [14, 86]). Thus, it was quickly realised that the methods developed would be useful for investigation of medical aspects of individual patients, and so patient-specific analysis was established and applied both in surgical planning (e.g. [85]), and in the understanding of patient-specific propensity to atherosclerosis (e.g. [89]), plaque vulnerability (e.g. [59, 72]), and aneurysm rupture (e.g. [80]). Although for many applications the arterial wall may be modelled as rigid, there are some where the elasticity of the wall is important, particularly in aortic aneurysms, where the wall becomes thin [6, 81]. If only the internal wall tensile stresses are required, it may be sufficient to simulate only the wall mechanics via finite element (FE) methods (e.g. [31]). However, sometimes the more complex simulation of fluid–structure interaction (FSI) may confer advantages for accurate determination of tensile stresses and/or fluid stresses in carotid stenosis [72] or aortic aneurysms [71]. In an example of turbulent flow in a thoracic aortic aneurysm, Tan et al. [71] found that better agreement with measured velocity profiles was achieved with FSI simulation which also incorporated a laminar-turbulent transition model. Comprehensive reviews of patient-specific modelling of arterial flow and cardiovascular mechanics can be found in Steinman [62] and Taylor and Figueroa [75].

While earlier studies using patient-specific modelling were focused on flow and related WSS parameters (e.g. [12, 35, 44, 54, 84]), patient-specific modeling has been refined and extended to include species transport. Because atherogenesis involves the ingress of, e.g. oxygen and LDL molecules into the wall, the simulation of mass transfer can offer important insights. Tarbell [74] gave an important review of arterial mass transport and its importance for atherogenesis. Following this lead, Sun et al. [65, 66] developed methods which could be applied in patient-specific studies. The transfer of oxygen, being a small molecule, is diffusive, whilst a large molecule like LDL is transported largely by the transmural flux. Moreover, WSS has an influence on both endothelial permeability and hydraulic conductivity. Sun et al. [65] developed the methodology for applying the shear-dependent transport mechanisms of LDL and oxygen initially in a single-layer wall model of an idealized stenosis in steady flow, showing the effects of the variation of WSS, which influences both permeability and hydraulic conductivity. The study was extended to the effect of transmural pressure on LDL transport in a multi-layered wall model in steady flow [66]. Subsequently the additional effects of pulsatile flow on LDL transport were demonstrated, and the methods required to deal with the additional complexity were introduced [67]. The methodology developed laid the basis for patient-specific studies: LDL and albumin transport in a single-layer wall model in a patient-specific RCA derived from a CT scan [64], and oxygen transport in an abdominal aortic aneurysm with intra-luminal thrombus [63]. It was found that the presence of thrombus could significantly impede oxygen transfer from the lumen to the aortic wall; even a 5 mm thick thrombus would be sufficient to cause hypoxia in the aortic wall. Subsequently, the influence of the release of haem-bound oxygen, previously neglected, on its transport to the wall of patient-specific retinal arteries

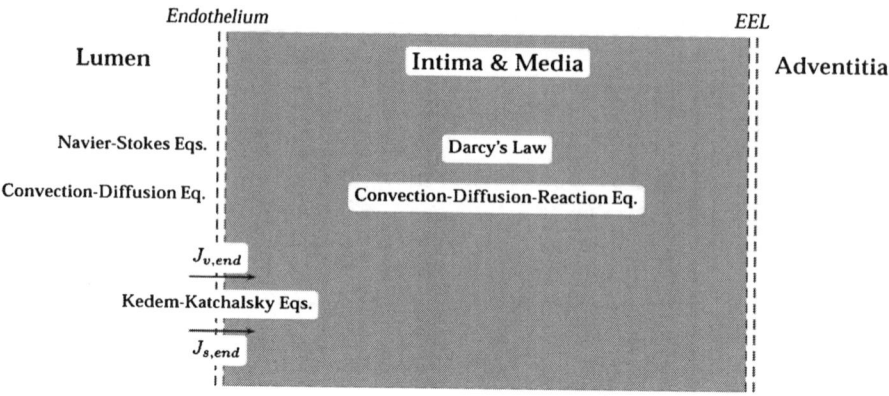

Fig. 1 A schematic diagram of a single-layered fluid-wall mass transport model

was reported by Liu et al. [34], showing how its effects contributed to the distribution of the dissolved oxygen and its diffusion into the wall. Detailed discussions of these patient-specific examples are given later in this chapter.

2 Computational Models

The transport of macromolecules in arteries is coupled with the bulk blood flow in the lumen and transmural flow in the arterial wall. Hence, fluid dynamics equations and solute transport equations are required for both the fluid domain (lumen) and the tissue domain (wall). In the context of arterial mass transfer, the wall usually refers to the intima and media, with the adventitia being considered as an outer boundary. A schematic diagram of a fluid-wall mass transport model is shown in Fig. 1. Details of the governing equations and boundary conditions are described in the following section.

2.1 Governing Equations for Flow and Mass Transfer

The laminar motion of pure fluids is completely described by the Navier–Stokes equations, which express Newton's law of the conservation of momentum in the fluid as it passes through a control volume and are combined with the equation of conservation of mass, otherwise known as the continuity equation. They are given below, in compact vector form, for an incompressible Newtonian fluid

$$\rho \frac{\partial u_l}{\partial t} - \mu \nabla^2 u_l + \rho(u_l \cdot \nabla)u_l + \nabla p_l = 0 \qquad (1)$$

$$\nabla \cdot u_l = 0 \qquad (2)$$

where subscript l is used to represent the lumen when applied to blood flow, u_l is the instantaneous fluid velocity vector at time t, p_l is the instantaneous static pressure relative to a datum level, and ρ_l and μ_l are respectively the fluid density and viscosity. Of course, blood is not a Newtonian fluid, in that the viscosity depends on local shear rate, which is incorporated in one of our examples given later.

Mass transfer in the lumen is described by the convection–diffusion equation as follows

$$\nabla \cdot (-D_l \nabla c_l + c_l u_l) = 0 \qquad (3)$$

where c_l is the solute concentration in the lumen, and D_l is the solute diffusivity in the lumen. This equation is coupled with the fluid flow governed by Eqs. 1 and 2.

For the tissue phase, the transmural flow in the arterial wall can de described by the Darcy's law

$$u_w - \nabla \cdot \left(\frac{\kappa_w}{\mu_p} p_w \right) = 0 \qquad (4)$$

$$\nabla \cdot u_w = 0 \qquad (5)$$

where subscript w represents the wall, u_w is the velocity vector of the transmural flow in the wall, p_w is the pressure in the wall, μ_p is the viscosity of blood plasma, and κ_w is the Darcian permeability coefficient of the wall. The coupled mass transfer in the wall can be described by the convection–diffusion-reaction equation as follows

$$\nabla \cdot (-D_w \nabla c_w + K_{\text{lag}} c_w u_w) = r_w c_w \qquad (6)$$

where c_w is the solute concentration in the arterial wall, D_w is the effective solute diffusivity in the arterial wall, K_{lag} is the solute lag coefficient, r_w is the consumption rate constant.

At the fluid-tissue interface (endothelium), the volume flux and solute flux are defined by the Kedem–Katchalsky equation [25] as follows

$$J_{v,\text{end}} = L_{p,\text{end}} (\Delta p_{\text{end}} - \sigma_{d,\text{end}} \Delta \pi_{\text{end}}) \qquad (7)$$

$$J_{s,\text{end}} = P_{\text{end}} \Delta c_{\text{end}} + (1 - \sigma_{f,\text{end}}) J_{v,\text{end}} \bar{c}_{\text{end}} \qquad (8)$$

where $J_{v,\text{end}}$ is the volume flux, $J_{s,\text{end}}$ is the solute flux, $L_{p,\text{end}}$ is the hydraulic conductivity of the endothelium, Δc_{end} is the solute concentration difference across the endothelium, Δp_{end} is the pressure drop across the endothelium, $\Delta \pi_{\text{end}}$ is the osmotic pressure difference across the endothelium, $\sigma_{d,\text{end}}$ is the osmotic reflection coefficient of the endothelium, $\sigma_{f,\text{end}}$ is the solvent reflection coefficient of the endothelium, P_{end} is the solute permeability of the endothelium, $\bar{c}_{\text{end}}$ is the mean interface concentration. In the cases described later, the osmotic pressure difference $\Delta\pi_{end}$ has been neglected to decouple the fluid dynamics from solute dynamics.

Finally, suitable boundary conditions need to be specified in order to obtain numerical solutions to the above equations. These include, for the fluid domain in the lumen, a given velocity and concentration profile at the inlet; a given pressure (either constant or time-varying depending on available data) and zero diffusive flux condition at the outlet. For the tissue domain in the wall, zero normal velocities can be assumed at the side boundaries together with an insulated condition for the transport equation, at the outer wall boundary (i.e. adventitia), a constant pressure along with a fixed concentration can be prescribed. Boundary conditions at the interface (endothelium) are provided by the Kedem–Katchalsky equations given above. Choice of model parameters and implementation of the basic modeling equations will be discussed in the specific cased presented later.

2.2 Imaging for Patient-Specific Model Reconstruction

The first requirement in patient-specific analysis is imaging of the region of interest under investigation. Usually the modality of interest or choice would be MRI, since it is safe and can provide both vessel anatomical (MR angiography, MRA) and velocity data (generally phase contrast—PC-MRI) for computer simulation. The available MR hardware advances over time and the MR sequences available are advancing continuously, particularly in anatomical imaging, so a sequence used in one investigation might have been replaced by the time another one is launched. This is not the place to describe MR in detail, since it is the most complex imaging modality. MR sequences involve the timing of a simultaneous set of transitory magnetic field gradients and radio frequency signals, which set parameters such as slice selection, navigation within the slice and 'flipping' of the spin axis of hydrogen protons in the blood or tissue. The size of the 'flip angle' determines attributes like the speed of acquisition; various refinements over the years have both reduced acquisition time and increased both temporal and spatial resolution [9, 41]. One sequence that has remained useful in MRA is 2D time of flight, which gives good contrast between blood and tissue. Multi-contrast weighted imaging has allowed definition of arterial wall components, particularly in atherosclerotic plaques [93].

Phase contrast velocity imaging makes use of transient positive and negative changes to the scanner's magnetic field in which moving hydrogen protons in the blood acquire different changes of phase in their spin from the fixed, or more slowly moving protons in the surrounding tissue. The change of phase is proportional to the velocity of the protons and hence of the substance being tracked, in this case blood and wall. The difference between them gives the blood velocity and the resolution is such as to give profiles of velocity components across the vessel, and in other directions if required [36].

The other 'non-invasive' modality, in the sense that it involves no harmful ionizing radiation, is ultrasound. Generally, B-mode ultrasound is the basis of anatomical imaging, whilst Doppler ultrasound is required for velocity acquisitions. Anatomical imaging depends on differences in ultrasonic impedance between tissues, such that at a boundary between them, ultrasound waves are partly reflected and partly transmitted. Depending on the speed of sound in each tissue, the timing of arrival of the reflected waves allows determination of the distance of the boundary from the transmitter. Thus, an image-based on these boundaries can be constructed [50]. Artery wall thickness, at least intima and media, can be detected and imaged with ultrasound (e.g. [3, 21]. Intra-vascular ultrasound (IVUS) is a newer invasive B-mode introduction, in which the ultrasound probe is mounted on a catheter, and has found application in coronary wall imaging [47, 56].

Doppler ultrasound makes use of the 'Doppler shift', i.e. the frequency change when the ultrasound signal is reflected from a moving target and will be positive or negative for targets moving towards or away from the target respectively. Neither PC-MRI nor Doppler ultrasound can give velocities very close to the wall, owing to artifacts, but MR gives better definition overall.

CT is an X-ray modality and, as such, involves ionizing radiation. Therefore, clinical necessity generally determines whether it may be used in individual circumstances. The CT scanner comprises rotating X-ray source or sources located opposite an array of detectors. The patient lies on a table, which is advanced through the rotating assembly, producing a spiral path of the X-ray source relative to the patient. As with MR, which is also a tomography modality, deconvolution of the received signals requires major computing facilities, which are part of the scanner assembly [22, 23]. As with MR, advances in CT capability, with multi-slice and rotating multi-detector array technology have increased spatial and temporal resolution. CT offers greater spatial resolution than MR, but generally lower temporal resolution. The blood-arterial wall boundary is clearly delineated, but wall thickness cannot be distinguished.

3 Specific Examples

3.1 LDL Transport in a Human Right Coronary Artery

The right coronary artery (RCA) originates from the aortic root and supplies the myocardium with oxygen and nutrients. As a large-medium sized artery, the diameter of the RCA is normally around 3–4 mm. Coronary artery atherosclerosis is responsible for acute coronary syndrome, myocardial infarction and sudden cardiac death. The early stages of the disease involve the accumulation of macromolecules, such as LDL, in the vessel wall. In this example, pulsatile flow in a patient-specific RCA coupled with LDL transport was simulated using the fluid-wall model

Fig. 2 Reconstructed geometry of a human right coronary artery based on CT images

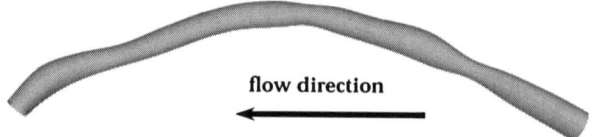

described in the preceding section. The RCA model was reconstructed based on multi-slice CT images acquired from a patient who had a clinically identified stenosis in his RCA, and a realistic flow waveform was applied as the upstream boundary condition. Experimental in vitro data [55] were used to derive the shear-dependent model for endothelial hydraulic conductivity, which is given by

$$L_p(|\tau_w|) = 0.392 \times 10^{-12} \ln(|\tau_w| + 0.015) + 2.7931 \times 10^{-12} \qquad (9)$$

where τ_w is the WSS. The relationship indicates that endothelial hydraulic conductivity increases with increasing WSS. Owing to the lack of suitable experimental data, a constant LDL permeability was assumed. Other model parameters for LDL transport include: diffusivity in the lumen $D_l = 2.867 \times 10^{-11}$ m²/s, effective wall diffusivity $D_w = 3.5 \times 10^{-12}$ m²/s, endothelial permeability $P_{end} = 4.84 \times 10^{-9}$ m/s, and consumption rate constant $r_w = -6.05 \times 10^{-4}$ 1/s. Values for D_w, P_{end} and r_w were determined using an optimization approach described by Sun et al. [66].

Figure 2 shows the reconstructed lumen surface of the RCA, corresponding to mid-diastole (75% into the duration of the cardiac cycle). The lumen inlet diameter (hydraulic diameter) of the RCA was 3.8 mm and the length of the RCA segment was about 75 mm. It has a mild stenosis (62% area reduction) near the proximal end, and two less obvious local constrictions with one in the middle and another towards the distal segment of the vessel. In the computational model, both the inlet and outlet were extended by five diameters in length to allow realistic velocity profiles to develop in the region of interest and to minimize the influence of artificial outlet boundary condition. Since the intima-media thickness cannot be resolved by CT, it was assumed to be 300 μm [24] throughout the vessel. The wall region was reconstructed by extruding the lumen surface in the normal direction by the assumed wall thickness. Here, the RCA wall was assumed to be rigid and its cardiac-induced dynamic motion was neglected.

To solve the system of partial differential equations, adequate boundary conditions need to be applied. At the inlet of the extended RCA, Womersley velocity profiles corresponding to a realistic coronary velocity waveform obtained using an electrocardiogram gated catheter delivered ultrasound Doppler probe (ComboWire, Volcano Co., Rancho Cordoval were applied. Based on this waveform, the mean inlet velocity was 0.1613 m/s and the resulting mean Reynolds number was 184. The Peclet number for LDL transport in the lumen was 2.14×10^7. The inlet concentration of LDL c_0 was assumed a normalised value 1, whease at the outer wall boundary, a constant concentration, $c_{adv}/c_0 = 0.016$, was prescribed [43]. The computational mesh consisted of 109,824 hexahedral elements in the fluid domain for solving the Navier–Stokes equations and 86,112 s order hexahedral elements with 704,965 nodes for solving the convection–diffusion equation. The tissue domain was subdivided into 66,240 hexahedral elements for solutions of Darcy's law and the convection–diffusion-reaction equation. The mesh density adopted was deemed sufficient to resolve the mass transfer boundary layer and to ensure grid independence of the computed concentration field. A fixed time step size of 0.002 s was chosen for the pulsatile flow simulation using an implicit time-marching scheme.

In this example, the Navier–Stokes equations were solved by means of a commercial CFD package, ANSYS CFX 10.0 (ANSYS, Inc., USA), while the rest of the model was implemented using Comsol Multiphysics 3.3 (Comsol AB, Sweden). One-way coupling of the two codes was achieved offline, which is sufficient for this problem since transmural flow in the wall and solute transport have little influence on the bulk flow in the lumen. To avoid potential loss of computational accuracy due to interpolation, the same mesh division was used in both codes.

Shown in Fig. 3 are contours of velocity magnitude and WSS in the RCA for steady flow (a) and pulsatile flow at different time points (b–f). The steady flow results are included here for comparison purpose. From the velocity contours, it is clear that flow is skewed towards the outer curvature which rotates along the vessel. Also, flow separation can be observed downstream of the stenosis near the proximal end of the RCA. The flow features are similar between the steady and pulsatile flow simulations, except that regions of slow flow and flow recirculation expand and contract with time during a pulse cycle. This is reflected in the instantaneous WSS plots where the sizes of low WSS zones change with time. Nevertheless, two distinct low WSS zones can be identified: one is located downstream of the stenosis near the proximal end, and the other is at the distal end of the vessel segment.

Based on the predicted instantaneous WSS values, shear-dependent endothelial hydraulic conductivities $L_{p,end}$ were calculated. As shown in Fig. 4, time-averaged shear-dependent hydraulic conductivity has a similar pattern to WSS with two distinct regions of low $L_{p,\,end}$ in correspondence to low WSS. Using the computed hydraulic conductivities, transmural flows were simulated. At a transmural pressure of 120 mm Hg, the predicted transmural velocities $J_{v,end}$ (see Fig. 5) are within the range of 2.0–3.5×10^{-8} m/s, which are consistent with experimental data on filtration velocities under the same transmural pressure [43]. It can be observed that endothelial hydraulic conductivity has a strong influence on transmural velocity: in the low $L_{p,\,end}$ regions, $J_{v,\,end}$ is considerably lower than that in other regions, indicating that the endothelium makes a major contribution to the overall resistance to transmural flow.

Using the fluid dynamics results discussed above, LDL transport was simulated. Contours of LDL concentration at the sub-endothelial surface (Fig. 6) show that LDL concentration is relatively high at locations where WSS is low, implying a focal accumulation of LDL in low WSS region. This is caused by a weaker

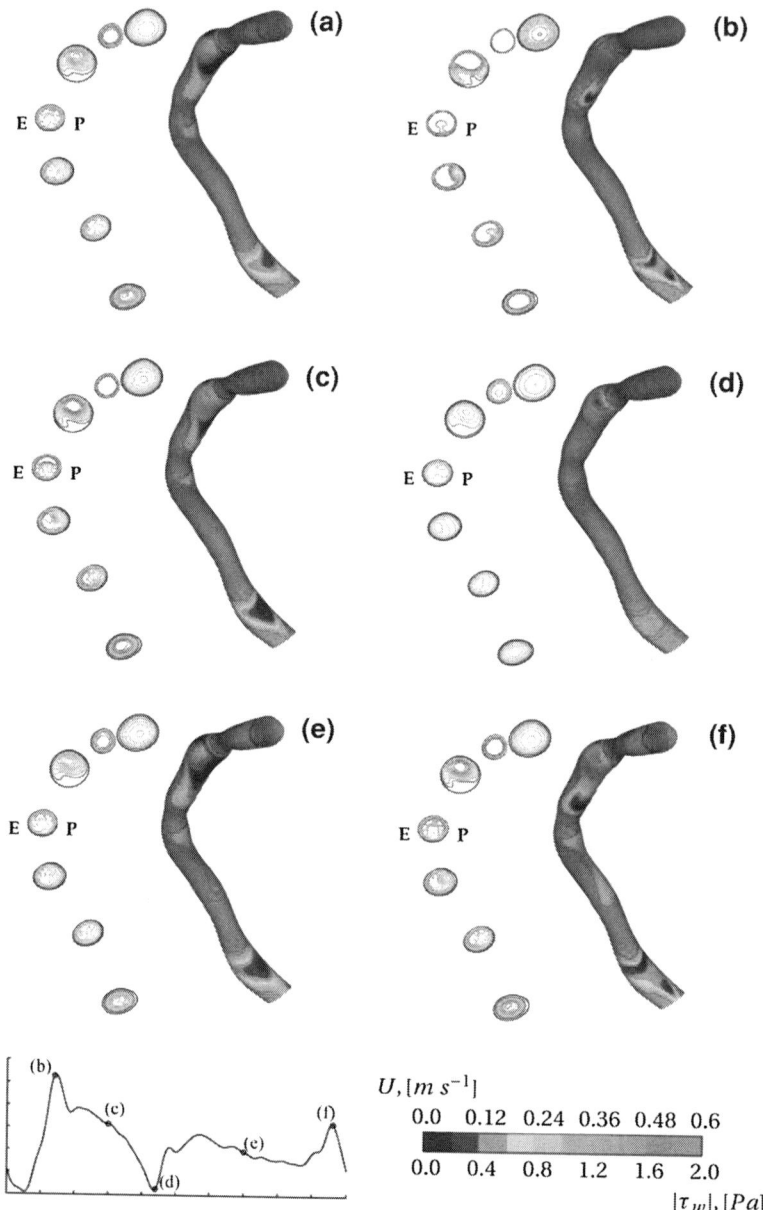

Fig. 3 Contours of velocity and WSS magnitude calculated from the steady flow simulation (**a**) and pulsatile flow simulation at different time points (**b–f**) in a human right coronary artery. E and P stand for epicardial and pericardial sides, respectively

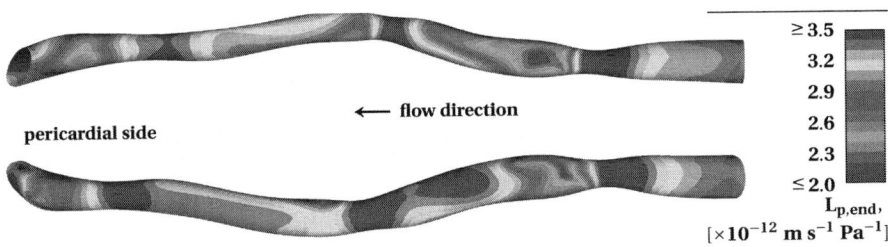

Fig. 4 Contours of endothelial hydraulic conductivity on the lumen surface of the right coronary artery

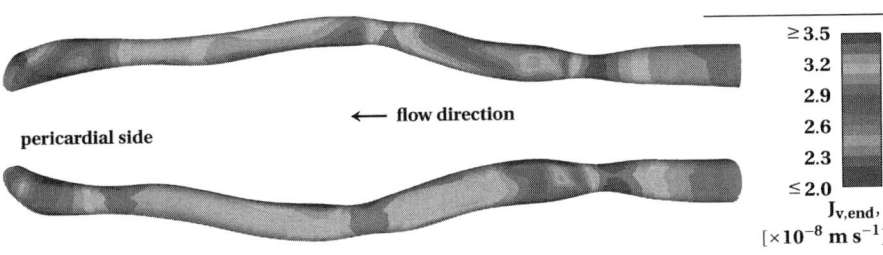

Fig. 5 Contours of transmural velocity in the right coronary artery at transmural pressure of 120 mm Hg

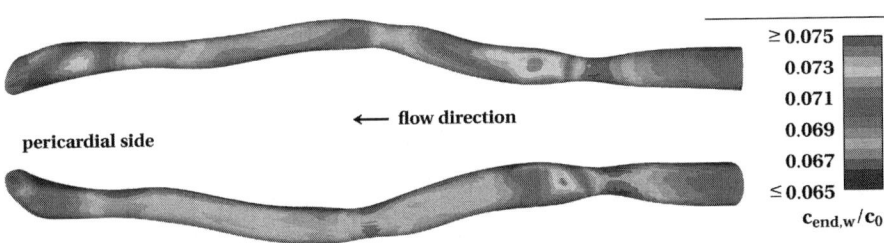

Fig. 6 Contours of LDL concentration at the sub-endothelial surface of the right coronary artery

convective clearance effect of the transmural flow in these areas. LDL transport in the arterial wall is convection-dominated and driven by the transmural flow. Owing to reduced transmural flow in the low WSS regions, convection of LDL from the sub-endothelial layer to the outer wall layer is weakened, resulting in local accumulation of LDL. On the other hand, the reduced transmural flow is caused by low hydraulic conductivity when the endothelium is subjected to low WSS. Analysis of the chain of events highlights that the endothelium plays an essential role in regulating arterial mass transport. It is worth noting that in the

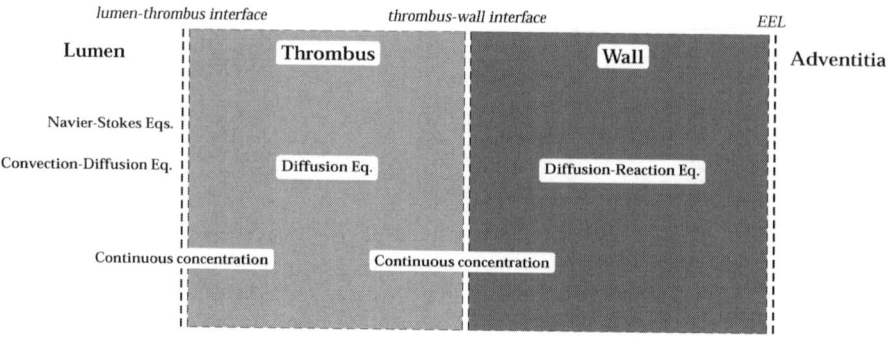

Fig. 7 A schematic diagram of the model formulation for oxygen transfer in a human abdominal aortic aneurysm

example shown here, a constant endothelial LDL permeability was assumed, and consequently, variation in trans-endothelial LDL flux was neglected. This also explains why the simulation results only showed a moderate elevation (10–15%) of LDL concentration in low WSS regions. If LDL permeability were to have similar shear-dependent characteristics to that of albumin which was reported by Kudo et al. [27], a much more significant LDL concentration elevation in the sub-endothelial layer would be expected in the low WSS regions. Sun et al. [64] investigated albumin transport in the same RCA geometry and found that albumin concentration in the sub-endothelial layer was around 100–150% higher in regions of low WSS when both the endothelial hydraulic conductivity and permeability were shear-dependent. Clearly, further experimental studies examining the effect of WSS on LDL permeability will be required in order to refine and improve the shear-dependent transport properties used in arterial mass transport models.

3.2 Oxygen Transport in an Abdominal Aortic Aneurysm

In this example, oxygen transport in a patient-specific abdominal aortic aneurysm (AAA) reconstructed from CT images was investigated using a fluid–thrombus–wall model that accounts for convection–diffusion in the lumen, diffusion in the thrombus and diffusion–reaction in the wall. The example differs from the first one in that: (1) oxygen transfer in the tissue domain is dominated by diffusion, hence convection can be ignored, and (2) an additional layer representing the introluminal thrombus (ILT) needs to be included in the model. A schematic diagram of the model formulation is given in Fig. 7. Briefly, blood flow was assumed to be steady, incompressible, laminar, Newtonian and hence described by the Navier–Stokes equations, whilst oxygen transfer in the lumen was coupled with the blood flow and governed by the convection–diffusion equation. Although oxygen in blood is carried both as a dissolved species and haem-bound in erythrocytes, the effect of oxygen binding to haemoglobin on oxygen concentrations in the lumen was neglected, as the focus here is on comparisons of relative transport over the aneurysm wall. For oxygen transfer in the arterial wall, the diffusion–reaction equation was applied. For trans-thrombus transport, it was assume that there was no significant plasma flow in ILT, which is justifiable due to the lack of driving force, hence oxygen transfer in ILT was described by the diffusion equation only. Since it is generally accepted that the formation of ILT causes, or is consequent upon, some degree of endothelium dysfunction [68], mass transfer resistance imposed by the endothelium can be ignored.

The AAA geometry was reconstructed from CT images by Leung [30]. Figure 8 shows a sample CT image and the reconstructed lumen and thrombus geometry of the AAA. It can be noticed that there is a local constriction at the proximal end of the aneurysm expansion, and the thrombus thickness is non-uniformly distributed, with the distal part of the AAA having the thickest thrombus (≥ 20 mm). The lumen diameter at the inlet is 16 mm and the length of the original aortic segment is approximately 120 mm. The intima–media thickness could not be measured from the CT images and was assumed to be 0.5 mm. The inlet and outlet of the reconstructed AAA model were extended by five diameters on each side (extensions not shown here) to minimise the effect of artificially imposed inlet and outlet boundary conditions.

Oxygen diffusivity was assumed to be $D_l = 1.6 \times 10^{-9}$ m^2/s in the lumen [53] and $D_w = 1.08 \times 10^{-9}$ m^2/s in the wall [7]. In addition, it was assumed that oxygen diffusivity in the thrombus was between D_w and D_l because ILT was found to be more porous than the wall [1]. Therefore, a mean value for oxygen diffusivity in the thrombus $D_t = (D_l + D_w)/2$ was chosen, and a sensitivity study was carried out to assess the influence of this parameter. The first-order oxygen consumption rate in the aortic wall was specified as $r_w = -4.89 \times 10^{-2}$ 1/s, which was obtained by solving a reverse one-dimensional diffusion–reaction problem using the corresponding data on oxygen partial pressures reported by Buerk and Goldstick [7]. The coupled fluid–thrombus–wall model for oxygen transfer was implemented by using a commercial FE package, Comsol Multiphysics 3.3 (Comsol AB, Sweden).

Since the flow was assumed to be steady and laminar in this example, a parabolic velocity profile with a mean velocity of 0.15 m/s was imposed at the extended inlet. This gave an inlet Reynolds number of 720, which is representative for flow in the abdominal aorta under resting condition. The corresponding Peclet number was 1.51×10^6. The inlet oxygen concentration was assigned a normalised value of 1, while a normalised value of 0.61 was specified at the outer wall boundary (media–adventitial interface). The latter was calculated based on oxygen tensions of 90 mm Hg at the inlet and 55 mm Hg in the adventitia, assuming a normal supply of oxygen to the outer layers of the aortic wall by vasa vasorum [7]. To investigate the importance of normal oxygen supply by the vasa vasorum, a sensitivity study was carried out by lowering the adventitial oxygen tension to 0.2 and 0.4 (normalised by the inlet value), respectively.

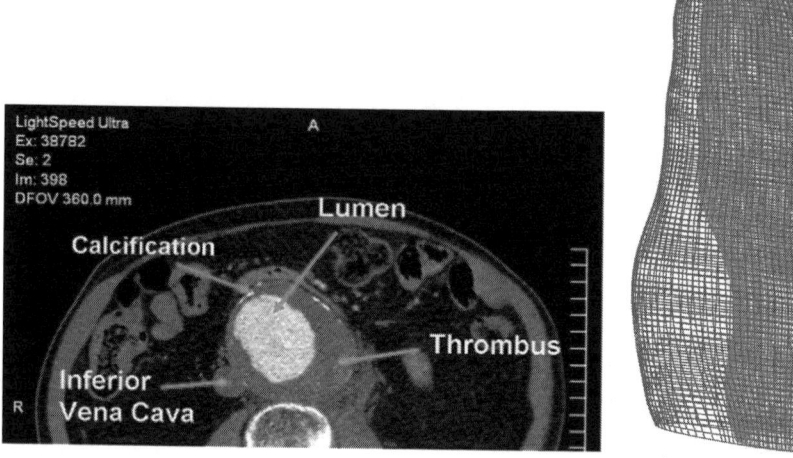

Fig. 8 A sample CT image (cropped) to show the abdominal aortic aneurysm (AAA) (*left*) and posterior view of the reconstructed AAA. The lumen surface is shown in *red* and the thrombus-wall interface in *grey*

The predicted flow pattern in the AAA is presented in Fig. 9 which shows the velocity magnitude contours and streamlines. It can be observed that the flow is skewed towards the posterior wall following the curvature and a local constriction at the proximal side of the AAA (refer to Fig. 8). This geometrical feature also results in a highly disturbed and swirling flow on the anterior side of the expansion, which is shown clearly by the streamline plot.

Figure 10 shows the oxygen concentration contours at the lumen–thrombus interface and the thrombus-wall interface. It is clear that oxygen concentrations on the luminal surface are high and vary in a narrow range (0.90–1.0). This is because ILT serves as an additional resistance to oxygen transport, thereby reducing oxygen efflux from the lumen. However, oxygen concentrations at the thrombus–wall interface are very low, with majority of the surface being covered by normalisaed oxygen concentrations below 0.1, compared with ≥ 0.9 at the lumen–thrombus interface. This means that the thrombus has caused an almost 80% reduction in oxygen concentration, and that a large proportion of the AAA wall is subject to some degree of hypoxia. The results demonstrate that the presence of ILT can significantly impede oxygen transport from the aortic lumen to the wall, thereby reducing oxygen concentration in the inner layer of the wall. Even if the haem-bound oxygen were included in this case, the effect of ILT

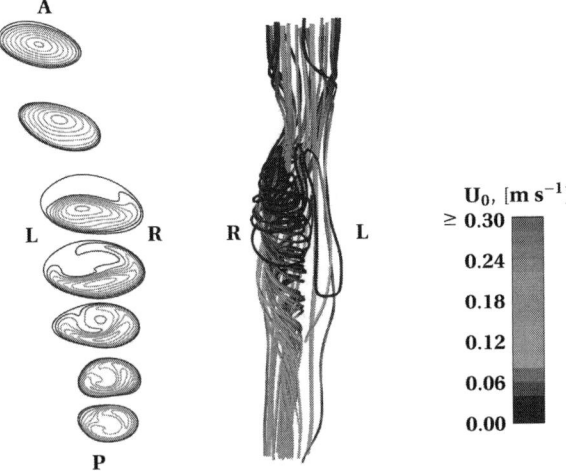

Fig. 9 Contours of velocity magnitude along the axial direction (*left*) and streamlines (*right*) in a human abdominal aortic aneurysm

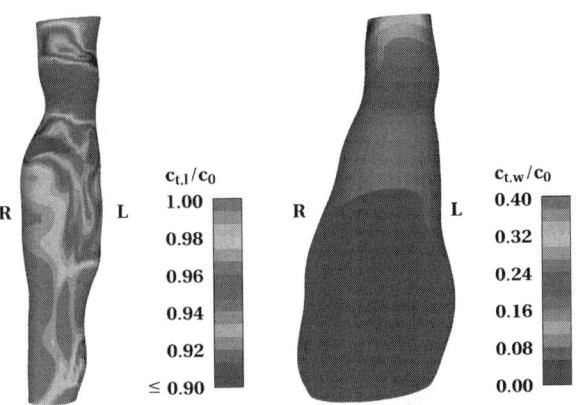

Fig. 10 Normalised oxygen concentration distributions at the lumen-thrombus interface (*left*) and thrombus-wall interface (*right*) on the anterior side of the abdominal aortic aneurysm model

would still be the same, because there would be little release of oxygen from the haemoglobin in blood.

To examine oxygen transfer within the wall and how this may be affected by ILT thickness, cross-sectional oxygen wall concentration profiles are shown in Fig. 11 at three different locations with ILT thickness of 0.7, 6.0 and 20.6 mm, respectively. It is clear that the highest oxygen concentration is at the outer wall where a normalised value of 0.61 was imposed as the outer wall boundary condition. It is interesting to note that wall concentration profiles at locations with ILT thickness of 6.0 and 20.6 mm differ only slightly, despite the large difference in ILT thickness; both show the lowest concentration within close proximity to the inner wall (thrombus-wall interface). But wall concentration profile at a location with ILT thickness of 0.7 mm is very different, with a much

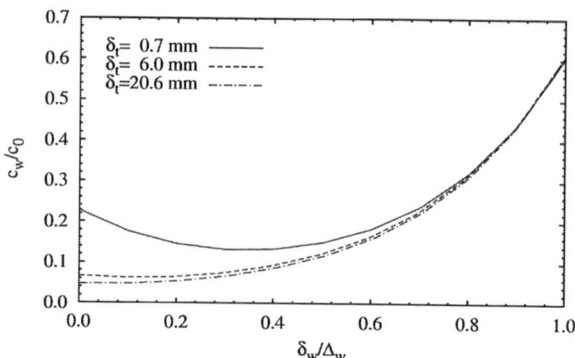

Fig. 11 Cross-sectional profiles of normalized oxygen concentration in the wall of a human abdominal aortic aneurysm at three different locations with thrombus thickness of 0.7 mm (*solid line*), 6.0 mm (*dashed line*) and 20.6 mm (*dashed dotted line*)

higher concentration level at the inner wall and the lowest concentration in the middle of the wall. These results suggest that oxygen required in the inner wall region is mainly provided by the lumen side if there is no thrombus or if the thrombus is very thin; but there exists a critical thrombus thickness, beyond which oxygen supply from the lumen is no longer sufficient to meet the demands of the wall. When this happens, oxygen concentration in the wall will depend on the normal function of vasa vasorum. Unfortunately, oxygen supplied by the vasa vasorum is mostly consumed within the wall before reaching the inner wall layer, causing hypoxia in some regions of the inner wall in AAA.

Sun et al. [63] found an inverse and non-linear relationship between oxygen concentration at the thrombus–wall interface and ILT thickness. They also determined the effects of oxygen diffusivity in the thrombus and oxygen concentration in the adventitia, and found that the predicted oxygen concentration in the wall was almost insensitive to oxygen diffusivity in the thrombus but highly sensitive to oxygen concentration in the adventitia.

3.3 Oxygen Transport in the Retinal Arteriolar Network

The retina, lying at the posterior fundal surface of the eye, has a high oxygen demand [82], and this makes it particularly vulnerable to vascular insults impairing oxygen and nutrient supply. Retinal vascular anatomy and network structure are adversely affected by systemic diseases, such as hypertension and diabetes. Hypertension is known to cause an increase of the length/diameter ratio of individual vessels and a narrowing of the bifurcation angles, resulting in potentially disadvantageous blood flow patterns [60]. Diabetic eye disease is one of the commonest causes of blindness in the UK. A number of studies have shown that generalized arteriolar narrowing and retinopathy are associated with increased risk of stroke, ischemic heart disease, heart failure, renal dysfunction, and cardiovascular mortality [87]. Therefore, in order to improve our understanding of the link

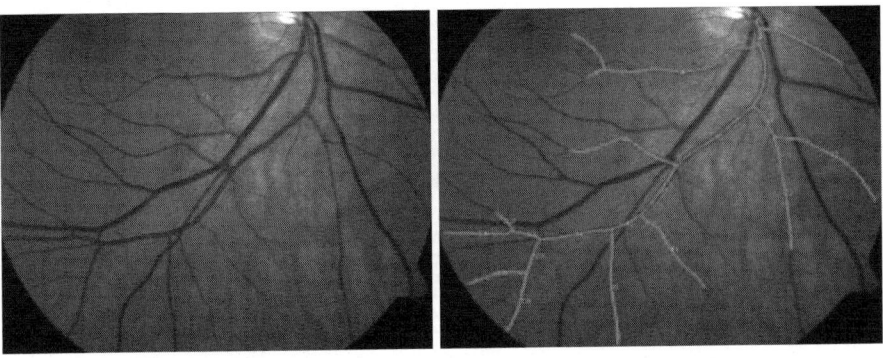

Fig. 12 A sample retinal image captured by a Zeiss FF450 + fundus camera (*left*) and the same image with arteries marked in red (*right*)

between systemic diseases and the retinal circulation, it is necessary to develop an approach to quantitatively determine blood flow and oxygen transport parameters in the retinal vascular network. This example differs from the previous examples in that it incorporates not only the effect of haemoglobin-bound oxygen but also the non-Newtonian properties of blood in microvessels.

In this example, two-dimensional models of the retinal arterial tree were reconstructed from retinal images acquired using a Zeiss FF450 + fundus camera (30° field of view). A sample original image and the corresponding arterial tree are shown in Fig. 12. Vessel segmentation included extraction of all vessels from the background and distinguishing arteries, which are thinner and lighter, from veins that were then erased from the image [39]. Owing to the limited imaging resolution and field of view, only a partial and visible retinal arteriolar network could be reconstructed from the digital image. To compensate for the lack of information on smaller peripheral vessels, asymmetric structured trees with self-similar binary bifurcations were generated and attached to each of the visible network outlets to represent smaller arterioles indistinguishable from the image. For the fractal tree, morphological parameters describing the asymmetric relationships between the radii of the parent and daughter vessels, such as junction exponent and asymmetry ratio, as well as vessel length to diameter ratio were required. These were derived from measurements made from the reconstructed retinal arterial network.

The retinal blood flow was assumed to be steady and governed by the Navier–Stokes equations which were coupled with the convection–diffusion equation for oxygen transfer in the arterioles. Since the diameters of the retinal arteries are less than 110 μm, the assumption of blood as an ideal homogenous Newtonian fluid would be invalid owing to the complex characteristics of blood, and its particulate nature has to be considered. In microcirculation, the apparent viscosity of blood depends on both haematocrit and vessel diameter, known as the Fahraeus-Lindqvist effect [51]. To account for this effect, the empirical viscosity model proposed by Pries et al. [52] was adopted, together with a constant haematocrit value of 0.45 owing to limited data on human retinal vessels [20]. The blood density was assumed to be 1050 kg/m³ and the walls were rigid.

For oxygen transfer, the reversible dynamic reaction between the oxygen carrier, haemoglobin, and free oxygen in the plasma, was incorporated by including both the free oxygen dissolved in plasma and the oxygen carried by oxyhaemoglobin. Coupled with the blood flow, the modified convection–diffusion equation is given by [34],

$$V_b \cdot \nabla P_b = \frac{\nabla D_p \left(1 + \frac{D_{Hb}}{D_p} \frac{C_{Hb} H_D}{\alpha_b} \frac{dS}{dP_b}\right) \nabla P_b}{1 + \frac{C_{Hb} H_D}{\alpha_b} \frac{dS}{dP_b}} \quad (10)$$

where P_b is the oxygen partial pressure in blood, α_b is oxygen solubility coefficient, C_{Hb} is the oxygen carrying capability of haemoglobin in blood, D_p is the free oxygen diffusion coefficient in plasma, and D_{Hb} is the diffusivity of oxyhemoglobin in red blood cells; the latter has been interpreted as a constant shear-augmented dispersion coefficient of a red blood cell [45]. H_D is the haematocrit, V_b is the velocity of blood, and S is the oxyhaemoglobin saturation function described by the Hill equation,

$$S = P_b^n / (P_b^n + P_{50}^n) \quad (11)$$

where P_{50} is the half partial pressure of oxygen saturation in haemoglobin (sO2) and n is the Hill exponent.

The Navier–Stokes equations together with the convection–diffusion equation described above were solved numerically via the commercial finite volume CFD solver, ANSYS CFX, supplemented by our own subroutines for the additional models incorporated here. Again, suitable boundary conditions are required to solve these equations and these include flow conditions and oxygen concentration at the inlet, as well as treatment of the outlets and the wall.

At the model inlet, a mean velocity of 7 cm/s was specified together with a flat velocity profile. This was based on color Doppler ultrasound measurements of blood velocities in the central retinal artery in healthy young subjects [18, 42]. The assumption of a flat velocity profile was reasonable since a fully developed profile could be established over a length of less than one diameter for the low Reynolds number considered here. For the mass transfer equation, a uniform oxygen partial pressure of 100 mm Hg (corresponding to oxygen saturation of 97.28% based on Eq. 11) was applied. This was derived from in vivo measurements in healthy volunteers which showed that oxygen saturation in retinal arterioles of around 100 μm in diameter was 97% [4, 17]. At the wall boundary, a continuous oxygen flux across the lumen wall interface was assumed, together with a constant permeability and wall-side oxygen partial pressure, P_w, which was set at 40 mm Hg according to experimental measurement of oxygen tension in the retinas of cat [8, 33] and other animals [5, 15, 92]. Assuming a constant P_w is a major simplification which was discussed by Liu et al. [34].

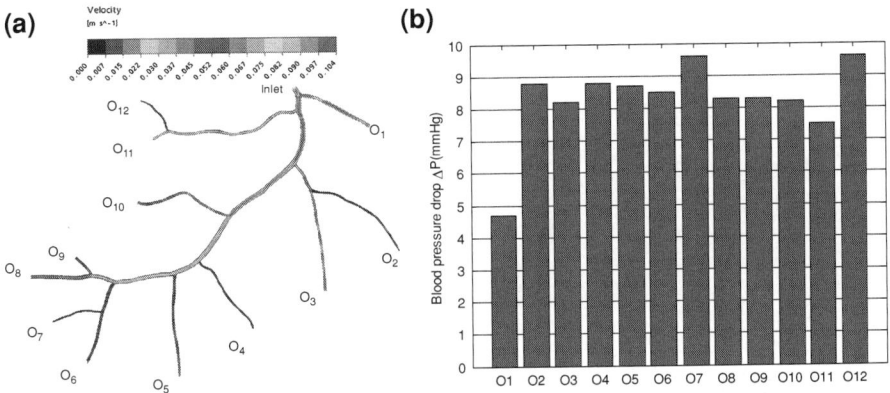

Fig. 13 Predicted velocity distribution (**a**) and pressure drop (**b**) in the retinal arterial tree of a normal subject

At the model outlets, whose diameters were in the range of 19–75 μm, it was assumed that retinal arterioles with outlets smaller than 30 μm in diameter would reach an equal pressure. Therefore, all the outlets in the reconstructed network with diameters greater than 30 μm were connected to an asymmetric structured tree generated for each outlet to represent its downstream vascular bed. The structured tree extended until the mean diameter of the major daughter vessel was <30 μm. The total resistance of each structured tree was calculated and pressure, which was derived from the total resistance of the fractal tree and corresponding volumetric flow, was specified as outflow boundary condition at each outlet.

Figure 13 shows the predicted blood velocity distribution and pressure drop in the reconstructed retinal arterial network of a normal subject, and the corresponding oxygen distribution is given in Fig. 14. The predicted flow profiles (not shown here) are approximately parabolic, which is in agreement with the in vivo velocity measurement made in human retinal vessels by Fourier domain Doppler optical coherence tomography [83] and would be expected at the low Reynolds numbers in these vessels. Predicted pressure drops between the inlet and each of the outlets are low and in the range of 4.5–10 mm Hg. From the oxygen distribution (Fig. 14), it is clear that oxygen concentration is high at the centre of the lumen and decreases towards the vessel wall. The drop of oxygen tension near the wall is caused by the combined effects of convection and diffusion of oxygen across the vessel, with the diffusion of dissolved oxygen at the lumen wall surface. Due to the efflux of oxygen, the average oxygen concentration decreases with the distance from the inlet, despite replenishment from the haemoglobin. Nevertheless, because of the dynamic reversible reaction between haemoglobin and free oxygen, when oxygen tension decreases in the vessel, the oxyhemoglobin dissociates to release more free oxygen into the plasma as a result of the

Fig. 14 Oxygen distribution in the retinal arterial tree of a normal subject

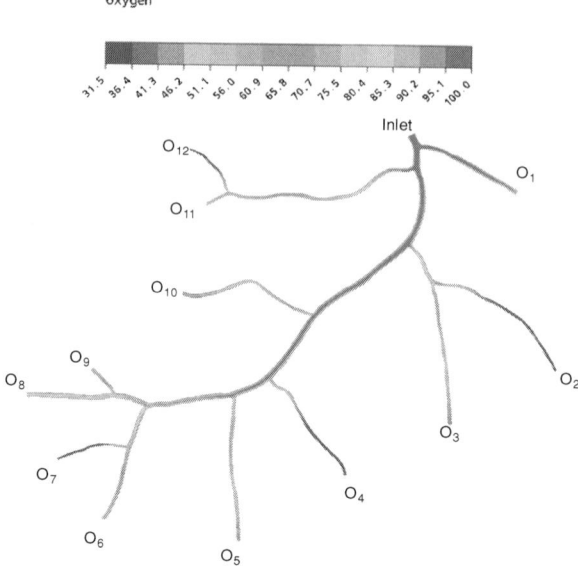

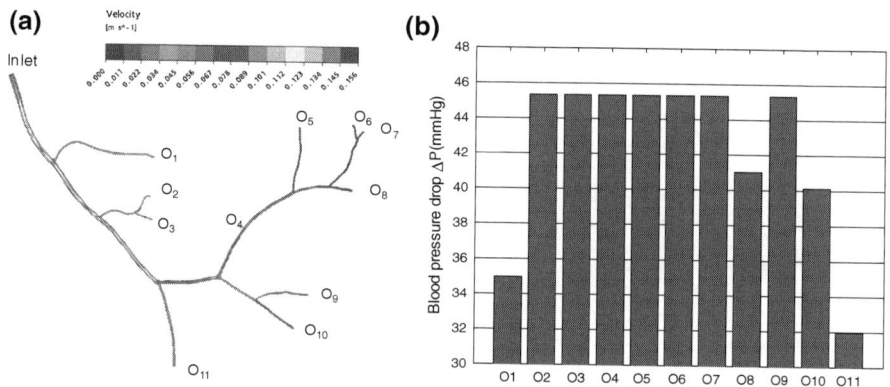

Fig. 15 Predicted velocity distribution (**a**) and pressure drop (**b**) in the retinal arterial tree of a hypertensive subject

relationship described by the oxyhaemoglobin saturation function, retaining a relatively stable oxygen concentration in the peripheral vessels.

Results for a hypertensive subject are given in Fig. 15 for flow distribution and pressure drop, and Fig. 16 for oxygen distribution. It is obvious that the hypertensive retinal arterial tree has fewer branches, higher asymmetry and larger length to diameter ratio than the normal retinal arterial tree presented earlier. These

Fig. 16 Oxygen distribution in the retinal arterial tree of a hypertensive subject

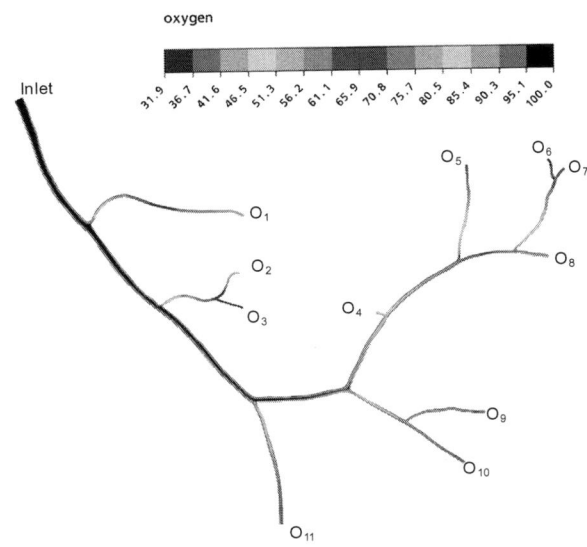

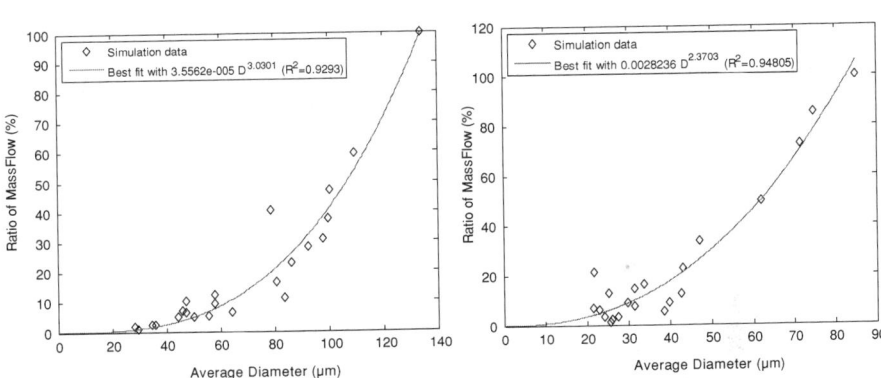

Fig. 17 Mass flow rate ratio verses vessel diameter in the normal (*left*) and hypertensive (*right*) retinal arteries

geometrical features have an effect on flow distribution and particularly pressure drops. As can be seen from Fig. 15b, pressure drops between the inlet and each of the 11 outlets vary in the range of 32–46 mm Hg, which are significantly higher than the pressure drops found in the normal retinal network. Oxygen distribution has a similar pattern to that in the normal retinal arterial tree.

Comparison of the relationships between fractional mass flow and vessel diameter for the normal and hypertensive retinal network is shown in Fig. 17. The best fit curves suggest that flow distribution in the normal retinal arterial tree is very close to that in an optimal vascular system whose flow (Q) and diameter (D) relationship follows the Murray's law, i.e. $Q \sim D^3$ [40]. On the other hand, flow distribution in the hypertensive retinal arterial tree deviates from the optimal vascular system owing to the morphological features altered by the disease. The implication is that the overall energy cost to maintain blood flow and wall metabolism in the retinal arteriolar network would be much higher for hypertensive than for normal subjects.

4 Concluding Remarks

The discussions and examples above have demonstrated some aspects of how far image-based computational modelling has come since the original ideas about the methodology were expressed. It is clear that the methodology and results obtained to date have shown great promise for its further application in both continuing the better understanding of arterial diseases, and in clinical decision making related to it, in tomorrow's medicine.

So far, to the authors' knowledge, there have been no direct patient-specific modelling links to genomics, but there have been examples of patient-specific molecular imaging, using PET-CT on AAA allied with FE analysis [91] and USPIO-MRI on carotid stenosis [73]. In the former, it was shown that in regions of high tensile stress, the expected rupture sites, the metabolic activity was raised, but we are unaware of any analysis of USPIO-MRI in conjunction with patient-specific analysis of flow and stress in carotid plaques. Meanwhile, investigations with endothelial cells of mouse aortas and human coronary arteries subjected to disturbed flow in vitro have allowed expression of bone morphogenic protein and related antagonists to be tracked in vulnerable sites [13]. These were important developments for the future of patient-specific analysis.

Acknowledgments The authors would like to thank the following people who have contributed to the data used here: Drs J. Leung, R. Torii, N. Witt, A. Wright, Professors A.D. Hughes and S.A. Thom.

References

1. Adolph, R., Vorp, A.D., Steed, D.L., Webster, M.W., Kameneva, M.V., Watkins, S.C.: Cellular content and permeability of intraluminal thrombus in abdominal aortic aneurysm. J. Vasc. Surg. **25**, 916–926 (1997)
2. Ariff, B.B., Glor, F.P., Crowe, L., Xu, X.Y., Vennart, W., Firmin, D.N., Thom, S.M., Hughes, A.D.: Carotid artery hemodynamics: observing patient-specific changes with amlodipine and lisinopril by using MR imaging computation fluid dynamics. Radiology **257**, 662–669 (2010)
3. Augst, A.D., Ariff, B., Thom, S.A.G., Xu, X.Y., Hughes, A.D.: Analysis of complex flow and the relationship between blood pressure, wall shear stress, anmd intima-media thickness in the human carotid artery. Am. J. Physiol. Heart Circ. Physiol. **293**, 1031–1037 (2007)

4. Beach, J.M., Schwenzer, K.J., Srinivas, S., Kim, D., Tiedeman, J.S.: Oximetry of retinal vessels by dual-wavelength imaging: calibration and influence of pigmentation. J. Appl. Physiol. **86**, 748–758 (1999)
5. Birol, G., Wang, S., Budzynski, E., Wangsa-Wirawan, N.D., Linsenmeier, R.A.: Oxygen distribution and consumption in the macaque retina. Am. J. Physiol. Heart Circ. Physiol. **293**, H1696–H1704 (2007)
6. Borghi, A., Wood, N.B., Mohiaddin, R.H., Xu, X.Y.: 3D geometric reconstruction of thoracic aortic aneurysms. BioMed. Eng. OnLine **5**(59), 1–13 (2006)
7. Buerk, D.G., Goldstick, T.K.: Arterial wall oxygen consumption rate varies spatially. Am. J. Physiol. Heart Circ. Physiol. **243**, H948–H995 (1982)
8. Buerk, D.G., Shonat, R.D., Riva, C.E., Cranstoun, S.D.: O2 gradients and countercurrent exchange in the cat vitreous humor near retinal arterioles and venules. Microvasc. Res. **45**, 134–148 (1993)
9. Bushong, S.C.: Magnetic Resonance Imaging: Physical and Biological Principles, 3rd edn. Mosby, St Louis MI (2003)
10. Caro, C.G., Fitz-Gerald, J.M., Schroter, R.C.: Atheroma and arterial wall shear-dependent mass transfer mechanism for atherogenesis. Proc. R. Soc. London **B177**, 109–159 (1971)
11. Cebral, J.R., Lohner, R.: Efficient simulation of blood flow past complex endovascular devices using an adaptive embedding technique. IEEE Trans. Med. Imaging **24**, 468–476 (2005)
12. Cebral, J.R., Yim, P.J., Löhner, R., Soto, O., Choyke, P.L.: Blood flow modeling in carotid arteries with computational fluid dynamics and MR imaging. Acad. Radiol. **9**, 1286–1299 (2002)
13. Chang, K., Weiss, D., Suo, J., Vega, J.D., Giddens, D., Taylor, W.R., Jo, H.: Bone morphogenic protein antagonists are coexpressed with bone morphogenic protein 4 in endothelial cells exposed to unstable flow in vitro in mouse aortas and in human coronary arteries: role of bone morphogenic protein antagonists in inflammation and atherosclerosis. Circulation **116**, 1258–1266 (2007)
14. Cheng, Z., Tan, F.P.P., Riga, C.V., Bicknell, C.D., Hamady, M.S., Gibbs, R.G.J., Wood, N.B., Xu, X.Y.: Analysis of flow patterns in a patient-specific aortic dissection model. J. Biomech. Eng. **132**, 051007 (2010)
15. Cringle, S.J., Yu, D.Y., Yu, P.K., Su, E.N.: Intraretinal oxygen consumption in the rat in vivo. Invest. Ophthalmol. Vis. Sci. **43**, 1922–1927 (2002)
16. Davies, P.F.: Flow-mediated endothelial mechanotransduction. Physiol. Rev. **75**, 519–560 (1995)
17. Delori, F.C.: Noninvasive technique for oximetry of blood in retinal vessels. Appl. Opt. **27**, 1113–1125 (1998)
18. Dorner, G.T., Polska, E., Garhofer, G., Zawinka, C., Frank, B., Schmetterer, L.: Calculation of the diameter of the central retinal artery from noninvasive measurements in humans. Curr. Eye Res. **25**, 341–345 (2002)
19. Groen, H., Gijsen, F., van der Lugt, A., Ferguson, M., Hatsukami, T., van der Steen, A.F.W., Yuan, C., Wentzel, J.J.: Plaque rupture in the carotid artery is localized at the high shear stress region: a case report. Stroke **38**, 2379–2381 (2007)
20. Iftimia, N.V., Hammer, D.X., Bigelow, C.E., Rosen, D.I., Ustun, T., Ferrante, A.A., Vu, D., Ferguson, R.D.: Toward noninvasive measurement of blood hematocrit using spectral domain low coherence interferometry and retinal tracking. Opt. Express **14**, 3377–3388 (2006)
21. Jackson, M., Wood, N.W., Zhao, S., Augst, A., Wolfe, J.H., Gedroyc, W.M.W., Hughes, A.D., Thom, S.A.M., Xu, X.Y.: Low wall shear stress predicts subsequent development of wall hypertrophy in lower limb bypass grafts. Artery Res. **3**, 32–38 (2009)
22. Kalender, W.A.: Computed Tomography: Fundamentals, System Technology Image Quality Applications. Publicis, Erlangen (2005)
23. Kalender, W.A.: X-ray computed tomography. Phys. Med. Biol. **51**, R29–R43 (2006)
24. Karner, G., Perktold, K., Zehentner, H.P.: Computational modeling of macromolecule transport in the arterial wall. Comput. Meth. Biomech. Biomed. Eng. **4**, 491–504 (2001)

25. Kedem, O., Katchalsky, A.: Thermodynamic analysis of the permeability of biological membranes to non-electrolytes. Biochim. Biophys. Acta **27**, 229–246 (1958)
26. Kilner, P.J., Henein, M.Y., Gibson, D.G.: Our tortuous heart in dynamic mode–an echocardiographic study of mitral flow and movement in exercising subjects. Heart Vessels **12**, 103–110 (1997)
27. Kudo, S., Ikezawa, K., Matsumura, S., Ikeda, M., Oka, K., Tanishita, K.: Effect of wall shear stress on macromolecule uptake into culyured endothelial cells. Tran. Jpn. Soc. Mech. Eng. Ser. B **64**, 367–374 (1998)
28. Lee, K.W., Wood, N.B., Xu, X.Y.: Ultrasound image-based computer model of a common carotid artery with a plaque. Med. Eng. Phys. **26**, 823–840 (2004)
29. Lehoux, S., Castier, Y., Tedgui, A.: Molecular mechanisms of the vascular responses to haemodynamic forces. J. Intern. Med. **259**, 381–392 (2006)
30. Leung, J.H.: Determination of flow patterns and stresses in patient-specific models of abdominal aortic aneurysm. PhD Thesis, Imperial College London, UK (2006)
31. Leung, J.H., Wright, A.R., Cheshire, N., Crane, J., Thom, S.A., Hughes, A.D., Xu, X.Y.: Fluid structure interaction of patient specific abdominal aortic aneurysms: a comparison with solid stress models. BioMed. Eng. OnLine **5**(33), 1–15 (2006)
32. Levick, J.R.: An Introduction to Cardiovascular Physiology, 5th edn. Hodder Arnold, London (2010)
33. Linsenmeier, R.A., Braun, R.D.: Oxygen distribution and consumption in the cat retina during normoxia and hypoxemia. J. Gen. Physiol. **99**, 177–197 (1992)
34. Liu, D., Wood, N.B., Witt, N., Hughes, A.D., Thom, S.A., Xu, X.Y.: Computational analysis of oxygen transport in the retinal arterial network. Curr. Eye Res. **34**, 945–956 (2009)
35. Long, Q., Xu, X.Y., Bourne, M., Griffith, T.M.: Numerical study of blood flow in an anatomically realistic aorto-iliac bifurcation generated from MRI data. Magn. Reson. Med. **43**, 565–576 (2000)
36. Lotz, J., Meier, C., Leppert, A., Galanski, M.: Cardiovascular flow measurement with phase-contrast MR imaging: basic facts and implementation. Radiographics **22**, 651–671 (2002)
37. Malek, A.M., Alper, S.L., Izumo, S.: Hemodynamic shear stress and its role in atherosclerosis. JAMA **282**, 2035–2042 (1999)
38. Marshall, I., Zhao, S.Z., Papathanasopoulou, P., Hoskins, P., Xu, X.Y.: MRI and CFD studies of pulsatile flow in healthy and stenosed carotid bifurcation models. J. Biomech. **37**, 679–687 (2004)
39. Martinez-Perez, M.E., Hughes, A.D., Stanton, A.V., Thom, S.A., Chapman, N., Bharath, A.A., Parker, K.H.: Retinal vascular tree morphology: a semi-automatic quantification. IEEE Trans. Biomed. Eng. **49**, 912–917 (2000)
40. Mayrovitz, H.N., Roy, J.: Microvascular blood flow: evidence indicating a cubic dependence on arteriolar diameter. Am. J. Physiol. Heart Circ. Physiol. **245**, H1031–H1038 (1983)
41. McRobbie, D.W., Moore, E.A., Graves, M.J., Prince, M.R.: MRI from Picture to Proton, 2nd edn. Cambridge University Press, Cambridge (2007)
42. Mendivil, A., Cuartero, V., Mendivil, M.P.: Ocular blood flow velocities in patients with proliferative diabetic retinopathy and healthy volunteers: a prospective study. Br. J. Ophthalmol. **79**, 413–416 (1995)
43. Meyer, G., Merval, R., Tedgui, A.: Effects of pressure-induced stretch and convection on low-density lipoprotein and albumin uptake in the rabbit aortic wall. Circ. Res. **79**, 532–540 (1996)
44. Milner, J.S., Moore, J.A., Rutt, B.K., Steinman, D.A.: Hemodynamics of human carotid artery bifurcations: computational studies with models reconstructed from magnetic resonance imaging of normal subjects. J. Vasc. Surg. **28**, 143–156 (1998)
45. Moore, J.A., Ethier, C.R.: Oxygen mass transfer calculations in large arteries. J. Biomech. Eng. **119**, 469–475 (1997)
46. Morris, L., Delassus, P., Walsh, M., McGloughlin, T.: A mathematical model to predict the in vivo pulsatile drag forces acting on bifurcated stent grafts used in endovascular treatment of abdominal aortic aneurysms (AAA). J. Biomech. **37**, 1087–1095 (2004)

47. Nissen, S.E., Yock, P.: Intravascular ultrasound: novel pathophysiological insights and current clinical applications. Circulation **103**, 604–616 (2001)
48. O'Brien, K.R., Myerson, S.G., Cowan, B.R., Young, A.A., Robson, M.D.: Phase contrast ultrashort TE: a more reliable technique for measurement of high-velocity turbulent stenotic jets. Magn. Reson. Med. **62**, 626–636 (2009)
49. Papathanasopoulou, P., Zhao, S., Köhler, U., Robertson, M.B., Long, Q., Hoskins, P., Xu, X.Y., Marshall, I.: MRI measurement of time-resolved wall shear stress vectors in a carotid bifurcation model, and comparison with CFD predictions. J. Magn. Reson. Imaging **17**, 153–162 (2003)
50. Pope, J.: Medical Physics: Imaging. Heinemann, Oxford (1999)
51. Pries, A.R., Secomb, T.W., Gaehtgens, P., Gross, J.F.: Blood flow in microvascular networks - experiments and simulation. Circ. Res. **67**, 826–834 (1990)
52. Pries, A.R., Secomb, T.W., Gaehtgens, P.: Biophysical aspects of blood flow in the microvasculature. Cardiovasc. Res. **32**, 654–667 (1996)
53. Rappitsch, G., Perktold, K.: Computer simulation of convective diffusion processes in large arteries. J. Biomech. **29**, 207–215 (1996)
54. Shojima, M., Oshima, M., Takagi, K., Torii, R., Nagata, K., Shirouzu, I., Morita, A., Kirino, T.: Role of the bloodstream impacting force and the local pressure elevation in the rupture of cerebral aneurysms. Stroke **36**, 1933–1938 (2005)
55. Sill, H.W., Chang, Y.S., Artman, J.R., Frangos, J.A., Hollis, T.M., Tarbell, J.M.: Shear stress increases hydraulic conductivity of cultured endothelial monolayers. Am. J. Physiol. **268**, H535–H543 (1995)
56. Slager, C.J., Wentzel, J.J., Schuurbiers, J.C., Oomen, J.A., Kloet, J., Krams, R., von Birgelen, C., van der Giessen, W.J., Serruys, P.W., de Feyter, P.J.: True 3-dimensional reconstruction of coronary arteries in patients by fusion of angiography and IVUS (ANGUS) and its quantitative validation. Circulation **102**, 511–516 (2000)
57. Slager, C., Wentzel, J.J., Gijsen, F.J.H., Schuurbiers, J.C.H., van der Wal, A.C., van der Steen, A.F.W., Serruys, P.W.: The role of shear stress in the generation of rupture-prone vulnerable plaques. Nat. Clin. Pract. Cardiovasc. Med. **2**, 401–407 (2005)
58. Slager, C.J., Wentzel, J.J., Gijsen, F.J.H., Thury, A., van der Wal, A.C., Schaar, J.A., Serruys, P.W.: The role of shear stress in the destabilization of vulnerable plaques and related therapeutic implications. Nat. Clin. Pract. Cardiovasc. Med **2**, 456–464 (2005)
59. Soloperto, G., Keenan, N.G., Sheppard, M.N., Ohayon, J., Wood, N.B., Pennell, D.J., Mohiaddin, R.H., Xu, X.Y.: Combined imaging, computational and histological analysis of a ruptured carotid plaque: a patient-specific analysis. Artery Res. **4**, 59–65 (2010)
60. Stanton, A.V., Wasan, B., Cerutti, A., Ford, S., Marsh, R., Sever, P.P., Thom, S.A., Hughes, A.D.: Vascular network changes in the retina with age and hypertension. J. Hypertens. **13**, 1724–1728 (1995)
61. Starmans-Kool, M.J., Stanton, A.V., Zhao, S., Xu, X.Y., Thom, S.A., Hughes, A.D.: Measurement of hemodynamics in human carotid artery using ultrasound and computational fluid dynamics. J. Appl. Physiol. **92**, 957–961 (2002)
62. Steinman, D.A.: Image-based computational fluid dynamics modeling in realistic arterial geometries. Ann. Biomed. Eng. **30**, 483–497 (2002)
63. Sun, N., Leung, J.H., Wood, N.B., Hughes, A.D., Thom, S.A., Cheshire, N.J., Xu, X.Y.: Computational analysis of oxygen transport in a patient-specific model of abdominal aortic aneurysm with intraluminal thrombus. Br. J. Radiol. **82**(1), S18–S23 (2009a)
64. Sun, N., Torii, R., Wood, N.B., Hughes, A.D., Thom, S.A., Xu, X.Y.: Computational modeling of LDL and albumin transport in an in vivo CT image-based human right coronary artery. J. Biomech. Eng. **131**, 021003-1-9 (2009b)
65. Sun, N., Wood, N.B., Hughes, A.D., Thom, S.A., Xu, X.Y.: Fluid-wall modelling of mass transfer in an axisymmetric stenosis: effects of shear-dependent transport properties. Ann. Biomed. Eng. **34**, 1119–1128 (2006)
66. Sun, N., Sun, N., Wood, N.B., Hughes, A.D., Thom, S.A., Xu, X.Y.: Effects of transmural pressure and wall shear stress on LDL accumulation in the arterial wall: a numerical study using a multilayered model. Am. J. Physiol. Heart. Circ. Physiol. **292**, H3148–H3157 (2007a)

67. Sun, N., Wood, N.B., Hughes, A.D., Thom, S.A., Xu, X.Y.: Influence of pulsatile flow on LDL transport in the arterial wall. Ann. Biomed. Eng. **35**, 1782–1790 (2007b)
68. Swedenborg, J., Eriksson, P.: The intraluminal thrombus as a source of proteolytic activity. Ann. NY. Acad. Sci. **1085**, 133–138 (2006)
69. Tan, F.P.P., Soloperto, G., Bashford, S., Wood, N.B., Thom, S., Hughes, A., Xu, X.Y.: Analysis of flow disturbance in a stenosed carotid artery bifurcation using two-equation transitional and turbulence models. J. Biomech. Eng. **130**, 061008-1-12 (2008)
70. Tan, F.P.P., Wood, N.B., Tabot, G.R., Xu, X.Y.: Comparison of LES of steady transitional flow in an idealized Stenosed Axisymmetric Artery model with a RANS transitional model. J. Biomech. Eng. **133**, 051001-1-12 (2011)
71. Tan, F.P.P., Torii, R., Borghi, A., Mohiaddin, R.H., Wood, N.B., Xu, X.Y.: Fluid-structure interaction analysis of wall stress and flow patterns in a thoracic aortic aneurysm. Int. J. Appl. Mech. **1**, 179–199 (2009)
72. Tang, D., Yang, C., Zheng, J., Woodard, P.K., Saffitz, J.E., Sicard, G.A., Pilgram, T.K., Yuan, C.: Quantifying effects of plaque structure and material properties on stress distributions in human atherosclerotic plaques using 3D FSI models. J. Biomech. Eng. **127**, 1185–1194 (2005)
73. Tang, T.Y., Muller, K.H., Graves, M.J., Li, Z.Y., Walsh, S.R., Young, V., Sadat, U., Howarth, S.P., Gillard, J.H.: Iron oxide particles for atheroma imaging. Arterioscler. Thromb. Vasc. Biol. **29**, 1001–1008 (2009)
74. Tarbell, J.M.: Mass transport in arteries and the localization of atherosclerosis. Ann. Rev. Biomed. Eng. **5**, 79–118 (2003)
75. Taylor, C.A., Figueroa, C.A.: Patient-specific modeling of cardiovascular mechanics. Annu. Rev. Biomed. Eng. **11**, 109–134 (2009)
76. Taylor, C.A., Draney, M.T., Ku, J.P., Parker, D., Steele, B.N., Wang, K., Zarins, C.K.: Predictive medicine: computational techniques in therapeutic decision-making. Comput. Aided Surg. **4**, 231–247 (1999)
77. Thorne, M.L., Rankin, R.N., Steinman, D.A., Holdsworth, D.W.: In vivo Doppler ultrasound quantification of turbulence intensity using a high-pass frequency filter method. Ultrasound Med. Biol. **36**, 761–771 (2010)
78. Torii, R., Wood, N.B., Hadjiloizou, N., Dowsey, A.W., Wright, A.R., Hughes, A.D., Davies, J., Francis, D.P., Mayet, J., Yang, G.Z., Thom, S.A., Xu, X.Y.: Stress phase angle depicts differences in coronary artery hemodynamics due to changes in flow and geometry after percutaneous coronary intervention. Am. J. Physiol. Heart Circ. Physiol. **296**, 65–76 (2009)
79. Varghese, S.S., Frankel, S.H., Fisher, P.F.: Modeling transition to turbulence in eccentric stenotic flows. ASME J. Biomech. Eng. **130**, 014503-1-7 (2008)
80. Vorp, D.A.: Biomechanics of abdominal aortic aneurysm. J. Biomech. **40**, 1887–1902 (2007)
81. Vorp, D.A., Vande Geest, J.P.: Biomechanical determinants of abdominal aortic aneurysm rupture. Arterioscler. Thromb. Vasc. Biol. **25**, 1558–1566 (2005)
82. Wang, L., Tornquist, P., Bill, A.: Glucose metabolism of the inner retina in pigs in darkness and light. Acta Physiol. Scand. **160**, 71–74 (1997)
83. Wang, Y., Bower, B.A., Izatt, J.A., Tan, O., Huang, D.: In vivo total retinal blood flow measurement by Fourier domain Doppler optical coherence tomography. J. Biomed. Opt. **12**, 041215 (2007)
84. Wentzel, J.J., Corti, R., Fayad, Z.A., et al.: Does shear stress modulate both plaque progression and regression in the thoracic aorta? Human study using serial magnetic resonance imaging. J. Am. Coll. Cardiol. **45**, 846–854 (2005)
85. Wilson, N.M., Arko, F.R., Taylor, C.A.: Predicting changes in blood flow in patient-specific operative plans for treating aortoiliac occlusive disease. Comput. Aided Surg. **10**, 257–277 (2005)
86. Wolters, B.J.B.M., Rutten, M.C.M., Schurink, G.W.H., Kose, U., de Hart, J., van de Vosse, F.N.: A patient-specific computational model of fluid0structure interaction in abdominal aortic aneurysms. Med. Eng. Phys. **27**, 871–883 (2005)
87. Wong, T.Y., McIntosh, R.: Systemic associations of retinal microvascular signs: a review of recent population-based studies. Ophthalmic Physiol. Opt. **25**, 195–204 (2005)

88. Wood, N.B., Weston, S.J., Kilner, P.J., Gosman, A.D., Firmin, D.N.: Combined MR imaging and CFD simulation of flow in the human descending aorta. J. Magn. Reson. Imaging **13**, 699–713 (2001)
89. Wood, N.B., Zhao, S.Z., Zambanini, A., Jackson, M., Gedroyc, W., Thom, S.A., Hughes, A.D., Xu, X.Y.: Curvature and tortuosity of the superficial femoral artery: a possible risk factor for peripheral vascular disease. J. Appl. Physiol. **101**, 1412–1418 (2006)
90. Xu, X.Y., Long, Q., Collins, M.W., Bourne, M., Griffith, T.M.: Reconstruction of blood flow patterns in human arteries. Proc. Inst. Mech. Eng. H. **213**, 411–421 (1999)
91. Xu, X.Y., Borghi, A., Nchimi, A., Leung, J., Gomez, P., Cheng, Z., Defraigne, J.O., Sakalihasan, N.: High levels of 18F-FDG uptake in aortic aneurysm wall are associated with high wall stress. Eur. J. Vasc. Endovasc. Surg. **39**, 295–301 (2010)
92. Yu, D.Y., Cringle, S.J., Su, E.: Intraretinal oxygen distribution in the monkey retina and the response to systemic hyperxia. Invest. Ophthalmol. Vis. Sci. **46**, 4728–4733 (2005)
93. Yuan, C., Hatsukami, T.: Chapter 23: MR plaque imaging. In: Gillard, J., Graves, M., Hatsukami, T., Yuan, C. (eds.) Carotid Disease: The Role of Imaging in Diagnosis and Management. Cambridge University Press, Cambridge (2007)
94. Zhao, S.Z., Ariff, B., Long, Q., Thom, S.A., Hughes, A.D., Xu, X.Y.: Inter-individual variations in wall shear stress and mechanical stress distributions at the carotid artery bifurcation of healthy humans. J. Biomech. **35**, 1367–1377 (2002)

Patient-Specific Modeling of Leg Compression in the Treatment of Venous Deficiency

Stéphane Avril, Pierre Badel, Laura Dubuis, Pierre-Yves Rohan, Johan Debayle, Serge Couzan and Jean-François Pouget

Abstract Leg compression is the process of applying external compression forces onto the human leg with stockings or socks, with the purpose of treating venous deficiency, recovering from a sporting competition, or simply as prevention against the economy-class syndrome for transcontinental flights. The objective of this book chapter is to present a computational approach for characterizing the biomechanical effects of compression onto the soft tissues of the leg. The originality of the approach is that it accounts for the inter-individual variability. After introducing the principal medical knowledge about venous deficiency and its treatment by compression therapy, a literature survey about computational biomechanical modelling regarding this topic is presented. The importance of patient-specific modelling is highlighted and the approach that we specifically developed for addressing this issue is further detailed. The main originality concerning the patient-specific numerical modelling is to account for the actual material properties of both the compression stocking and biological soft tissues inside the leg. Suitable identification methods, based on image warping and model updating from MRI scans (for the internal leg tissues), are developed for retrieving these material properties. Eventually, the local pressure induced by compression in the tissues surrounding the veins is determined for different subjects. Perspectives are presented for integrating the effects of compression therapy onto the blood flow itself and for improving the design of compressive garments.

S. Avril (✉) · P. Badel · L. Dubuis · P.-Y. Rohan · J. Debayle
Centre for Health Engineering, Ecole Nationale Supérieure des Mines,
CNRS UMR 5146, 158 cours Fauriel, 42023 Saint-Étienne cedex 2, France
e-mail: avril@emse.fr

S. Couzan · J.-F. Pouget
Clinique Mutualiste, 3 rue le verrier, 42013 Saint-Étienne, France

1 Introduction

Veins normally contain about 80% of the total volume of blood in the systemic vascular system. This role of stocking blood was considered for a long time as the only role of veins. Blood flow in the veins is however a topic of research that deserves as much attention as hemodynamics in the large arteries. Indeed, it has been estimated that more than 50% of women and 25% of men are affected by diseases related to chronic venous insufficiency (CVI), confirming that CVI is a major public health issue. CVI is a medical condition where the veins cannot pump enough oxygen-poor blood back to the heart. This may occur after deep vein thrombosis (DVT) or phlebitis. The latter is also a common complication in hospitalised patients. A serious outcome is pulmonary thrombo-embolism, which is implicated in 10% of all hospital deaths.

The effects of CVI can grade from heavy legs and varicose veins to phlebetic lymphedema, chronic swelling of the legs and ankles, and increased risks of ulcers [1]. Although the effects of CVI are well known, the causes are rather complex, which makes them difficult to determine.

Compression therapies have been demonstrated to be effective non-operative options to relieve symptoms associated with venous disorders in the human lower limb, such as reducing venous hypertension and improving venous blood return [1, 2]. Elastic compression is supposed to help the venous return, to decrease venous pressure, to prevent venous stasis and deterioration of venous walls, and to efficiently relieve aching and heavy legs. This simple, but very efficient therapy, is now also extended to recover from a sporting competition, or for instance as a prevention against the economy-class syndrome for transcontinental flights.

Now, the main limitation to develop new stocking products is that the biomechanical effects of compression therapy have not been fully elucidated. The role of computational models can be essential to address these issues.

The objective of this book chapter is to present a computational approach accounting for the inter-individual variability in modelling the biomechanical effects of compression therapy onto the leg, focusing especially on the calf muscle. After introducing the principal medical knowledge about venous deficiency and its treatment by compression therapy, a literature survey about computational biomechanical modelling on this topic is presented. The importance of patient-specific modelling is highlighted and a dedicated original approach is further detailed for addressing this issue.

2 Anatomy and Physiology

A large number of troubles at the origin of CVI occur in the leg and this chapter, as most of the literature dedicated to venous deficiency, will be focused on the region of the calf muscle.

This section is aimed at presenting the principal medical knowledge about venous deficiency and its treatment by compression therapy. More details can be found in [1] and [2].

2.1 Anatomy of the Calf Venous System

It is important to dissociate in the venous system two types of veins:

1. the primary collecting veins which are passive, thin-walled reservoirs that are very distensible. Most are supra-fascial, surrounded by loosely bound alveolar and fatty tissue that is easily displaced. These supra-fascial collecting veins can dilate to accommodate large volumes of blood with little increase in back pressure, so that the volume of blood sequestered within the venous system at any moment can vary by a factor of two or more without interfering with the normal function of the veins.
2. secondary veins or conduits which benefit of the outflow from primary collecting veins. They have thicker walls and are less distensible. Most of these veins are sub-fascial and are surrounded by tissues that are dense and tightly bound.

These veins may be grouped in 3 (sub)-systems:

1. the superficial venous system, which is a complicated and extremely variable network of interconnecting veins. Many superficial collecting veins deliver their blood into the great and small saphenous veins, which deliver most of their blood into the deep system through the sapheno-femoral junction SFJ and the sapheno-popliteal junction (SPJ).
2. the deep venous system which receives the blood before its way back to the right atrium of the heart.
3. the system of perforating veins which drains the blood from the superficial system to the deep veins. Venous valves prevent reflux of blood from the deep veins into the superficial system.

2.2 The Calf Muscle Pump

The passage of blood upward from the feet against gravity depends on a complex array of valves and pumps. Muscle pumps of the calf and foot provide the motive force for venous return. This is frequently called the calf muscle pump or musculo-venous pump and is thought to function as the peripheral heart. Each segment of the calf muscle pump works in the same way.

When the gastrocnemius and soleus muscles contract, they expel more than 60% of venous blood into the large popliteal vein. In the resting state, veins of the calf fill at 1–2 mm per second—actively by compression of the sole of the foot and passively during muscle relaxation. It takes in average 70 s to refill 90% of the blood volume back into the calf and 31 s in average to restore hydrostatic pressure. However, these time durations can vary dramatically between individuals.

Ambulatory venous pressure (AVP) is the "gold standard" test of the efficiency of the calf musculo-venous pump. It is performed by placing a small needle into one of the veins on the back of the foot and connecting the needle to a blood pressure measurement machine. The test has three parts:

1. The subject is then asked to stand up and the standing venous pressure is measured.
2. The subject is asked to perform ten heel raise exercises to work the musculo-venous pump and the AVP is recorded.
3. The subject is asked to rest again in the standing position and the rate at which the ambulatory pressure returns to the standing pressure is measured, called the refilling time.

In a normal subject the standing venous pressure is around 90 mmHg (depending on their height). During exercise this should fall to around 30 mmHg and after exercise this should only rise slowly over half a minute or so back to the standing pressure [3].

2.3 Venous Dysfunction

Venous dysfunction can result from primary muscle pump failure, from venous obstruction (thrombotic or nonthrombotic), or from venous valvular incompetence, which may be segmental or may involve the entire length of the vein. Venous return is impaired and it can trigger different possible abnormal AVP:

- The pressure does not fall normally during exercise, which indicates that the calf pump is not working effectively.
- The pressure rises rather than falls during exercise, which indicates that the deep veins are occluded.
- The AVP returns to the standing pressure too quickly, which indicates reflux in either the deep or superficial veins due to absent or damaged valves or due to valves made incontinent because of dilation.

The effects of venous dysfunction can grade from heavy legs and varicose veins to phlebetic lymphedema, chronic swelling of the legs and ankles, and increased risks of ulcers.

2.4 Treatments by Compression Therapy

Graduated compression stockings (GCS), used as one of the important compression therapies, have been demonstrated to be an effective non-operative option to relieve symptoms associated with venous disorders in the human lower limb, such as reducing venous hypertension and improving venous blood return. Elastic compression is supposed to help the venous return, to decrease venous pressure, to prevent venous stasis and deterioration of venous walls, and to efficiently relieve aching and heavy legs. This simple, but very efficient therapy, is now also extended to recover from a sporting competition, or for instance as a prevention against the economy-class syndrome for transcontinental flights.

In case of DVT, medicinal anticoagulants, such as low molecular weight heparin, are often used in conjunction with GCS. Intermittent pneumatic compression of the calf or foot can also be applied [4].

Scientific design of compression stockings can not only enhance their medical functions, but also bring wearers comfort sensory perceptions. Comfort is an important issue. With the principle of graduated compression from the ankle to the knee, it is required to apply significant pressures at the ankle. Such high pressures at the ankle tend to be questioned and novel designs of compression stockings, with larger pressures at the level of the calf muscle, are now developed. Several medical studies based on the analysis of venous return with MRI and Doppler ultrasounds, have confirmed that large pressures at the calf are more efficient for treating moderate venous insufficiency [5-10].

Now, the main limitation to develop new stocking products with targeted biomedical actions is that the biomechanical effects of compression therapy have not been fully elucidated. The role of computational models can be essential to address these issues.

3 Review of Computational Studies in Biomechanics Dedicated to Venous Deficiency and Compression Therapy

Computational modelling was developed by numerous authors in the domain of venous deficiency and compression therapy. A review is presented further. Generally two different objectives are sought:

1. to understand venous return and the mechanisms of muscular draining, in order to address its possible dysfunction and treatments.
2. to study and characterize the effects of elastic compression, possibly to improve them, either for a better comfort or for a better efficiency.

3.1 Modelling Blood Flow in Veins

A very general approach is to model the blood flow in veins by modelling them as rigid or collapsible pipes [11-15].

3.1.1 Modelling Steady Flow

It has often been considered that the blood flow in veins is driven by the Hagen–Poiseuille law [16] which states that the mean flow in a cylindrical rigid pipe is laminar and follows the rule:

$$Q = -\pi R^4/8\eta(\Delta P/L)$$

where Q is the mean flow rate, R is the radius of the pipe, η is the dynamic viscosity of the fluid, ΔP is the pressure difference between the inlet and the outlet and L is the length of the pipe.

This model is consistent with the application of external graduated compression: larger pressures are applied at the ankle inducing a negative pressure gradient ΔP. However, it suffers from several limitations: it does not consider the action of tissues surrounding the veins and it does not consider the deformability of veins.

Regarding the latter point, an important theory has been developed since Shapiro and Kamm in 1979 [17]: the theory of collapsible tubes. Indeed, vessel collapse is most readily seen in the veins, such as in the veins of a hand raised above the level of the heart or in the jugular vein when a person is standing erect. Collapse is also observed in the arteries when high external pressures are applied, such as when an artery is compressed by a sphygmomanometer cuff during blood pressure measurement.

Flow through a collapsible tube has been extensively studied in the laboratory. Kamm and Shapiro describe the collapse process, in the presence of an external pressure gradient, as observed in their experiments [17]. They consider the tube collapse as consisting of three phases which they describe as the initial transient phase, the quasi-steady phase, and the viscous drainage phase. The initial transient is identified as a period of flow acceleration giving the initial peak observed in the output flow. The quasi-steady emptying is the period where the output flow drops from its initial peak and where there is the establishment of a quasi-steady throat in the region of minimum cross sectional area. Viscous drainage occurs as the region of collapse increases and the remaining fluid in the tube is forced out as a new equilibrium configuration is approached. An important aspect of modelling flow in collapsible tubes is an accurate description of the pressure/volume relationship, known as the tube law. More details about this theory may be found in the original papers from the authors [17].

Using the theory of collapsible tubes, Kamm was able in 1982 [18] to explore the detailed effects of different modes of compression including (i) uniform compression, the simultaneous application of uniform pressure over the entire lower leg, (ii) graded compression, the application of uneven pressure, maximum at the ankle and minimum at the knee, and (iii) wavelike compression, a wave of compression proceeding from the ankle toward the knee. These numerical results indicate that the effectiveness of uniform compression is severely compromised by the formation of a flow-limiting throat at the proximal end of the compression cuff that reduces both the rate at which blood is discharged from the lower leg and the total blood volume removed. Both of these detrimental effects can be avoided by the use of either wavelike or graded compression. Both alternated methods are shown to produce more uniform augmentation of volume flow rate, flow velocity and shear stress, throughout the entire lower leg.

3.1.2 Computer Fluid Dynamics Modelling

The theory of collapsible tubes is only 1D and several authors have shown the importance of considering 3D modelling of the blood flow in veins for analysing the effects of compression therapy.

Cros and coworkers [13] had to resort to 3D computer fluid dynamics (CFD) for modelling the flow at confluent junctions, showing that confluence effects cannot be neglected when evaluating pressure decreases.

Despite the importance of the flow rate in venous deficiency, another factor has been investigated recently: the wall shear stress (WSS). It is believed that WSS is a factor that regulates blood vessel structure and influences the development of vascular pathology. A group from Imperial College, London, UK, developed a computational methodology for estimating the local values of WSS [14].

The purpose was to estimate WSS in individual vessels of the venous circulation of the calf and quantify the effects of elastic compression based on change of vessel geometry and velocity waveform. The great saphenous vein and either a peroneal or posterior tibial vein have been imaged in healthy subjects using magnetic resonance imaging (MRI), with and without the presence of a grade 1 medical stocking. Flow through image-based reconstructed geometries was numerically simulated for both a range of steady flow rates and ultrasound derived transient velocity waveforms, scaled to give a standardised time averaged flow rate. For steady flow the stocking produced an average percentage increase in mean WSS of approximately 100% in the great saphenous vein. The percentage increase in the peroneal/posterior tibial veins varied from 490 to 650%. The average minimum value of WSS in all vessels without the stocking was less than 0.1 Pa. With the stocking this was increased to 0.7 Pa in the great saphenous and 0.9 Pa in the peroneal/posterior tibial veins.

The main downside of these CFD models is that veins are modelled with rigid walls. Patient-specific geometry is digitized, with and without compression. Interesting prospects would be to model the wall deformation under the action of internal blood pressure and external muscle pressure. For that, fluid structure interaction (FSI) should be considered.

Another important aspect which is rather absent from the computational approaches is the role of the valves. It is believed that a complete 3D CFD approach should include both the role of valves on the flow and the effect of wall deformation.

3.2 Modelling of Stocking and Skin Pressure Distribution

Mechanical interaction between skin and sock is an important factor affecting wearing comfort. A crucial factor in the compression therapy, pressure itself has not been understood sufficiently. Therefore, how the pressure performances and how to develop new stocking products with more satisfactory pressure profiles becomes an urgent matter.

3.2.1 The Laplace Law

The basic Laplace law provides the pressure P induced, on a cylinder of radius r, by a medium of width b stretched on it with a force F which can depend on the cylinder radius for non-linear constitutive materials:

$$P = F/rb$$

It can be noticed that this basic law expresses a pressure which is proportional to the curvature. Thus an increase in the radius of curvature, with a constant tension, is accompanied by a pressure decrease. In practice, the perimeter of the leg increases naturally upwards. Therefore by preserving a constant tension a decreasing pressure is obtained.

Gaied and coworkers [19] showed that experimental pressures measured using sensors are in good agreement with the Laplace law. They also showed that the curvature of the leg can vary significantly from one location of the leg to another, inducing large local pressure variations, with zero pressure at concave zones.

3.2.2 Finite Element Structural Modelling

Application of the Laplace law requires some knowledge about the local tension of the stocking. It is sometimes more relevant to model directly the stocking deformation and its contact with the skin. Modelling the contact between stockings and the leg was achieved by different authors [19–25]. The rationale behind such studies is that the increase of the intramuscular pressure is found to be either lower or higher than the pressures measured between the skin and the stocking at the middle of the calf [26]. This discrepancy is due to the pressure exerted by stockings which is not uniform around the leg.

For instance, You and coworkers [21] calculated the distribution of garment pressures on the surface of human body using elastic human body and mass-spring clothing model under large deformation situation. Zhang and coworkers [22] introduced a finite element (FE) approach to model the contact between the foot and sock. A biomechanical foot model consisting of bones and soft tissues and an orthotropic and elastic sock model were constructed for simulating the contact process. By using the model, they simulated the process of wearing a sock, which consisted of two kinds of materials. The stress and pressure distributions in the sock as well as that in the foot were predicted. The comparison between the pressure measurements at several geometrically characteristic points and the predicted pressures confirmed that the model was able to simulate the sock wearing condition and predict the pressure exerted by socks. It was also demonstrated that the skin pressure depended on the curvature of the contact surface and the stress in the fabric. The developed FE model allowed the authors to carry out parametric analysis on socks of different styles and materials with a relatively quick and easy way and to provide guidance to the functional design of socks.

Dai and coworkers [25] developed a 3D biomechanical mathematical model to simulate and predict accurately the mechanical interaction between human lower limb and compression stockings. They showed that the external pneumatic compression of the lower legs is an effective prophylaxis against deep vein thrombosis.

3.3 Modelling the Response of Deep Soft Tissues to Compression

Modelling blood flow in the veins under compression requires a precise knowledge of the pressure applied to the veins by the tissues surrounding it. Downie and coworkers [14] have shown that the increase of this pressure when the leg is compressed can significantly reduce the cross sectional area of the veins and this may induce larger wall shear stress in the blood.

Though the contribution of elastic compression is shown to be clinically efficient [27–29], the actual mechanisms of elastic compression and its biomechanical effects on internal tissues are still partially unknown.

Measurement of the intramuscular pressure in the compressed leg has been achieved several times [3, 26]. However, the effects remain unclear and the very intrusive measurements are fairly difficult to achieve at a large scale. This may explain that computational analyses of the response of deep tissues to compression have become an important field of research in biomechanics.

Applications of computational biomechanics to leg compression were initiated by Dai and coworkers [25] using a patient-specific geometry for FE analyses. The purpose was to develop a three-dimensional biomechanical model to simulate and predict accurately the mechanical interaction between human lower limb and compression stockings in conjunction with experimental measurements of the skin pressure and properties of the textile. The established mathematical model can predict and visualize the skin pressure magnitude and distribution applied by different compression stockings without practical wearing, which provided the authors with more effective engineering design and scientific evaluation approach.

The main difficulty related to the development of the computational models of the compressed leg lie in the determination of relevant constitutive equations for the tissues. Generally, biological soft tissues are assumed hyper-elastic. By medical images and inverse methods, the material properties of soft tissues are identified for instance in different organs [30–39] among which the compressed leg

[38, 39]. In such studies, natural mechanical loading is used to identify constitutive parameters. FE models of the undeformed tissues are built using medical images of the non-compressed tissues. Using a second set of images taken in the compressed state, constitutive properties input in the FE model are calibrated [9, 38, 39]. It gives access to the internal pressure distribution of the patient-specific leg.

3.4 New Challenges

It can be remembered from this literature survey that numerical modelling has already contributed significantly to understanding the mechanisms of venous deficiency and the effects of its treatment by medical compression.

The new challenges offered to computational mechanics are now:

1. to develop FSI approach taking into account both the blood flow in the veins and the deformations of the surrounding tissues. Modelling the effect of muscular contraction on the venous flow with and without compression is especially challenging.
2. to develop patient-specific modelling of the biomechanical effects of compression for adapting the treatments to the morphology and pathology of each subject.

The latter has been addressed recently by the current authors. A specific methodology was developed. It is presented in the following section along with practical results and future developments.

4 Patient-Specific Modelling of Compressed Deep Tissues

Downie and coworkers [14] showed that the deformation of the veins as a result of the static external compression is very subject-dependent. The shape and location of the bones, the architecture and mechanical properties of the muscles and other soft tissues, are highly variable and can significantly affect the deformation of the deep veins, which will in turn affect the flow rate in the deep veins. Due to the great inter-individual variability of these factors this justifies the development of patient-specific models for characterizing how the pressure applied by the stocking to the skin is transmitted towards the deep veins through the fat and muscle tissues. Such a patient-specific approach is proposed in this section.

4.1 Method for Setting up a Patient-Specific FE Model

The methodology proposed here is aimed at setting up patient-specific FE models about the effect of elastic compression onto the leg. The models are fully patient-specific, which means that:

Fig. 1 Different steps of mesh construction

1. the geometry of the model is deduced from 3D medical images of the subject.
2. the properties of the materials constituting the model are adjusted according to an image of the deformed leg for matching the actual properties.
3. the boundary conditions are determined so as to reproduce the actual mechanical action of the compression stocking onto the leg of the subject.

4.1.1 Geometry and Meshing

Image Acquisition and Processing

3D CT-scan images of the leg are acquired on patients or subjects according to procedures that have been approved by the ethical review council of the University Hospital of Saint-Etienne, France. The CT-scans are acquired with and without elastic compression in seated position. The elastic compression used is the BVSport® Pro Recup sock. The images resolution is $512 \times 512 \times 376$ voxels and the voxel size is set to about $1 \times 1 \times 1$ mm. After different steps of rigid registration, the images are segmented by thresholding in three regions: adipose tissue (skin and fat), muscles and bones (tibia and fibula), which are shown in Fig. 1. Segmentation is achieved using the ImageJ software, the thresholds are prescribed manually.

Mesh Generation

The soft tissues are meshed with around 300,000 tetrahedral linear elements and 70,000 nodes. The mesh is created after image segmentation using the Avizo® software. Three quality criteria of the surface mesh must be satisfied:

1. Aspect Ratio (< 10). This computes the ratio of the radii of the circumcircle and the incircle for each triangle.
2. Dihedral Angle (> 11). For each pair of adjacent triangles, the angle between them at their common edge will be computed.
3. Tetra Quality (< 20). This computes the ratio of the radii of the circumsphere and the inscribed sphere for the tetrahedron which would probably be created.

The surface mesh is smoothed in the horizontal plane following a five degree Fourier descriptor [9, 40]. This step is aimed at clearing local irregularities of the mesh on the surface, these irregularities occurring due to the segmentation process applied here on a relatively coarse grid of voxels. The tetramesh is created from surface mesh and is exported to the Abaqus® software for FE computations.

The whole procedure is shown for a subject in Fig. 1.

4.1.2 Boundary Conditions

All the degrees of freedom along the contours of the bones are fixed, as both bones are assumed to be infinitely rigid (their elastic properties are nearly 1,000 times larger than those of soft tissues).

An inhomogeneous pressure is applied on the external surface of the leg. Unlike usual graduated compression socks, the maximal compression of the socks used for this study is at calf and not at ankle, which has been shown to fit comfort requirements and recovering efficiency after sport efforts [8]. The pressure increases linearly from ankle (less than 1 kPa in mean) to knee (3.5 kPa in mean).

In the patient-specific model, local pressure values are obtained from the Laplace law, neglecting the curvature along the height of the leg and only considering the circumferential curvature:

$$P = T/r = k\,\varepsilon/r$$

with P the pressure, T the sock tension, r the local curvature radius of the leg in the horizontal plane (curvature in the vertical plane is neglected), k the fabric stiffness, and ε the sock strain along the circumferential direction. The latter is derived knowing leg and sock perimeters from the CT-scans.

The fabric stiffness denoted k is measured at the sock's ankle and calf using a non-destructive system based on the NF G30-102 standard [41]. This system is used to reproduce loading conditions similar to those observed when the sock is worn and measure the fabric stiffness k. The set up is illustrated in Fig. 2.

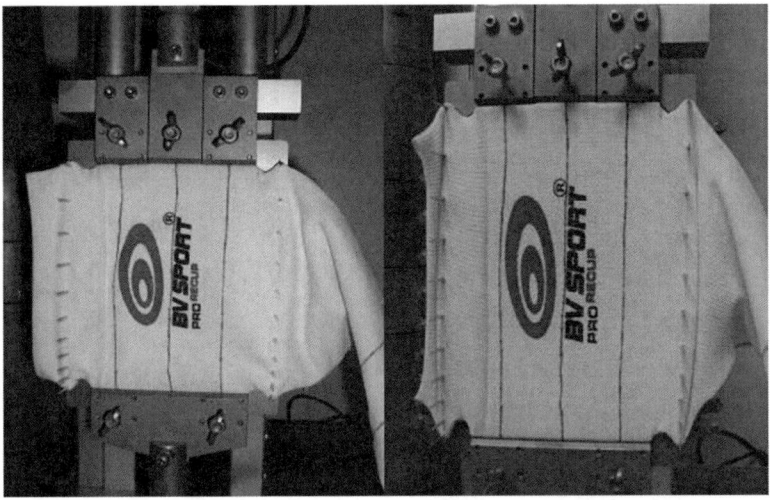

Fig. 2 Picture of the tensile test used to characterize the fabric stiffness

The value of k is estimated with an accuracy of 14%. It depends only on the vertical coordinate, i.e. it is constant in horizontal planes. It is also constant from the ankle to height 60 mm, then increases linearly up to height 140 mm and is constant again in the calf region.

The local curvature radius r used for the Laplace law is computed from the CT-scans using image analysis of the leg wearing compression socks.

Friction of the sock on the leg is neglected, which means that no tangential stress is considered. Another approach would be to include the sock in the model and to simulate its application onto the leg. This approach has been developed in two-dimension and results proved that the distribution of the applied pressure fits the one predicted by the Laplace law [16]. It was decided to use the Laplace law in three dimensions in order to reduce the complexity of the problem posed by the contacts between the sock and the leg.

4.1.3 Material Properties

Choice of the Constitutive Equations

Soft tissues (adipose and muscle tissues) are defined as isotropic, compressible and hyper-elastic materials. A Neo–Hookean strain energy function is used as constitutive equations [9, 39]. This energy function models the compression behaviour of the equivalent material: aponeurosis, veins and arteries are included in the muscle material and skin and some veins are included in fat. Anisotropy of the muscles is not considered, its influence is marginal for the response of the leg to compression, which is mainly a uniaxial loading. The contractile behaviour of the muscle is not considered either as the analysis is focused here on the quasi-static passive response of the leg to compression.

The strain energy function can be written as:

$$W = C_{10}(I_1 - 3) + K_v[(J-1)/2 - \ln(J)]$$

where $J = \det(\mathbf{F})$ and $I_1 = J^{-2} \operatorname{Tr}({}^t\mathbf{F}/\mathbf{F})/3$ with $\mathbf{F}$ being the transformation gradient and Tr the trace of a tensor or matrix.

The behaviour of each tissue is driven by two parameters:

- C_{10}^m and K_v^m for the muscle tissue.
- C_{10}^f and K_v^f for the fat.

These constitutive parameters have to be identified by an inverse method for ensuring patient-specificity of the constitutive parameters. The inverse approach is detailed in the following paragraph.

Characterization of the Leg Deformation Induced by Compression

In 29 horizontal cross sections (one centimetre apart), Fourier descriptors that fit the contours are determined in the volume images of the leg before and after wearing the compressive sock. For each of the 29 cross sections, 2 Fourier descriptors are computed:

- 1 for the external contour of the leg.
- 1 for the contour separating the muscle compartments and the adipose tissue.

Using these contours, two scalars are computed at the 29 horizontal cross sections of the leg:

1. the cross section area of the whole leg at a given height numbered k, denoted $Al(k)$ before wearing the sock and denoted $al(k)$ after wearing the sock.
2. the cross section area of the muscle compartments at height k, denoted $Am(k)$ before wearing the sock and denoted $am(k)$ after wearing the sock.

For example, Al and al are plotted for the whole height of the leg in Fig. 3. The gap between Al and al is used to quantify the deviation of volume images between the compressed and uncompressed state. The larger this gap, the larger the deformation of the leg induced by the compression (Fig. 4).

Identification of Soft Tissue Properties

In order to determine the material properties of soft tissues, an identification method based on the bounded Levenberg–Marquardt optimisation algorithm (BLVM) is coded in Matlab®. The cost function to minimize is built up from the

Fig. 3 Definition of the cross section areas of the leg and of muscles

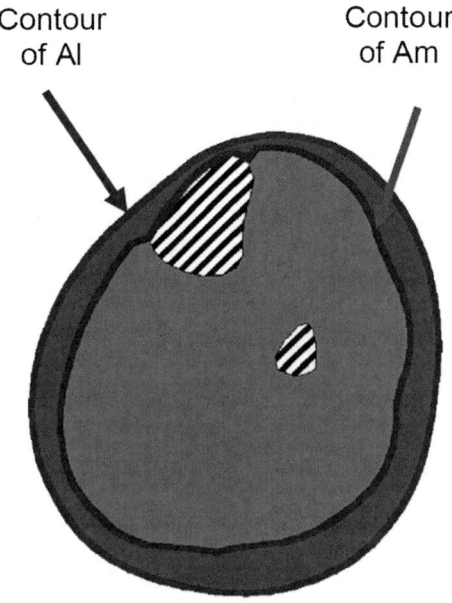

principles presented above. From the FE model of compressed leg, the action of compression onto the cross section areas of the leg is computed numerically. The response only depends on the undetermined material parameters C_{10}^m, C_{10}^f, K_v^m and K_v^f. Let X denote the vector set of unknown material parameters. For a given set of X, let $al_X(k)$ denote the cross section area of the leg at height number k in its deformed geometry computed by the FE model with the set of parameters X. Similarly, let $am_X(k)$ denote the cross section area of the muscle compartments at height k in the deformed state computed by the FE model with the set of parameters X (Fig. 5).

Eventually, the choice of material parameters is ranked by the following cost function:

$$C(X) = \Sigma_k [al(k) - al_X(k)]^2 + [am(k) - am_X(k)]^2$$

It is assumed that the actual values of the set of material parameters, denoted Y, are such that:

$$am_Y = am \text{ and } al_Y = am,$$

So that:

$$C(Y) = 0.$$

In practise, due to experimental artefacts, we can only try to minimize C but not to zero it. The method for identifying the actual material parameters is to solve the following minimization problem:

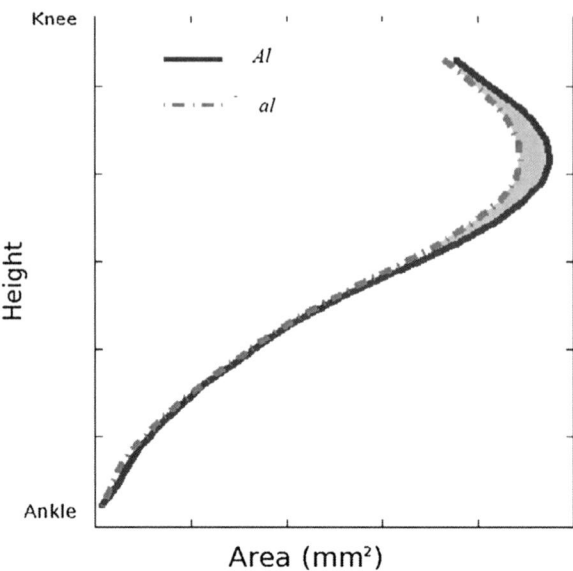

Fig. 4 Cross section areas of the leg varying across the height for a given subject before and after wearing a compressive sock

$$Y = \text{Arg } \min_X(C(X))$$

The BLVM algorithm is used to solve this minimization problem since BLVM is dedicated to least-square minimization problems. It requires computing the gradient of C with respect to the parameters, which is performed by backward finite differences.

The global principle of the method is schemed in Fig. 6. Two stopping criteria have been used: one threshold on the value of C (i.e. good quality of identification is reached), and another one on the norm of the parameter increments (i.e. no more improvement can be reached).

4.2 Obtained Results for Different Subjects

The preceding approach was applied to compute the distribution of pressure inside the tissues of the leg and to characterize inter-individual variations.

Three-dimensional CT-scans of the right leg were acquired on volunteers without venous pathology, following informed consent and according to a protocol approved by the local institutional ethics committee. The scans are acquired with and without elastic compression in seated position. The used elastic compression is the BVSport® Pro Recup sock, the size of the sock is chosen in agreement with the perimeter of the leg at the calf location, in order to deliver theoretically an average pressure of 3.5 kPa ≈ 25 mmHg at the calf.

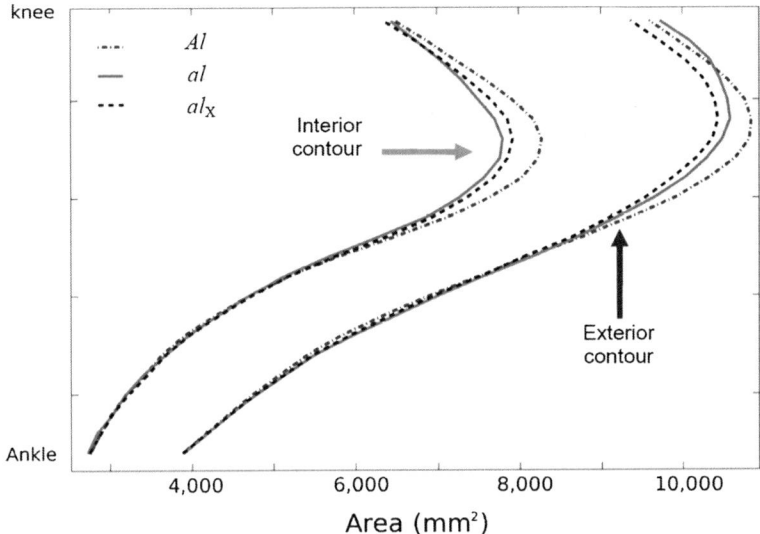

Fig. 5 Cross section areas before and after minimizing C

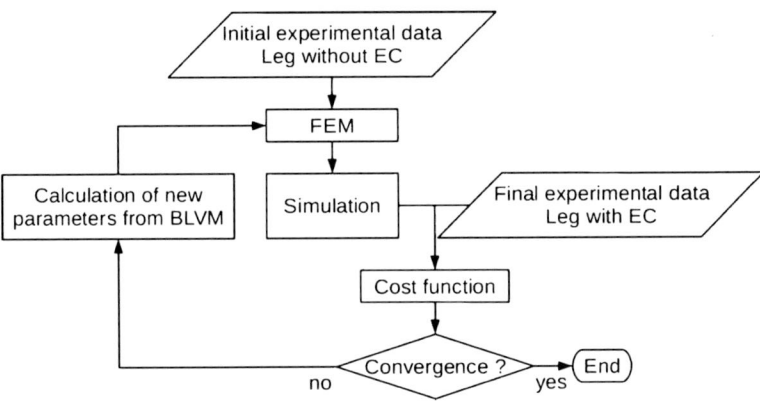

Fig. 6 Flowchart of the identification approach

We present here the results for three different leg morphologies, whose cross sections at the calf are shown in Fig. 7.

The obtained results are analysed in terms of pressure distribution. The hydrostatic pressure in the soft tissues of the leg, computed for each of the 3 patient-specific models, is shown in Fig. 8 at 4 different cross sections. The hydrostatic pressure in the tissues is an important mechanical parameter for different reasons:

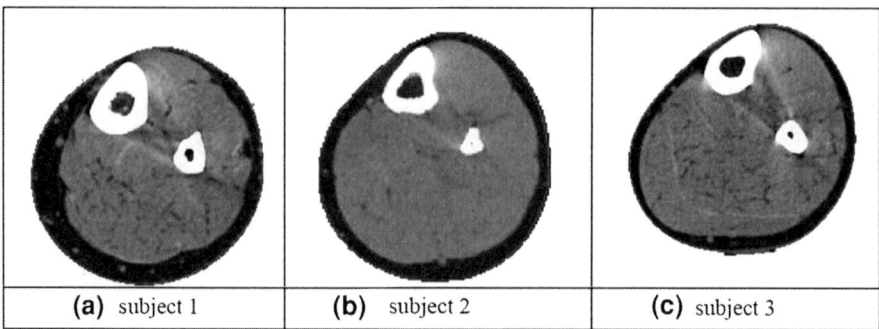

Fig. 7 Cross section image of three different legs on which the patient-specific modelling approach has been applied for characterizing the distribution of pressure inside the leg

- in the deep tissues, the hydrostratic pressure of the tissue is the pressure outside the veins induced by the compression. The circumferential stresses in the walls of the veins have to balance the difference between the blood pressure inside the vein, and the hydrostatic pressure in the tissues surrounding the veins.
- in the skin and adipose tissue, the pressure can be directly correlated to comfort aspects as these zones have many sensorial receptors. An important aspect is also the effect of the hydrostatic pressure on the development of ulcers [11].

The main conclusions regarding these results are the following:

1. the average pressure applied to the leg is transmitted towards the deepest tissues without significant attenuation, the pressure being always of 20 mmHg = 2.5 kPa in the deepest tissues. Compared to the blood pressure in the veins in the standing position (about 100 mmHg), this pressure is not large enough for affecting significantly the calibre of the veins. However, in the supine position, the pressure is relatively significant and may induce some important reduction of the vein calibre even for the deep veins. This was confirmed by in vivo measurements involving echodoppler [7] and/or MRI measurements [42].
2. the main spatial variations in the local distribution of the pressure in a given subject at a given cross section are induced by the morphology, especially the local variations of the radius of curvature. For an average pressure of 25 mmHg delivered by the sock onto the skin, local pressure of 55 mmHg may occur in the region of small radii of curvature. This is important to consider because the local pressure can exceed the blood pressure in some veins for someone in the supine position. This may induce the collapse of the vein wall, occlude the vein and obstruct the blood flow.
3. large inter-individual variations may occur depending on the individual morphology and tissue compliance. Legs with larger proportion of fat tend to be rounder after compression, inducing a more uniform distribution of the pressure.

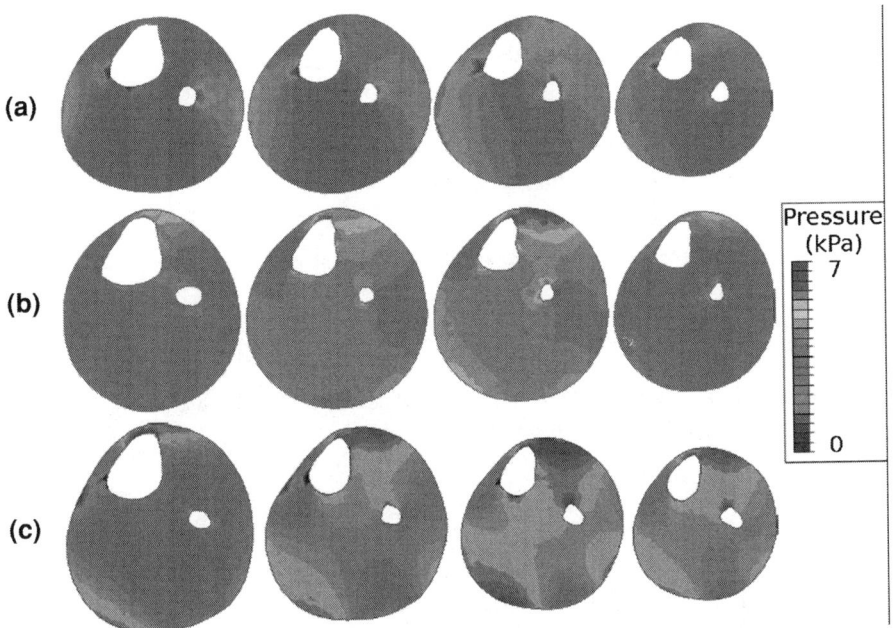

Fig. 8 Pressure distribution computed with the patient-specific approach for 3 subjects at 4 different cross sections of their leg

These results have important consequences on the design of compressive garments. The study is continuing on a large number of subjects for confirming these trends.

5 Conclusions and Perspectives

In this chapter, after presenting the state of the art regarding finite element modelling of elastic compression and its effects for treating venous deficiency, a novel methodology dedicated to set up patient-specific models of compressed legs has been presented. It has permitted to compute pressure distribution inside the leg tissues for different subjects and the methodology is now ready for applications to clinical study and as a design tool for the compressive garments.

Fundamentally, the approach is still under improvement, especially on the following points:

- patient-specific modelling of blood return in the compressed leg, combining computational fluid dynamics and our solid mechanics models in a FSI approach;

- taking into account the contractile behaviour of muscles for modelling the effect of compression during sport;
- modelling the pressure around a vein and modelling the evolution of varicose veins under compression;
- tying up the local pressure in soft tissues to their remodelling and possible development of ulcers around varicose veins.

It is expected that these improvements open the way to important clinical applications, especially regarding the development of patient-specific compressive treatments for deep thrombosis and/or ulcers.

Acknowledgments The authors are grateful to the funding partners involved in this study: The Rhône-Alpes regional Council through a CIBLE2008 grant and the French National Agency of Research through a ANR-08-JCJC-0071 grant. The authors are also grateful to the BVsport Company for supplying materials to this study.

References

1. Bergham, J.J.: The Vein Book. Elsevier, New York (2007)
2. www.phlebology.org
3. Neglén, P., Raju, S.: Differences in pressures of the popliteal, long saphenous, and dorsal foot veins. Twelfth Annual Meeting of the American Venous Forum, Phoenix, Ariz, 3–6 Feb 2000. The Society for Vascular Surgery and the American Association for Vascular Surgery, a Chapter of the International Society for Cardiovascular Surgery (2000)
4. Morris, R.J., Woodcock, J.P.: Intermittent pneumatic compression or graduated compression stockings for deep vein thrombosis prophylaxis? A systematic review of direct clinical comparisons. Ann. Surg. **251**(3), 393–396 (2010)
5. Ducrozet P.: Un nouveau concept de contention-compression veineuse évalué par un essai randomisé en double aveugle : contention progressive versus contention classique dans l'insuffisance veineuse légère : l'étude BOOSTER. Thèse de médecine de Saint Etienne n°44, Dec 2004
6. Pouget, J.F., Geedroyc, W.M.W., Prüfer, M.: Nuclear magnetic resonance (NMR) study of leg venous drainage in men with and without elastic compression socks (traditional and new bvsport concept). Int. Angiol. **19**, 41 (2000)
7. Couzan, S., Prüfer, M., Ferret, J.M.: Un nouveau concept de contention-compression : apport de l'écho-Doppler couleur avec prise des pressions veineuses et de l'IRM. Phlébologie **55**, 159–171 (2002)
8. Couzan, S., Pouget, J.F., Prüfer, M.: Study of the athletes venous system by Doppler scan with pressures measurement and the perfecting of a new elastic compression concept. Int. Angiol. **19**, 41 (2000)
9. Dubuis, L., Avril, S., Debayle, J., Badel, P.: Identification of the material parameters of soft tissues in the compressed leg. Computer Methods in Biomechanics and Biomedical Engineering, 2011, (in press)
10. Couzan, S., Assante, C., Laporte, S., Mismetti, P., Pouget, J.F.: Booster study: comparative evaluation of a new concept of elastic stockings in mild venous insufficiency. Presse Med. **38**(3), 355–361 (2009)
11. van Donkelaar, C.C., Huyghe, J.M., Vankan, W.J., Drost, M.R.: Spatial interaction between tissue pressure and skeletal muscle perfusion during contraction. J. Biomech. **34**(5), 631–637 (2001)

12. Guesdon, P., Fullana, J.M., Flaud, P.: Etude expérimentale du drainage musculaire. C. R. Mec. **335**(4), 207–212 (2007)
13. Cros, F., Flaud, P., Dantan, P.: A digital model for the venous junctions. Comput. Meth. Biomech. Biomed. Eng. **5**(6), 421–429 (2002)
14. Downie, S., Raynor, S., Firmin, D., Wood, N., Thom, S., Hughes, A., Parker, K., Wolfe, J., Xu, X.Y.: Effects of elastic compression stockings on wall shear stress in deep and superficial veins of the calf. Am. J. Physiol-Heart C. **294**(5), H2112–H2120 (2008)
15. Aubert, J.T., Bassez, S., Louisy, F., Ribreau, C.: Accélérations dans le réseau veineux du membre inférieur au cours de la marche stationnaire. Paper presented at: XVeme Congrès Francais de Mécanique, pp. 3–7. Centre Universitaire Marseille Saint-Charles, Marseille (2001)
16. Fung, Y.C.: Biomechanics: Mechanical Properties of Living Tissues. Springer, Berlin (1993)
17. Kamm, R.D., Shapiro, A.H.: Unsteady flow in a collapsible tube subjected to external pressure or body forces. J. Fluid Mech. **95**(1), 1–78 (1979)
18. Kamm, R.: Bioengineering studies of periodic external compression as prophylaxis against deep vein thrombosis–part I: numerical studies. J. Biomech. Eng. **104**(1), 87–95 (1982)
19. Gaied, I., Drapier, S., Lun, B.: Experimental assessment and analytical 2D predictions of the stocking pressures induced on a model leg by medical compression stocking. J. Biomech. **39**, 3017–3025 (2006)
20. Lee, W.C.C., Zhang, M.: Using computational simulation to aid in the prediction of socket fit: A preliminary study. Med. Ing. Phys. **29**, 923–929 (2007)
21. You, F., Wang, J.M., Liao, G.J.: The simulation of elastic human body deformation and garment pressure with moving Mesh method. Stud. Comp. Intell. **55**, 289–300 (2007)
22. Zhang, M., Dai, X., Li, Y., Cheung, J.M.: Computational simulation of skin and sock pressure distributions. Stud. Comp. Intell. **55**, 323–333 (2007)
23. Partsch, H., Partsch, B., Braun, W.: Interface pressure and stiffness of ready made compression stockings: comparison of in vivo and in vitro measurements. J. Vasc. Surg. **44**(4), 809–814 (2006)
24. Li, Y., Zhang, X., Yeung, K.: A 3D Biomechanical model for numerical simulation of dynamic mechanical interactions of bra and breast during wear. Sen'I Gakkaishi **59**(1), 12–21 (2003)
25. Dai, X., Liu, R., Li, Y., Zhang, M., Kwok, Y.: Numerical simulation of skin pressure distribution applied by graduated compression stockings. Study Comp. Intell. **55**, 301–309 (2007)
26. Maton, B., Thiney, G., Ouchene, A., Flaud, P., Barthelemy, P.: Intramuscular pressure and surface EMG in voluntary ankle dorsal flexion: influence of elastic compressive stockings. J. Electromyogr. Kines. **16**(3), 291–302 (2006)
27. Musani, M., Matta, F., Yaekoub, A., Liang, J., Hull, R., Stein, P.: Venous compression for prevention of post-thrombotic syndrome: a meta-analysis. Am. J. Med. **123**(8), 735–740 (2010)
28. Oduncu, H., Clark, M., Williams, R.: Effect of compression on blood flow in lower limb wounds. Int. Wound J. **1**(2), 107–113 (2004)
29. Amsler, F., Willenberg, T., Blttler, W.: In search of optimal compression therapy for venous leg ulcers: a meta-analysis of studies comparing divers bandages with specifically designed stockings. J. Vasc. Surg. **50**(3), 668–674 (2009)
30. Gefen, A.: Stress analysis of the standing foot following surgical plantar fascia disease. J. Biomech. **35**, 629–637 (2002)
31. Portnoy, S., Yizhar, Z., Shabshin, N., Itzchak, Y., Kristal, A., Dotan-Marom, Y., Siev-Ner, I., Gefen, A.: Internal mechanical conditions in the soft tissues of a residual limb of a trans-tibial amputee. J. Biomech. **41**, 1897–1909 (2008)
32. Linder-Ganz, E., Shabshin, N., Itzchak, Y., Gefen, A.: Assessment of mechanical conditions in sub-dermal tissues during sitting: a combined experimental-MRI and finite element approach. J. Biomech. **40**, 1443–1454 (2007)

33. Gefen, A., Chen, J., Elad, D.: A biomechanical model of Peyronie's disease. J. Biomech. **33**, 1739–1744 (2000)
34. Oomens, C., Bressers, O., Bosboom, E., Bouten, C., Bader, D.: Can loaded interface characteristics influence strain distributions in muscle adjacent to bony prominences? Comput. Meth. Biomech. Biomed. Eng. **6**(3), 171–180 (2003)
35. Chung, J.H., Rajagopal, V., Nielsen, P.M.F., Nash, M.P.: A biomechanical model of mammographic compressions. Biomech. Model Mechanobiol. **7**(1), 43–52 (2008)
36. Azar, F., Metaxas, D., Schnalli, M.: Methods for modeling and predicting mechanical deformations of the breast under external perturbations. Med. Image Anal. **6**(1), 1–27 (2002)
37. Vogl, T.J., Then, C., Naguib, N., Nour-Eldin, A., Larson, M., Zangos, S., Silber, G.: Mechanical soft tissue property validation in tissue engineering using magnetic resonance imaging. Acad. Radiol. **17**, 1486–1491 (2010)
38. Bouten, L., Drapier, S.: In vivo identification of soft biological tissues using MR imaging. Eur. J. Comp. Mech **18**(1), 21–32 (2009)
39. Avril, S., Bouten, L., Dubuis, L., Drapier, S., Pouget, J.F.: Mixed experimental and numerical approach for characterizing the biomechanical response of the human leg under elastic compression. ASME J. Biomech. Engrg. **132**(3), 031006 (2010)
40. Zahn, C.T., Roskies, R.Z.: Fourier descriptors for plane closed curves. IEEE Trans. Comput. **c-21**(3), 269–281 (1972)
41. Le comité technique CEN/TC205: Bas médicaux de compression. Prénorme Européenne, Comité Européen de normalisation CEN, 2001
42. Partsch, H., Mosti, G., Mosti, F.: Narrowing of leg veins demonstrated by magnetic resonance imaging (MRI). Int. Angiol. **29**(5), 408–410 (2008)

Part IV
Airways

Scan-Based Flow Modelling in Human Upper Airways

Perumal Nithiarasu, Igor Sazonov and Si Yong Yeo

Abstract In this chapter, an overview of a scan-based modelling technique to investigate the air flow and heat transfer in a human upper respiratory system is presented. The scan-based modelling process includes image segmentation, producing a valid mesh for analysis and flow modelling. All the three aspects are briefly discussed in this work.

1 Introduction

The fundamentals of air flow through a human respiratory tract and the associated forces exerted by the air motion are the essential ingredients for understanding many airway-related problems in humans, such as sleep apnoea and other sleeping disorders and nasal flow obstruction. The problems associated with the human respiratory tract may include asthma [1], airway stenosis [2], obstructive sleep apnoea [3, 4], throat cancer [5], nasal airway blockage [6] and chronic obstructive pulmonary disease [7]. The upper human airways have been recently receiving serious attention from the patient-specific numerical modelling community due to the fact that sleep apnoea, throat cancer and nasal airway blockage are becoming

P. Nithiarasu (✉) · I. Sazonov · S. Y. Yeo
Computational Bioengineering Group, College of Engineering,
Swansea University, Swansea SA2 8PP, UK
e-mail: P.Nithiarasu@swansea.ac.uk

I. Sazonov
e-mail: I.Sazonov@swansea.ac.uk

S. Y. Yeo
e-mail: genyeoius@gmail.com

more prevalent in developed countries. The modelling trend in this area is changing and many recent studies use realistic geometries to investigate flow in human upper airways. To study any of the disorders or to understand the fundamental aspects of human upper airway fluid dynamics, it is essential to create a problem that is as close to reality as possible. Here, scan-based modelling approaches provide a realistic platform for modelling human upper airway fluid dynamics. Such an approach is particularly suited for human upper airways as the large inter-human variability of geometries due to age, gender, ethnic origin and the state of the human upper airway makes the use of simplified and ideal geometries redundant.

The general structure of any scan-based human upper airway flow modelling framework or pipeline consists mainly of image segmentation, meshing and solution stages. The image segmentation stage depends heavily on the type of available image and its resolution. The image processing is normally followed by a meshing stage in which a domain discretisation is carried out. The connection between the image processing and meshing is established via a geometry definition step. Majority of the efforts during the mesh generation process goes towards establishing a valid and high quality surface mesh that is a close representation of the geometry. The surface meshing in general is followed by the boundary layer mesh construction and wall meshing (if applicable). The boundary layer and arterial wall volume discretisations are followed by automatic volume meshing of the central flow domain of the human upper airway. The flow solver is then used along with appropriately generated boundary conditions to complete the pipeline. The solver may include problems with rigid geometry, flow and structure coupling or turbulence. In the present chapter, our focus is on all the three stages of such a pipeline.

One of the main challenges in the computational modelling of human upper airway flow is the accurate reconstruction of the geometry. Anatomically accurate geometric models of the nose and middle airways are essential for realistic flow simulations and analysis. The anatomical information used to reconstruct the geometries are usually provided in the form of medical image datasets (scans) from imaging modalities such as computed tomography (CT) and magnetic resonance (MR) imaging. Manual reconstruction of the geometries can be tedious and time consuming. There is also the issue of variability between the geometries extracted manually by different individuals, and variability of geometries extracted by the same individual on different occasions. In order to allow the computational flow modelling to be efficiently applied as a diagnostic or predictive tool, the amount of user intervention required in the process should be reasonably small so that a large number of geometries can be rapidly processed. Therefore, a robust and efficient method that can be used to accurately segment the geometry from medical image datasets can be very useful and advantageous in the modelling process [8].

Assuming that a reasonably accurate reconstruction of the geometry is possible, the next stage of the pipeline would be to generate a surface mesh or meshes to accurately discretise the surface or surfaces of the airway geometry. Unlike well defined standard engineering geometries, the reconstructed subject-specific, human upper airway geometries are often defined by binaries. Thus, an alternative approach is required to that of the standard engineering applications [9].

The analysis of air flow through subject-specific upper human airway geometries is currently one of the widely researched topics of patient-specific modelling [10]. Fluid flow modelling through static geometries is more or less well established except for large scale simulations at high flow rates and extremely complex geometries. However, correct application of flow boundary conditions is still an issue for both steady and unsteady state problems. On the other hand, topics such as fluid-structure interaction has also been studied but with little or no success. The main problem in fluid-structure interaction studies is the lack of availability of subject-specific, in vivo material properties and difficulties associated with imaging and segmentation of walls. One way of overcoming the material property issue is via transiently registering the image and imposing the wall motion as a boundary condition. However, if accurate wall stress distribution is of interest, imposing wall motion may not be the solution.

Before discussing the scan-based flow modelling in some detail, it is essential to provide a brief overview on the anatomy and various disorders of human upper airways. This is essential to give a purpose to the flow modelling procedure presented in this chapter. Thus, a brief introduction to the human upper airway anatomy and disorders is presented in Sect. 2. Although the focus of the chapter is scan-based modelling procedure of flow in human upper airways, the problems of practical interest here are generic flow structure and ventilation of inhaled air. Thus, in Sect. 3, we provide a brief literature survey on these two aspects of human upper airways. The idea here is not to present a serious critique of the available literature but to provide a summary of literature on modelling aspects and results in the two areas mentioned. Section 4 describes a deformable model based segmentation method (first part of the scan-based modelling pipeline) for human upper airway reconstruction and the meshing methods and cosmetics used in the present study are explained in Sect. 5 (second part of the pipeline). In Sect. 6, the flow modelling steps, including boundary condition generation, boundary condition cosmetics, governing flow equations and their solution method are provided (third and final part of the pipeline). Four examples, including a model validation problem, are discussed in Sect. 7 and finally, Sect. 8 draws some conclusions and lists the challenges and unresolved problems.

2 Anatomy and Disorders of the Human Upper Airways

2.1 Anatomy

To understand the need for, and importance of, patient-specific human upper airway flow simulation, it is necessary to understand the anatomy of the airway and the common effects of disease on airflow. Figure 1 shows a diagram of a

Fig. 1 Anatomy of human upper airways

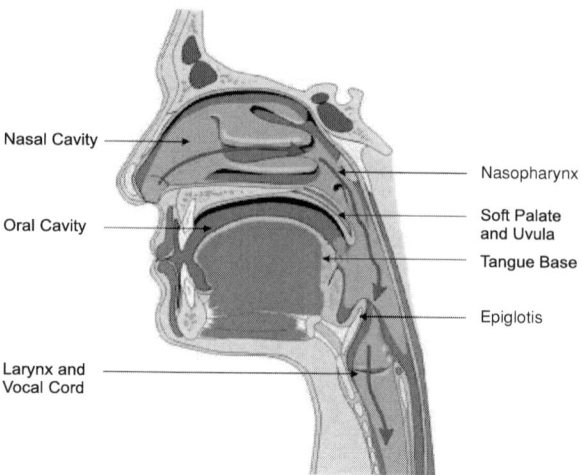

vertical cross section (sagittal) of a typical human upper airway. It is usually divided into anatomically distinct sections, based on anatomical, physiological or pathological considerations. We have divided the upper airway into six segments: the nasal cavities, nasopharynx (including the inferior portion or velopharynx), oropharynx, laryngopharynx, larynx and trachea [4]. The nasal cavities extend from the external nares (nostrils) to the posterior choanae, where the two cavities open into the nasopharynx. The lower part of the nasopharynx (velopharynx) lies behind the soft palate and is highly flexible and constantly changes in shape. It is a common site of snoring. The oropharynx extends from soft palate to the level of the tongue base and the laryngopharynx from tongue base to larynx. The larynx is a valve separating the pharynx and the trachea. The trachea (windpipe) extends from the larynx into the chest and lungs. The walls of the upper airway are formed mostly of soft tissues, including over 20 muscle groups, adipose tissue (fat), tonsil tissue and a flexible mucosal lining. The muscles actively constrict and dilate the lumen of the upper airway [11–13]. In the nasal cavities, larynx and trachea, the walls also contain more rigid cartilaginous and bony structures. A large number and wide range of cross-sectional measurements of the airway have been made [4]. The minimum airway size in the awake state occurs at one of the three levels: the anterior nares (nasal valve), the velopharynx (behind the soft palate) and at the junction of larynx and trachea (the subglottis) [14–16]. During sleep, the two narrowest portions are at the velopharynx and in the oropharynx at tongue base level. These two sites are of great interest as likely areas of soft tissue collapse of the airway, causing snoring and, most importantly, obstructive sleep apnoea. This makes the oropharynx the site of most interest for modelling. The posterior and lateral walls contain the superior, middle and inferior constrictor muscles, while the anterior wall comprises soft palate and tongue base. The laryngeal and tracheal parts of the airway are much less variable during sleep. However, in the awake state there are significant changes in glottic (laryngeal) area. During quiet breathing the vocal cords open during inspiration, to widen the airway, and close a little during expiration. During phonation the vocal cords are closely approximated with near complete closure of the airway.

2.2 Human Airway Disorders

The upper airway is susceptible to many diseases. We will focus on three common disorders, each of which predominantly affects one section of the airway.

The first problem of interest is that of vocal cord paralysis. This may be caused by cancer in the chest or neck or the result of surgical trauma to nerves supplying the larynx. Good quality, loud sound can only be produced when the vocal cords are closely approximated. A vocal cord, paralysed in a lateral position, results in a gap during phonation, producing a weak, breathy voice. It is not usually possible to re-innervate the larynx but insertion of a prosthesis to push the paralysed cord into an optimal position for vocalizing, can allow excellent voice production. The difficulty is that, too small a prosthesis will result in a poor voice and, too large, will narrow the airway, causing breathlessness. The final decision must be a compromise between voice and airway. We believe that preoperative modelling will allow more accurate prediction of the effects of prosthesis size on both voice and airway. Although the fluid-structure interaction in this problem is important, the mechanism is normally simple. Vocal cords move up and down and laterally and medially in a systematic way and a prescribed motion may be sufficient to determine an optimal flow rate. The geometry is not as complex as that of the nasal passages although scan resolution is still an issue.

The second, increasingly common, problem is that of sleep apnoea. This has received more attention than the previous problem due to the larger numbers affected and the concern that sleep apnoea may be an important factor in some road accidents and in early death from hypertension and heart disease. However, patient-specific modelling of upper airway collapse has so far not been successfully attempted. The major challenge in this problem is that of tackling the combination of fluid forces, structural movement and the neuromuscular activities contributing to airway collapse. Coupling passive material with fluid is difficult but neuromuscular actions make the problem even more complex. On the positive side, sleep apnoea is routinely investigated, providing high-quality digital audio-visual data during controlled sleep studies.

The third problem is of interest is the nasal blockage. This problem is more prevalent than the first two and often it requires surgery. The common causes of nasal blockage includes medically treatable conditions such as common cold, bacterial sinusitis, allergy, sensitivity to dust, smoke, pollution and other irritants. Mechanical causes of nasal blockage include anatomic abnormalities such as a deviated septum, nasal polyps, obstructed sinuses, over-enlarged turbinates and obstructing adenoids. One very common reason for nasal blockage is a deviated septum. The septum is the wall that divides the nose down the middle, into a right and left side. It is made of cartilage and bone and has a mucous membrane lining on both sides. When the septum is straight, it simply acts as the divider of the nose and allows for streamlined airflow and easy breathing. If it is deviated or twisted, it can cause nasal obstruction. The septum can twist to the right and block the right side, and then come around further back in the nose and twist to the left to block the left side as well. If the septum is causing nasal obstruction, only surgery (septoplasty) can correct it. For an optimal planning of septoplasty, a surgical flow simulation a priori to the surgery may be useful.

In all the three problems, the unifying factor is the study of patient-specific fluid dynamics. Irrespective of other factors involved, it is essential to develop a procedure allowing repeatable fluid dynamics studies to be performed. This chapter is a demonstration of such a patient-specific fluid dynamics study in a human upper airway. Before describing of the various stages of a modelling framework, we summarise the state of the art of the flow modelling in the following subsection.

3 A Brief Literature Survey

There are many fundamental studies available on human upper airway fluid dynamics, but many of them are carried out on a non-realistic geometry [17]. The basic human airway studies have generally concentrated on simplified geometries assuming, for instance, perfectly circular cross sections of the upper airway or circular triple bifurcation geometries [18–20]. Collapsible tubes have also received significant attention in recent years [21, 22]. It is assumed that some of the theory of collapsible tubes may be extended to help the understanding of upper airway closure. Other basic studies include large eddy simulation (LES) of flow in simplified human airways [23]. The majority of applications in the literature concentrate around the topic of particle transport in upper and lower human airways. Studies on this topic include application to spray dynamics and smoke particles [24–33]. In these studies, in addition to a fluid dynamics algorithm, an efficient way of tracking the particles is essential. The particle tracking studies reported the use of a mixture of simplified geometries and real geometries extracted from real human airways. The fluid dynamic studies on realistic upper human airway geometries are emerging fast [34–42]. However, focus of these studies is the flow analysis and they have given little attention to either reconstruction of the geometries or mesh generation.

The inclusion of the nasal cavity into a computational model of flow through a human upper airway has been restricted by the ability of an imaging technique to capture the narrow passages in the nasal cavity in detail and the meshing technique to generate mesh inside a complex geometry. However, recent growth in the number of published studies on nasal cavity flow [35–42] shows some advances in medical imaging techniques, digital image processing and mesh generation methods. However, high quality CT images are still an issue due to radiation dosage associated with increase in image resolution. Reconstruction and meshing of the airway geometries, even now, needs a significant amount of effort and time. Only a very limited number of works on human airway flow studies include nasal cavity and trachea. Among them the work of Jeong et al. [36] and Mylavarapu et al. [43] are notable. Another influential factor on the flow pattern in the human conducting airway is the inlet flow conditions. It is often easy to assume a flow rate and adjust the inlet flow conditions to match the flow rate if only one inlet exists, such as the oral cavity. In the case of nasal cavity, there are two independent inlets. Often these nares are not equally sized and thus equally dividing the flow between the nares may be questionable even for steady flow cases.

In the past, the majority of the published studies on respiratory flow were considered to be laminar by assuming sedentary or quiet normal breathing characterised by low inlet velocity. Recently, this trend has been reversed and most publications now account for turbulent flow. Turbulent flow can be simulated using three different approaches, namely, direct numerical simulation (DNS), large eddy simulation (LES), and Reynolds averaged Navier-Stokes (RANS). DNS solves Navier-Stokes equations directly where all length and time scales are resolved. This approach is computationally very expensive and therefore its application currently is limited only to low-to-moderate Re flows. In LES, large scale turbulent eddies are resolved directly while the unresolved small-scale eddies being represented by simple models. A number of papers have applied LES techniques to resolve turbulence in human respiratory flows [23, 43–45]. Often both DNS and LES studies need specially arranged structured meshes to obtain a correct solution. This is possible on model human airways but on subject-specific geometries, this needs very large computing resources. RANS models on the other hand still have the advantage of being a fast method with reasonable accuracy. Although the RANS models do not resolve turbulent eddies, the dynamics of the turbulent eddies are predicted using semi-empirical models. Thus, RANS has been widely used in engineering applications.

Different RANS models have been used by different authors as closures. Among them the one-equation model proposed by Spalart-Allmaras [17, 43, 46, 47] is appealing for solving flow in complex geometries. This is due to the fact that this model is easy to implement, not expensive and naturally lend itself for extending it to more advanced detached eddy simulation (DES). Several studies have been carried out to validate the use of RANS models in respiratory flow with the experimental results. Stapleton et al. [48] presented a numerical simulation of particle deposition in an idealised model of mouth and throat geometry. They state that the characteristics of flow in a complex mouth and throat geometry cause two equation turbulence models to perform poorly. In a recently published article by Mylavarapu et al. [43], the validation was carried out on a subject-specific human upper airway geometry. Comparing between the computed and measured wall static pressure they found that the standard $k - \omega$ model provides the best agreement with the experimental measurements and outperforms the LES and the Spalart-Allmaras models. This finding contradicts the popular belief that LES is more accurate and reliable than

RANS. Our previous experience on both model and realistic upper human airways clearly shows that Spalart-Allmaras model performs better [17, 47, 49].

The human lungs operate under a very tight physiological condition. Any small variation above or below its normal working condition makes the lungs susceptible to pathological conditions that may impair normalcy. In order to keep the lungs healthy, inspired ambient air must be filtered and conditioned to the aveolar condition (i.e., fully saturated at body temperature) before entering the lungs. Studies by Webb [50], Cole [51] and Inglestedt [52] show that the heat and mass transfer associated with air conditioning process is primarily conducted in the nasal cavity.

In the past, relationship between airflow pattern and air conditioning in a nose was predominantly investigated using experimental studies via in vivo temperature measurements. However, in vivo measurement is severely restricted by two difficulties. The first one is the difficulties associated with obtaining good spatial and time resolution. The second one is the complexity of nasal cavity severely limiting access to all parts of the nose. However, recent advances in numerical modelling tools [53–56] currently allows the heat transfer process in a human respiratory system to be studied in detail. In general, the numerical modelling studies can be grouped into two different categories.

The first and most popular one assumes that the conditioning of the inspired ambient air is carried out primarily in the nasal cavity. Therefore, most authors considered only nasal geometry in their study on respiratory transport phenomena [39, 57–60]. The early published works on nasal cavity heat transfer employed simplified geometrical model and two-dimensional simulations at various transverse cross sections of the nose [57].

The second category of problems arise from the observation that there is a close relationship between respiratory heat and mass transfer and bronchoconstriction in patients suffering from exercise-induced asthma and the increase in risk of infection in patients with intubated airways. Only a limited number of publications belong to this class. They mainly concern with the intra-thoracic region of the respiratory system and disregard the nasal cavity from their model. The most notable are the works of Ingenito et al. [61], Zhang and Kleinstreuer [62] and Tawhai and Hunter [63]. Ingenito et al. [61] used finite difference methods to solve the heat transfer between the airway and the submucosal layer. Their airway model was based on a symmetrical dichotomously branching system. Zhang and Kleinstreuer [62] conducted a numerical simulation of heat and mass transfer in a human upper airway. The model of airways used in their study was a simplified human cast replica, consisting of a representative of extra-thoracic airway from oral cavity to trachea and a symmetric four-generation upper bronchial tree model. The steady 3D airflow are simulated for laminar as well as locally turbulent flow conditions using low-Re (LRN) $k - \omega$ model. Using both symmetric and anatomically consistent airway model geometries Tawhai and Hunter [63] simulated the thermofluid dynamics in the intubated airway from the trachea (Horsfield order 0) down to Horsfield order 6.

Most of the published articles on respiratory heat transfer, in particular those that only consider nasal geometry, assume laminar flow condition due to low peak velocities and low *Re* [57, 59], though some turbulence modelling have been attempted. This assumption is supported by several experimental studies on nasal cavity flow which show that for normal breathing condition with an airflow rate of less than 180 cm^3/s on one side of cavity, the flow is predominantly laminar over most of the cavity. Even though some instability was experienced, they did not develop into a fully turbulent flow [64, 65]. However, if the domain of interest is beyond the nasopharyngeal region of the airway, even under normal breathing condition, the airflow distal to this region is highly turbulent with turbulent intensity ranging between 5 and 20%, as pointed out by Lin et al. [66].

Despite the fact that a large number of works are being carried out in the area of modelling airflow through the human upper airway, the preprocessing stages still consume a large part of the analysis time. This is due to the fact that the geometry is complicated and the scan quality of the human upper airways can be very low. The flow analysis part is also not without difficulties. The questions about boundary conditions and turbulence are still open to interpretation. However, the preprocessing stages are mundane procedures but pose a great deal of challenge. Thus, the next two sections of this chapter are devoted to the first two stages (preprocessing) of the scan-based modelling framework.

4 Image Segmentation

Recent advances in medical imaging technology and digital image processing permit reconstruction of anatomically realistic geometries of airways. The anatomical information used to reconstruct the geometries is usually provided in the form of medical image data sets (scans) from imaging modalities such as computed tomography and magnetic resonance imaging. Typical resolution of such scans is around 0.5–1 mm which is close to the thickness of nasal bones and passages and small bronchi. Higher resolution CT scans are still an issue due to radiation dosage. The MRI resolution depends directly on the strength of magnetic field, but its resolution is poorer at the interface between the air and tissue than the CT scans. Thus, one of the main problems in airway geometry reconstruction is working with low resolution scans.

Direct manual reconstruction of the geometries can be tedious and time consuming. Some commercial softwares (AMIRA, MIMICS, etc) can be used if the scan resolution is good, and the user is prepared to intensively interact with the segmentation. Therefore, a robust and efficient method that can be used to accurately and automatically segment a geometry from medical image data sets can be very useful and advantageous in the modelling process. Here, we adopt our recently developed level set-based image segmentation technique [8]. A brief overview of this method and implementation details are given below.

Let the 3D gray level image be described by function $I(\mathbf{x})$ where $\mathbf{x} = [x, y, z]^T \in \mathcal{D}$ is a point in the image domain $\mathcal{D}$. Let Ω be an object to be segmented. We employ the level set method in which the object boundary $\partial\Omega(t)$ is defined through a level set function $\Phi(\mathbf{x}, t)$:

$$\partial\Omega(t) = \{\mathbf{x} : \Phi(\mathbf{x}, t) = 0\}.$$

The most often used PDE to compute $\Phi(\mathbf{x}, t)$ is (cg., [67])

$$\partial\Phi/\partial t = \alpha g(\mathbf{x})\kappa(\mathbf{x}, t)\|\nabla\Phi\| - (1 - \alpha)(\mathbf{F}(\mathbf{x}) \cdot \nabla\Phi) \qquad (1)$$

where α is a tuning parameter, $g(\mathbf{x}) = 1/(1 + \|\nabla I\|)$ is the edge stopping function, $\kappa(\mathbf{x}, t) = \nabla \cdot \hat{\mathbf{n}}$ is the mean curvature of the surface $\Phi = \text{const}$, (here $\hat{\mathbf{n}} = \nabla\Phi/\|\nabla\Phi\|$ is the unit normal vector to that surface), $\mathbf{F}(\mathbf{x}) = [F_x, F_y, F_z]^T$ is the flow function determined by image I. The flow function which is derived from image data act as an external force on the deformable model, and it is the most critical part in this kind of deformable model formulation.

In our approach the geometrical potential force (GPF) proposed in [8] is used:

$$\mathbf{F}(\mathbf{x}) = G(\mathbf{x}) \cdot \hat{\mathbf{n}}, \qquad G(\mathbf{x}) = P.V. \iint_{\mathbf{x}' \in \mathcal{D}} \frac{\mathbf{x} - \mathbf{x}'}{\|\mathbf{x} - \mathbf{x}'\|^{\lambda+1}} \cdot \nabla I(\mathbf{x}') d\mathbf{x}'. \qquad (2)$$

Here G is the scalar potential represented by the convolution $K * \nabla I$ where $K = P.V.(\mathbf{x}/\|\mathbf{x}\|^{\lambda+1})$ and $P.V.$ denotes the principle value. This force is always normal to the active surface $\Phi(\mathbf{x}, t) = 0$. A discrete analogue of the convolution kernel is $K(\mathbf{x}) = \mathbf{x}/\|\mathbf{x}\|^{\lambda+1}$ if $\mathbf{x} \neq \mathbf{0}$ and $\mathbf{0}$ if $\mathbf{x} = \mathbf{0}$. The fastest way to evaluate convolution (2) and compute the potential G is to apply the 3D Fast Fourier Transform (FFT) method.

Parameter λ depends on the domain dimension: $\lambda = 3$ for a 3D domain. It is possible to show that in the 2D case with $\lambda = 2$ this force coincides with a magnetic-active force [68], but in contrast to magnetic force the GPF can be defined and applied to images of any dimension (even to 4D, i.e., time varying 3D scans).

To set the initial condition we have to define the initial surface S_0 somewhere in the vicinity of the studied airway. An initial level set function is defined as a signed distance

$$\Phi(x_i, y_j, z_k, 0) = \mathsf{D}[S_0]$$

from the initial surface S_0. The signed distance can be computed efficiently using the algorithm described in [69]. At the boundaries of the image domain the zero Neumann boundary conditions are imposed.

Equation (1) is solved numerically by the finite difference method: using upwind and downwind differences for the flow $\mathbf{F} \cdot \nabla\Phi$ and central differences for other derivatives. Advancing in time is performed by the forward Euler method.

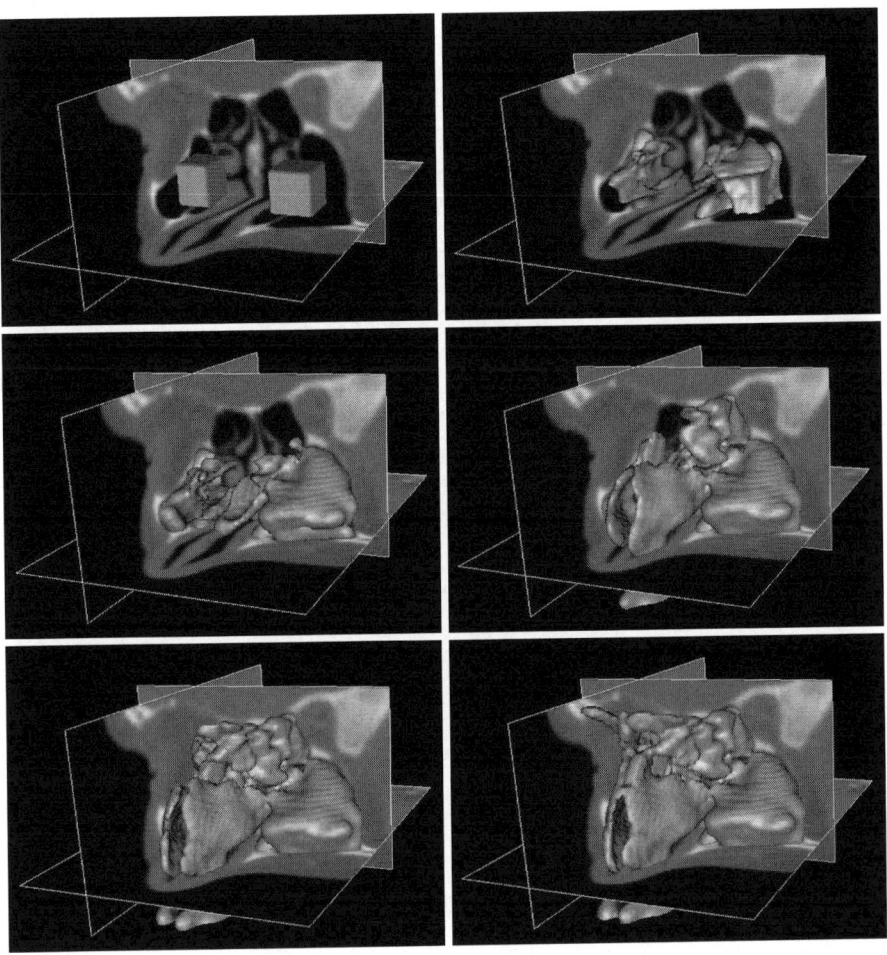

Fig. 2 Automatic segmentation of the human upper airway–from *left* to *right*, *top* to *bottom*: initial level set surface, intermediate stages of the level set evolution, and the converged deformable model

Compared with the other well known segmentation methods, the proposed one is more stable to topological complexity of the object, to image noise, to the initialization. It also shows much smaller leakage at the fuzzy edge. See details in Ref. [8]. The segmentation of the nasal part of upper airways using the GPF deformable model is depicted in Fig. 2. The resolution of the CT scan used is $0.877 \times 0.877 \times 1.0$ mm (1.0 mm is the distance between the slices). The binary level-set function obtained from the segmentation is used in the surface mesh generation procedure discussed in the following section.

5 Mesh Generation

Next step is to discretize the segmented domain, i.e., to build a mesh of polyhedron elements (as usual tetrahedron or hexahedron, sometime hybrid, for example tetrahedron with prismatic elements near the surface). We use a tetrahedron mesh with triangular surface elements in our simulation. The mesh should approximate the geometry of airway passages with appropriate accuracy and its elements should have proper size to resolve flow patterns and good shape parameters to minimize computational errors.

At present, powerful mesh generation methods (surface and volume) exist for traditional engineering applications such as aerospace and process engineering [70–72]. In such applications, the object boundaries may be defined and described analytically or piecewise analytically. However, in subject-specific biomedical geometries, the surfaces are not well-defined and analytical description is not easy. Thus, it is clear that the main difference between the domain discretization of a well defined object and patient-specific geometry is in building the surface mesh.

5.1 Surface Meshing

After the segmentation stage, the domain of interest may be represented as a binary image (1 for voxels belonging to the object and 0 for others) and as a level-set function set on the voxel-grid or as a cloud of points [73]. In our study, we use first two forms.

5.1.1 Obtaining Initial Mesh

The easiest way to build a surface mesh of triangles from a binary object is a marching cube method (MC) [74]. The surface elements thus generated have the edges of the voxel as element edges and half the size of the voxels. Their quality may be satisfactory but the surface may not be smooth, i.e., it is ragged and contains stair-like features (see Fig. 3a).

5.1.2 Correcting Non-manifold Edges and Nodes

If the size of a flow passage is close to the voxel size, the MC method can give an inconsistent surface mesh in which two different surfaces are in contact by sharing a common node or a common edge (Fig. 4). Such nodes or edges are referred to as non-manifold nodes or edges.

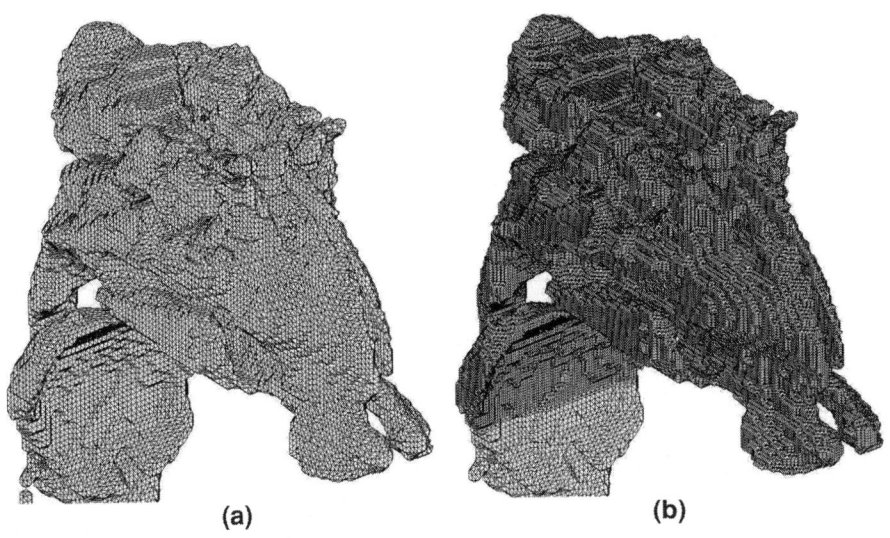

Fig. 3 Nasal cavity of a human upper airway: **a** Initial mesh. **b** After doubling the resolution of the nasal cavity part

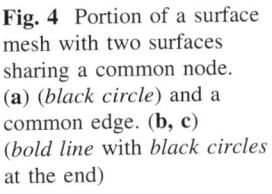
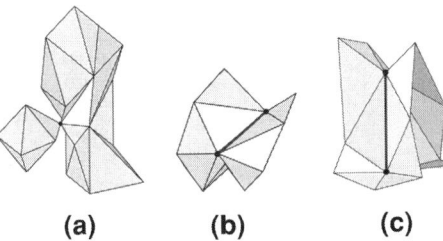

Fig. 4 Portion of a surface mesh with two surfaces sharing a common node. (**a**) (*black circle*) and a common edge. (**b, c**) (*bold line* with *black circles* at the end)

In a consistent and closed surface mesh, every edge is a manifold edge, and exactly two faces share it. If the number of faces sharing an edge exceeds two, then this is a non-manifold edge. If non-manifold edges are isolated, the mesh can be corrected. For example, by removing all faces containing a non-manifold edge (or a point) and then by triangulating the formed gaps, a valid mesh can be obtained. But in the nasal part, a rather complicated situation occurs: several non-manifold edges are normally grouped together at a location.

5.1.3 The Double-Resolution Method to Correct Grouped Non-manifold Edges

To overcome these difficulties associated with geometrical complexities, we artificially double the scan resolution of the complex part of the geometry. Then, we interpolate linearly the binary entries onto a grid with twice the original resolution. This gives a smoother level set function field with some level-set function values in between zero and unity.

Then an advanced version of MC method [75] can be applied to the grey-scale image on a higher resolution grid, which gives a mesh containing much smaller number of non-manifold edges and points, and therefore can be easily handled by the correction methods described previously.

Figure 3a shows an initial mesh of an upper human airway obtained by applying the standard MC method to the original binary image. The image is then split into the nasal part and the tracheal part of the airway. Followed by this, the double resolution technique is applied only to the nasal part of the mesh where the geometry is the most complex (Fig. 3b). The double resolution technique produces a mesh with essentially reduced number of non-manifold edges and nodes and thus it can be easily corrected.

After the non-manifold edges and nodes are corrected, the nasal part of the surface mesh is stitched to the mesh of the remaining part via a stitching method described in [76] to produce a single mesh. The resulting input mesh is shown in Fig. 3b.

5.1.4 Clipping the Domain

For numerical simulation, we need to clip off the extra parts of the mesh at the exits (bronchi) (Fig. 5a) and triangulate the clipped plane outlets. The exits are clipped approximately orthogonal to the bronchi's wall (Fig. 5b). The so-called 'stitching' method [76] is then employed to triangulate the outlet planes. In the nasal cavity part, the nasal inlets are triangulated via marching cube method to form horizontal planes.

5.1.5 Surface Mesh Cosmetics

The surface mesh obtained at this stage is not of a proper quality in terms of element size, element shape, surface smoothness. Therefore, the surface mesh needs some mesh 'cosmetics' which creates a surface mesh for the same domain with much better element quality at a prescribed element size. There are essentially three procedures used here to improve the mesh quality: (1) mesh smoothing, (2) edge swapping, and (3) edge splitting/contraction (see Fig. 6).

Mesh Smoothing

In the mesh smoothing methods, the nodes are moved along the surface (without changing the mesh topology) in order to increase the element quality (Fig. 6a).

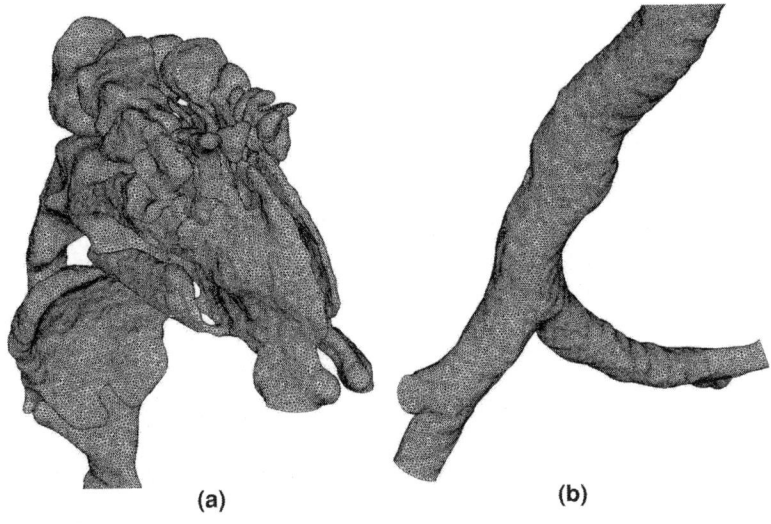

Fig. 5 The thorax part of the upper airways before (**a**), and after (**b**) cropping

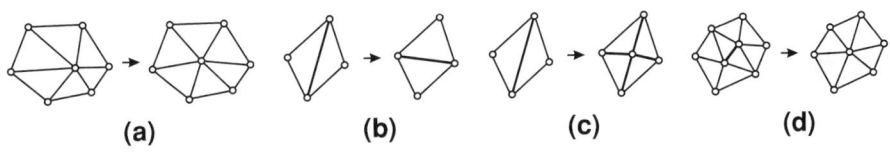

Fig. 6 Surface mesh cosmetics methods: **a** Mesh smoothing. **b** Edge swapping. **c** Edge splitting. **d** Edge contraction

The Taubin smoothing [77] is preferred over the more commonly used Laplace smoothing to avoid continuous shrinkage of convex objects.

To diminish the possible deviation of the surface from its initial position, a constraint is added, i.e., if displacement of a point from its initial position computed from the Taubin smoothing equation (see [77]) is greater than some maximally allowed value, h_{max}, then the point is displaced by h_{max} only in the direction given by the Taubin smoothing equation. Natural choice of maximal distance h_{max} is half of a voxel size: then every node will remain within its initial voxel, and the accuracy of the object boundary segmentation is not violated.

Topology Based Edge Swapping

Improving the mesh topology is very likely to have a beneficial effect on the elements quality if combined with smoothing. The edge swapping approach based on an analysis of the topology is proposed in [78]. The nodal index d_p is defined as the number of contiguous nodes to node p. Its optimal number is six if the element

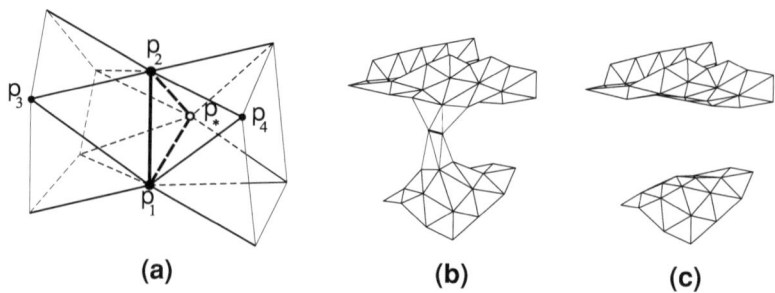

Fig. 7 Contraction of edge (p_1, p_2) results in appearing a non-manifold edge (**a**), Fragment of a mesh with a narrow vertical 'tunnel' before (**b**), and after (**c**) the correction

size is small compared to the radius of the local curvature of the surface. The aim of the topology based edge swapping is to reduce the deviation of d_p from the optimal number. The edge is to be swapped (Fig. 6b) if swapping reduces the so-called mesh relaxation index $U = \sum_p (d_p - 6)^2$.

Edge Splitting/Contraction

If the mesh contains very long or/and short edges and the average element size is not close to the value required by the flow solver, then edge splitting and contraction are very useful procedures. If the edge is too long, it can be divided by inserting a new node approximately at the midpoint of the edge as shown in Fig. 6c. If an edge is too short, it can be contracted into a single node located approximately at the midpoint of the edge as shown in Fig. 6d. One of the problems is placing the new node onto the object boundary which is not defined explicitly. If not accurately placed, a systematic shrinkage of convex surface will occur. Some methods, described based on the surface interpolation can be found in [79]. After this operation cosmetics should be performed: topology based swapping of edges containing the new node and then Taubin smoothing of a new node and node contiguous to it.

Correcting Geometric Complexities

In complicated geometries such as a nasal cavity, the contraction of some edges makes the mesh inconsistent, i.e., a non-manifold edge can appear. In this part of the airway, the width of the passages and the width of the bones between them are comparable in size to that of the scan resolution. Thus, the noise in the scans and also the presence of some mucus (e.g. snivel) in the airways result in a significant number of narrow 'tunnels' to appear in the initial mesh (see Fig. 7). These narrow tunnels can be removed during the edge contraction procedure which should be added to the non-manifold edge checking. If such edge appears and is found, the local mesh correction procedure described above (*Correcting non-manifold edges and nodes*) can be applied. The result is shown in Fig. 7c.

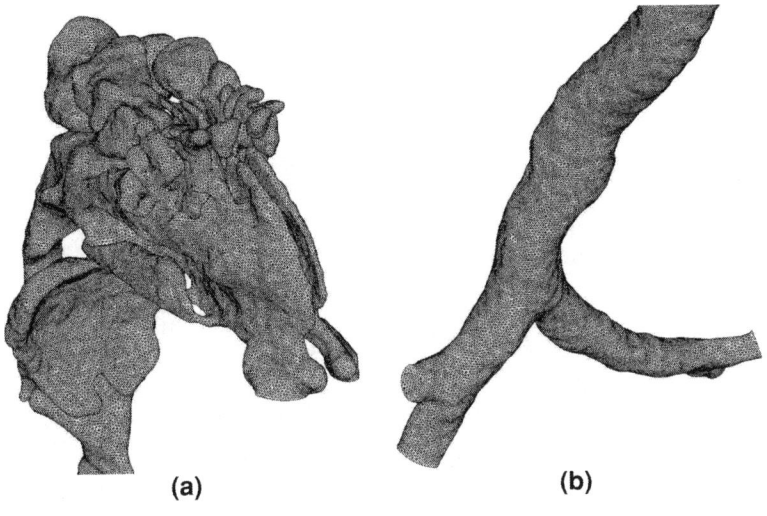

Fig. 8 Surface mesh after mesh cosmetics in its head part (**a**), and in its thoracic part (**b**)

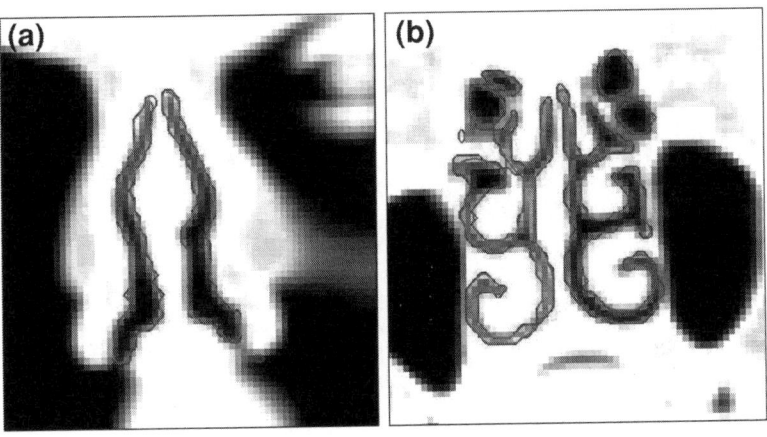

Fig. 9 Vertical slide, initial mesh generated (*blue line*), mesh after edge contacting and smoothing (*red line*): **a** Vestibulum nasi. **b** Intranasal part

5.1.6 Surface Mesh Generated

Result of mesh cosmetics procedures is shown in Fig. 8. Two vertical cuts of the CT scans are shown in Fig. 9. Here the cut of the initial and final surfaces are depicted. A vertical cut across the nasal part of the airway (both surface and volume meshes) is shown in Fig. 10.

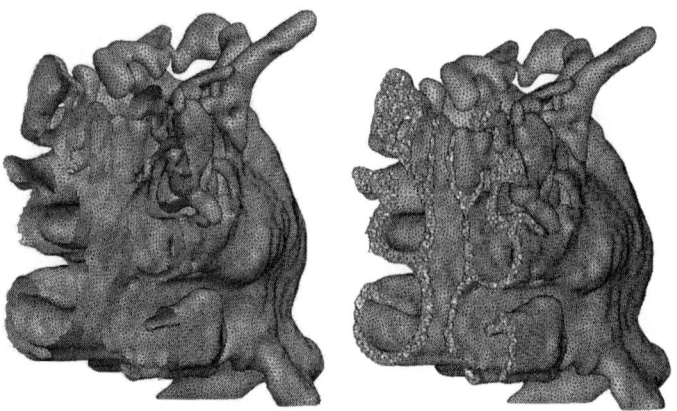

Fig. 10 Vertical cut of the nasal part of the surface (*left*) and volume (*right*) meshes

5.2 Volumetric Mesh

After the high quality surface mesh is generated, the inner part of the domain can be triangulated by tetrahedra. To generate the volumetric mesh, an in-house developed code [80] based on a Delaunay triangulation to insert the points is used. Additional 3D smoothing technique based on 3D Lloyd's [81] iterations is also applied along with standard 3D mesh cosmetics.

6 Flow Solution

6.1 Governing Equations

The flow in upper airways is considered to be turbulent and incompressible. For turbulent incompressible flow computations, the Reynolds-averaged Navier-Stokes equations are expressed in conservation form as
Mean continuity

$$\frac{\partial}{\partial x_i}(\rho \bar{u}_i) = 0 \tag{3}$$

Mean momentum

$$\frac{\partial}{\partial t}\bar{u}_i + \frac{\partial}{\partial x_j}(\bar{u}_i \bar{u}_j) = -\frac{1}{\rho}\frac{\partial}{\partial x_i}p + \frac{1}{\rho}\frac{\partial}{\partial x_j}\left(\tau_{ij} + \tau_{ij}^R\right) \tag{4}$$

where β is an artificial compressibility parameter, $\bar{u}_i, i = 1, 2, 3$ are the mean velocity components, p is the mean pressure, ρ is the density, τ_{ij} is the mean laminar shear stress tensor given as

$$\tau_{ij} = \nu \left(\frac{\partial \bar{u}_i}{\partial x_j} + \frac{\partial \bar{u}_j}{\partial x_i} - \frac{2}{3} \frac{\partial \bar{u}_k}{\partial x_k} \delta_{ij} \right) \quad (5)$$

The Reynolds stress tensor, τ_{ij}^R, as introduced by Boussinesq's assumption, has the form

$$\tau_{ij}^R = -\overline{u_i' u_j'} = \nu_T \left(\frac{\partial \bar{u}_i}{\partial x_j} + \frac{\partial \bar{u}_j}{\partial x_i} - \frac{2}{3} \frac{\partial \bar{u}_k}{\partial x_k} \delta_{ij} \right) \quad (6)$$

where ν is the kinematic viscosity of the fluid, ν_T is the turbulent eddy viscosity and δ_{ij} is the Kronecker delta.

6.2 Spalart-Almaras Turbulence Model

The SA [46] model was first introduced for aerospace applications and now it is commonly used in incompressible flow calculations. The SA model is a one-equation model, which employs a single scalar equation and several constants, to model turbulence. We find this model both to be fast and accurate for incompressible and internal flows compared with other RANS models.

Using SA model, the eddy viscosity ν_T is a function of turbulent viscosity $\hat{\nu}$ given by $\nu_T = f_{\nu 1} \hat{\nu}$ in which

$$f_{\nu 1} = \frac{X^3}{X^3 + c_{\nu 1}^3} \quad \text{and} \quad X = \frac{\hat{\nu}}{\nu}$$

The transport of turbulent viscosity $\hat{\nu}$ is governed by

$$\frac{\partial \hat{\nu}}{\partial t} + \frac{\partial}{\partial x_j}(\bar{u}_j \hat{\nu}) = c_{b1} \hat{S} \hat{\nu} + \frac{1}{\sigma} \left[\frac{\partial}{\partial x_i}(\nu + \hat{\nu}) \frac{\partial \hat{\nu}}{\partial x_i} + c_{b2} \left(\frac{\partial \hat{\nu}}{\partial x_i} \right)^2 \right] - c_{w1} f_{w1} \left(\frac{\hat{\nu}}{y} \right)^2 \quad (7)$$

where y is distance from the wall. Parameter $\hat{S}$ is given by

$$\hat{S} = S + \left(\frac{\hat{\nu}}{k^2 y^2} \right) f_{\nu 2} \quad \text{where} \quad f_{\nu 2} = 1 - \frac{X}{1 + X f_{\nu 1}}$$

and $S = |\nabla \times \bar{\mathbf{u}}|$ is the magnitude of vorticity. Parameter f_{w1} and the constant c_{w1} are, respectively, given by

$$f_w = g \left[\frac{1 + c_{w3}^6}{g^6 + c_{w3}^3} \right]^{\frac{1}{6}} \quad \text{and} \quad c_{w1} = \frac{c_{b1}}{k^2} + \frac{(1 + c_{b2})}{\sigma},$$

where $g = r + c_{w2}(r^6 - r)$ and $r = \hat{v}/(\hat{S}k^2 y^2)$.

The constants are $c_{b1} = 0.1355$, $\sigma = 2/3$, $c_{b2} = 0.622$, $k = 0.41$, $c_{w1} =$, $c_{w2} = 0.3$, $c_{w3} = 2$ and $c_{v1} = 7.1$.

6.3 Boundary Conditions

In reality, respiratory flow is driven by the pressure difference created by the deformation of lung tissue. Owing to unavailability of the patient-specific information on outlet pressure evolution during a breathing cycle, standard boundary conditions in which inlet velocity profile and constant pressure at the outlet are used widely in published works. The inlet velocity profile can be applied directly at the nares or at inlet of an extension tube attached to the nares. Although prescribing pressure difference is a better approximation to the physical process, the correct flow rate can only be achieved by iteratively adjusting the pressure difference.

To create a physiologically realistic inlet boundary condition, Doorly and co-workers [40–42] included an external face of a subject. In [42], they used this model as a reference to compare two of the most commonly used inflow boundary conditions, namely, imposing inlet velocity profile directly at the nares and using extension tubes. They found that the use of a tapered extension pipe resulted in a better approximation although this increased the computational cost.

To simplify the geometry, reconstruction and mesh generation process, we impose parabolic velocity profiles applied directly on both nares. The parabolic profile is generated by solving the Poisson equation at the inlet boundaries. Here, every inlet is treated using local co-ordinates such that the x and y axes are in the plane of the inlet and the z co-ordinate is orthogonal to this plane. With this transformation, the equation for the velocity component normal to the plane reads as,

$$\nabla_\perp^2 \bar{\mathbf{u}}_n |_{\Gamma_{inlet}} = -\frac{1}{\mu} \partial_z p |_{\Gamma_{inlet}} \tag{8}$$

where $\nabla_\perp^2 = \partial_x^2 + \partial_y^2$ is in-plane laplace operator, $\partial_z p|_{\Gamma_{inlet}}$ is the pressure gradient assumed to be constant across the inlet plane and $\mu = \rho \nu$ is the dynamic viscosity. The integration is performed by the finite element method using the existing triangulation of inlets. In case of circular pipe, this method gives the parabolic Poiseuille velocity profile and in the present study it gives a natural generalization of the profile for more complicated inlet shapes. This way of implementing the inlet boundary conditions clearly simplifies the problem without compromising accuracy.

6.4 Numerical Scheme

The CFD algorithm used here has been tested in the past for various laminar and turbulent flows. The method is based on the characteristic based split CBS algorithm originally proposed by Zienkiewicz and Codina [82].

The algorithm is based on the fractional step method to stabilize pressure and a characteristic-based approach to stabilize discrete convection operators. This combination, along with a local time stepping, was found to be robust [83, 84]. The method has been employed, both in its fully explicit and semi-implicit forms, in the past. We have used the artificial compressibility-based method here [85, 86].

In order to solve the incompressible Navier-Stokes equations fully explicitly in a parallel computing environment, Artificial Compressibility (AC) method is used in conjunction with the CBS scheme: continuity Equation 3 is rearranged as

$$\frac{1}{\beta^2}\frac{\partial}{\partial t}p + \frac{\partial}{\partial x_i}(\rho \bar{u}_i) = 0 \qquad (9)$$

where β is an artificial compressibility parameter [85]. The value of β is computed locally such that the convective and viscous characteristic time step sizes are represented by using the following relation:

$$\beta = \max\{\varepsilon, u_{\text{conv}}, u_{\text{diff}}\} \quad \text{where} \quad u_{\text{conv}} = \sqrt{u_i u_i}, \quad u_{\text{diff}} = \nu/h.$$

Here ε is a small constant used to avoid β approaching zero, u_{conv} and u_{diff} are the convection and diffusion characteristic velocities, respectively, and h is the local element size.

The method solves the incompressible Navier-Stokes equations in three steps. In the first step, an intermediate momentum field is solved. In the second step, pressure is computed and in the third step, the momentum field is corrected. The one-equation SA turbulence model is added as a fourth step. The first three steps of the CBS scheme is used in its semi-discrete form may be summarized as

Step 1: intermediate momentum

$$\Delta \tilde{U}_i = \Delta t \left[-\frac{\partial(\bar{u}_j U_i)}{\partial x_j} + \frac{\partial \tau_{ij}}{\partial x_j} + \frac{\partial \tau_{ij}^R}{\partial x_j} + \frac{\Delta t}{2}\bar{u}_k \frac{\partial}{\partial x_k}\frac{\partial(\bar{u}_j U_i)}{\partial x_j} \right]^n \qquad (10)$$

Step 2: pressure

$$p^{n+1} = p^n - \Delta t(\beta^n)^2 \left[\frac{\partial U_i^n}{\partial x_i} + \theta_1 \frac{\Delta \tilde{U}_i}{\partial x_i} - \Delta t \theta_1 \frac{\partial}{\partial x_i}\frac{\partial p^n}{\partial x_i} \right] \qquad (11)$$

Step 3: momentum correction

$$\Delta U_i = U_i^{n+1} - U_i^n = \Delta \tilde{U}_i - \Delta t \frac{\partial p^n}{\partial x_i} \qquad (12)$$

Step 4: transport of turbulence variable

$$\hat{v}^{n+1} = \hat{v}^n + \Delta t \left[-\frac{\partial}{\partial x_i}(\hat{v}\bar{u}_i) + c_{b1}\hat{S}\hat{v} + \frac{1}{\sigma}\left(\frac{\partial}{\partial x_i}(v+\hat{v})\frac{\partial \hat{v}}{\partial x_i} + c_{b2}\frac{\partial \hat{v}}{\partial x_i}\frac{\partial \hat{v}}{\partial x_i}\right) \right.$$
$$\left. - c_{w1}f_{w1}\left(\frac{\hat{v}}{y}\right)^2 \right] + \frac{\Delta t^2}{2}\bar{u}_j\left[\frac{\partial}{\partial x_j}\left(\frac{\partial}{\partial x_i}(\bar{u}_i\hat{v})\right)\right] \quad (13)$$

Here superscript n mean the value is taken at time $t_n = \Delta t(n-1)$, Δt is the time step, $U_i^n = \rho \bar{u}_i(t_n)$ is the mass flux, $\tilde{U}_i = U_i^n + \Delta \tilde{U}_i$ is its intermediate value.

Parameters $\theta_{1,2}$ should be in the ranges $\theta_1 \in [0.5, 1]$, $\theta_2 \in [0, 1]$. For the explicit scheme, employed in this study, $\theta_2 = 0$. The equations are derived from the observation that the time discretization is carried out along the characteristic. A simple approximate backward integration gives the equations with some extra convection stabilization terms (last term in the right-hand side at Step 1). These extra terms are consistent and reduce oscillations, due to the standard Galerkin discretization of convective terms. These higher-order terms are very important when the time discretization is explicit. The artificial compressibility parameter β is defined locally, based on the local velocity scales. Further details on the selection of β may be found in [85]. The scalar transport turbulence equations are also subjected to the characteristic time discretization, similar to Step 1. Once the semi-discrete form is available, the standard Galerkin spatial discretization follows [85, 87].

7 Results

In this section we describe computational results of the numerical framework described in Sects. 3, 4 and 5. The first example presented in the following subsection is to demonstrate the accuracy of flow solver on an idealized geometry. This is followed by steady flow through a truncated and full scan-based geometries.

7.1 A Model Human Airway Problem

An idealized human upper airway model is considered first to verify the steady-state calculations described in [88]. This reference provides some experimental data for this idealized geometry. Although identical reproduction of the experimental configuration is difficult, we have reproduced an approximate configuration from the geometrical data given in [88]. The geometry used here was constructed from the 13 cross sections provided by Heenan et al. [88]. The sections were linearly connected to obtain the full geometry. An unstructured mesh with about one million tetrahedral elements to discretize the domain is generated (Fig. 11a).

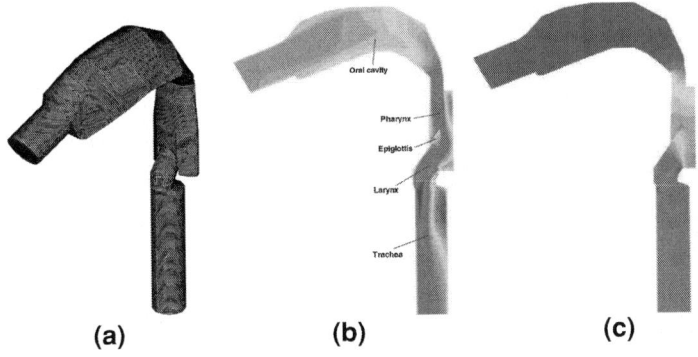

Fig. 11 a Unstructured mesh for an idealized human upper airway model. Computed flow at the flow rate = 37.35 L/min. **b** u_3 velocity component. **c** Pressure

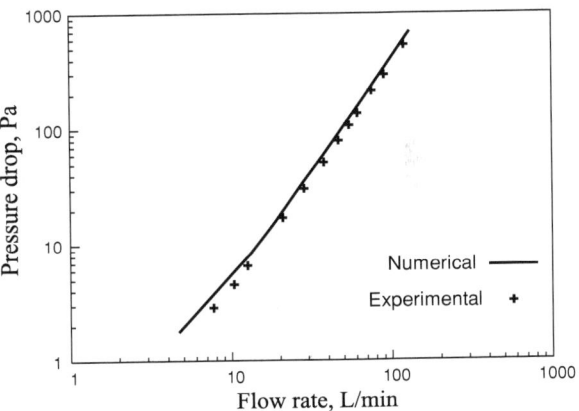

Fig. 12 Pressure drop in the flow through an idealized human upper airway. Solid line is for computed data, crosses—for the experimental data from [88]

Figure 11b and c show some sample qualitative results at a flow rate of 37.35 L/min. The u_3 velocity distribution clearly shows the acceleration as the flow enters the pharynx and also when it leaves the pharynx through the epiglottis. As a result of the complex nature of the geometry, some slow moving, recirculatory regions are also noted, near the entrance to pharynx. It is also observed that the majority of the pressure drop occurs as the fluid enters the pharynx and also as it passes the epiglottis. The oral cavity shows almost no pressure change. The qualitative results obtained compare well with the experimental results given in [88].

To further increase the confidence in the flow solver, the pressure drop calculations were carried out for different flow rates. The pressure drop is calculated as the difference between the average pressure values at the inlet and exit. Figure 12 shows the comparison of pressure drop values obtained from the present numerical study and the experimental data. As seen the agreement between the two

7.2 Inhalation Studies in the Central Part of Human Upper Airway

Inhalation studies are important for determining the flow behaviour during drug deliveries and forced inhalation. In this study, the influence of nasal part is disregarded and only the central part of the airways is considered. The area at the inlet of the patient-specific geometry is approximately 222 mm^2. The properties of air are assumed to be constant. A kinematic viscosity of 1.69×10^{-5} m^2/s and density of 1.2 kg/m^3 are used in the calculations. The flow rate in the human airways normally varies between 15 and 45 L/min depending on the state of the subject. The inhalation flow is studied by prescribing a velocity based on the assumed flow rate. In order to accurately represent the velocity boundary conditions, it may be essential to consider the flow through the nasal passages (see next section).

The mesh sensitivity study shows that, at a Re (based on 100 mm) of 15,000, there is a difference in the pressure drop of about 2.5% between meshes with half and one million linear tetrahedral elements. To obtain as accurate results as possible, we have employed one million element mesh throughout this section. At the inlet of the geometry, a constant velocity value is assumed depending on the flow rate.

The qualitative results, at an inhalation flow rate of 33.8 L/min, are presented in Figs. 13 and 14. The vector plots in Fig. 13c clearly show a recirculation region just below the vocal folds. Although the geometry looks fairly straight, the recirculation makes the physics more interesting. As expected, the flow accelerates near the tongue base (oropharynx) as it passes into the laryngopharynx (see Fig. 13b). The flow stabilizes and reaches a nearly developed state as it moves towards the bifurcation, within the trachea.

Figure 14a, b shows the pressure distribution. The maximal and minimal pressure are 72.8 and −33.1 Pa, respectively. The total pressure drop is 40.9 Pa. Once again, all the action takes place near the laryngopharynx. The flow recirculation below the epiglottis also creates a negative wall pressure distribution at the posterior surface of the airway. However, the pressure at the anterior remains positive. The negative pressure created by the recirculation and the naturally narrow portion of oropharynx are responsible for the large pressure drop. As in the model airway discussed in the previous section, here also the pressure changes are almost nil in the oral cavity.

The wall shear stress distribution in Fig. 14c, d also shows a trend in accordance with the flow distribution. The maximum wall shear stress occurs near the narrow portion of the airway. It is also noticed that the wall shear stress is low near the recirculation on the posterior surface of the airway. However, the accelerating

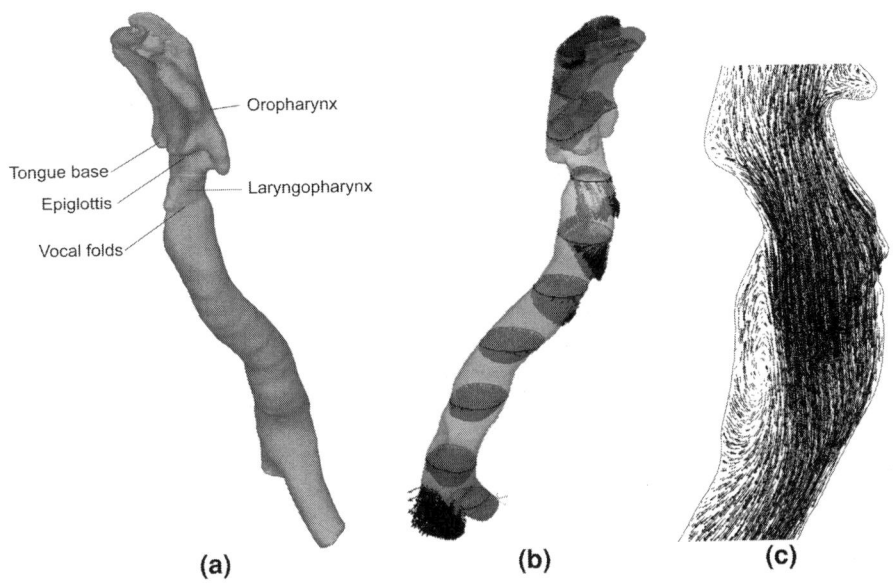

Fig. 13 Airway geometry. **a** Velocity field at an inhaling flow rate of 33.8 L/min. **b** Velocity profiles. **c** Recirculation below the vocal folds

downward velocity along the anterior wall opposite to the recirculation increases the velocity gradient and increases the wall shear stress.

Table 1 presents comparisons of the pressure drop for the model airway presented in the previous section and the patient specific airway studied in this subsection for three different flow rates. Although a very close match between the two is not anticipated, the difference between the pressure drops is not extremely high. The difference is about 8% at 50.6 L/min, about 12.1% at 33.8 L/min and 17.9% at 16.9 L/min. There are two major differences between the model and patient specific upper human airways. The first is that, although similarity exists between the two, they are not geometrically identical. The second difference is that the model airway has about a 16% longer flow path than the patient specific airway. This second difference is clearly reflected in the higher pressure drop obtained on the model human upper airway.

Figure 15 shows the wall pressure and shear stress distributions along the anterior and posterior walls of the airway up to the bifurcation. From the pressure distribution, it is clear that the major pressure drop occurs between −620 and −632 mm with the minimum pressure reported near −631 mm. This is the location just above the vocal folds in the laryngopharynx. The negative pressure difference obtained at 16.9 L/min is almost nil and, as the flow rate increases, the pressure drop noticed in the laryngopharynx also increases. The negative pressure created near the vocal folds triggers a flow recirculation near the posterior surface, as the flow leaves the laryngopharynx, into the larynx and trachea. The recirculation near

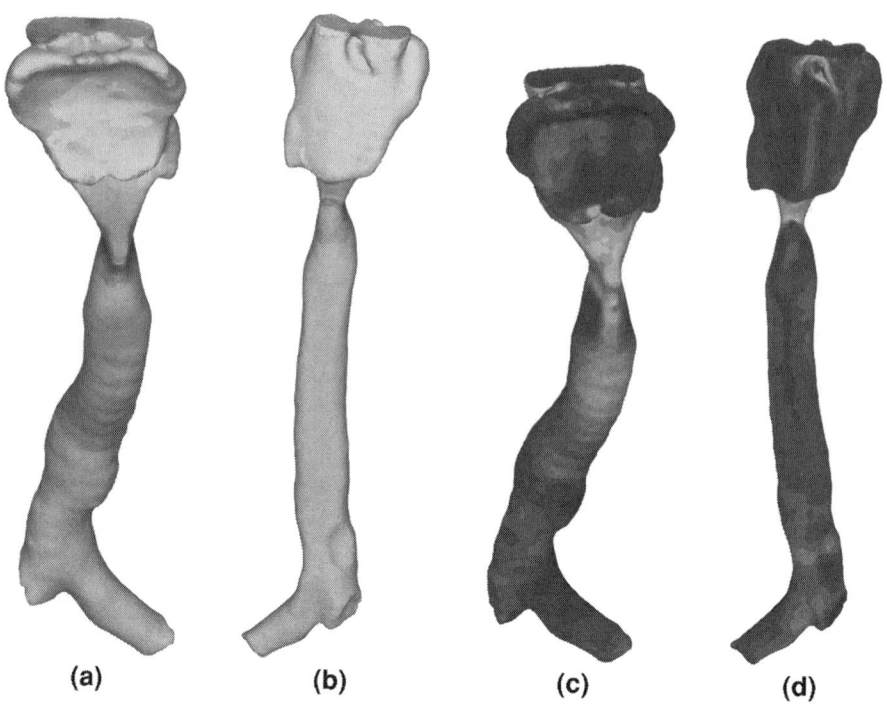

Fig. 14 Pressure (**a, b**) and wall shear stress distribution. (**c, d**) at a flow rate of 33.76 L/min. (**a, c**) anterior view. (**b, d**) posterior view

Table 1 Pressure drop (Pa) comparison of numerical calculations of model and patient specific human upper airways

Flow rate	Model airway (present)	Cast model [89]	Patient specific airway (present)
16.9	12.89	–	10.58
33.8	47.73	44.00	41.95
50.6	108.45	–	99.746

the posterior wall results in flow acceleration near the anterior wall, with an increase in velocity gradients along the anterior wall. This is clearly represented by the higher wall shear stress distribution on the anterior wall between −650 and −700 mm. In addition, the shear stress peaks at both the anterior and posterior walls coincide with the negative peak pressure value. The combination of the negative pressure difference and the peak shear stress value makes the laryngopharynx location an important area for airway collapse investigation. It is noticed that both the pressure difference and wall shear stress distributions are not very smooth. This is mainly due to the rapid changes in the surface contours. A higher mesh resolution may marginally increase the smoothness of the wall shear stress

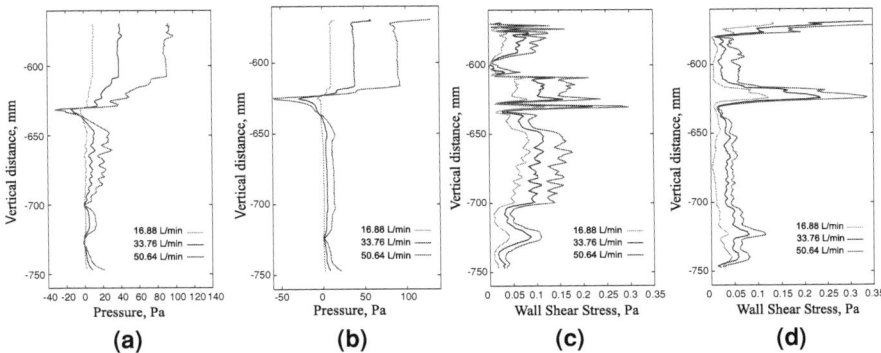

Fig. 15 Wall pressure (**a, b**) and shear stress (**c, d**) distributions along the anterior (**a, c**) and posterior (**b, d**) wall up to bifurcation. The data is extracted approximately along the centre of the anterior surface

distribution. Even on structured meshes, the shear stress distribution is expected to have rapid changes all along the surface [24].

The patient specific study has shown that the pressure drop is large close to the narrow portion just below the oropharynx. It has also shown that the wall shear stress is very large near the narrow portion of the pharynx. All this clearly indicates that the oropharynx and laryngopharynx are the two major areas of interest for developing an appropriate fluid-structure model for airway collapse (sleep apnoea).

7.3 Full Upper Airway Flow Modelling

This section focuses on the impact of including nasal cavity on airflow through a human upper respiratory tract. Here we compare the flow in the airways with and without the nasal cavity. But we do not simply cut the nasal cavity and compare with geometry studied in Sect. 7.2 as the inlet boundary effect can be essential. So to minimize the influence of the closely located inlet, an extension tube is attached at the inlet boundary.

Thus we are working with two geometries. Geometry I includes nasal cavity, pharynx, larynx, trachea and two generations of airway bifurcations below trachea as show in Fig. 16a, b. To build Geometry II, Geometry I is truncated just below the nasopharynx. Note, that the space occupied by uvula is one of the geometrical features that needs special attention in the truncated geometry. This space divides the air flow in the oropharynx region into two. In the geometry considered in Sect. 7.2, the distance between the inlet boundary and this separation region is very close and as a consequence directly imposing velocity profile on the inlet boundary can cause instability. That is why we have to add the artificial prolongation as

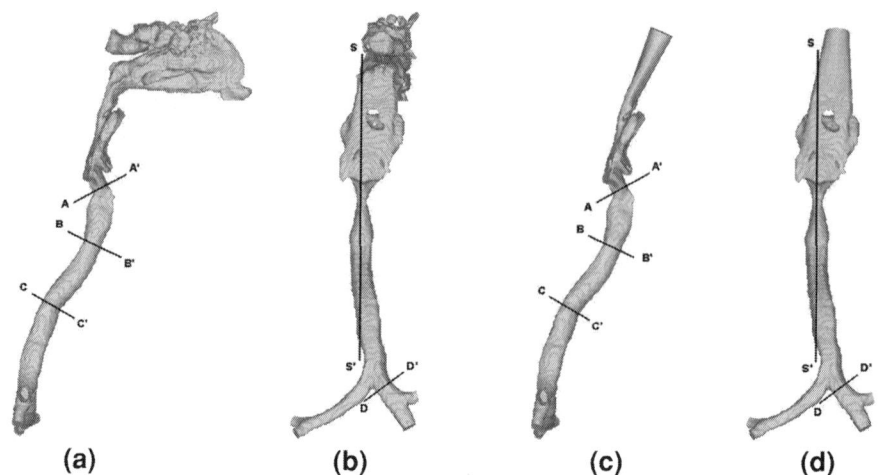

Fig. 16 Airways geometries studied: Geometry I (**a, b**). Geometry II (**c, d**). Sagital view (**a, c**). posterior view (**b, d**). Locations of different cross sections are indicated where computational results are displayed

shown in Fig. 16c, d. In Geometry II, a tube of the constant cross-sectional area (but variable shape) is added. The cross-sectional shape gradually varies from the circular one (at the inlet) to the real cross-section of the airways in the cut-plane. The direction of the cut-plane is approximately orthogonal to the axis of the airways at this location. The axis of the extension tube passes through the bary-centre of the local cross-section.

In the general population, the left and right nares has a slight geometrical difference and it varies from person to person. Therefore, in addition to the differences in flow pattern due to nasal cavity inclusion, we have also studied the effects of prescribing two different inlet flow boundary conditions on Geometry I. In the first case, the flow rate is divided equally between each side of the nares and in the second problem flow rate is divided proportionally with respect to the area of the nares (area weighted flow rates). In all studies reported here the air kinematic viscosity is $v = 1.69$ cm^2/s. For the first study, the flow rate is $Q = 15$ L/min. This value is chosen based on the values used in the Sect. 7.2. In the second study, we consider a flow rate of $Q = 4$ L/min.

First of all, we discuss the differences in flow results obtained between the geometry with and without the inclusion of the nasal cavity. Figure 17 shows flow patterns at various sections along the length of the airway for both the geometries. Figure 18 shows the sagital view of the flow pattern along the length. Both the cross sections and the sagital section used for generating the flow patterns are defined in Fig. 16. The vorticity magnitude contours for Geometry I is between 100 and 1,500 while for Geometry II is between 200 and 2,600. For the sake of comparison, the area weighted flow rate boundary condition is used when the geometry with nasal cavity is employed.

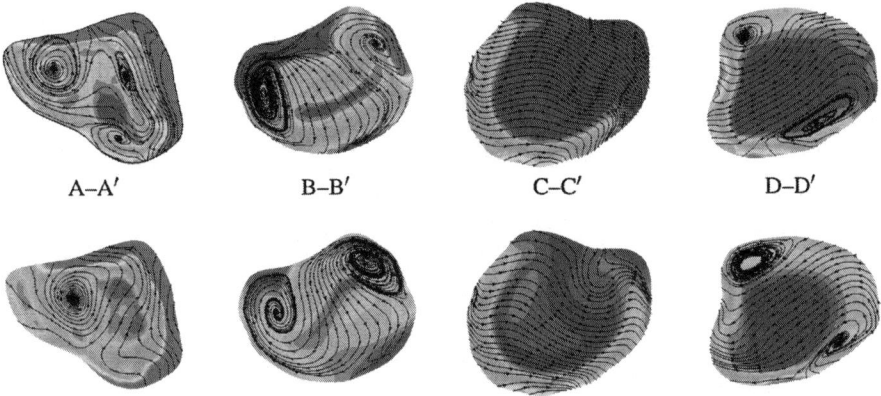

Fig. 17 Contours of vorticity magnitude and streamlines at different cross-sections: Geometry I (*top row*), Geometry II (*bottom row*)

Fig. 18 Contours of vorticity magnitude and streamlines at S–S' section: **a** Geometry I. **b** Geometry II

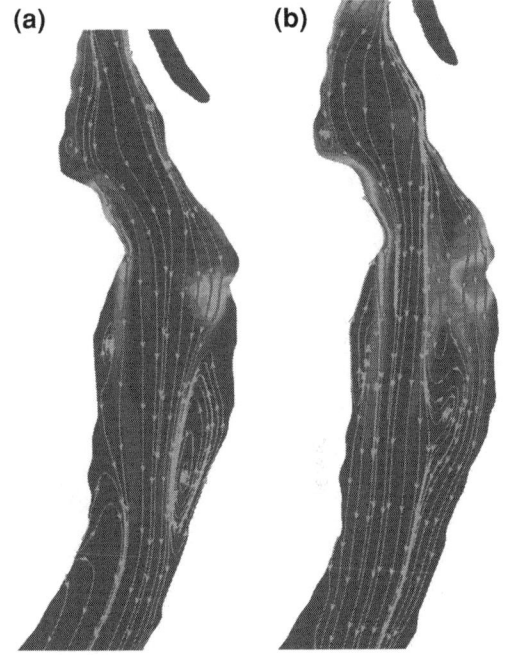

The first noticeable impact of the inclusion of the nasal cavity is that in Geometry I, more than 90° direction change causes the flow to recirculate and mix more than the Geometry II. In addition, inclusion of nasal cavity makes the flow to develop more secondary flow vortices in the region superior to the larynx. Distal to the larynx differences in the secondary flow are less pronounced. This can be

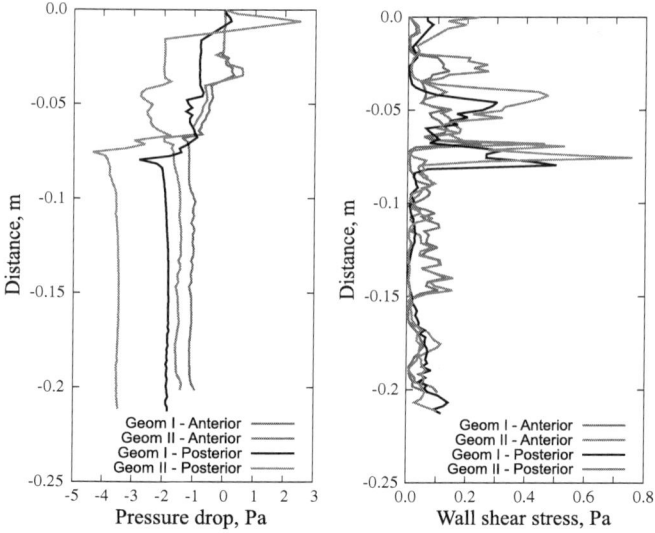

Fig. 19 Comparison of pressure and WSS distributions along the anterior and posterior walls between two geometries. Zero pressure value at tracheal entrance is obtained by subtracting the average pressure at that section

clearly seen in Figs. 17 and 18. Contours of velocity and streamlines at different cross-sections for both Geometry I and Geometry II are depicted in Fig. 17. Examination of these plots reveals that the differences in the pattern of secondary flows between the two geometries is significant up to cross section B–B'. However, the differences in sagital flow pattern depicted in Fig. 18 are substantial. The inclusion of nasal geometry induces the formation of stronger recirculation region in the anterior part of the trachea as opposed to the only weak posterior one when the nasal geometry is excluded from computation. The differences in the primary flow pattern bring serious implications to modelling the airflow without the nasal cavity [47, 90]. In addition to predicting wrong locations of airway collapse, this can lead to wrong location of particle deposition, when incomplete geometries are used in particle inhalation studies. This also means that the location of negative pressure will be predicted in two completely different locations with and without the nasal cavity.

Figure 19 shows the pressure and wall shear stress (WSS) distributions along the anterior and posterior walls of the airway. As seen, the differences between the two geometries are clearly visible. The pressure values are scaled (subtracted) to the average value at the location of the truncation. These plots show a very similar trend published in our previous work on truncated geometries [47]. However, the difference here is that the geometry with nasal passage has a less dramatic pressure drop at the laryngeal part of the geometry. This may be the result of extra mixing and more controlled flow of air, due to the inclusion of the nasal part. Due to the

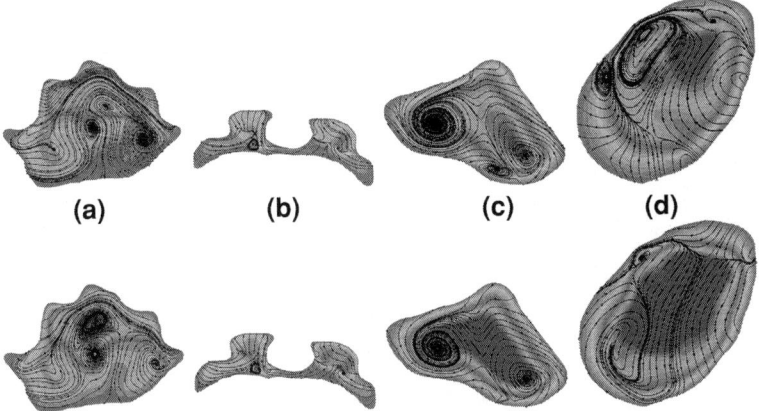

Fig. 20 Contours of velocity magnitude and streamlines of human upper airway flow by imposing equally (*top row*) and proportionally (*bottom*) distributed flow rate between the nares. **a** Nasopharyngeal section. **b** Oropharyngeal section. **c** Laryngeal section. **d** Tracheal section

uvula obstruction in Geometry II facing the inlet flow head on, the pressure shows dramatic drop close to vocal folds. The peak WSS distribution in both cases appear close to the epiglottis and qualitatively in agreement with previous studies.

The rest of this section is devoted to investigating the differences induced by the two boundary conditions applied on Geometry I. Figure 20 shows the results that describe the differences between the two flow boundary conditions. These figures are plotted using the same scale. As expected, imposing equally divided inlet flow rate (BC-1) at both nares (resulting in prescribing higher inlet velocity distribution in one nares than the other) causes the flow to skew towards the side with higher inlet velocity as shown Fig. 20 (top). Imposing area weighted flow rate between the two nares (BC-2) causing the maximum velocity to approximately align with the centerline of the airway as shown in Fig. 20 (bottom). For the sake of simplicity, we will refer to the case with equally distributed inlet flow rate as BC-1 and proportionally distributed or area weighted flow rate condition as BC-2.

Figure 20a shows the flow pattern on the nasopharyngeal section. While the same number of vortices appear in both cases, BC-1 and BC-2, the locations and their strengths are different. In BC-1 case, the two major vortices appear on the left and right side of the cross-section and the minor vortex is in the middle. This is clearly caused by the fact that the applied inlet velocity is higher on the right side than at the left. In BC-2 case the two major vortices are located inline in the middle of the airway and a minor vortex appears on the righthand side. The nasopharyngeal section is the region where the differences in flow pattern is most pronounced. By inspecting Fig. 20b, one can conclude that there is no apparent difference in flow pattern at the oropharyngeal section. At the larynx, the results display different number of vortical structures appear on BC-1 and BC-2 as depicted in Fig. 20c. Even for a non-symmetrical geometry, at this part of airway,

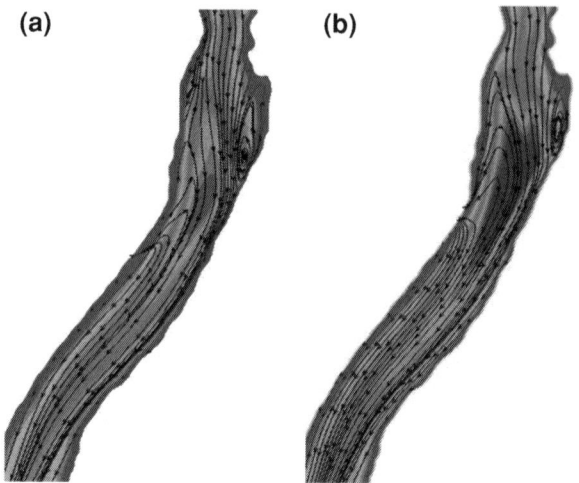

Fig. 21 Flow pattern in the S–S′ (longitudinal) cross-sections indicated in Figure reffig:geometries: Geometry I (**a**), and Geometry II (**b**)

imposing BC-2 results in nearly symmetrical secondary flows. Further down at the proximal trachea, the differences in flow pattern is still visible as shown in Fig. 20d.

Figure 21 also clearly shows the differences between the two flow boundary conditions. Due to the unequal velocity distribution between the nares, BC-1 develops a more mixed flow with two vortices at both anterior and posterior walls of the trachea. On the other hand a more uniform nature of the flow with BC-2 is again observed.

Figure 22 shows the pressure and WSS distributions for both the boundary conditions. Since BC-1 with equally distributed flow between the nares results in unequal velocity distribution between them, it significantly affects the symmetry of pressure and WSS distributions between the two sides of the geometry. This is clearly evident in Fig. 22, especially in the nasal cavity area. With BC-1 the pressure drop between the two sides of the nasal cavity are significantly different. This is also true for the WSS distribution. This is the result of one side of the nasal cavity being smaller than the other and equal flow distribution results in enhanced velocity values in the smaller sized nasal inlet. The pressure and WSS distributions for a weighted area based inlet flow conditions (BC-2) are very similar between both sides of the nasal cavity. In both cases the pressure drop is significant across the epiglottis and vocal folds.

7.4 Numerical Simulation of the Heat Transfer in the Nasal Cavity

In this section, heat transfer in a subject-specific human upper airway geometry is numerically investigated. The flow is assumed to be steady, inspiratory flow for quiet normal breathing condition (tidal volume $V_T = 0.5$ L and flow rate

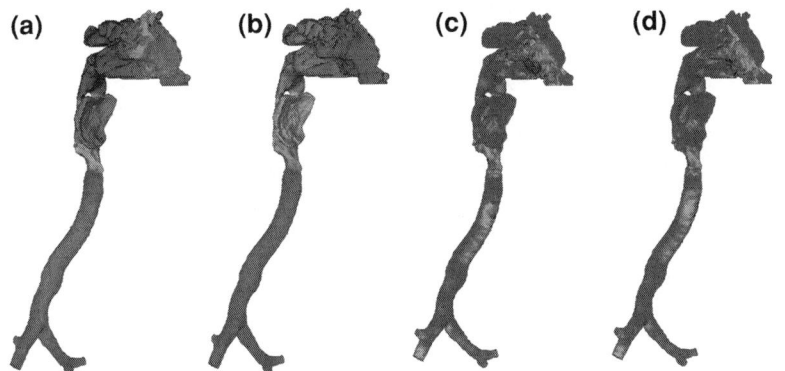

Fig. 22 Differences in pressure (**a**, **b**) and WSS (**c**, **d**) between the two inlet flow boundary conditions: BC-1 (**a**, **c**) and BC-2 (**c**, **d**)

$Q = 15$ L/min). Since all the heat transfer aspects can be studied without any loss of accuracy using a steady state assumption, transient results are not presented. In the first case considered, inspired air temperature is assumed to be substantially lower than the temperature of the airway wall and the second one uses an inspired air temperature that is slightly higher than the wall temperature. The temperatures of the inspired air used are $T_\infty = -30°C$ and $T_\infty = 45°C$ for the first and the second cases respectively. The temperature of the wall is assumed to be $T_{\text{wall}} = 37°C$ and constant.

The boundary conditions for the flow problem are as follows. On the wall, a no-slip boundary condition of $u_i = 0|_{\Gamma_{\text{wall}}}$ is imposed. Parabolic velocity profiles are applied directly on both nares as inflow boundary conditions. These parabolic profiles are generated by solving Poisson's equation on the inlet boundary surface. If every inlet/outlet is transformed to a local co-ordinate system such that x_1 and x_2 axes are in the plane of the inlet and the x_3 co-ordinate is orthogonal to this plane, then the equation for the velocity component normal to the plane reads as

$$\frac{\partial}{\partial x_\alpha}\frac{\partial u_n}{\partial x_\alpha}\bigg|_{\Gamma_{\text{inlet}}} = -\frac{1}{\mu}\frac{\partial p}{\partial x_3}\bigg|_{\Gamma_{\text{inlet}}} \qquad \alpha = 1, 2 \tag{14}$$

where u_n is the magnitude of the velocity normal to the inlet plane and $\partial p/\partial x_3|_{\Gamma_{\text{inlet}}}$ is the pressure gradient assumed to be constant across the inlet plane. The integration is performed by the FE method using the surface triangulation of inlet/outlets. In case of circular pipe, this method gives the parabolic Poiseuille velocity profile and for non-circular cross sections this method provides a natural generalization. This way of implementing the inlet/exit boundary conditions clearly reduces the computation required for extension tubes often used by other researchers. The outlet boundary condition is imposed by applying $p = 0|_{\Gamma_{\text{outlet}}}$.

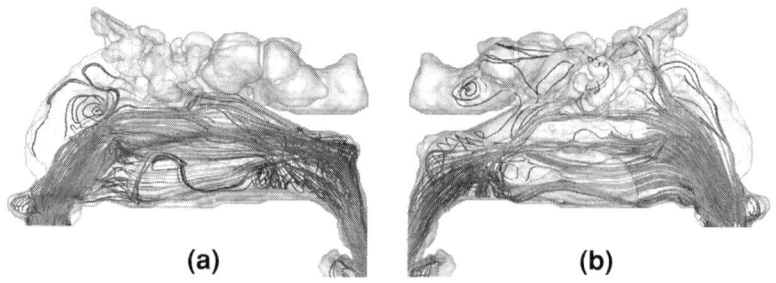

Fig. 23 Streamlines of the flow the colour indicates the magnitude of the velocity: **a** Left cavity. **b** Right cavity

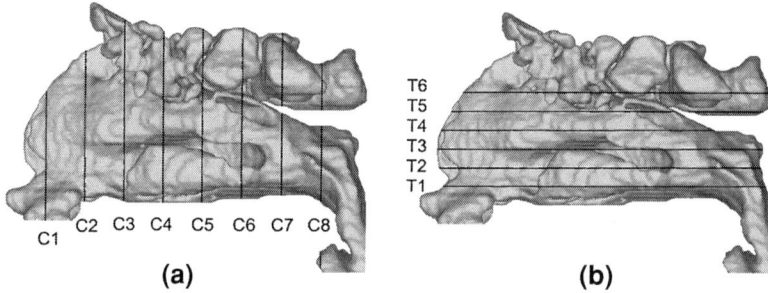

Fig. 24 Location of the visualisation planes: **a** Coronal direction. **b** Transversal direction

Figure 23 shows the stream traces released from the right and left nostrils respectively. The flow is characterized by the formation of flow recirculations at the anterior part of nasal vestibule and atrium of both left and right sides of the cavity and another one at the nasopharyngeal region. Another prominent flow feature is that only very small proportion of air flows through the olfactory slit and meatuses. Majority of the flow is close to the nasal septum. This result is in agreement with the experimental results conducted by Kelly et al. [91] and Doorly and co-workers [65]. The only difference is that the circulation in the nasopharyngeal region did not appear in the experimental results. This is due to the fact that the anatomy of every subject differs. Both experiments were conducted using a unilateral model of a nasal cavity. However, both sides of the cavity are included in the present numerical model. The fact is that a certain degree of asymmetry between left and right cavity makes the flow to rotate and form a recirculation in the confluence area, where the flow from both sides meet.

To analyse the results, the velocity and temperature fields are plotted on several coronal and transversal planes. The locations of these planes are depicted in Fig. 24. Plane C1, C2, C3, C4, C5, C6, C7, and C8 are located, respectively at 1, 2, 3, 4, 5, 6, 7, and 8 cm from the left most point of the nasal geometry as shown.

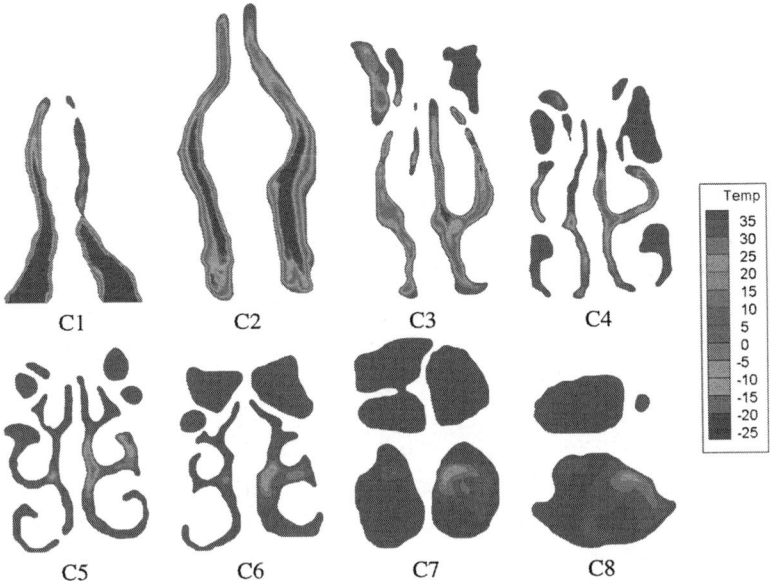

Fig. 25 Distribution of temperature for $T_\infty = -30°C$ in the coronal planes

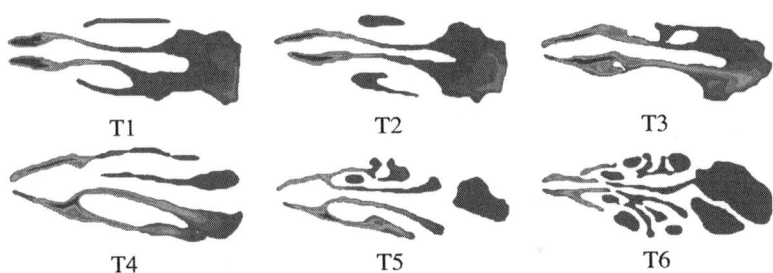

Fig. 26 Distribution of temperature for $T_\infty = -30°C$ in the transversal planes. Color-scale is the same as in Fig. 25

For the transversal planes, planes T1, T2, T3, T4, T5 and T6 are located respectively at 0.5, 1.0, 1.5, 2.0, 2.5 and 3.0 cm from the lowest point at the nasal floor.

The first noticeable result from the numerical simulation of heat transfer process is that the air temperature has reached the body temperature as it passes the nasopharynx irrespective of the ambient temperature of the inspired air. Between 85 and 90% of the heat transfer happens inside the nasal cavities. Temperature distribution plots for $T_\infty = -30°C$ and $T_\infty = 45°C$ are presented in Figs. 25 and 26. These figures demonstrate that the nasal interior structures are effective in air-conditioning.

In the first case at steady state, the temperature of the inferior region of the nasal cavity (T1 and T2 planes) is very close to wall temperature with exception of the nasal vestibule region, main nasal cavity and the distal nasopharyngeal region. In the middle region (T3, T4 and T5 planes), the region of low temperature in the left side of the cavity extend further to the distal part of the nasopharyngeal region while in the right side, the air temperature is very close to the wall temperature. Superior to this region (T6 plane), only the air in the nasal cavity and left superior meatus has temperature lower than the wall temperature. The air inside the sinuses has the same temperature as the wall.

As seen the results of heat transfer calculation show that by the time the inspired air reaches the nasopharyngeal region its temperature is within 10–15% of the lung operating temperature. In both sides of the nasal cavity, the region between 2–4 cm from the most anterior part of the nares is the region with high temperature gradient. This region lay between the nasal valve and the anterior part of the middle meatus. Any changes in the geometry of the nasal interior structure in this region is very likely to severely affect the heat transfer capacity of the nose. This finding confirms the in vivo measurement that, under quiet normal breathing, the majority of the heat transfer processes is happening in the nasal cavity and shows the effectiveness of the nasal internal structures to condition the inspired air temperature closer to the lung operating temperature.

8 Conclusion

An overview of scan based modelling of human upper airways has been provided in this chapter. Although various stages of the modelling framework, such as image segmentation, meshing and flow solver have been documented here, these stages have room for substantial improvements. The key to the future development is creating a procedure that is robust, fast and accurate. Due to the nature of scan based modelling techniques and the requirement for dealing with a large number of geometries, a fully or semi-automated system is essential to make progress. Among the various stages of the modelling framework, the first two preprocessing stages pose a real challenge. The flow analysis challenges include modelling turbulence, high resolution transient calculations and correct fluid-structure interaction models.

Acknowledgements This work was funded by EPSRC, UK grant D070554. The authors also acknowledge the help and support of Dr. P.H. Saksono.

References

1. Ebden, P., Jenkins, A., Houston, G., Davies, B.H.: Comparison of 2 high-dose corticosteroid aerosol treatments, beclomethasone dipropionate (150 ug day) and budesonide (1600 ug day), for chronic astma. THORAX **41**, 869–874 (1986)
2. Hawkins, D.B., Luxford, W.M.: Laryngel stenosis from endotracheal intubation–a review of 58 cases. Ann. Otol. Rhinol. Laryngol. **89**, 454–458 (1980)

3. Wright, J., Johns, R., Watt, I., Melville, A., Sheldon, T.: Health effects of obstructive sleep apnoea and the effectiveness of continuous positive airways pressure: a systematic review of the research evidence. Br. Med. J. **314**, 851–860 (1997)
4. Ayappa, I., Rapoport, D.M.: The upper airway in sleep: physiology of the pharynx. Physiol. Rev. **7**, 9–33 (2003)
5. De Boer, M.F., McCormick, L.K., Pruyn, J.F.A., Ryckman, R.M., van den Borne, B.W.: Physical and psychosocial correlates of head and neck cancer: a review of the literature. Otolaryngol. Head Neck Surg. **120**, 427–436 (1999)
6. Constantian, M.B., Clardy, R.B.: The relative importance of septal and nasal valvular surgery in correcting airway obstruction in primary and secondary rhinoplasty. Plast. Reconstr. Surg. **98**, 38–54 (1996)
7. Kulish, V.: Human Respiration. Anotamy and Physiology, Mathematical Modeling, Numerical Simulation and Applications. WIT Press, Southampton (2006)
8. Yeo, S.Y., Xie, X., Sazonov, I., Nithiarasu, P.: Geometrically induced force interaction for three-dimensional deformable models. IEEE Trans. Image Process. **20**(5), 1373–1387 (2011)
9. Sazonov, I., Yeo, S.-Y., Bevan, R.L.T., Xie, X., van Loon, R., Nithiarasu, P.: Modelling pipeline for subject-specific arterial blood flow–a review. Int. J. Numer. Methods Biomed. Eng. (2011). doi:10.1002/cnm.1446
10. Saksono, P.H., Nithiarasu, P., Sazonov, I., Yeo, S.-Y.: Computational flow studies in a subject-specific human upper airway using a one-equation turbulence model. Influence of the nasal cavity. Int. J. Numer. Methods Eng. **87**, 96–114 (2011)
11. Nishino, T., Hiraga, K.: Coordination of swallowing and respiration in unconscious subjects. J. Appl. Phsiol. **70**, 988–993 (1991)
12. Fouke, J.M., Teeter, J.P., Strohl, K.P.: Pressure–volume behaviour of the upper airway. J. Appl. Phsiol. **61**, 912–918 (1986)
13. Strohl, K.P., Fouke, J.M.: Dialating forces on the upper airway of anethetized dogs. J. Appl. Phsiol. **58**, 452–458 (1985)
14. Verneuil, A., Berry, D.A., Kreiman, J., Gerratt, B.R., Ye, M., Berke, G.S.: Modelling measured glottal volume velocity waveforms. Ann. Otol. Rhinol. Laryngol. **112**, 120–131 (2003)
15. Brancatisano, T., Collett, P., Engel, L.: Respiratory movements of the vocal cords. J. Appl. Phsiol. **54**, 1269–1276 (1983)
16. Renotte, C., Bouffioux, V., Wilquem, F.: Numerical 3d analysis of oscillatory flow in the time-varying laryngeal channel. J. Biomech. **33**, 1637–1644 (2000)
17. Nithiarasu, P., liu, C.-B., Massarotti, N.: Laminar and turbulent flow through a model human upper airway. Commun. Numer. Methods Eng. **23**, 1057–1069 (2007)
18. Wilquem, F., Degrez, G.: Numerical modeling of steady inspiratory airflow through a three-generation model of the human central airways. J. Biomech. Eng. Trans. ASME **119**, 59–65 (1997)
19. Martonen, T.B., Zhang, Z.Q., Yu, G.Q., Musante, C.J.: Three-dimensional computer modeling of the human upper respiratory tract. Cell Biochem. Biophys. **35**, 255–261 (2001)
20. Zhao, Y., Brunskill, C.T., Lieber, B.B.: Inspiratory and expiratory steady flow analysis in a model symmetrically bifurcating airway. J. Biomech. Eng. Trans. ASME **119**, 52–58 (1997)
21. Grotberg, J.B., Jensen, O.E.: Biofluid mechanics in flexible tubes. Ann. Rev. Fluid Mech. **36**, 121–147 (2004)
22. Marzo, A., Luo, X.Y., Bertram, C.D.: Three-dimensional collapse and steady flow in thick-walled flexible tubes. Fluids Struct. **20**, 817–835 (2005)
23. Luo, X.Y., Hinton, J.S., Liew, T.T., Tan, K.K.: Les modelling of flow in a simple airway model. Med. Eng. Phys. **26**, 403–413 (2004)
24. Li, W.-I., Perzl, M., Heyder, J., Langer, R., Brain, J.D., Englmeier, K.-H., Niven, R.W., Edwards, D.A.: Aerodynamics and aerosol particle deaggregation phenomena in model oral-pharyngeal cavities. J. Aerosol Sci. **27**, 1269–1286 (1996)
25. Hofmann, W.: Modeling techniques for inhaled particle deposition: the state of the art. J. Aerosol Med. **9**, 369–388 (1996)

26. Balashazy, I., Heistracher, T., Hofmann, W.W.: Air flow and particle deposition patterns in bronchial airway bifurcations: the effect of different cfd models and bifurcation geometries. J. Aerosol Med. **9**, 287–301 (1996)
27. Kaufman, J.W., Scherer, P.W., Yang, C.C.G.: Predicted combustion product deposition in the human airway. Toxicology **115**, 123–128 (1996)
28. Li, W.-I., Edwards, D.A.: Aerosol particle transport and deaggregation phenomena in the mouth and throat. Adv. Drug Delivery Rev. **26**, 41–49 (1997)
29. Gemci, T., Corcoran, T.E., Yakut, K., Shortall, B., Chigier, N.: Spray dynamics and deposition of inhaled medications in the throat. ILASS-Europe 2001, Zuirch 2–6 (2001)
30. Gemci, T., Corcoran, T.E., Chigier, N.N.: A numerical and experimental study of spray dynamics in a simple throat model. Aerosol Sci. Technol. **36**, 18–38 (2002)
31. Sheu, T.W.H., Wang, S.K., Tsai, S.F.: Finite element analysis of particle motion in steady inspiratory airflow. Comput. Methods Appl. Mech. Eng. **191**, 2681–2698 (2002)
32. Martonen, T.B., Zhang, Z., Yue, G., Musante, C.J.: 3D particle transport within the human upper respiratory tract. Aerosol Sci. **33**, 1095–1110 (2002)
33. Hofmann, W., Golser, R., Balashazy, I.: Inspiratory deposition efficiency of ultrafine particles in a human airway bifurcation model. Aerosol Sci. Technol. **37**, 988–994 (2003)
34. Ma, B., Lutchen, K.R.: An anatomically based hybrid computational model of the human lung and its application to low frequency oscillatory mechanics. Ann. Biomed. Eng. **34**, 1691–1704 (2006)
35. Croce, C., Fodil, R., Durand, M., Sbirlea-Apiou, G., Caillibotte, G., Papon, JF., Blondeau, JR., Coste, A., Isabey, D., Louis, B.: In vitro experiments and numerical simulations of airflow in realistic nasal airway geometry. Ann. Biomed. Eng. **34**(6), 997–1007 (2006)
36. Jeong, S.-J'., Kim, W.-S., Sung, S.-J.: Numerical investigation on the flow characteristics and aerodynamic force of the upper airway of patient with sleep apnea using computational fluid dynamics. Med. Eng. Phys. **29**, 637–651 (2007)
37. Brouns, M., Jayaraju, S.T., Lacor, C., De Mey, J., Noppen, M., Vincken, W., Verbanck, S.: Tracheal stenosis: a flow dynamics study. J. Appl. Phsiol. **102**, 1178–1184 (2007)
38. Jayaraju, S.T., Brouns, M., Verbanck, S., Lacor, C.: Fluid flow and particle deposition analysis in a realistic extrathoracic airway model using unstructured grids. J. Aerosol Sci. **38**, 494–508 (2007)
39. Garcia, G., Bailie, N., Martins, D., Kimbell, J.S.: Atropic rhinitis: a cfd study of air conditioning in the nasal cavity. J. Appl. Phsiol. **103**, 1082–1092 (2007)
40. Doorly, D.J., Taylor, D.J., Franke, P., Schroter, R.C.: Experimental investigation of nasal airflow. Proc. Inst. Mech. Eng. Part H: J. Eng. Med. **222**, 439–453 (2008)
41. Doorly, D.J., Taylor, D.J., Schroter, R.C.: Mechanics of airflow in the human nasal airways. Respir. Physiol. Neurobiol. **163**(1-3), 100–110 (2008)
42. Taylor, D.J., Doorly, D.J., Schroter, R.C.: Inflow boundary profile prescription for numerical simulation of nasal airflow. J. R. Soc. Interface **7**, 515–527 (2010)
43. Mylavarapu, G., Murugappan, S., Mihaescu, M., Kalra, M., Khosla, S., Gutmark, E.: Validation of computational fluid dynamics methodology used for human upper airway flow simulations. J. Biomech. **42**(10), 1553–1559 (2009)
44. Mihaescu, M., Murugappan, S., Kalra, M., Khosla, S., Gutmark, E.: Large Eddy Simulation and Reynolds-Averaged Navier-Stokes modeling of flow in a realistic pharyngeal airway model: an investigation of obstructive sleep apnea. J. Biomech. **41**(10), 2279–2288 (2008) JUL 19
45. Jin, H.H., Fan, J.R., Zeng, M.J., Cen, K.F.: Large eddy simulation of inhaled particle deposition within the human upper respiratory tract. J. Aerosol Sci. **38**(3), 257–268 (2007)
46. Spalart, P.R., and Allmaras, S.R.: A one-equation turbulence model for aerodynamic flows. AIAA paper 92-0439, (1992)
47. Nithiarasu, P., Hassan, O., Morgan, K., Weatherill, N.P., Fielder, C., Whittet, H., Ebden, P., Lewis, K.R.: Steady flow through a realistic human upper airway geometry. Int. J. Numer. Methods Fluids **57**, 631–651 (2008)

48. Stapleton, KW., Guentsch, E., Hoskinson, MK., Finlay, W.H.: On the suitability of k-epsilon turbulence modeling for aerosol deposition in the mouth and throat: a comparison with experiment. J. Aerosol Sci. **31**(6), 739–749 (2000) JUN
49. Nithiarasu, P., Liu, C.-B.: An explicit characteristic based split (cbs) scheme for incompressible turbulent flows. Comput. Methods Appl. Mech. Eng. **195**, 2961–2982 (2006)
50. Webb, P.: Air temperatures in respiratory tracts of resting subjects. J. Appl. Physiol. **4**, 378–382 (1951)
51. Cole, P.: Some aspects of temperature, moisture and heat relationships in upper respiratory tract. J. Laryngol. Otol. **67**, 449–456 (1953)
52. Inglestedt, S.: Studies on conditioning of air in the respiratory tract. Acta Oto-Laryngologica **46**(s131), 1–80 (1956)
53. Radaelli, A.G., Peiro, J.: On the segmentation of vascular geometries from medical images. Int. J. Numer. Methods Biomed. Eng. **26**(1), 3–34 (2010)
54. Mut, F., Aubry, R., Löhner, R., Cebral, J.R.: Fast numerical solutions of patient-specific blood flows in 3D arterial systems. Int. J. Numer. Methods Biomed. Eng. **26**, 73–85 (2010)
55. Gee, M.W., Foerster, Ch., Wall, W.A.: A computational strategy for prestressing patient-specific biomechanical problems under finite deformation. Int. J. Numer. Methods Biomed. Eng. **26**, 52–72 (2010)
56. Nithiarasu, P., Löhner, R.: Special issue on patient-specific computational modelling. Int. J. Numer. Methods Biomed. Eng. **26**, 1–2 (2010)
57. Naftali, S., Schroter, R., Shiner, R., Elad, D.: Transport phenomena in the human nasal cavity: a computational model. Ann. Biomed. Eng. **26**, 831–839 (1998)
58. Lindemann, J., Keck, T., Wiesmiller, K., Sander, B., Brambs, H., Rettinger, G., Pless, D.: A numerical simulation of intranasal air temperature during inspiration. Laryngoscope **114**, 1037–1041 (2004)
59. Naftali, S., Rosenfeld, M., Wolf, M., Elad, D.: The air-conditioning capacity of the human nose. Ann. Biomed. Eng. **33**, 545–553 (2005)
60. Elad, D., Wolf, M., Keck, T.: Air-conditioning in the human nasal cavity. Respir. Physiol. Neurobiol. **163**, 121–127 (2008)
61. Ingenito, E., Solway, J., McFadden, E. Jr, Pichurko, B., Cravalho, E., Drazen, J.M.: Finite difference analysis of respiratory heat transfer. J. Appl. Physiol. **61**, 2252–2259 (1986)
62. Zhang, Z., Kleinstreuer, C.: Species heat and mass transfer in a human upper airway model. Int. J. Heat Mass Transfer **46**, 4755–4768 (2003)
63. Tawhai, M.H., Hunter, P.J.: Modelling water vapor and heat transfer in the normal and the intubated airways. Ann. Biomed. Eng. **32**, 609–622 (2004)
64. Chung, S.K., Kim, S.K.: Digital particle image velocimetry studies of nasal airflow. Respir. Physiol. Neurobiol. **163**, 111–120 (2008)
65. Doorly, D.J., Taylor, D.J., Franke, P., Schroter, R.C.: Experimental investigation of nasal airflow. Proc. Inst. Mech. Eng. Part H: J. Eng. Med. **222**, 439–453 (2008)
66. Lin, CL., Tawhai, M.H., McLennan, G., Hoffman, E.A.: Characteristics of the turbulent laryngeal jet and its effects on airflow in the human intra-thoracic airways. Respir. Physiol. Neurobiol. **157**, 295–309 (2007)
67. Xu, C., Prince, J.L.: Snakes, shapes, and gradient vector flow. IEEE Trans. Image Process. **7**(3), 359–369 (1998)
68. Xie, X., Mirmehdi, M.: MAC: Magnetostatic active contour model. IEEE Trans. Pattern Anal. Mach. Intell. **30**(4), 632–647 (2008)
69. Felzenszwalb, P.F., Huttenlocher, D.P.: Distance transforms of sampled functions. Cornell Comput. Inform. Sci. TR2004–1963 (2004)
70. Peirò, J.: Surface grid generation. In: Thompson, J. F., Soni, B.K, Weatherill, N.P. (eds.) Handbook of Grid Generation, CRC Press, LLC
71. Tremel, U., Deister, F., Hassan, O., Weatherill, N.P.: Automatic unstructured surface mesh generation for complex configurations. Int. J. Numer. Methods Fluids **45**(4), 341–364 (2004)

72. Wang, D.S., Hassan, O., Morgan, K., Weatherill, N.P.: EQSM: an efficient high quality surface grid generation method based on remeshing. Comput. Methods Appl. Mech. Eng. **195**, 5621–5633 (2006)
73. Frey, P.J.: Generation and adaptation of computational surface meshes from discrete anatomical data. Int. J. Numer. Methods Eng. **60**, 1121–1133 (2004)
74. Lorensen, W.E., Cline, H.E.: Marching cubes: a high resolution 3D surface construction algorithm. Comput. Graphics **21**(4), 163–169 (1987)
75. Chernyaev E.V.: Marching cubes 33: construction of topologically correct isosurfaces. Technical report Cern CN 95-17, CERN (1995)
76. Sazonov, I., Wang, D., Hassan, O., Morgan, K., Weatherill, N.P.: A stitching method for the generation of unstructured meshes for use with co-volume solution techniques. Comput. Methods Appl. Mech. Eng. Corresp. **195**(13–16), 1826–1845 (2006)
77. Taubing, G.: A signal approach to fair surface design. In Proceedings of SIGGRAPH (August 1995) pp. 351–358 (1995)
78. Frey, W.H., Field, D.A.: Mesh relaxation: a new technique for improving triangulation. Int. J. Numer. Meth. Eng. **31**, 1121–1133 (1991)
79. Löhner, R.: Regridding surface triangulations. J. Comput. Phys. **126**, 1–10 (1996)
80. Weatherill, N.P., Hassan, O.: Efficient three-dimensional delaunay triangulation with automatic point creation and imposed boundary constraints. Int. J. Numer. Methods Eng. **37**, 2005–2039 (1994)
81. Du, Q., Wang, D.: Tetrahedral mesh generation and optimization based on CVT. Int. J. Numer. Methods Eng. **56**, 1355–1375 (2003)
82. Zienkiewicz, O.C., Codina, R.: A general algorithm for compressible and incompressible flow–part i. The split, characteristic-based scheme. Int. J. Numer. Methods Fluids **20**(8-9), 869–885 (1995)
83. Codina, R., Owen, H.-C., Nithiarasu, P., Liu, C.-B.: Numerical comparison of cbs and sgs as stabilization techniques for the incompressible navier-stokes equations. Int. J. Numer. Methods Eng. **66**, 1672–1689 (2006)
84. Massarotti, N., Fausto, F., Lewis, R.W., Nithiarasu, P.: Explicit and semi-implicit cbs procedures for incompressible viscous flows. Int. J. Numer. Methods Eng. **66**, 1618–1640 (2006)
85. Nithiarasu, P.: An efficient artificial compressibility (ac) scheme based on the characteristic based split (cbs) method for incompressible flows. Int. J. Numer. Methods Eng. **56**, 1815–1845 (2003)
86. Nithiarasu, P., Mathur, J.S., Weatherill, N.P., Morgan, K.: Three-dimensional incompressible flow calculations using the characteristic based split (cbs) scheme. Int. J. Numer. Methods Fluids **44**, 1207–29 (2004)
87. Zienkiewicz, O.C., Taylor, R.L., Nithiarasu, P.: The Finite Element Method for Fluid Dynamics. Elsevier Butterworth-Heinemann, Burlinton (2005)
88. Heenan, A.F., Matida, E., Pollard, A., Finlay, W.H.: Experimental measurements and computational modeling of the flow in an idealized human oropharynx. Exp. Fluids **35**, 70–84 (2003)
89. Kleinstreuer, C., Zhang, Z.: Laminar-to-turbulent fluid-particle flows in a human airway model. Int. J. Multiphase Flow **29**, 271–289 (2003)
90. Malve, M., Perez, A., lopez Villalobos, JL., Ginel, A., Boblare, M.: FSI analysis of the coughing mechanism in a human trachea. Ann. Biomed. Eng. **38**(4), 1556–1565 (2010)
91. Kelly, J.T., Prasad, A.K., Wexler, A.S.: Detailed flow patterns in the nasal cavity. J. Appl. Physiol. **89**, 323–337 (2000)

A Decission Support System for Endoprosthetic Patient-Specific Surgery of the Human Trachea

Olfa Trabelsi, Angel Ginel, Jose L. López-Villalobos,
Miguel A González-Ballester, Amaya Pérez-del Palomar
and Manuel Doblaré

Abstract This chapter presents a patient specific decision support system for helping surgeons in endoprosthetical tracheal surgery. The system is based on a set of 3D finite element numerical simulations of the human trachea during swallowing with and without prosthetic insertion. The associated code uses the constitutive behavior derived from a set of experimental and histological tests performed for its main components: smooth muscle and cartilage. A complete set of numerical results is derived from those simulations using a parametric approach relying on a robust design of experiments and on a automatic mesh adaptation process. These results feed a tool for statistical analysis of human swallowing without and with prosthesis that, together with the response surface method, allows

O. Trabelsi · A. Pérez-del Palomar · M. Doblaré (✉)
Group of Structural Mechanics and Materials Modelling (GEMM).
Aragón Institute of Engineering Research (I3A), University of Zaragoza,
Betancourt Bldg. María de Luna s/n, 50018 Zaragoza, Spain
e-mail: mdoblare@unizar.es

O. Trabelsi
e-mail: olfa@unizar.es

A. Pérez-del Palomar
e-mail: amaya@unizar.es

A. Ginel · J. L. López-Villalobos
Thoracic and Neumology Service, Hospital Virgen del Rocío,
Av. Manuel Siurot s/n, 41013 Seville, Spain
e-mail: aginel@yahoo.es

J. L. López-Villalobos
e-mail: jllopezvillalobos@hotmail.com

M. A González-Ballester
ALMA IT Systems, Vilana 4B, 4-1, 08022 Barcelona, Spain
e-mail: miguel.gonzalez@alma3d.com

getting a set of response functions for the most important variables of the problem (displacements and principal stresses) for every point of the trachea in terms of a set of patient-specific parameters that define the geometry of the trachea and prosthesis. This provides the thoracic surgeon a fast, accurate and simple tool for predicting the stress state of the trachea and the reduction in the ability to swallow after implantation thus helping in taking decisions during pre-operative planning of tracheal interventions. This tool is also useful for performing patient specific studies of pathological swallowing and for the design of new types of prostheses.

1 Introduction

One of the main tools for surgical intervention in many types of application (orthopaedics, vascular, respiratory or digestive systems) are endoproshteses, both permanent or temporary. They have been considered as a real breakthrough in less invasive treatments, with an incredible rate of success [1, 2]. However, its generalized use for every patient still implies complications [3, 4]. Therefore, a careful planning of the associated surgery before the intervention is needed to decide the type, size, location, angulation and method of insertion of the endoprosthesis for each specific patient with the objective of ensuring a proper fixation, minimizing short and long-term damage in the tissues and, of course, to get the correct functioning of the organ, as well as to minimize the duration of the operation and postoperative recovery. The many parameters involved in this type of surgery and the associated outcomes have made the surgeon experience the main (and usually unique) tool for taking such decisions in pre-operative planning. In the last years, however, and thanks to the continuous improvement of testing protocols and computer simulation, Biomechanics in general, and computational Biomechanics, in particular, have become powerful tools for prosthesis design, tissue behaviour simulation, biological process modelling, and pathologies follow-up [5–7]. In addition, computer simulation allows checking new hypotheses for biological systems, establishing new lines of experiments and formulating new theories and lines of thinking for complex biomedical phenomena [8].

In this context, many medical problems associated with surgery, trauma, and rehabilitation have been identified, conceptualized and numerically solved [9, 10]. It is true that studies "in vivo" allow a more objective valorisation of traumatic disorders as well as the effects of the different methods used in their treatment, but, alternatively, numerical simulation permits studying biomechanical problems in normal and pathological circumstances caused by traumatisms or degenerations provoked, for example in the trachea, by a prosthetic implantation or growth of granulomas.

One of the main difficulties of simulation within Biomechanics is the extreme complexity of the behaviour of biological materials. Indeed, there is not a sufficiently contrasted database for modelling the behaviour of tissues like those forming the human trachea (the cartilaginous tissue or the smooth muscle), which are still subject of multiple studies due to their special characteristics of heterogeneity, anisotropy and non-linear stiffness, as well as to the relation with age, loads and particular anatomy of each individual. In addition, the geometrical no-linearity that always lead to large displacements and finite strains and the problems associated to complex contacts between organs or between organs and prostheses have to be also taken into account [11]. Despite these complexities, the development of accurate constitutive models has permitted, nowadays, to model the behaviour of biological materials with reasonable accuracy within Finite Element frameworks, thus gradually improving the ability to simulate real clinical situations, including the variation of factors that change from one patient to another.

One of the clinical fields in which surgical planning has become a real need is tracheal surgery. Tracheal stenosis, which is the most common tracheal injury [12], can be considered as a pressure necrosis sometimes caused by a tracheotomy intubation with unsuitable pressure. Prolonged ischemia and infection cause necrosis of the tracheal wall and deterioration of the cartilaginous structure through formation of granulation tissue [13]. Many methods are available to treat tracheal stenosis as tracheal dilation, excision, anastomosis or reconstruction, but after treatment, the stenosis may reappear (recidives) especially if it was longer than two centimetres [13–16]. In 1964 and thanks to Montgomery [17], tracheal stents were invented with the objective of creating a support on which the lesion scar can be modelled, and then removed once the scaring process is stabilized. Since then, the stents have had an unceasingly evolution, from the simple tube design to much more complex geometries [18, 19], and from the static silicone stent to the dynamic or expandable metallic stents [20–22]. These interventions, however, can bear, in the post-operating period, many problems such as migration of the stent, development of granulation tissue at the edges of the stent with overgrowth of the tracheal tissue, or accumulation of secretions inside the prosthesis. The most popular stents used in the management of tracheal obstructive lesions are probably the silicone ones, and in particular the Dumon stent (Novatech; Aubagne, France) [19], which is considered to be the most effective airway stent presently available due to its low cost and safety [23, 24].

Despite the clinical interest of the implantation of tracheal stents, most of the works have been focused on obtaining experimental and clinical results of tracheal stents insertion [12–26]. There are few studies on the stent-tracheal tissues interaction based on computational models. In addition, most of the numerical studies have been based on analytical models, simulating an isolated tracheal ring formed by cartilage and muscular membrane [27]. A finite element model of the lamb's windpipe was constructed to predict the behaviour of the trachea before airway ventilation treatment [28]. Gustin [29] analyzed the influence of endotracheal prosthesis implantation through an axisymetric FE model. Wiggs et al. [30] presented a finite element model without the cartilaginous structure to analyse the mechanism of mucosal folding that occured after smooth muscle shortening and which explains the difference in airway narrowing between asthmatic and control airways.

The most common clinical consequence after endoprosthesis implantation is the formation of granular tissue at the ends of the endoprosthesis. Despite the more than 100 models of stents available [18], the problem of formation of granulation tissue in the tracheal wall surrounding the implant has not been resolved yet [21]. Therefore, the objective of this chapter is to describe a decision support system based on experimental, numerical and statistical surveys with a first clinical verification that may help the thoracic surgeon to choose the position and appropriate dimensions of the prosthesis for each patient in an optimal time, with the objective of reducing the damage generated in the living tissues after placement of the implant. With this aim, we analyzed the mechanical behaviour of the main constituents of the human trachea to perform computational models reproducing its functional behavior under physiological conditions and after prosthetic implantation. Among the movements that trachea carries out, swallowing seems to drive harmful consequences for the tracheal tissues surrounding the prosthesis. Swallowing is a complex movement which varies from one person to another, and depends on many physiological parameters. These parameters have a huge influence on the ascendant displacement of the trachea (swallowing capacity), and on the muscular force exerted by each individual to achieve the swallowing movement. When a prosthesis is inserted, additional factors intervene in these values and on the stress state of the trachea during swallowing. This is the reason why we have concentrated in analyzing the effect of prosthesis implantation into the biomechanical behaviour of the trachea during swallowing, performing several numerical simulations, based on physiological tracheal geometries of specific individuals and their real swallowing trajectories.

This chapter consists of an introduction and four sections distributed as follows: Sect. 2 exposes the histologies made on the principal tissues of the trachea (cartilage and the smooth muscle) and the protocols of the uniaxial tensile tests performed to characterize the passive properties of these tissues. The mechanical properties of the silicone of the Dumon stent are also described. Section 3 describes the different steps for obtaining patient swallowing trajectories, tracheal and prostheses finite element models, and the different boundary conditions and materials used to perform the simulations; in Sect. 4 the simulation results both for physiological and pathological swallowing (with the Dumon prosthesis and the Mongomery T-tube) are presented. Swallowing simulations of several real patients with tracheal geometries adapted to each case are performed and clinically validated to justify those computational results. Section 5 includes a statistical study of the human swallowing in presence of the Dumon prosthesis. A code for mesh adaptation to patient-specific geometries was implemented and used to complete a robust experimental design with the aim of detecting the main factors influencing the stress state of the trachea and its capacity to swallow. The results derived from that statistical study are compared to those of the finite elements simulations. The decision support system developed to optimize cost, effort and time for planning tracheal prosthesis implantation is finally presented.

2 Experimental Characterization of the Mechanical Passive Behaviour of the Tracheal Tissues

To characterize the mechanical behaviour of a living tissue, two techniques are usually needed: histology, to analyze the underlying structure, and mechanical testing, to get the actual constitutive behavior. Among the several tissues that form the trachea, the cartilage rings and the smooth muscle are the principal components responsible for the global mechanical behaviour of this essential component of the respiratory system. The mechanical function of the cartilaginous structure is to maintain the trachea open during the respiratory movements supporting the interthoracic pressure. The contraction of the smooth muscle and the transmural pressure generate bending in these cartilages, collapsing them to regulate the air flow by modulating the diameter of the airway.

The mechanical characteristics of the human trachea have not been completely determined. Few studies have been made to better understand the behaviour of its components [31–33]. With respect to the cartilage, in some of these works it is considered as a non-linear material displaying higher strength in compression than in extension [33], but in most of them, the isolated tracheal cartilage ring is considered as a linear elastic material [34, 35]. In Rains's work [36], the stress-strain relations obtained from slices of human tracheal cartilage were considered linear for deformations up to 10% of the initial length of the samples. In fact, negligible hysteresis and no residual strain were found in specimens tested up to a maximal applied strain of 10%. Beyond this limit, strain hysteresis and residual strain increased progressively [31]. The same assumption was used by Kempson [37], where the response of the articular cartilage was considered linear for deformations up to 5% of the initial length of the samples. This hypothesis was also used in several analytical models [27, 38], as well as in some finite elements simulations [28, 39].

With respect to the mechanical properties of the tracheal muscle, most of the developed works dealt with its plasticity, stiffness and extensibility. Gunst et al. [40] treated the effects of its length on the stiffness and the extensibility at contractile time activation. Cumin et al. [41] analyzed its mechanical response to oscillatory conditions and Stephens et al. [42] analyzed the influence of the temperature on the force-velocity relationships. However, none of them analyzed the mechanical properties of this membrane that makes possible the collapse of the rings and therefore the pressure balance inside the trachea.

In this section, a histological analysis was made on cartilaginous and muscular tracheal samples taken from an individual aged 46 to relate the microstructure of the different tissues with their mechanical behaviour. Then, the experimental tests carried out for cartilage and smooth muscle samples are described. Cartilage rings obtained from the autopsy of individuals aged 79 and 82, without tracheal pathologies were tested using a cyclic uniaxial tensile test confirming their quasi-linear behaviour. Uniaxial quasi-static tensile tests on human tracheal smooth muscle were also made to characterize its passive response along the two preferential directions of anisotropy corresponding to the two orthogonal families of muscle fibers confirmed in the histology. The obtained experimental curves were then analyzed and the material parameters that appear in the Holzapfel strain-energy density function [43] were fitted, thus considering this material as non-linearly hyperelastic.

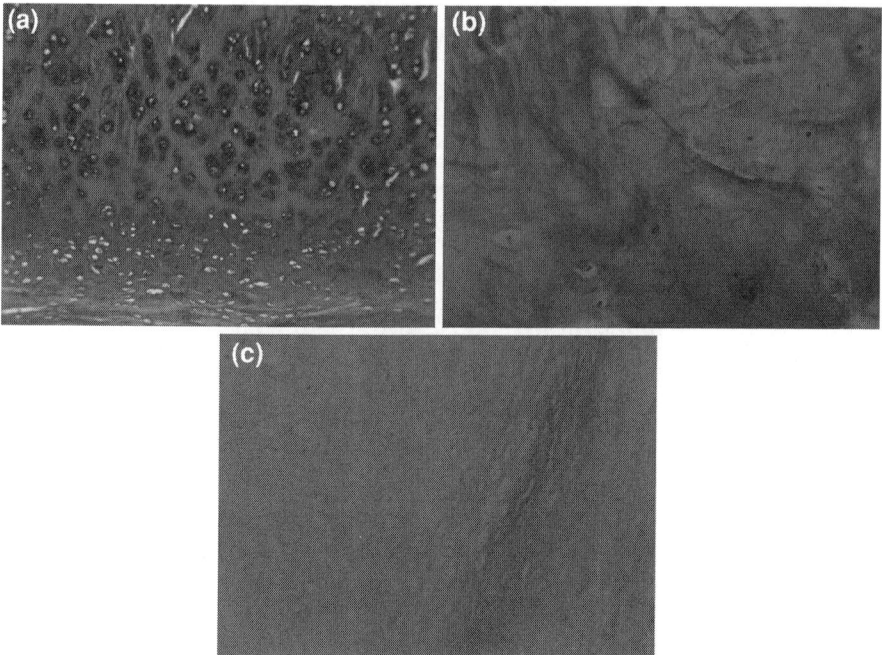

Fig. 1 Human cartilage histology. It can be seen that there is not a clear orientation of the collagen fibers. **a** X10, Haematoxylin and eosin stain: nucleus in *black* and rest of cells in *purple*. **b** X40, Masson's Trichrome stain: nucleus in *black* and collagen in *blue*. **c** X20, Van Gieson's Trichrome stain: nucleus in *black* and collagen in *red*

2.1 Tracheal Histology

In order to choose an appropriate constitutive behavior for the different tissues, related to their actual microstructure, a histological study of both the cartilage rings and smooth muscle was made on samples cut from the human trachea of an individual aged 46 with no tracheal disorder. Muscular samples and cartilaginous rings were conserved in formol, fixed first in formalin 10% during 24 h at room temperature and then in 70° alcohol. The samples were next embedded and cut in slices of 5 μm thickness in the required directions. Then, they were dyed with different compounds such as Haematoxylin and eosin (to see nucleic acids and ergastoplasm), Masson and Van Gieson's Trichrome stain (to see connective tissues and fibers). After dehydration, the samples were observed using an inverted microscope (Axiovert 40 CFL-ZEISS). The histology demonstrates that cartilage is arranged in an interlaced network of fine fibrils of collagen oriented towards the surface and become dispersed and disorientated once they approach the center of the cartilage (Fig. 1). This was clearly described by Roberts et al. [32] in their study on the ultra-structure of the human cartilage.

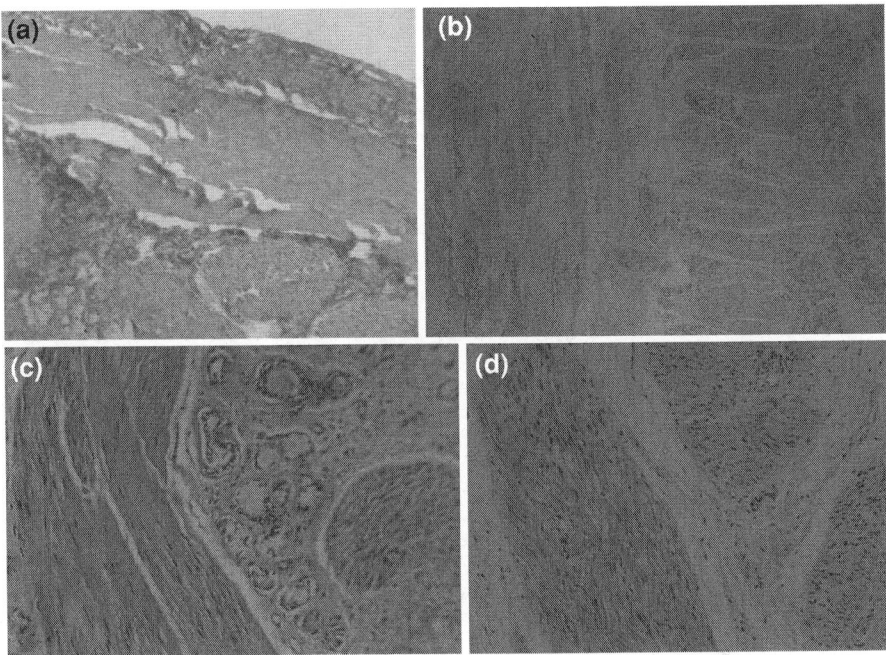

Fig. 2 Histology of human tracheal smooth muscle. The two directions of fiber bundles: one running longitudinally and other orthogonally are seen in both pictures. **a** Transverse cut of the human smooth muscle, X10 Van Gieson's Trichrome: muscle fibers in *red*. **b** Longitudinal cut of the human smooth muscle, X2 Haematoxylin and eosin: muscle fibers in *red*. **c** Longitudinal cut of the human smooth muscle, X10, Haematoxylin and eosin, fibers in *pink*. **d** Transverse cut of the human smooth muscle, X10, Haematoxylin and eosin stain, fibers in *pink*

Referring to samples of smooth muscle, cuts in the longitudinal and transverse directions demonstrate the existence of two families of muscle fibers. Figure 2 shows clearly how these fibers are oriented in the two cutting directions, being the transversal family the thicker and stronger and inserted at the level of the cartilage edges to create a transverse connection between them, so that their contraction bends the cartilage, altering the cross-section of the trachea [44] according to the functional needs. On the other hand, the weaker longitudinal fibers are important in other movements of the trachea such as swallowing [44].

2.2 Experimental Characterization of Tracheal Mechanical Properties

Fibered soft tissues are usually modeled as incompressible, non-linearly hyperelastic materials, being their constitutive behavior derived in isothermal processes, as is usually the case, from the definition of an appropriate local strain-energy

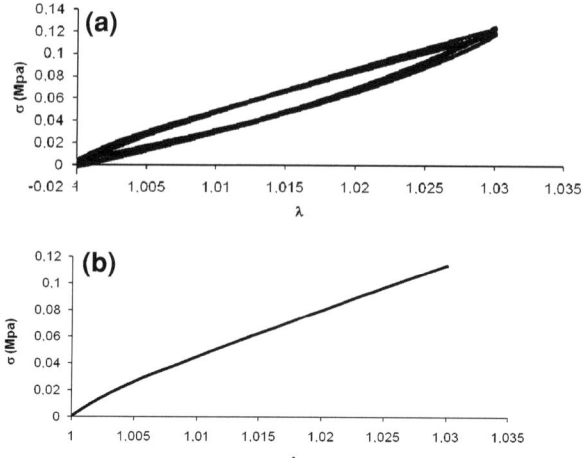

Fig. 3 Stress-stretch curve obtained in a cyclic tension test performed on the central part of the second human tracheal ring. **a** Stress-stretch curve obtained from a cyclic test performed on the central part of the human tracheal ring. **b** Loading part of the stress-stretch curve obtained from the tensile test performed on the human tracheal cartilaginous ring

density function Ψ at each point $\mathbf{X}$, that depends on both the local deformation tensor (the Green $\mathbf{C}$ or Lagrange $\mathbf{E}$ tensors are usually used in finite strain formulations) and the directions of the different fiber families, each defined by a unit reference vector $\mathbf{a}_0^i$ that identifies the initial preferential direction of each fiber bundle i [45, 46]. For incompressible or quasi-incompressible materials as soft tissues, a decoupled form of the strain-energy density function is usually stated. In fibered materials this may be written as:

$$\Psi(\mathbf{C}, \mathbf{A}^i) = \Psi_{vol}(J) + \overline{\Psi}(\mathbf{C}, \mathbf{A}^i) = \Psi_{vol}(J) + \overline{\Psi}(\bar{I}_1, \bar{I}_2, \bar{I}_4^i) \quad (1)$$

where Ψ_{vol} and $\overline{\Psi}$ are scalar functions describing the volumetric and isochoric responses of the material, respectively. $\bar{I}_1$ and $\bar{I}_2$ are the standard invariants of the right Cauchy-Green tensor $\mathbf{C}$ associated with the isotropic material behavior, and the pseudo-invariant $\bar{I}_4^i$ characterizes the constitutive response of the fibers on the direction (i).

$$\bar{I}_1 = tr(\mathbf{C}) \quad (2)$$

$$\bar{I}_2 = \frac{1}{2}[(tr(\mathbf{C}))^2 - tr(\mathbf{C}^2)] \quad (3)$$

$$\bar{I}_4^i = \overline{\mathbf{C}} : \mathbf{A_i} \quad (4)$$

$$\mathbf{A_i} = \mathbf{a}_0^i \otimes \mathbf{a}_0^i \quad (5)$$

From the tensile tests on human tracheal cartilages up to 5% of strain, quasi-linear experimental curves of stress-stretch were obtained (Figs. 3 and 4); therefore an elastic neo-Hookean model was used to fit the experimental data. In this case, the isochoric strain energy density function is defined as:

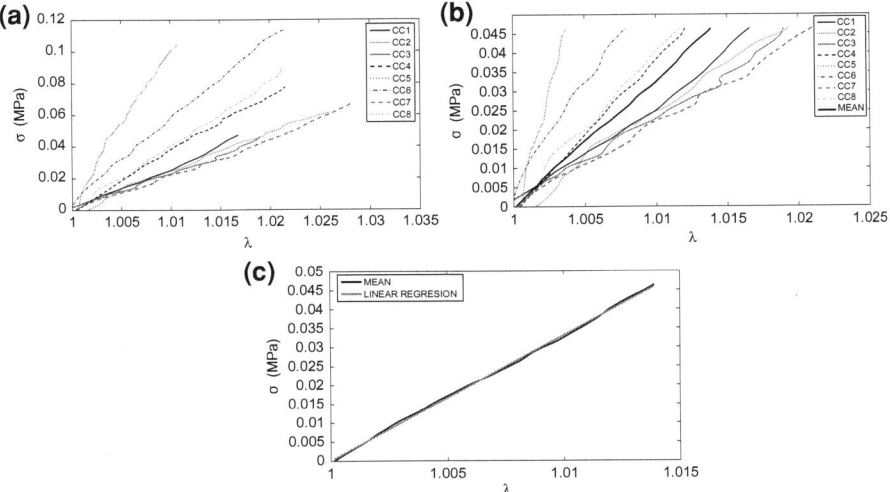

Fig. 4 Experimental cartilage curves and Neo-Hookean regression of the mean. **a** Curves of tensile tests on human tracheal cartilage. **b** Curves of tensile test on human tracheal cartilage and the mean. **c** Neo-Hookean regression of the mean of the experimental cartilage curves: $C_{10} = 0.557\,\text{MPa}, E = 3.334\,\text{MPa}, \nu = 0.494$ and $R = 0.999$

$$\Psi_{iso}(\bar{I}_1) = C_{10}(\bar{I}_1 - 3) \tag{6}$$

where $C_{10} > 0$ is a stress-like material parameter (fitted from the experimental curves).

Referring to the smooth muscle, and considering the results of the histology which determined the existence of two orthogonal families of fibers, the exponential Holzapfel strain-energy density function [47] was used.

$$\bar{\psi}\left(\bar{I}_1, \bar{I}_2, \bar{I}_4^1, \bar{I}_4^2\right) = C_{10}(\bar{I}_1 - 3) + \frac{k_1}{2k_2}\left\{\exp\left[k_2\left(\bar{I}_4^1 - 1\right)^2\right] - 1\right\} \\ + \frac{k_3}{2k_4}\left\{\exp\left[k_4\left(\bar{I}_4^2 - 1\right)^2\right] - 1\right\} \tag{7}$$

where the isotropic behavior of the smooth muscle matrix is described by the first neo-Hookean term; k_1 and k_2 are constants related to the first family of fibers, and k_3 and k_4 correspond to the second family (all these parameters will be determined by fitting the Holzapfel's exponential function [45, 48] to the experimental results). In the next lines, we explain the mechanical testing protocols and the experimental results obtained to compute these material parameters that will be subsequently used in the numerical simulation.

The samples taken from human tracheas were conserved in a 0.9% physiological saline solution and stored at $-70°C$. The postmortem autolysis was minimized by immediate freezing of the tissue and by performing the tests soon after the samples were thawed. For the mechanical tests of the cartilage and the muscle, pieces of sandpaper were bound to the samples to avoid sliding between the tips of the samples and the clamps of the tensile machine. To determine the initial positioning of the sample and to calibrate the video-extensometer, two references were drown on the central part of the samples as markers for the longitudinal deformation measurement [47]. The tracheal tissues were mounted in an Instron MicroTester 5,548 machine to perform the tensile tests, with a precision of 0.0001 N and 0.001 mm in force and displacement respectively. An Instron video-extensometer with an Elecentric focus of 55 mm able to collect data at 10 Hz was used to register the longitudinal deformation between the two markers. During the test, the humidity and temperature were maintained constant to avoid the dehydration of the biological tissues in order to reproduce their natural environment. Ten samples of cartilage were tested after dissecting them from the surrounding perichondrium and soft connective tissues. From each cartilaginous ring only the straight central part was conserved to neglect the curvature of the ring. The specimens of the cartilage measured 2.1 ± 0.3 mm in thickness, 6.6 ± 3 mm in length and 6 ± 3.5 mm in width. The distance between the two reference points was about 3.6 ± 0.7 mm.

Regarding the tracheal smooth muscle, samples were cut in transverse (5 samples) and longitudinal directions (5 samples). As done for the cartilage samples, the same protocol for calibration and test preparation was followed. The distance between the reference dots in the muscular tissue was 4.2 ± 0.47 mm for the longitudinal samples and 3.84 ± 0.97 mm for the transversal ones. The dimensions of the longitudinal muscular samples were 1.2 ± 0.77 mm, 14.66 ± 3 mm and 6 ± 3.5 mm for length, width and thickness, respectively. Those along the transversal direction were 1.62 ± 0.2 mm, 12.42 ± 1.73 mm and 5.22 ± 1.04 mm respectively.

For the cartilages and muscle tissue, different testing protocols were performed. For cartilage, samples were preconditioned 5 times under a speed of 0.2 mm/min until a deformation of 1% of the initial length of the sample, and then a cyclic test was accomplished at a constant speed of 0,1 mm/min until getting a deformation of 5% of the initial length of the samples, at room temperature. The imposed strain rate makes reference to the one used by Korhonen in his tests on articular cartilage [49]. For the smooth muscle, as the mechanical behavior is different, the samples were first preconditioned by applying 10 cycles at a rate of 3 mm/min until a 5% of deformation, and then monotonically tested at 1 mm/min until getting 20% of deformation. This loading protocol is similar to the one performed by Holzapfel [47] to test the smooth muscle of coronary arteries.

As a result of the experimental study, Fig. 3 illustrates the experimental curves obtained from the cyclic tests made on the central part of tracheal cartilage rings. It can be observed that, during the loading step, the cartilage has a quasi-linear response, but when unloaded, some hysteresis is noticed. This is explained by the intervention of water which provides this visco-elastic behavior to cartilage during relaxation. The cartilage rings work under strains well below the value of 5%, so only the loading phase (Fig. 3b) until this strain of 5% was taken into account.

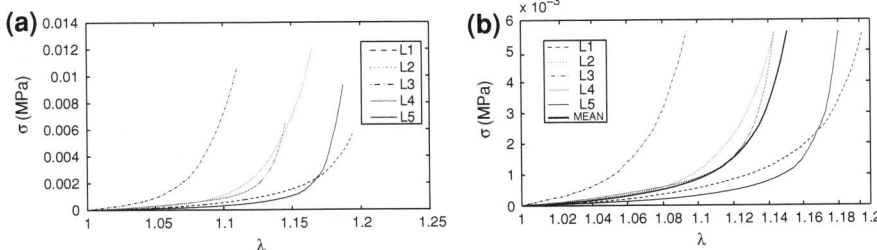

Fig. 5 Experimental curves of the human tracheal smooth muscle and mean curves obtained by uniaxial extension tests along the longitudinal (*L*) direction. **a** Experimental curves of longitudinal human smooth muscle cuts (*L*). **b** Mean of the experimental curves of longitudinal human tracheal smooth muscle cuts (*L*)

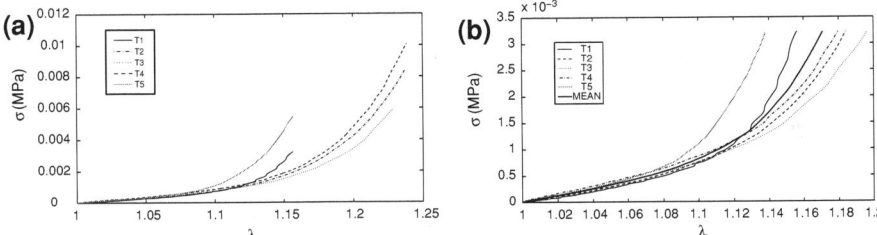

Fig. 6 Experimental curves of the human tracheal smooth muscle and mean curves obtained by uniaxial extension tests along the transversal (*T*) direction. **a** Experimental curves of transverse human smooth muscle cuts (*T*). **b** Mean of the experimental curves of transverse human tracheal smooth muscle cuts (*T*)

A code was implemented in Matlab 7.1 to get the mean of all cartilage curves for the same level of stress. The value of C_{10} in Eq. 6 (see Fig. 4) was then computed for this curve getting a value of $C_{10} = 0.56$ MPa. The relationship between the Neo-Hookean fitting and the experimental mean stress-strain curve was then assessed by means of the Pearson's correlation coefficient, resulting into a value of $R = 0.99$ (Fig. 4). Assuming elastic behavior and quasi-incompressibility, the corresponding value of the total Young modulus was computed as 3.33 MPa.

Regarding the tracheal muscle, the histology revealed two orthogonal families of fibers (Fig. 2). In order to characterize its passive mechanical properties, two steps were made: once the matrix stiffness C_{10} was computed, the longitudinal and transverse fiber's parameters (k_1, k_2, k_3 and k_4) were estimated. From the experimental data, the means of the transverse and longitudinal curves were obtained (Figs. 5, 6), and the linear range (in our case up to 5% of strain) that essentially characterizes the behavior of the matrix was chosen. In this range, a linear fitting was applied to the mean curves using the Neo-Hookean strain-energy

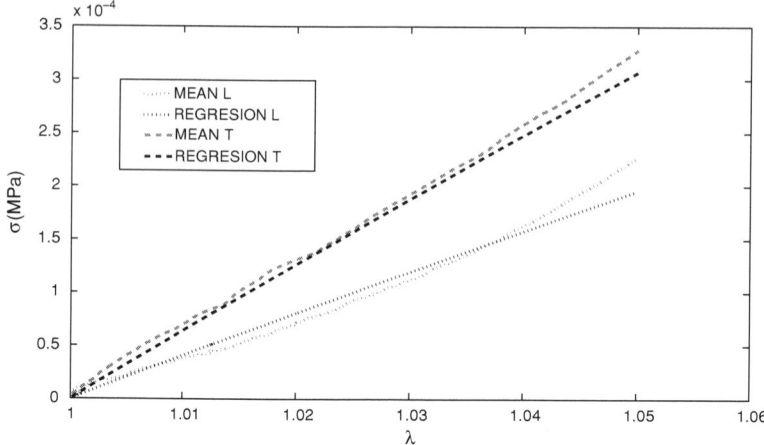

Fig. 7 Neo-Hookean fitting of the strain interval 0–5% along the longitudinal (*L*) and transversal (*T*) directions of the mean experimental curves: $C_{10}(L) = 0.684$ KPa and $C_{10}(T) = 1.074$ KPa

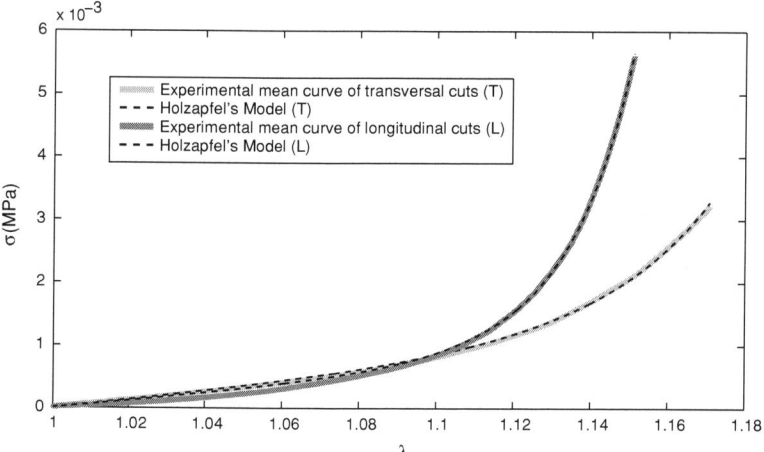

Fig. 8 Regression of the longitudinal (*L*) and transversal (*T*) mean curves using Holzapfel's function: $C_{10} = 0.877$ KPa, $k_1 = 0.154$ KPa, $k_2 = 34.157$, $k_3 = 0.347$ KPa, $k_4 = 13.889$ and $R = 0.999$

density function to determine the stiffness of the matrix C_{10}. The mean of the obtained transversal and longitudinal stiffnesses was $C_{10} = 0,877$ KPa (Fig. 7). To characterize the anisotropic behavior of the muscle, the means of longitudinal and transversal muscle curves were fitted using the Holzapfel strain-energy density function adapted to a material with two families of fibers (Fig. 8). The "fminsearch" function available in Matlab ® v7.1. was used for this fitting process. For the longitudinal direction, the parameters of the Holzapfel strain-energy density resulted $k_1 = 0.154\,\text{KPa}$ and $k_2 = 34.157$, while for the transversal direction, the values of these parameters were $k_3 = 0.347\,\text{KPa}$, and $k_4 = 13.889$ ($R = 0.999$) (Fig. 8).

Finally, some of the many limitations of this work deserve to be mentioned. All soft biological tissues are viscoelastic to some extent. However, in this case, its mechanical behavior was simplified to be purely hyperelastic. Cartilage rings in the trachea are very rigid elements (the physiological strain range is well under 5%) that bend under muscle contraction, being negligible the contribution of its biphasic nature to the overall behavior of the trachea. Also, the uniform directional dispersion of the collagen fibers in tracheal cartilage rings drove us to use an isotropic Neo-Hookean model homogenizing the overall stiffness of both matrix and fibers. The value obtained for the equivalent Young Modulus, assuming incompressibility, for cartilage rings was $E = 3.33$ MPa which is in the range found in the literature. For instance, Lambert et al. [35], using a photographic technique, reported that the elastic modulus was in the range between 2.5 and 7.7 MPa for human cartilage tracheal rings. Rains et al. [31] performed tensile tests of tracheal cartilage samples to examine the hysteresis behavior of this material, getting elastic moduli from 1.8 ± 2.1 to 15 ± 5.1 MPa, depending on the age of the patient. Roberts et al. [32] tested samples of cartilage from different depths of the tracheal rings obtaining a range of variation from 1 to 20 MPa.

Regarding the tracheal muscle, the histological analysis showed that the matrix is mainly composed of two orthogonal families of fibers, fact that was only occasionally mentioned [44] but not considered in other studies made on this tissue [28, 50, 51]. The longitudinal fibers stiffen the tissue during movements like swallowing, while the transversal ones work mainly during breathing and coughing. In order to investigate the non-linearity of the tracheal smooth muscle on the two directions of fibers, the Holzapfel strain energy function was used to fit the experimental data. As far as the authors knowledge, this is the first study performed in the human tracheal muscle; only Sarna et al. [50] developed similar studies for canine tissue. In their work, different non-linear material models were considered to study the muscular contraction, but their conclusions cannot be extrapolated to our case not only due to the different species but also because they measured the mechanical behavior making artificial contraction with a very high strain rate.

Only tracheal samples from elder people were analyzed, when it is known that biological tissues loose water and collagen with age changing the mechanical properties, so an experimental campaign with a wider range of ages should be performed to get age-independent conclusions. Regarding the range of deformation, the cartilage was tested only in tension up to 5%; in this range, a quasi-linear behavior was obtained. To generalize the constitutive law to other strain intervals, including collapse of the tracheal rings, additional tests for a wider range of deformation should be performed [33]. For the tracheal smooth muscle, only its passive behavior under quasi-static conditions was investigated. It would be important to include viscoelasticity and perform biaxial tests to better approximate the anisotropy and the combined contribution of the two families of fibers.

Moreover, there is a large dispersion in the experimental curves, fact that can be related to using samples from different individuals. Therefore additional tests would be necessary to draw more accurate conclusions. Despite these limitations, the results presented herein have to be seen as a first approach to characterize the behavior of the human tracheal tissues, thus improving our understanding of the mechanical properties of tracheal cartilage and smooth muscle, and allowing reproducing the physiological tracheal behavior by means of computational models.

3 Description and Construction of the Finite Element Model

3.1 Reference Geometrical Model

In this section, a complete FE model of the human trachea based on the previous experimental study of the mechanical behaviour of the principal tracheal constituents: cartilage and smooth muscle, is presented. The first step to build such a model corresponds to defining the geometry. For the reconstruction of the physiological geometry of the human trachea, DICOM files provided by the Hospital Virgen del Rocío in Seville (Spain), derived from a Computational Tomography of a 70 years old healthy man, were used. From these files, a sweeping of transversal cut images in several superior regions of the patient (head and thorax) was obtained. The first step in any image analysis is the segmentation process, that is the identification of the boundaries of the organs under study and the interfaces between different tissues and structures. The accuracy of this segmentation depends on the specific application and the quality of the images available [52]. In our case, the images provided a clear picture of the internal cavity of the trachea filled with air (Fig. 9a). The detection of the outer surface and therefore the thickness of the wall was not that easy. For this reason, a non automatic segmentation of the CT scan was accomplished to determine the real geometry of the trachea and to distinguish between the muscle membrane, with lower density, and the cartilaginous structures (thyroid, cricoid and rings) with higher density. MIMICS software [53] was used for this purpose, getting a complete 3D geometrical model of the trachea that will be subsequently used in our numerical simulations.

Slight variations of the tracheal thickness are usually detected along the tracheal axis (starting from the third cartilaginous ring after the cricoids), with an average of 3 mm. Consequently, we decided to use the segmentation of DICOM images defining the internal surface of the trachea and obtain the external one by an offset equal to the constant thickness obtained from the cartilage structure. The segmented geometry such defined was then stored in a file written in STL format that defines the surfaces in terms of a mesh of triangles. This STL file was subsequently translated into IGES format (Initial Graphics Exchange Specification) and also to a

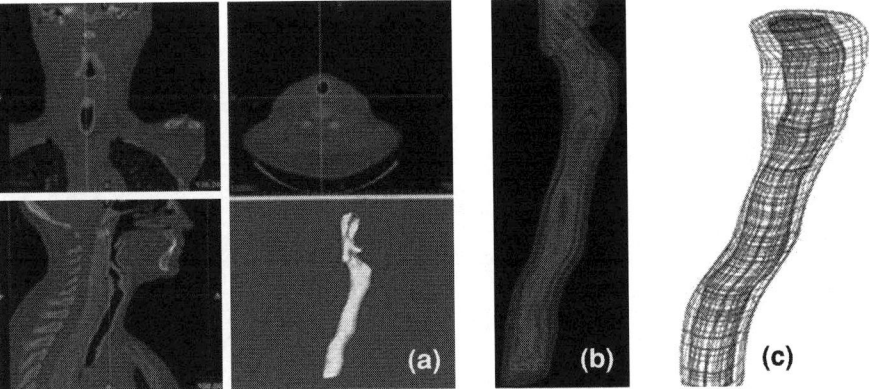

Fig. 9 Construction of the 3D finite element model of the human trachea. **a** Segmentation and 3D-extrusion of the external and internal surfaces of the airway using MIMICS 10. **b** Polylines that define the shape of the external and internal airway membranes. **c** External and internal tracheal surfaces IGES files

neutral format that allows digital interchange between CAD systems (Fig. 9). Rhinoceros software [54] was used to make this conversion from STL to IGES. This IGES file of the external and internal tracheal surfaces was then used to construct a structured finite element volume mesh with hexahedral elements (see Fig. 10b), using PATRAN (MSC, Santa Ana, USA), a well-known standard pre and post finite elements processor. IDEAS (Integrated Design and Engineering Analysis Software, Siemens, PLM, Plano, USA) was used to define the different groups corresponding to the different components of the trachea. The final complete model is composed of 28,350 finite elements (8,100 quadratic shell elements corresponding to the surfaces and 20,250 hexahedral elements) subdivided in 5 groups: external and internal membranes, the tracheal smooth muscle, the cartilaginous rings including the cricoids and thyroid, and finally the annular ligaments (see Fig. 10). This first geometrical model of the human trachea, considered as the referential one, measured around 130 mm in length, 3 mm in thickness and had 14 mm of internal diameter (mean values); it was composed of 20 cartilaginous rings in alternation with 20 annular ligaments, the cricoids, the thyroids, the smooth muscle and the internal and external membranes.

The objective of this numerical approach is to analyze the influence of the implantation of a tracheal endoprosthesis in the overall behaviour of the trachea. Therefore, the prosthesis has to be accurately located inside the tracheal wall. Influenced by his own profesional experience, the thoracic surgeon decides which kind of prosthesis is most suitable for each patient. Nowadays, the most common and used prostheses are the silicone ones, mainly the Montgomery T-tube and the Dumon endoprosthesis. There are different designs for the tracheal Dumon endoprosthesis [19] as the TD and TF Dumon Tracheal Stents; nevertheless their geometry is usually close to a simple cylinder with several diameters and

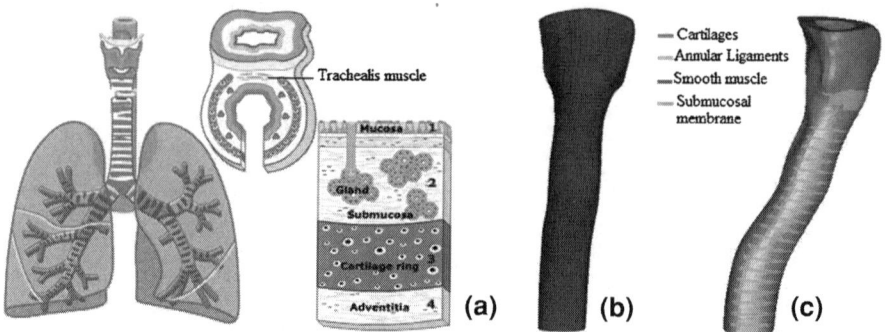

Fig. 10 Construction of the 3D finite element model of the human trachea. **a** Tracheal anatomy [55]. **b** Structured hexahedral tracheal mesh. **c** Different groups constituting the human trachea

Fig. 11 Dumon endoprosthesis geometrical model. **a** Dumon prosthesis. **b** Meshed geometrical model

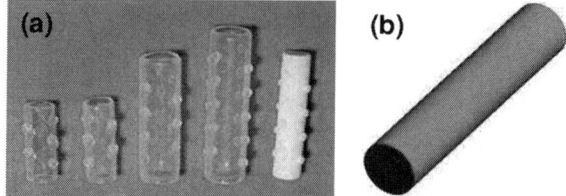

lengths [18]. The prosthesis diameter has to be large enough to fit inside the trachea remaining fixed to the wall avoiding its migration. Since our aim is focused on the behaviour of the trachea, this prosthesis was then approximated by a simple cylindrical surface (small thickness compared to the tracheal one: 1 vs 3 mm) with length and diameter derived from the commercial catalogue of NOVATECH DUMON tracheal and bronchial stents [24]. This simple geometry was then easily meshed by means of IDEAS employing bilinear quadrilateral shell elements (1,600 Elements) (see Fig. 11). In the reference model, a length of $L = 50$ mm and a thickness of $e = 1$ mm were used, while the diameter "d" is adapted to the patient tracheal diameter D through the process explained in the next section.

The same approach and similar simplifications were also used to model the Montgomery T-tube. This prosthesis consists of two perpendicular branches, the main one (I) fits to the interior of the trachea while the second one (II) is used to maintain the incision open for air exchange with the outside. The dimensions of the Montgomery T-tube geometrical model were taken from the Bostom Medical's catalog (see Fig. 12b). The geometrical model was approximated by two perpendicular cilyndrical surfaces. The longest one with 60 mm in length (principal branch) was meshed with 1,362 bilinear quadrilateral shell elements (see Fig. 13). The orthogonal tube (secondary brach E) had 12 mm in length. Both branches cross just at the mid-length of the principal brach. The length of the stent is

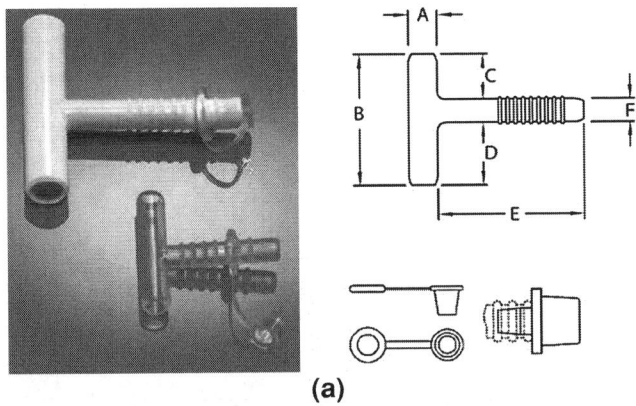

Fig. 12 Boston medical products catalogue of the montgomery T-tube [56]. **a** Montgomery T-tube. **b** Catalogue of the montgomery T-tube

generally cut to adapt the specific geometry of patient's trachea and stenosis. To avoid problems of convergence caused by excessive distortion of the finite elements, the geometrical final diameter of the hole (incision) D_2 was fixed to 8 mm regardless of the patient, which represents the smallest dimension (F) (see Fig. 12) of the T-tube's Montgomery catalog. To obtain the geometrical model of the perforated trachea to introduce the Montgomery T-Tube, the same geometrical model of the healthy trachea was used. This model was then perforated on its frontal face between the second and the third cartilaginous rings, where surgeons usually make the incision. The diameter of the perforation was chosen equal to the diameter of the Montgomery tube (II) F = 8 mm (see Fig. 13) [56]. As the lines used to generate the internal and external surfaces are parallel to the tracheal axis, the presence of holes perturbs the continuity of the surfaces and make finite element meshing complicated. To solve this problem, the surfaces were divided in parts, separately "meshable" and then joined to the rest of the surfaces node by node (see Fig. 14). PATRAN was used again to mesh the geometrical model of the perforated trachea (see Fig. 13), which was finally composed of 16,795 elements (11,800 hexaedral elements and 4,995 membrane elements).

Fig. 13 Mesh of the perforated trachea and a T-tube implant

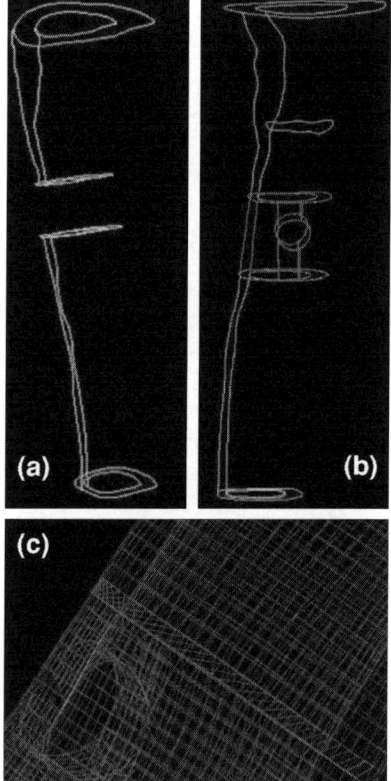

Fig. 14 Process to obtain a meshable geometry of the perforated trachea

3.2 Process of Mesh Adaptation

Since the principal objective of this work is the development of a tool for surgeons to predict the consequences of implantation of a tracheal prosthesis in a specific patient, it is necessary to create patient specific meshes in an automatic manner. With this aim, an automatic adaptation process of the reference mesh described above was implemented.

From the CT images of each patient trachea, STL files for the tracheal inner and outer surfaces are automatically obtained using the ALMA 3D software [57]. Then, the CVMLCPP free code (open-source C++ software library (LGPL)) [58] is used to voxelize the two surfaces of the trachea. The CVMLCPP code voxelizes a given geometry of triangles and stores the voxel representation in a 3D-matrix. In our case, voxelization of the internal and external surfaces is performed independently in two cubes of voxels with dimensions $128 \times 128 \times 256$. These voxelized geometries will be the "templates" to adapt the original referential mesh obtained as described in the previous section. After several attempts, to make the adaptation process easier and more efficient, a smoothing process was applied to that original model preserving the geometry of the tracheal section. The thyroidal cartilage is the biggest irregularity of the referential model, so some modifications were made to soften the structure maintaining the same lumen section. To adapt the original geometry to the specific dimensions of a patient, the trachea reference length was scaled. Since, the voxelized Z dimension (principal tracheal axis) consists of 256 voxels, 256 slices (0.6 mm average) are considered to compare the sections of the two geometries in each (X, Y) plane. In each of these slices, every node of the referential geometry is translated in the (X, Y) plane to adjust its position to that of its reciprocal point in the patient-specific geometry, thus keeping the total number of nodes for each Z section (see Fig. 15). The obtained adapted geometry usually presents geometrical irregularities from one slice to the other. To avoid such problem that may cause mesh element distortion, the (X, Y) coordinates of the section center of gravity in each cut are adjusted to the mean of the coordinates of the section centers of the anterior and posterior slice. This process of voxelization and geometry adaptation is done for both internal and external surfaces separately, so, although these two surfaces have almost the same principal axis (centers of gravity of the corresponding cross-sections), due to the voxelization is process, these section-centers may differ by a small value. Therefore, at the time of joining the two surfaces of the trachea, the sections-centers of the external surface are moved to the ones of the internal surface considered then as the master one.

This process of mesh adaptation is made through an in house computational code implemented in C^{++} that preserves the same hexahedral structural mesh, same element labels, node labels, groups of elements (cartilaginous structure, smooth muscle, annular ligaments, internal and external membranes) (see Fig. 16). After completion of this surface adaptation process, the difference in coordinates for each mesh node between the referential and the target geometries are saved in a

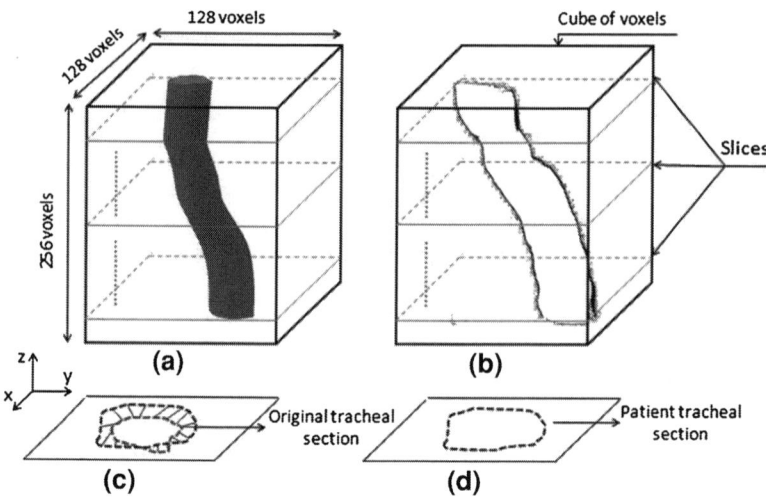

Fig. 15 Voxelization and process of mesh adaptation. **a** Voxelization of the internal surface of the original model. **b** Voxelization of the internal STL file of the patient trachea. **c** In a section slice, (X, Y) displacements are applied to each node of the reference section to adapt it to the target section. **d** Final coincidence of the original and patient-specific tracheal sections

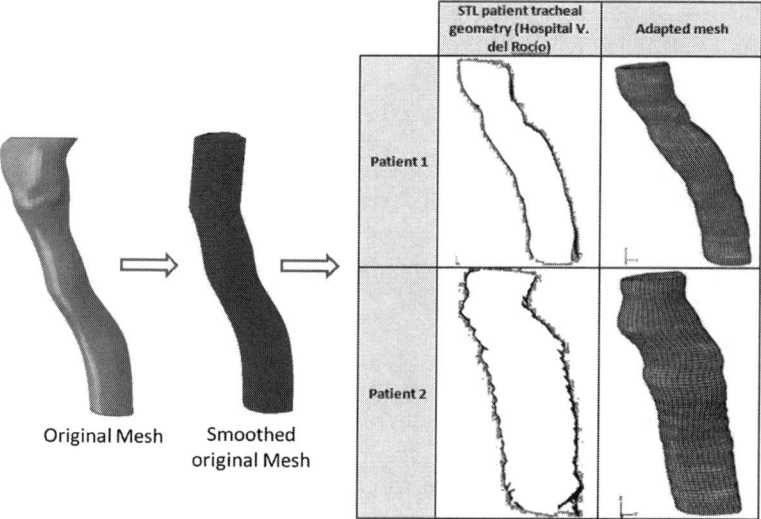

Fig. 16 Some examples of mesh adaptation

text file. Using these values as imposed boundary displacements, the whole volumetric mesh of the referential trachea is analyzed using the commercial code ABAQUS, getting the final position of the internal nodes of the relaxed final model from the computed displacements of those nodes and, with it, the adapted patient-specific tracheal geometry (see Fig. 16). It is important to remark again that we have a bijective mapping between any mesh in nodes (labels and topological location), elements, materials and groups that will be essential in the statistical analysis described in the following sections as well as in the response surface method used to approximate the mechanical variables associated to each point of the tracheal domain.

3.3 Materials: Constitutive Models and Parameters

To obtain a numerical model that fits the mechanical behaviour of the human trachea, the results of the previous experimental study were used [59]. To recapitulate, the Young modulus (E) of the human tracheal cartilaginous structure was estimated by tensile tests up to 5% of deformation while histology demonstrates that tracheal cartilage is a material in which the collagen fibres are randomly distributed. So an elastic isotropic material with a Young Modulus E = 3.33 MPa and a Poisson coefficient $v = 0.49$ were finally considered to be introduced in our model. Referring to the smooth muscle, and considering the results of the histology which determines the existence of two orthogonal families of fibres, the Holzapfel strain-energy density function, adapted to two families of fibres and quasi-incompressible materials, was used. The corresponding exponential function was fitted to the experimental data, and k_1, k_2, k_3 and k_4, the constants related to the first and second family of fibres, were computed. For the longitudinal direction, the parameters of the Holzapfel strain-energy density function (SEDF) resulted $k_1 = 0.154$ KPa and $k_2 = 34.157$, while for the transversal direction these values were $k_3 = 0.347$ KPa and $k_4 = 13.889$. To characterize the behavior of the matrix, data up to 5% of strain of the smooth muscle curves were linearly fitted using the Neo-Hookean strain-energy density function and a value for C_{10} of 0.877 KPa was obtained (see Sect. 2 and [59]). Therefore, an anisotropic model with two orthogonal families of fibers is proposed to approximate the tracheal smooth muscle behaviour being this the model used in our computational model of the human trachea.

Since the final goal of this work is to construct a valid tool for thoracic surgeons, computationally efficient, some additional simplifications regarding the materials behaviour were adopted. For this regard, the muscle was also analyzed as an isotropic hyperelastic Neo-Hookean material with equivalent mechanical properties. As the swallowing is a movement that occurs mainly in the lengthwise direction of the trachea, only the contribution of the longitudinal fibers was then taken into account by fitting only the part of the experimental strain-stress curve where the fibers are stretched (in this case for $\lambda > 1.14$) (see Fig. 17). This part of the curve is adjusted to a Neo-Hookean function obtaining a stiffness value C_{10} equal to 32.7 KPa. To validate the use of this approximation, a comparison of the results derived from the anisotropic smooth muscle hypothesis and those computed using the isotropic assumption was made.

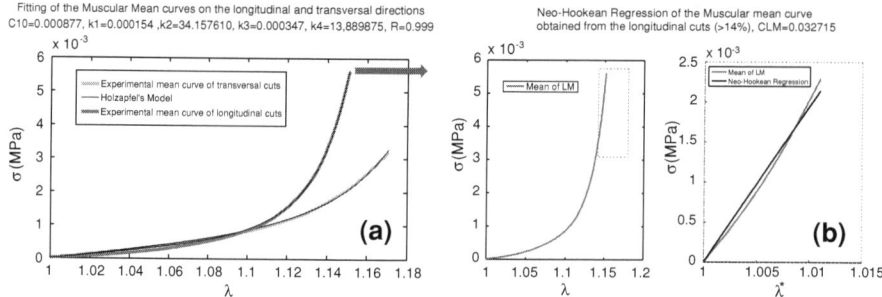

Fig. 17 Obtention of the isotropic properties of the tracheal smooth muscle. **a** Regression of the transversal and longitudinal mean curves using the Holzapfel's function. **b** Fitting with a Neo-Hookean function of the region of the mean strain-stress curve up to 14% of strain obtained from experimental data of longitudinal smooth muscle cuts (LM)

Regarding the remaining tracheal components (the annular ligaments, the submucosa or internal membrane and the adventitia), no experimental studies were made on these human tissues to determine their mechanical properties. Wang et al. [60] made a study about the mechanical properties of the tracheal mucosal membrane in rabbits; they concluded that mucosa was a collagen fibred tissue mainly oriented in the longitudinal direction whose stiffness was found between 8 and 18 KPa. Wiggs et al. [30] studied the mechanism of the airway mucosal folding and considered that the lumen-bordering layer of the trachea was stiffer than the less organized extracellular matrix between the smooth muscle and this layer, and considered it as an isotropic linear elastic (Neohookean) material. In our tracheal model, the annular ligament and the external and internal membranes were considered as isotropic hyperelastic materials with a stiffness parameter of $C = 32.7$ KPa and a Poisson ratio of $v = 0.3$ equivalent to the muscular tissue. Finally the silicone composing the stents was assumed to be an elastic material with a Young modulus of 1 MPa and a Poisson's coefficient of 0.28 [61–64].

3.4 Loads and Boundary Conditions for Swallowing Simulations

According to the clinical experience, from all functions that the trachea accomplishes, swallowing and coughing are those which most likely generate post-operative complications. The movement of the larynx and the elevation of a pathological trachea with a prosthetic insertion during swallowing cause damage on the surrounding tracheal tissues due to the accumulated effects of the relative displacement between prosthesis and trachea, and the pressure applied by the implant on the tracheal wall. In our study, the movement of the trachea during swallowing was then simulated. For this purpose, ABAQUS 6.8, a general purpose commercial finite element program was used applying appropriate boundary conditions.

Swallowing is a complex process that, in addition to driving food, permits the elimination of rhino-pharyngeal secretions, and the contents of an eventual esopharyngeal reflux. During swallowing 3 phases are distinguished: lip-mouth time, pharyngeal time and esophagus time [55]. Swallowing movement involves a series of coordinated motor events to propel liquid or food bolus from the mouth to the esophagus through the oral and pharyngeal cavities. During swallowing, the trachea ascends accompanying the larynx producing a non-homogeneous elongation, maximal in the cranial segments, while the inferior remain practically fixed. This movement can be compromised when a prosthesis is inserted because this element stiffens the overall behaviour of the trachea. This explains the inefficiency of the glottis to close that has been observed sometimes after placement of tracheotomy stents and also the high incidence of prosthesis migration [24, 25].

A variety of swallowing measures have been developed to quantify the physiological temporal organization of the laryngeal closure [65, 66], the upper esophageal sphincter opening [67] or bolus transit [68], and some studies have been made to evaluate the effects of aging and bolus characteristics on the temporal sequence of swallowing events [69]. The common conclusion is that the temporal relations between structural movements to obtain normal swallowing are not fixed but can be systematically affected by bolus volume and viscosity, and age [70]. Swallowing movement measures (spatial and temporal) vary depending on the methodology, the technique, and the chosen reference point. Most studies have analyzed the movement of the tongue [70, 71], jaw [72] or hyoid bone [73] to have spatial temporal measurements of swallowing using generally videofluorography, a sequence of X-ray images onto fluorescent screen [69, 73, 74]. Honda et al. [75] demonstrated that the information given by dynamic imaging of swallowing obtained by an open-configuration Magnetic Resonance Imaging (MRI) scanner was comparable to the one given by the videofluorographic images. Other researchers employed electromyography (EMG) to identify the occurrence of swallowing, to describe swallowing physiology, and to treat impaired swallowing function in dysphagic patients [76]. Electromyograms from the anterior tongue, suprahyoid muscles and the submental muscle were also recorded to have information about the swallowing process [71, 77]. Current multi-sensory non-invasive portable systems capable to detect spontaneous swallowing are based on electromyograms that record voltage potentials generated by the throat muscles [78]. Despite its increasing use, limited information is available regarding its reliability to quantify swallowing [76]. New techniques are continuously appearing to analyze swallowing such as electromagnetic midsagital articulography [72] of the mandibular jaw or the custom-built magnetoelastic sensors designed for insertion in the larynx to measure the force of transverse deformation of the thyroarytenoid and cricothyroid muscles during deglutition [79, 80].

For our study, the measurement of the tracheal swallowing movement was made in the Hospital Virgen del Rocío in Seville, using a Phillips Integris V Medical system with high-speed DICOM image interface [81]. Patients are asked to drink a suspension of barium sulfate and fluoroscopy images are taken as the barium is swallowed. The patients are asked to swallow the barium a number of

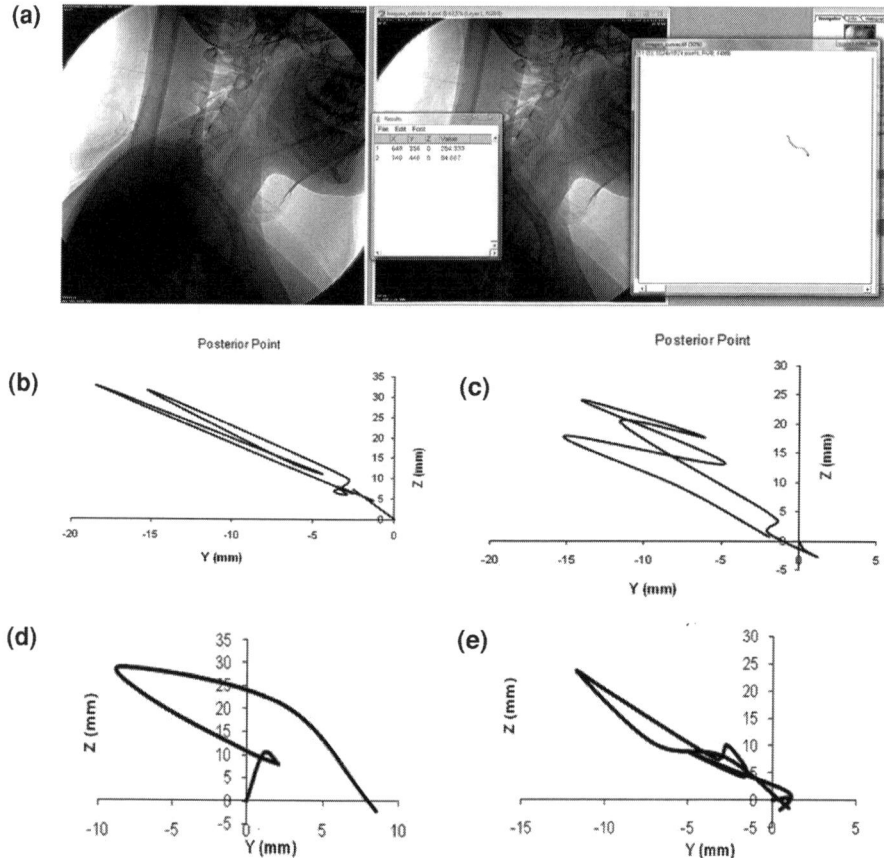

Fig. 18 Swallowing trajectories **a** Measurement of the displacement of the thyroid cartilage during swallowing. **b** Trajectory of the thyroid during physiological swallowing of patient (A). **c** Trajectory of the thyroid during swallowing of patient (A) after insertion of a Dumon prosthesis. **d** Trajectory of the thyroid during physiological swallowing of patient (B). **e** Trajectory of the thyroid during swallowing of patient (B) after insertion of a Dumon prosthesis

times, whilst standing in different positions, to assess the 3D structure as well as possible. The total displacement that underwent the trachea during this process can be measured marking in each image the superior edge of the thyroid cartilage, which is the only perceivable and valid reference to follow the movement in the 2D computerized tomography images (Fig. 18a).

Measurement of the different displacements of each of the cartilaginous structures was not possible, since the tracheal rings could not be distinguished in the fluoroscopic images; therefore, the obtained movement was only the one associated to the thyroid in the sagital plane. Taking into account that the mobility of the trachea in the transversal plane is almost negligible due to the esophagus position and neck shape and muscles, this movement is acceptable for our purpose. The swallowing trajectories of two men, one aged 28 with 75 kg and the other aged 62 and with 83 kg, were recorded before and after implantation of the Dumon prosthesis. Images were performed with the subjects seated while swallowing oral contrast medium from a cup. In the following, these patients will be named as Patient (A) and Patient (B) respectively. Both patients underwent a prolonged intubation (more than 15 days) that caused them a bad healing and left a fibrous stenosis. In our computational model of the human trachea during swallowing, the movement of the tracheal inferior end, at the level of the bifurcation was restrained, whereas in the superior end, displacements in the sagital plane (Y, Z) corresponding to the curves obtained by means of the physiological and pathological fluoroscopic images of the two patients were imposed. In the case of the Montgomery T-tube, since the swallowing trajectories were only detected from a Dumon prosthesis, the results of the physiological swallowing simulations were used to estimate the muscle forces necessary to swallow. The two cases of patient (A) and patient (B) are then taken as references and the corresponding forces (F_y, F_z) found during physiological deglutition will be used to simulate swallowing in presence of a Montgomery T-tube. This force ($F_z = 13.6$ N and $F_y = -3.11$ N distributed in a time ramp on 270 nodes that represent the thyroid surface of the trachea) is considered as the minimum one exerted by the tracheal elevator muscles to achieve the swallowing movement and will be taken as reference in all the subsequent pathological swallowing simulations, despite the prosthesis.

Finally, to simulate the endoprosthesis implantation process of the Dumon or Montgomery tubes, a prosthesis with a diameter smaller than the one chosen is located inside the trachea in the specified position. The prosthesis model is then deformed radially by imposing an increasing radial displacement on all its nodes until reaching the actual final dimensions of the prosthesis (see Figs. 19 and 23). This first step in the simulation allows reproducing the real stress distribution in the tracheal wall due to prosthesis insertion.

4 Computational Results

4.1 Simulation of Physiological Swallowing

The trajectory of the tracheal swallowing movement of patient (A) before inserting the implant (Fig. 18b) was included as imposed displacements in our geometrical model of the human trachea, considering the anisotropy of the smooth muscle and therefore the mechanical effect of its two orthogonal fiber families. Figure 18b shows that to achieve the experimental swallowing movement, the trachea has to elevate 33 mm in (Z), the principal axis of the trachea, and move 18 mm in (−Y). The movement of the trachea in the Y-direction is limited by the presence of the esophagus. The vertical displacement distribution (Z) is shown in Fig. 20a, while

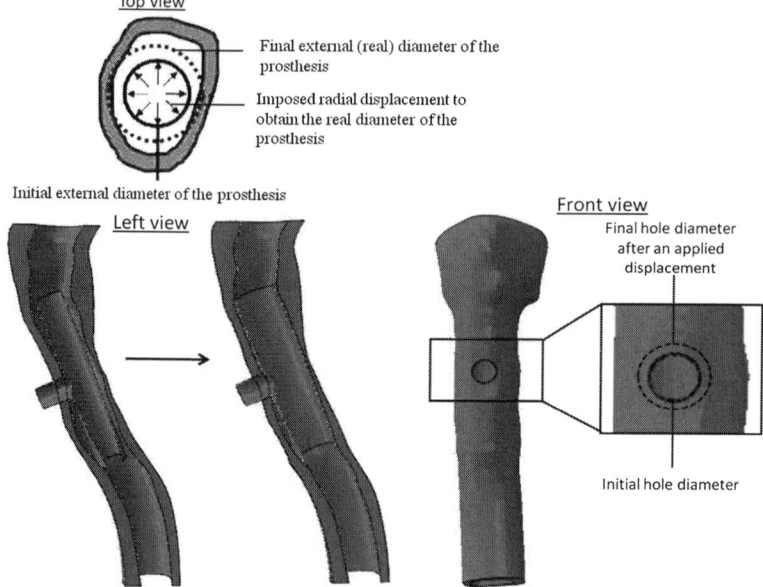

Fig. 19 T-tube insertion model

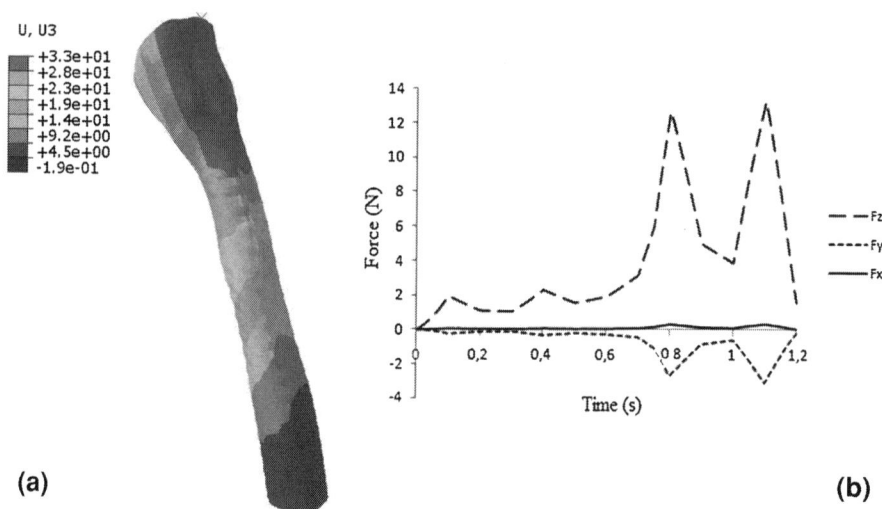

Fig. 20 Swallowing simulation of patient A before surgery using a two-fiber families constitutive model for the smooth muscle. On the *left*, displacements (mm) in the longitudinal direction; on the *right*, reaction forces (N)

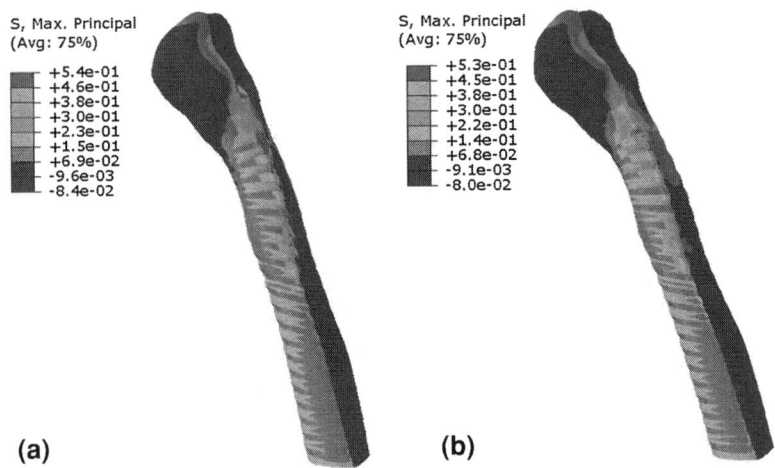

Fig. 21 Comparison between the maximal principal stress distribution during swallowing simulation of Patient A before surgery using anisotropic and isotropic models for the smooth muscle. On the *left*, stress distribution (MPa) using the anisotropic model for the muscle. On the *right*, stress distribution (MPa) using the isotropic model for the muscle

Fig. 20b represents the reaction force associated to the imposed displacement that may be identified with the force of the tracheal elevating muscles. This value was around 13 N in the direction (Z), and a minimum of 3.11 N in the direction ($-Y$). It can be seen that, when a patient swallows, two force peaks appear. These peaks are due the difficulty in swallowing the "barium liquid" in only one cycle.

The same simulation of the tracheal movement for Patient (A) during swallowing was again analyzed but now under the simplifying hypothesis of isotropic behavior of the smooth muscle. Comparing both behaviors, the distribution of principal maximal stress in both models is very similar; in fact the maximal value of the stress using a Neo-Hookean muscle was 0.53 MPa (Fig. 21b), located on the tips of the superior cartilaginous rings, while in the model with anisotropic muscle the maximum value of stress was 0.54 MPa (Fig. 21a) appearing in the same region.

These results validate to some extent the idea of using a simplified material model for the smooth muscle (isotropic hyperelastic), which allows an important reduction of the computational cost (the time spent to solve the whole model with isotropic smooth muscle was 3,386 s, whereas the one corresponding to an anisotropic muscle was 19,155 s). All the following simulations use the simplified isotropic Neo-Hookean model for the tracheal smooth muscle.

The same simulations were performed for Patient (B) (Fig. 18d). In this case, and again before surgery, the trachea reaches a maximal displacement on the principal tracheal axis (Z) of 28.94 mm with a maximum reaction force of 11.68 N, whereas in the ($-Y$) direction, the trachea had a maximal displacement

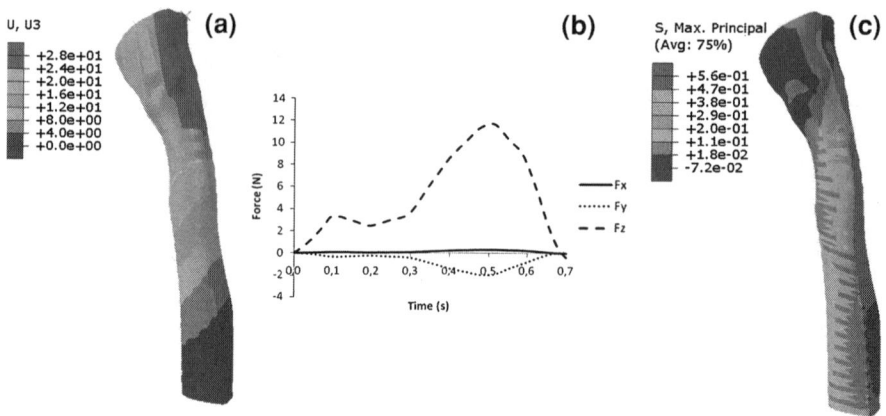

Fig. 22 Swallowing movement of Patient B before surgery. **a** Displacements in the vertical direction (mm). **b** Reaction forces (N). **c** Maximal principal stress (MPa)

of 8.5 mm in the (−Y) direction with an action force equal to 1.92 N. Again, the maximal stress that appears in this case was 0.56 MPa (Fig. 22b), situated in both tips of the cartilaginous rings of the upper half of the trachea.

4.2 Simulation of Swallowing with a Dumon Prosthesis

To numerically simulate the endoprosthesis implantation process, as commented: a first analysis step is performed starting from a smaller prosthesis than is located inside the trachea in the specified position, so that no contact between the internal surface of the trachea and the prosthesis is established, then the prosthesis is radially deformed to its actual diameter (see Fig. 23). With this new configuration, the stress distribution in the tracheal wall due to prosthesis insertion is obtained. The case of a Dumon prosthesis with initial diameter of d = 10 mm, inserted into the tracheal model diameter with D = 14 mm is shown in Fig. 23.

In a second simulation step, the pathological swallowing trajectories are imposed as known displacements to simulate the movement. For Patient (A), the insertion of a Dumon tube makes him losing capacity to swallow defined as the tracheal displacement in the vertical direction (Z). Analyzing Fig. 18b and c, it can be seen that the maximum displacement in the direction Z is 24.04 mm, whereas it was 33 mm before the insertion of the implant. The same conclusion applies to Patient (B); during his physiological swallowing movement, the displacement of the trachea in the direction Z was 28.94 mm whereas during pathological swallowing decreases to 23.47 mm (see Fig. 18d and e) which means a loss close to 19% of his capacity to swallow after implantation of the Dumon prosthesis. Figure 24 shows the displacements of the swallowing movement and its equivalent reaction force for

A Decission Support System for Endoprosthetic Patient-Specific Surgery 309

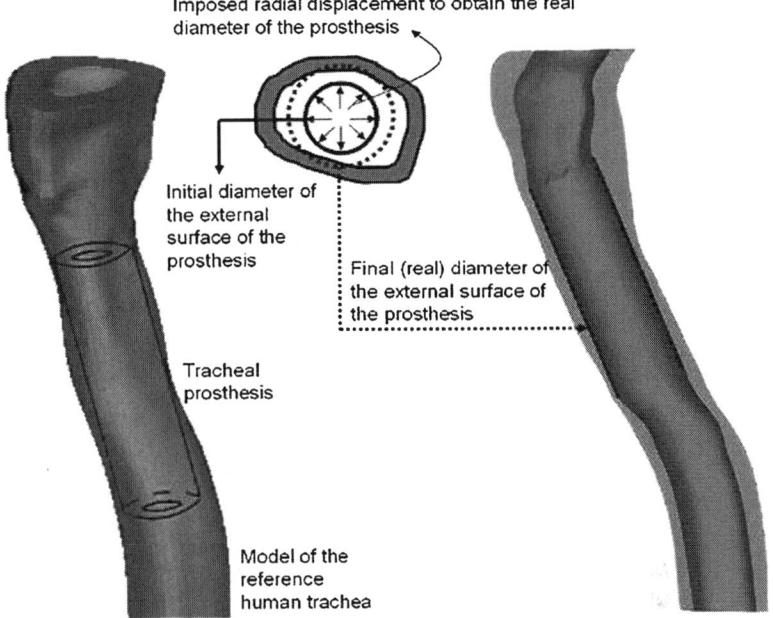

Fig. 23 Implantation of a tracheal Dumon prosthesis. First a smaller diameter stent is placed inside the trachea and then this prosthesis is deformed to reach its actual diameter

patient A with Dumon prosthesis. An equivalent maximum force of 14.05 N is obtained to cover the maximal 24.04 mm. This means that with a 3.3% of force increase, the patient looses 27% of its physiological capacity to swallow. Regarding the stresses distribution (Fig. 24c), the implanted trachea drives to a maximal principal stress of 0.54 MPa which is almost equal to the one before surgery (0.53 MPa). Again, this maximum is located on the superior border of the contact trachea/prosthesis. For Patient (B), trachea undergoes a displacement of 23.47 mm after surgery, with a maximum resulting reaction of 14.78 N. The maximum principal stress is 1.09 MPa also located at the upper part of the trachea in contact with the prosthesis (Fig. 25). Therefore, Patient (B) after surgery needs 26.5% higher force on the principal axis of the trachea to swallow, while looses 19% of his ability to swallow. This may be also related with the distribution of maximal principal stresses which are doubled with respect to the presurgery state. We can conclude from these simulations that patients almost use the same axial force to swallow, despite being implanted or not, with an average of 13.5 N which can be considered as the physiological force necessary to close the glottis during swallowing, although they clearly loose part of their ability to swallow after prosthesis insertion (Table 1).

Using the mesh adaptation code, the finite element models of 3 patients (E1, E2, E3) were obtained to simulate their virtual swallowing with inserted Dumon stents.

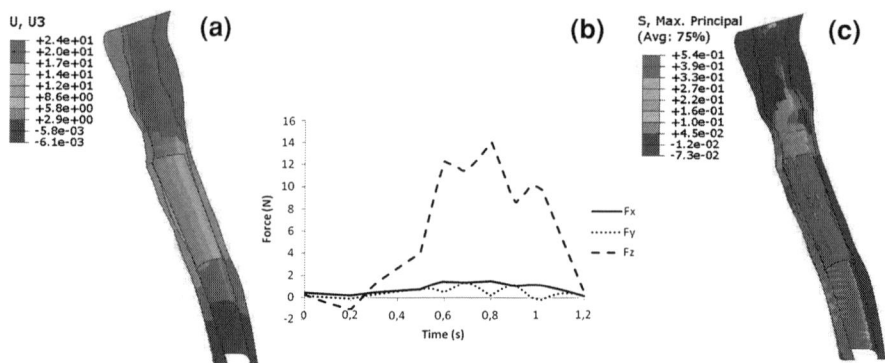

Fig. 24 Swallowing movement of Patient A after prosthesis implantation. **a** Displacements in the vertical direction (mm). **b** Reaction forces (N). **c** Maximal principal stress (MPa)

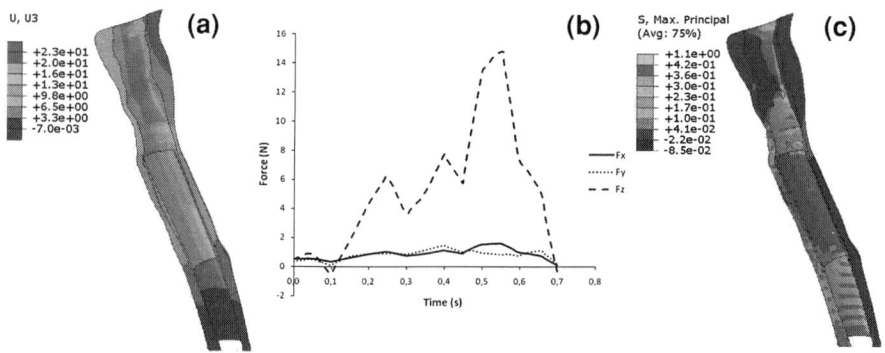

Fig. 25 Swallowing movement of Patient B after surgery. **a** Displacements in the vertical direction (mm). **b** Reaction forces (N). **c** Maximal principal stress (MPa)

Table 1 Displacement (Uz), reaction force (Fz), and principal maximum stress (PMS) for patients A and B before and after surgery

	Without stent (before surgery)			With stent (after surgery)		
	Uz (mm)	Fz (N)	PMS (MPa)	Uz (mm)	Fz (N)	PMS (MPa)
Patient (A)	32.68	13.6	0.5	24.04	14.05	0.54
Patient (B)	28.94	11.68	0.56	23.47	14.78	1.09

Due to the difficulties found in obtaining their swallowing trajectories, results of Patient A and B swallowing simulations were exploited. Using the reference model inclined $-18.43°$ with respect to the vertical direction, the minimum force in the direction Z obtained to achieve swallowing was 11 N (patient B), whereas

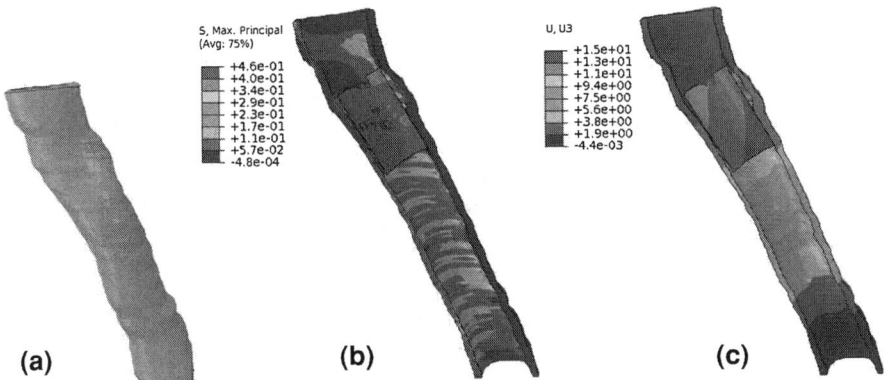

Fig. 26 Patient E1. **a** FE tracheal model adapted to the patient tracheal geometry. **b** Maximal principal stress distribution (MPa) during swallowing. **c** Maximal displacement in the direction Z (mm)

the minimum force in the direction Y was −3 N (patient A), which is equivalent to a physiological swallowing force module equal to 11.4 N. Then, with the agreement of the surgeons, we assumed that natural swallowing can be accomplished with a minimum force of 10 N. The same boundary conditions applied to the trachea and the stent were used, but now instead of imposing the displacements corresponding to the swallowing trajectory, external forces were applied with a constant magnitude of F = 10 N with components in the Z and Y directions dependent on the tracheal angle with respect to the vertical direction (Z). In all cases, the diameter, length and position of the prosthesis were defined by the surgeons and adapted to each patient.

For patient (E1), the dimensions of the Dumon prosthesis used were 15 × 30 [diameter (mm) × length (mm)]. The corresponding results are shown in Fig. 26, getting a maximum displacement in the vertical direction of 15 mm and a maximal stress of 0.46 MPa located in the upper contact of the stent border and the tracheal wall. The dimensions of the stent for patient E2, were 16 × 40, being the displacement and stress distribution ae shown in Fig. 27. The maximal displacement in the direction Z was 19 mm and the maximal stress 0.74 MPa, located in the upper contact between the stent border and the tracheal wall. This patient presents also a clear stress concentration in the contact zone between the lower border of the prosthesis and the trachea. For the last case of patient E3, the stent dimensions were 13 × 40. The maximal displacement in the direction Z achieved during the simulation was 19 mm and the maximal stress 0.74 MPa (see Fig. 28), located again in the upper contact between the prosthesis's border and the tracheal wall.

These patients all had a clinical follow-up from the day of their arrival to the hospital. Images of the tracheal endoscopy of each of these patients were used to verify and clinically validate the results of the numerical simulations. Patient E1

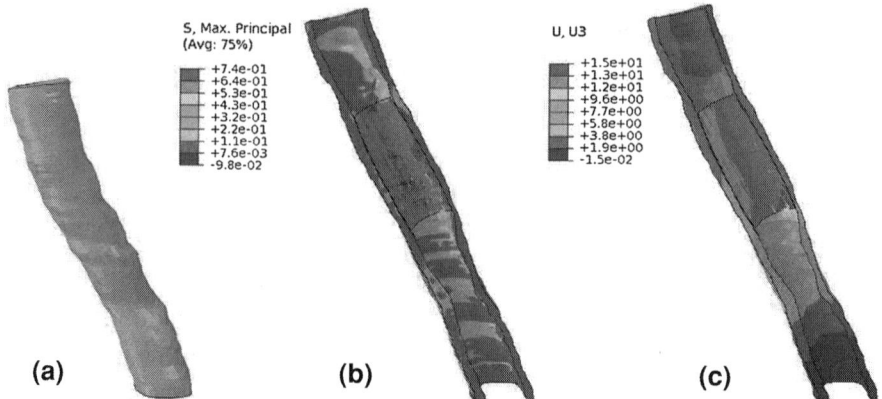

Fig. 27 Patient E2. **a** FE tracheal model adapted to the patient tracheal geometry. **b** Maximal principal stress distribution (MPa) during swallowing. **c** Maximal displacement in the direction Z (mm)

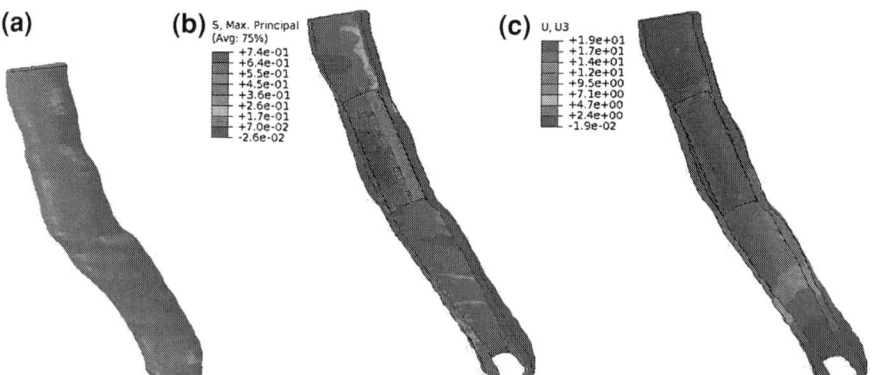

Fig. 28 Patient E3. **a** FE tracheal model adapted to the patient tracheal geometry. **b** Maximal principal stress distribution (MPa) during swallowing. **c** Maximal displacement in the direction Z (mm)

was admitted to the emergency service with a complicated epileptic state and a tracheal stenosis. Two months after the surgical intervention and the prosthetic implantation, a big granuloma in the upper contact prosthesis/trachea appeared in the endoscopic images during the postoperative revision, while in the lower contact a granuloma formation seems to be starting (see Fig. 29). This supports the results found in our simulations (see Fig. 26). Surgeons had to make two other interventions on this patient to eliminate the granuloma and then to re-locate the stent that migrated toward the vocal cords almost four months after the second intervention (see Fig. 30). Patient E2 images show the formation of granulomas in

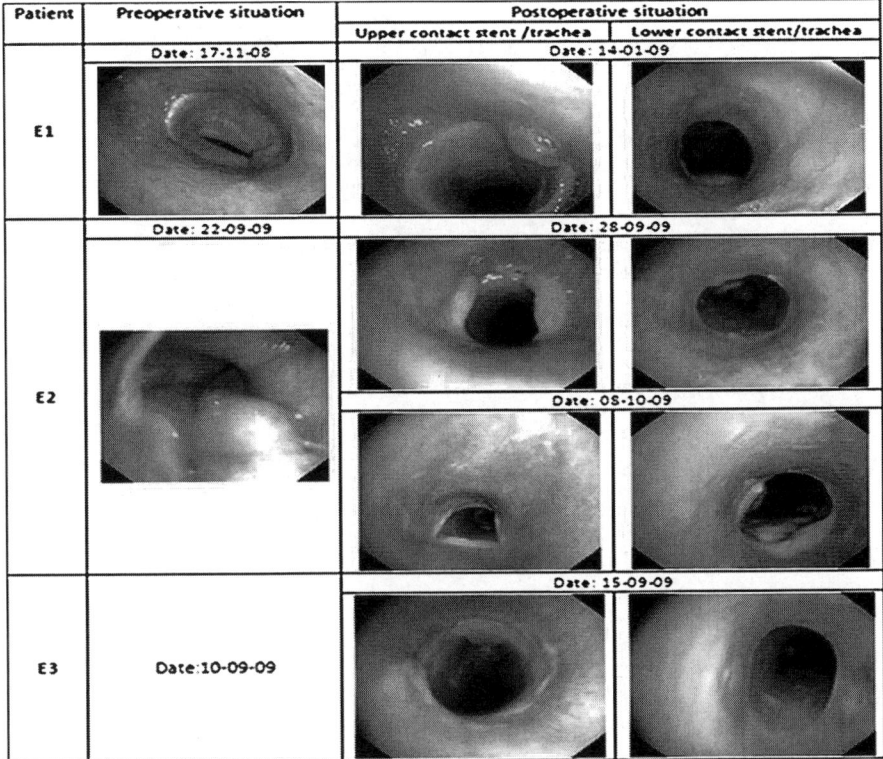

Fig. 29 Clinical history of patients E1, E2 and E3

the upper part of the stent/trachea contact one week after the implantation of the prosthesis. Two weeks after the intervention, granulomas appear in the tissue surrounding the two edges of the prosthesis. The computational results reached also similar conclusions (see Figs. 27 and 29). In the case of patient E3, a few days after prosthetic implantation, the endoscopic images show the beginning of granulomas formation in the upper area of the contact, which was also demonstrated by the computational simulations.

The common conclusion for all the three patient simulations is that the high values of maximal stress that appears in the trachea during one swallowing cycle (mostly located in the upper contact zone between the border of the stent and the internal surface of the trachea) can be the cause of the granuloma formation shown in the endoscopic images presented in Fig. 29. The prosthesis migration toward the vocal cords, that the patient E1 presented can also be a consequence of the low contact pressure between the stent and the tracheal wall. No information was provided about patient tracheal displacement during swallowing due to medical ethics.

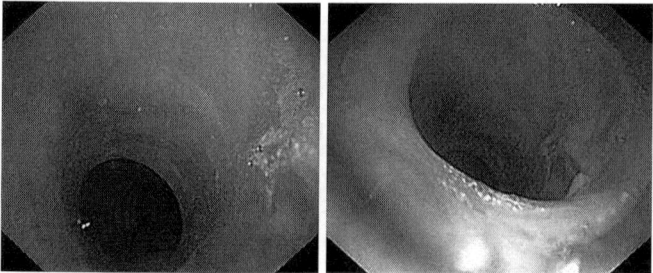

Fig. 30 Migration of patient E1 prosthesis

4.3 Simulation of Swallowing with a Montgomery T-tube Implant

The equivalent forces of patient (A) found in the sagittal plane to achieve physiological swallowing were then applied to the perforated model of the trachea with an inserted T-tube. The corresponding results of the simulation are shown in Fig. 31. During healthy swallowing, patient (A) reaches 33 mm in (Z) and −18 mm in (Y), while with an inserted Montgomery T-tube, the same patient loses 75% of its displacement in (Z) (8.01 mm) and 63% in (Y) (−6.63 mm). The maximal principal stress obtained during the swallowing process was 1.6 MPa surrounding the hole and in the upper contact with the tracheal wall. This stress level is three times higher than the one obtained in physiological swallowing without the T-tube (0.53 MPa).

4.4 Conclusions and Discussion

As intermediate conclusions of this section, we can mention that a first complete finite element model of the human trachea based on experimental results and complex mechanical behaviour models for the tracheal tissues has been presented and then simplified in order to optimize the computational cost. The stress distribution in the tissue has been analyzed during swallowing, before and after implantation of two types of tracheal prostheses. It has been shown that the presence of silicone stents (mainly the Montgomery T-Tube) in the human trachea may generate high values of stresses that would lead to possible damage in the tracheal wall. The level of stress obtained after prosthesis implantation was almost twice the physiological one and was generally located in the tracheal wall tissue in contact with the superior border of the stent. This implication has also been shown in other studies [13], whereas other works referred to the granulation formation at both ends of the stent [21, 24]. In addition to this increase of stress values, patients loose about 20% of their natural ability to swallow as a consequence of tracheal stiffening due to stent implantation.

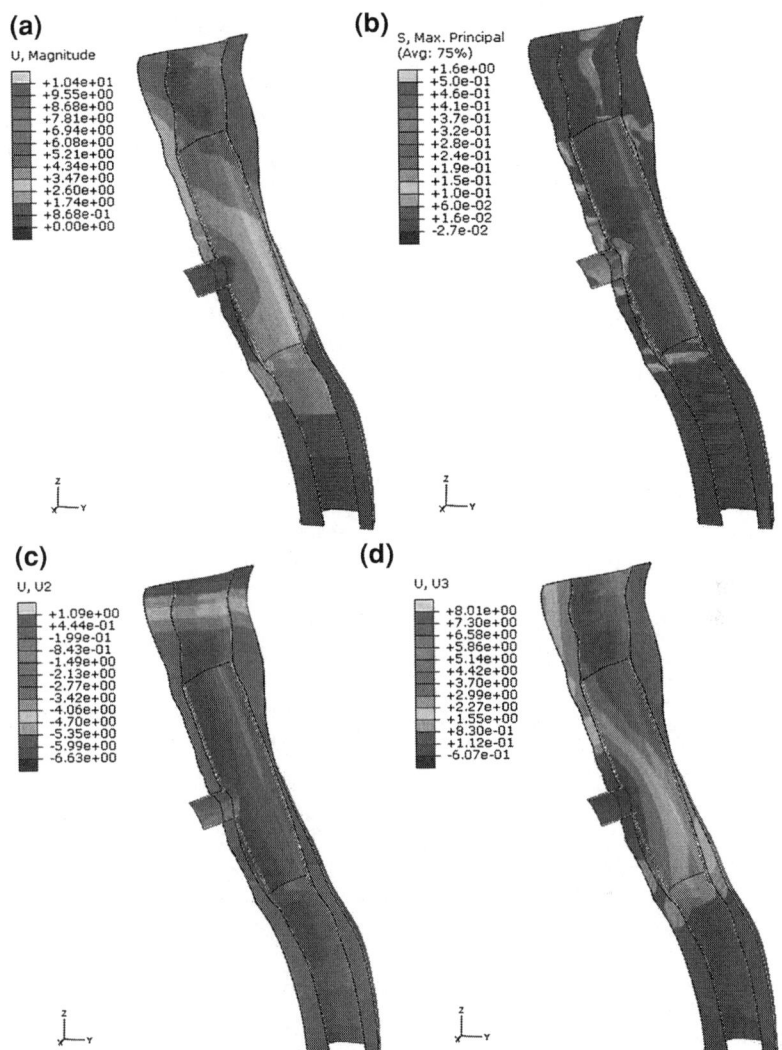

Fig. 31 Simulation of the swallowing process of patient (A) with a Montgomery T-Tube. **a** Amplitude of the displacement during swallowing (mm). **b** Principal stress distribution (MPa). **c** Displacements in Y-direction (mm). **d** Displacements in Z-direction (mm)

The used swallowing boundary conditions were based on two real cases of patients before and after implantation. The swallowing movement obtained from fluoroscopic images was obtained by marking the thyroid cartilage in the sagittal plane, tracking its position. Taking into account that the mobility of the trachea in the transversal plane is negligible, this movement would represent the overall movement of the trachea during swallowing. To be able to validate the simulated swallowing movement, the measurement of the movement of each cartilaginous ring would be valuable, but till the moment these rings cannot be distinguished in the fluoroscopic images. Another important result that can be deduced from these swallowing simulations is that the reaction force related to the imposed displacement is maintained in a small range of values which allows using the force as input obtaining the displacement and with that estimating the loss of the patient's ability to swallow as a result of the simulation.

Regarding the endoprosthesis itself, a rough approximation has been made assuming it as a perfect cylindrical surface. The Dumon prosthesis is actually a perfect cylinder; however it has several studs to help fixing it to the internal wall of the trachea. This effect has not been considered, but the friction of the prosthesis-internal wall interface ($\mu = 0.5$) was high enough to prevent migration.

5 Decision Support System for Endoprosthetic Human Tracheal Preoperative Planning Based on Robust Design of Experiments

The objective of this work is to get a fast, simple and efficient tool for helping surgeons in deciding the type, dimensions, location and angulation of a tracheal endoprostesis for each specific patient. The previous models and associated simulations may accomplish this goal; however, there are still some drawbacks that still have to be overcome. The first is the complexity of the tool itself. Surgeons are not specialists in developing FE models, taking very technical decisions and dealing with problems such as convergence. The second problem is computational time. Even with the reduction in computer cost achieved with the simplification of the material behaviour, still the time needed is too high to allow the specialist to test many hypotheses or to vary parameters as needed when using this tool efficiently. Finally, there are too many parameters to vary in the problem and the individual and combined effect of each of them on the final outcome (displacement and stresses) has not been assessed. To solve these problems, a statistical parametric study using the bases of robust design of experiments is implemented. From this, the effect of each of the main parameters of the implant will be analyzed and a set of correlation functions will be obtained from that statistical study to compute, although approximately, the main variables of the problem in a very short time. The description of these additional tools will be the aim of this latter section.

A Robust Design of Experiments is performed to understand which product and/or process design parameters are critical to achieve the proposed outcome with minimal variation [82–84]. Designing the experiment involves the determination of which factors and values should be varied during the experiments and which response will be measured. One simple and effective albeit long approach is to use a factorial design which gives all possible combinations of factor levels (possible combinations of k factors, each with n levels, is n^k) [84]. Its main disadvantage is the number of experiments that have to be done; therefore an alternative is needed in order to make many-parameter experiments manageable. Several alternatives exist such as Montgomery's method [85] or Berger and Maurer method [86]. But the approach that has become increasingly popular for its simplicity and effectiveness is the use of Taguchi's Method [84]. In the 40s, Genichi Taguchi introduced several new statistical concepts which proved to be valuable tools in the subject of quality improvement [86]. The Taguchi's method was introduced in 1980 in several American industries and resulted a significant improvement in product and manufacturing process design. Many of the statistical methods proposed by Taguchi, such as the use of signal-to-noise ratios, orthogonal arrays, linear graphs, and accumulation analysis have still room for improvement [87, 88]. The bases for Taguchi's method are the orthogonal arrays, which impose a selection of the factor levels to perform the fewest possible runs. Generally, four steps are carried out to analyze the information got from the experiments:

- Estimation of factors effects in the average response.
- Quantification of each factor influence on the response.
- Analysis of interactions between factors.
- Prediction using a mathematical model of the experiment and determination of the optimal factors values for a given response.

The following statistical study is based on the principles of Robust Experimental design using the Taguchi's Method. This methodology enables obtaining information about quantitative analysis as the percentage of influence of each factor (ANOVA) and qualitative analysis as interactions between factors.

5.1 Experimental Design

To evaluate the consequences of stent implantation and to accomplish simulations, the swallowing trajectory before and after a prosthetic insertion for each patient is needed. Patients come usually to emergency services with breathing problems, which make impossible to ask them gulp down substances and record swallowing movement. Therefore, to find a way to standardize swallowing for every patient, we propose to substitute the displacement trajectory of the trachea during swallowing by equivalent forces applied by the elevator muscles. The effect of these forces depends on the properties of the airway, so it varies for each patient. However, the simulations results of physiological swallowing of two patients based on real swallowing trajectories and presented in the previous section and in [11] showed that the minimal forces exerted by the muscles involved in swallowing were 11.68 N in the vertical direction (Z) and -3.11 in the antero-posterior direction (Y), in a tracheal model with an angle of $-18.43°$ with respect to the vertical direction. In the statistical study presented herein, we assume the hypothesis that the force applied by the muscles involved in swallowing is always constant, regardless sex, age, weight and height of the patients. It is then assumed that the physiological swallowing can be accomplished by a minimum force of 10.5 N despite the tracheal inclination.

We present herein a statistical study of swallowing movement with Dumon prosthesis to analyze and study the parameters that have the most important influence on the stress state of the trachea and on the ability to swallow when the prosthesis is inserted. The capacity to swallow (maximal vertical displacement) is one of the important variables to take into account a direct consequence of prosthesis implantation. This relation between maximum tracheal displacement in the (Z) direction and the prosthetic intervention is demonstrated in [11]. Therefore, the tracheal displacement in (Z), U_Z, is considered one main output variable. The damage caused by the presence of a stent within the trachea is manifested in form of granulomas that usually appear at the edges of the prosthesis in contact with the tracheal wall. These granulomas are a direct result of the "hammer blows" effect that the stent exerts on the trachea and the relative friction that appears in the contact zone prosthesis/trachea. Thus, a concentration of stress can be at the origin of pain and fibrous tissue formation; for this reason the maximum principal stress S_{PM} that arises in the trachea during swallowing is also considered a second main output variable.

In collaboration with the surgeons and based on the results of the previous section and [11], several experimental designs were made (L_9, L_9, L_{16}) with several variables and levels. After a first analysis, depending on the percentage of influence of each variable on the output ones (less than 5%), some of them were just taken as fixed. Other were also considered fixed for their direct dependence on the input variables or for geometrical reasons. These are the following:

- The tracheal length (L_T) is measured from the vocal cords to the carina and taken iqual to 130 mm [89].
- The thickness of the trachea (e_T) was fixed to 3 mm.
- The stenosis diameter $\phi_e = \phi_T - 2$, which traduces the surgical intervention made on the stenosis before the prosthesis implantation. Surgeons confirm that they let around one millimeter of the stenosis in radius to use as a support for the stent.
- The prosthesis length $L_P = L_e + 2$, exceeds one millimeter each border of the stenosis. It is supposed that L_P cannot exceed the interval $[L_T \times 10\%, L_T \times 90\%]$. The 10% of the bottom of the trachea is excluded to have margin from the carina, and the upper 10% is not taken into account to remain sufficiently far from the vocal cords.

Then the factors with higher influence on the output variables were selected. These parameters were the diameter of the prosthesis (ϕ_P), the length of the stenosis (L_e), the position of stenosis (P_e), the tracheal inclination (α) and the penetration of the prosthesis in the trachea (P_n) (see Fig. 32). The level of each of these variables is selected based on anatomical, physiological and commercial references. For example, the value of the stent diameter is defined by the manufacturers, being available in integer values generally between 12 and 18 mm. The levels of ϕ_P are chosen according to the most used Dumon tube dimensions available in the market,

Fig. 32 Dependent and independent variables

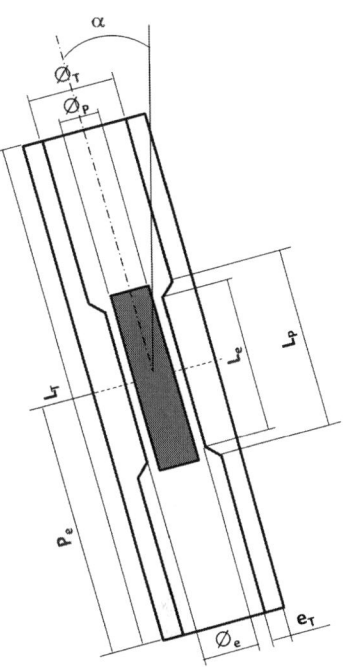

which are 14–17 mm. A stenosis with a minimal length of $L_e = 20$ mm is considered [90]. Stenosis can affect any part of the trachea, but as the length of the prosthesis is conditioned by the interval $[L_T \times 10\%, L_T \times 90\%]$, L_e is included in the interval $[(L_T \times 10\%) - 1, (L_T \times 90\%) - 1]$. As the tracheal length is fixed to 130 mm, stenoses of 25, 45, 65 and 85 mm in length were tested. The position of the stenosis is measured from the bottom of the trachea to the middle of the stenosing zone; therefore, P_e is included in the interval $\left[(L_T \times 10\%) + 1 + \frac{L_e}{2}, L_T \times 90\% - 1 - \frac{L_e}{2}\right]$ where L_e is the longest dimension of the stenosis (85 mm). Therefore and as the tracheal length is fixed, P_e varies in the interval [56.5, 73.5], and the chosen levels are 56.5, 62.5, 68.5 and 73.5 mm. The tracheal inclination (α) is the angle between the tracheal longitudinal mean axis and the vertical direction (Z). In this study, it is always taken negative due to the chosen reference. No values could be found in litterature on the range of this angle, but anatomically human trachea is never vertical for the shape of the neck and the relative position of the bellows and the skull. Ragavan et al. [91] worked on the interactions of airway oscillation, tracheal inclination and mucus elasticity on cough clearance, and they took 0, 15, 30 and 45 as values of tracheal angles. In this design, the tracheal angle measured respect to the vertical was chosen equal to 12, 17, 22 and 27°. The prosthesis penetration P_n is defined as the difference between the prosthesis and the stenosis diameters supposed to be always within the interval [0, 1] mm. Thus P_n levels were taken equal to 0.2, 0.4, 0.6 and 0.8 mm. With the chosen independent factors (ϕ_P, L_e, P_e, α, P_n) and with 4 levels each, the appropriate orthogonal Taguchi's array is the L_{16}.

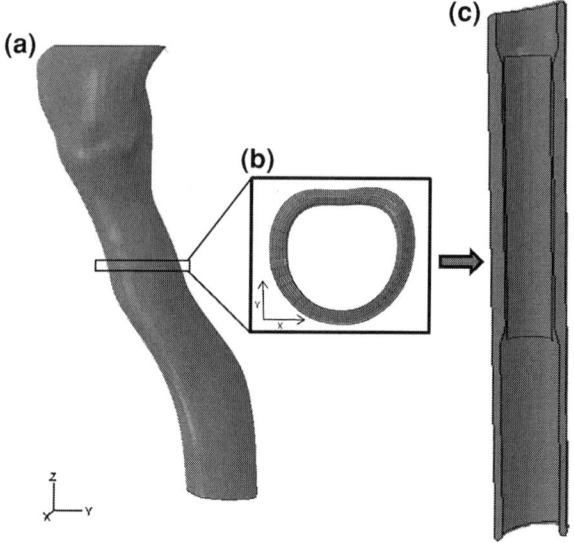

Fig. 33 Process to mesh the statistical model. **a** Reference mesh. **b** Ring of elements taken from the reference model. **c** Meshed statistical model

For each experiment to run, a finite element model based on a simplified geometry was generated. A whole ring was chosen from our reference mesh used in Sect. 3 and in [11, 92], to preserve the real tracheal section in the statistical models. A program implemented in C^{++} was implemented to replicate elements of the ring in the vertical direction (Z) until getting a total length of 130 mm. The obtained geometry is simpler than the real one but the mesh is stil hexahedrical and structured. To vary the length of the obtained mesh, the z-coordinate of the nodes is scaled to the required dimension. In the stenotic region, the "x" and "y" coordinates are scaled to reduce the tracheal diameter 2 mm (see Fig. 33). The 16 models to run were simulated with the mechanical properties of each tracheal tissue used in [59]:

- Cartilaginous structure: considered as elastic material with Young Modulus $E = 3.33$ MPa and Poison coefficient $v = 0.49$.
- Smooth muscle, internal and external membranes, and annular ligaments: approximated to isotropic hyperelastic materials with a stiffness of $C = 32.7$ KPa and a coefficient of Poisson $v = 0.3$.
- The Dumon silicone tube is simulated as an elastic material with $E = 1$ Mpa and $v = 0.28$.

The boundary conditions used in these virtual experiments are:

- Null displacements at the bottom layer of the trachea, at the level of bifurcation.
- The initial model of the prosthesis always had a diameter 2 mm less than the final one. In a first step, a radial displacement of 1 mm is imposed to the nodes of the prosthesis driving to the stenosis/prosthesis contact.

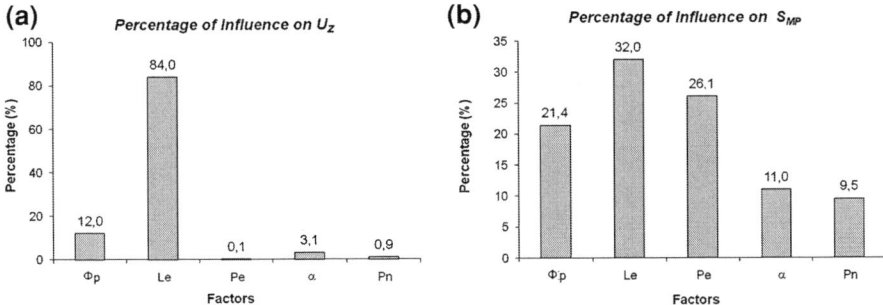

Fig. 34 Percentage of influence of the different factors on the output variables. **a** Percentage of influence of the different factors on U_Z. **b** Percentage of influence of the different factors on S_{MP}

- The force exerted by the swallowing muscles and estimated by the simulations in [11] and in Sect. 3, has a module of 10.5 N. This force is then decomposed in "Z" and "Y" depending on each value of inclination angle, and applied at each node of the top layer of the tracheal model.

All simulations were made in the same way and using the same contact option. The friction coefficient was set to 0.5 sufficient to prevent relative sliding between trachea and prosthesis.

5.1.1 Analysis of the Statistical Data

The percentage of influence of each factor on the capacity to swallow and on the Maximal Principal tracheal stress during swallowing is presented in Fig. 34, illustrating the importance of each selected variable. Some factors as the stenosis position, the tracheal inclination and the prosthesis penetration have a negligible influence on the vertical displacement of the trachea during swallowing U_Z (percentage of influence <5%), but they keep having an important influence on the tracheal stress distribution S_{MP}. A smooth variation of the different influence percentages on S_{MP} was detected, which makes all five factors determinant for the Robust Design of Experiments (see Fig. 34). As a summary, the stenosis length, and the prosthesis or tracheal diameter are the only factors that have an important influence on the patient swallowing capacity, whereas all the five factors have almost a similar effect on the stress distribution.

The relation between the output variables and each of the considered factors is shown in Figs. 35 and 36. U_Z decreases non linearly when ϕ_P increases, and as ϕ_P is directly related to ϕ_T by the relation $\phi_T = \phi_P - P_n$, the relation between the vertical displacement and ϕ_T is similar. The same happens with the stenosis length, U_Z has a decreasing linear behavior when L_e increases. In the chosen range, P_e seems to have no influence on U_Z which presents almost a horizontal curve (see Fig. 35), whereas the vertical displacement has a non-monotonic behavior

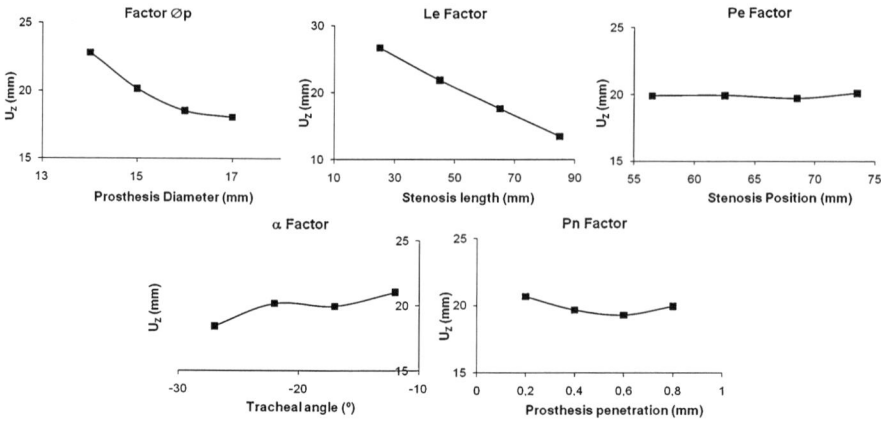

Fig. 35 Effects of the different factors on U_Z

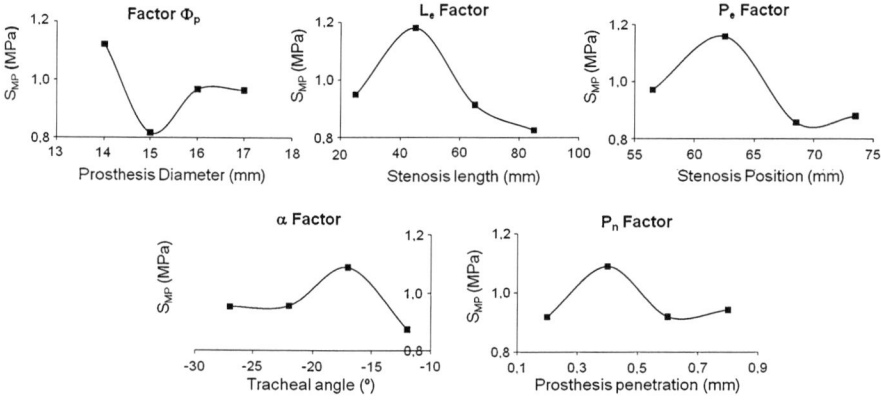

Fig. 36 Effects of the different factors on S_{MP}

when related to α and P_n. Regarding S_{MP} non-monotonic relations were obtained with ϕ_P (or ϕ_T), L_e, P_e, α and P_n.

The effect of the different factors on U_Z does not interact in the selected ranges, but two critical cases have to be discussed with detail. Firstly, as the stenosis length is the main factor affecting the U_Z behavior, curves of interaction $L_e - \phi_P$ (see Fig. 37) present a high risk of intersection for higher values of L_e, but L_e values are limited by surgical conditions (upper and lower 10% of the tracheal length shouldn't be covered by the prosthesis). The other critical case corresponds to the interaction between P_n and P_e. In this case, a high probability of intersection is detected for higher values of P_n, but as P_n, P_e and α have a percentage of influence less than 5% on U_Z, these interactions may be neglected. Regarding the

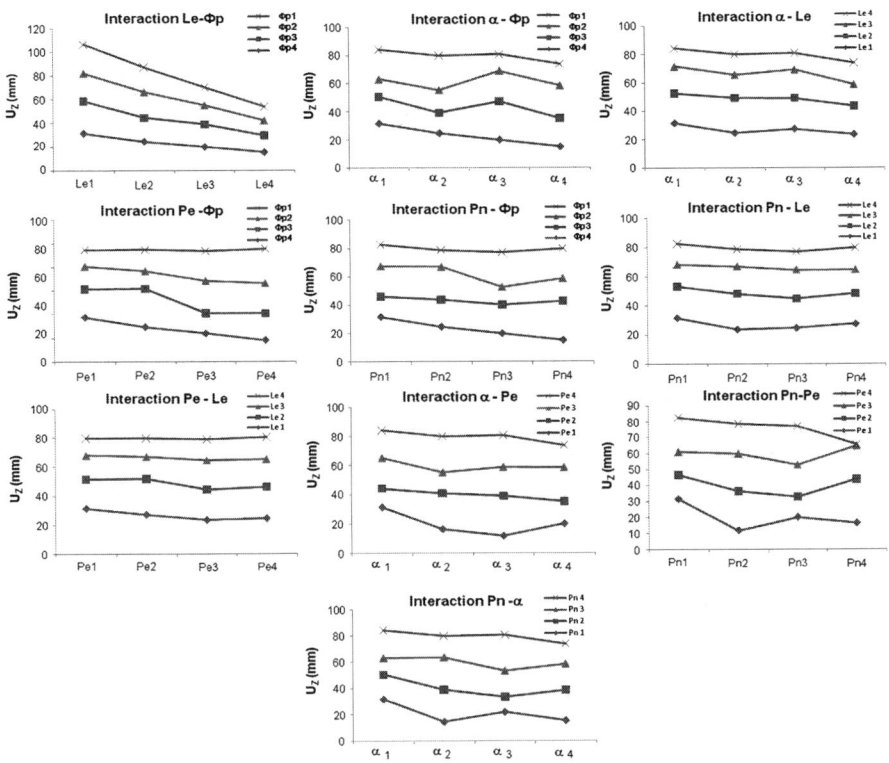

Fig. 37 Interaction between factors on U_Z

interaction of the different factors on S_{MP}, only synergystic interactions appear in Fig. 38, with almost parallel curves which guarantees that no intersection is expected in the physiological and commercial ranges.

5.2 Regression Models

The objective of this section is to generate a simplified numerical model linking the independent input variables with higher effect on each output variable relevance, determined after the application of the experimental design. In this work, a quadratic regression is used but formulated in a multiple linear matrix regression formulation, where the product of factors and the squares of factors are considered as variables of the multiple linear regression approach. This study applies the multiple linear regression method proposed by Neter et al. in 1996 [93], that formulates the expression of a variable y, in terms of the independent variables $x_1, x_2, ..., x_n$. The multiple linear regression is written as:

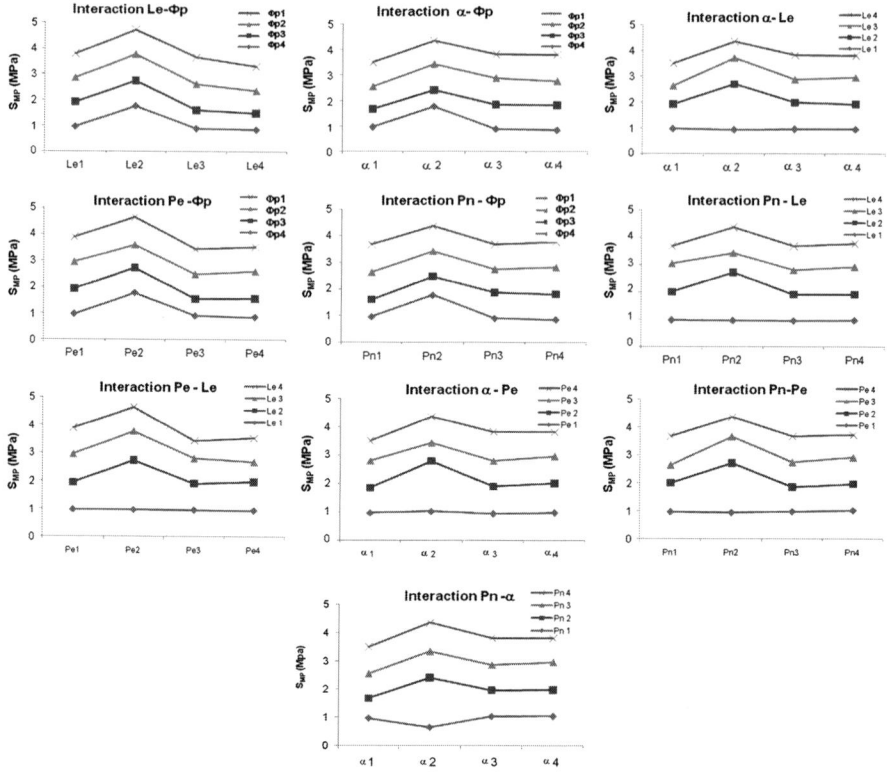

Fig. 38 Interaction between factors on S_{MP}

$$y_i = \beta_0 + \sum_{j=1}^{k} \beta_j X_{ij} + \varepsilon_i \qquad (8)$$

equivalent to the matrix form:

$$\mathbf{y} = \mathbf{X}\boldsymbol{\beta} + \boldsymbol{\varepsilon} \qquad (9)$$

where $\mathbf{y} = \begin{pmatrix} y_1 \\ y_2 \\ \vdots \\ y_n \end{pmatrix}$ is an $n \times 1$ vector of dependent variable observations, $\mathbf{X}$ an $n \times p$ design matrix, $\mathbf{X} = \begin{pmatrix} 1 & x_{11} & x_{12} & \cdots & x_{1,p-1} \\ 1 & x_{21} & x_{22} & \cdots & x_{2,p-1} \\ \vdots & \vdots & \vdots & \cdots & \vdots \\ 1 & x_{n1} & x_{n2} & \cdots & x_{n,p-1} \end{pmatrix}$, $\boldsymbol{\beta} = \begin{pmatrix} \beta_0 \\ \beta_1 \\ \vdots \\ \beta_{p-1} \end{pmatrix}$ a $p \times 1$

vector of regression parameters, and $\boldsymbol{\varepsilon} = \begin{pmatrix} \varepsilon_1 \\ \varepsilon_2 \\ \vdots \\ \varepsilon_n \end{pmatrix}$ the $n \times 1$ vector of additive errors.

The Least-Squares is used to fit a regression line to the data $\{\mathbf{x_i}, \mathbf{y_i}\}_{i=1}^n$, where $\mathbf{x}_i = \{x_{i,1}, \ldots x_{i,p-1}\}$. Thus, the objective is to find the regression coefficient estimates $\hat{\boldsymbol{\beta}}$ that minimizes the criterion

$$Q(\boldsymbol{\beta}) = (\mathbf{y} - \mathbf{X}\boldsymbol{\beta})^T(\mathbf{y} - \mathbf{X}\boldsymbol{\beta}) = \sum_{i=1}^n (y_i - X_i\boldsymbol{\beta})^2 \qquad (10)$$

Taking derivatives with respect to $\boldsymbol{\beta}$, and setting these equal to 0, the normal equations are obtained:

$$\frac{dQ}{d\boldsymbol{\beta}} = 2\mathbf{X}^T(\mathbf{y} - \mathbf{X}\boldsymbol{\beta}) = 0 \Longrightarrow (\mathbf{X}^T\mathbf{X})\boldsymbol{\beta} = \mathbf{X}^T\mathbf{y} \qquad (11)$$

To solve for $\boldsymbol{\beta}$, the inverse of $\mathbf{X}^T\mathbf{X}$ is applied to both sides of the previous equation so: $\hat{\boldsymbol{\beta}} = (\mathbf{X}^T\mathbf{X})^{-1}\mathbf{X}^T\mathbf{y}$.

The independent factors, their percentage of influence on the output variable, the behavior of the output variable when they vary and their interactions govern the predictive answer equation. As the capacity to swallow U_Z depends only on L_e and ϕ_P (see Fig. 34), the U_Z equation should be written as:

$$F_{U_Z} = \beta_0 + \beta_1 \phi_P + \beta_2 L_e + \beta_3 \phi_P^2 + \beta_4 L_e^2 + \beta_5 \phi_P L_e \qquad (12)$$

whereas the function that predicts the Maximal Principal Stress S_{MP} depends on all the five factors, their squares and their interactions (see Fig. 34), so it is written as:

$$\begin{aligned} F_{S_{MP}} = &\beta_0 + \beta_1 \phi_P + \beta_2 L_e + \beta_3 P_e + \beta_4 \alpha + \beta_5 P_n + \beta_6 \phi_P^2 + \beta_7 L_e^2 + \beta_8 P_e^2 + \beta_9 \alpha^2 \\ &+ \beta_{10} P_n^2 + \beta_{11} \phi_P L_e + \beta_{12} \phi_P P_e + \beta_{13} \phi_P \alpha + \beta_{14} \phi_P P_n + \beta_{15} L_e P_e \\ &+ \beta_{16} L_e \alpha + \beta_{17} L_e P_n + \beta_{18} P_e \alpha + \beta_{19} P_e P_n + \beta_{20} \alpha P_n \end{aligned} \qquad (13)$$

A matrix model linking the information generated by the different computational simulations coming out from the Design of Experiments was generated. Using Minitab® program [94], two procedures based on robust design were applied: Stepwise Regression and Best Subsets Regression. The stepwise model-building technique [95–97] involves:

Vars	R-Sq	R-Sq (adj)	Mallows C-p	S	Φ_p	L_e	Φ_p^2	L_e^2	Φ_p-L_e
1	74,1	73,3	49,2	4,0603					X
1	59,2	57,9	95,1	5,1015		X			
2	87,1	86,3	11,5	2,9122	X	X			
2	85,6	84,7	16,2	3,0795		X	X		
3	89,6	88,6	5,8	2,6548	X	X	X		
3	89,0	87,8	7,8	2,7399	X	X			X
4	90,7	89,4	4,5	2,5578	X	X	X		X
4	90,4	89,0	5,5	2,6043	X	X	X	X	
5	90,9	89,2	6,0	2,5805	X	X	X	X	X

(a)

Step	1	2	3	4
Constant	37,47	53,59	80,28	118,71
Φ_p-L_e	-0,0197	-0,0154	0,0263	0,0205
T-Value	-9,57	-8,16	2,24	1,82
P-Value	0,000	0,000	0,033	0,079
Φ_p		-1,36	-3,21	-9,32
T-Value		-4,45	-5,57	-3,48
P-Value		0,000	0,000	0,002
L_e			-0,61	-0,53
T-Value			-3,59	-3,26
P-Value			0,001	0,003
Φ_p^2				0,24
T-Value				2,33
P-Value				0,027
S	4,06	3,22	2,74	2,56
R-Sq	74,13	84,21	88,95	90,69
R-Sq(adj)	73,32	83,19	87,85	89,41
Mallows C-p	49,2	20,3	7,8	4,5

(b)

Fig. 39 Regression models of U_Z using Minitab®. **a** Best subsets regression. **b** Stepwise regression

- Identifying an initial model.
- Iteratively "stepping", that is, repeatedly altering the model at the previous step by adding or removing a predictor variable in accordance with the "stepping criteria".
- Terminating the search when stepping is no longer possible, that is the stepping criterion is fulfilled, or when a specified maximum number of steps has been reached.

The Best Subsets Regression helps to determine which predictor (independent) variables should be included in a multiple regression model. It enables comparing the full model (containing all the independent variables) and subset models (containing different subsets of independent variables) [98]. This method involves examining all possible combinations of predictor variables. It uses the indicator (R^2) to compare the models and choose the best. First, all models with only one predictor variable are checked and the two models with the highest accuracy (R^2) are selected. Then all models with two predictor variables are checked and the two models with the highest (R^2) are chosen. This process continues until all combinations of all predictors variables have been taken into account [99]. To determine the best model arising from the collected information, two indicators were used: the standard deviation (S), and the quadratic correlation (R-Sq). The best regression corresponds to the lowest (S), while (R-Sq) should tend to 100%. As shown in the results (Fig. 39), several regressions drove to low (S) and high (R-Sq). From all of them, we selected the model with the least number of independent variables, getting the following regression equations:

$$F_{U_Z} = 118.71 - 9.32\phi_P - 0.53L_e + 0.24\phi_P^2 + 0.0205\phi_P L_e \tag{14}$$

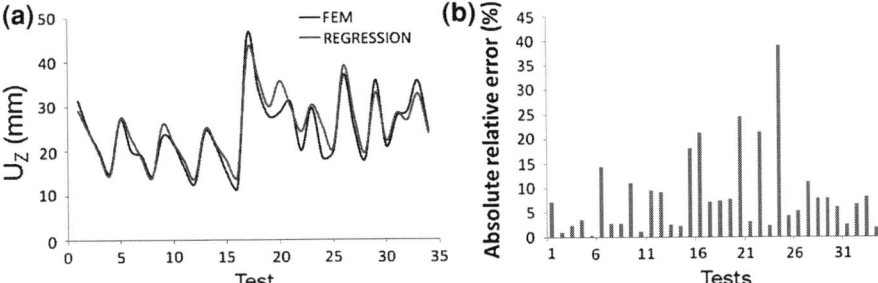

Fig. 40 Comparison between FE and regression results for U_Z **a** Comparison between FE and regression results for U_Z. **b** Absolute relative error between FE and regression results for U_Z

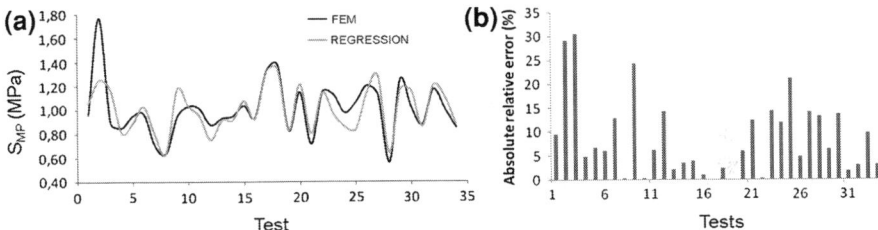

Fig. 41 Comparison between FE and regression results for S_{MP}. **a** Comparison between FE and regression results for S_{MP}. **b** Absolute relative error between FE and regression results for S_{MP}

$$\begin{aligned}F_{S_{MP}} = &-1.63 + 0.309\phi_P + 0.0121L_e + 0.0813P_e + 0.126\alpha - 4.69P_n - 0.0229\phi_P^2 \\ &- 0.000223L_e^2 - 0.000758P_e^2 - 0.00089\alpha^2 - 1.16P_n^2 + 0.00176\phi_P L_e \\ &- 0.00011\phi_P P_e - 0.00692\phi_P \alpha + 0.247\phi_P P_n - 0.000360L_e P_e \\ &- 0.000193L_e \alpha + 0.00510L_e P_n - 0.00122P_e \alpha + 0.0381P_e P_n + 0.0331P_n \alpha\end{aligned}$$
(15)

The quadratic correlation related to this function was 72.5% with a standard deviation of 0.18%.

Figures 40 and 41 show the comparison between the Finite Element results for U_Z and S_{MP} of the 34 experiments solved in all Experimental Designs and their respective values calculated using the regressions equation of F_{U_Z} and $F_{S_{MP}}$ respetively. The relative error found between the FEA results of U_Z and those found with the regression equations varies from 0.32 to 39%, while the relative error between the FEA S_{MP} results and the one found with the regression equations fluctuates between 0.04 and 30.6%.

Three additional cases different from those included in the Experimental Designs were run to verify both the stress and vertical displacement regression

Table 2 F_{U_z} and $F_{S_{MP}}$ verification

	Φp(mm)	Le(mm)	Pe(mm)	α(°)	Pn(mm)	Uz(mm) Fem	Uz(mm) FUZ	Relative error Uz (%)	SMP (Mpa) Fem	SMP (Mpa) FSMP	Relative error SMP (%)
Case1 1.3	14	90	60	–	15	0.5	11.74	13.4	14.1	0.75	0.76
Case 2 14.1	12	30	90	–	26	0.9	35.97	32.91	8.5	1.06	0.91
Case 3 6.2	13	35	80	–	19	0.7	28.75	28.89	0.4	1.12	1.19

functions (see Table 2). Values of different factors used in these cases of verification were chosen within the intervals defined in the Robust Design process. The relative error calculated from the difference between vertical displacement deduced from Finite Element Analysis and the F_{U_z} regression function presents a minimum of 0.4% and a maximum of 14.1%; Similarly for the Maximal Principal Stress, this relative error resulted between 6.2 and 14.1%.

5.3 Conclusions

The aims of the statistical procedure presented here are:

- Getting a quantitative idea about the main influential input variables, their degree of influence and the effect of their interactions both in the movement of the trachea and its stress state during swallowing and after prosthetic insertion.
- Avoiding the use of commercial programs that, although more accurate, are expensive and complex to use, which can be annoying for surgeons.
- Providing to surgeons a simple, fast and efficient tool to help them planning surgical prosthetic implantation in an optimal way and reduced time.

Although the relative error between the results of the statistical study and the simulations performed with the FE program may reach values close to 40% which is considered high, this only happens at certain points and special situations. So we consider that the main goal of this work, as it is constructing a tool that allows the thoracic surgeons to quantify the behavior of a patient-specific trachea after prosthesis implantation with the possibility of varying the dimensions and stent position getting reasonably accurate results in a very short time, has been accomplished. However, there are still some important aspects that deserve further study. For example, as for any other statistical tool, the regression equations should be additionally "trained" and the factors ranges increased. Also, the robust experimental design was accomplished by means of FE models with simplified tracheal geometry (the tracheal geometry was smoothed and approximated to a simple cylindrical shape although the tracheal section was conserved) and a simplified material model (the smooth muscle was considered as isotropic hyperelastic and not as anisotropic). Finally, the swallowing force obtained in the computational study in Sect. 3 was used as boundary condition to simulate the pathological swallowing in the robust design, which means that the force exerted by the elevator muscles to achieve swallowing is considered constant independently of the patient, gender, age, weight, etc. This hypothesis should be also discussed.

Acknowledgments The support of the Instituto de Salud Carlos III under the CIBER initiative is greatly acknowledged.

References

1. Ritt, M.J.P.F., Stuart, P.R., Naggar, L., Beckenbaugh, R.D.: The early history of arthroplasty of the wrist from amputation to total wrist implant. J. Hand Surg. Br. **19**(6), 778–782 (1994)
2. Fattori, R., Piva, T.: Drug-eluting stents in vascular intervention. Lancet **361**(9353), 247–249 (2003)
3. Zakaluzny, S.A., Lane, J.D., Mair, E.A.: Complications of tracheobronchial airway stents. Am. Acad. Otolaryngology-Head Neck Surg. **128**, 478 (2003)
4. Norwood, S., Vallina, V.L., Short, K., Saigusa, M., Fernandez, L.G., McLarty, J.W.: Incidence of Tracheal Stenosis and other late complications after percutaneous Tracheostomy. Ann. Surg. **232**, 233–241 (2000)
5. Fung, Y.C.: Biomechanics: Mechanical Properties of Living Tissues. 1st edn. Springer, Heidelberg/New York (1990)
6. Garzon Alvarado, D.A., Garcia, J.M., Doblare, M.: A Reaction-diffusion model for long bone growth. Biomed. Environ. Sci. **8**(5), 381–395 (2009)
7. Hernandez, B., Pena, E., Pascual, G., Rodriguez, M., Calvo, B., Doblare, M., Bellon, J.: Mechanical and histological characterization of the abdominal muscle. A previous step to modelling hernia surgery. J. Mech. Behav. Biomed. Mater. **4**(3), 392–404 (2011)
8. Enderle, J., Blanchard, S.M., Bronzino, J.: Introduction to Biomedical Engineering. Biomedical Engineering, 1st edn. Elsevier, Amsterdam (2000)
9. Malve, M., Perez-del Palomar, A., Trabelsi, O., Lopez-Villalobos, J., Ginel, A., Doblare, M.: Modeling of the fluid structure interaction of a human trachea under different ventilation conditions. Int. Commun. Heat Mass Transf. **38**(1), 10–15 (2011)
10. Bayod, J., Losa-Iglesias, M., Becerrode Bengoa-Vallejo, R., Carlos Prados-Frutos, J., Jules, K., Doblare, M.: Advantages and Drawbacks of Proximal Interphalangeal Joint Fusion Versus Flexor Tendon Transfer in the Correction of Hammer and Claw Toe Deformity. A Finite-Element Study. J. Biomech. Eng. Trans. ASME **132**(5), 051002, (2010)
11. Trabelsi O., Pérez Del Palomar A., Mena Tobar A., López-Villalobos J.L., Ginel A., Doblaré M.: FE simulation of human trachea swallowing movement before and after the implantation of an endoprothesis. Appl. Math. Model. (2011)
12. Grillo, H.C., Donahue, D.M., Mathisen, D.J., Wain, J.C., Wright, C.D.: Postintubation tracheal stenosis, treatment and results. J. Thorac. Cardiovasc. Surg. **109**(3), 486–492 (1995)
13. Huang, C.J.: Use of the silicone T-tube to treat tracheal stenosis or tracheal injury. Ann. Thorac. Cardiacovasc. Surg. **7**, 192–196 (2001)
14. Rob, C., Bateman, G.: Reconstruction of the trachea and cervical oesophagus; preliminary report. Br. J. Surg. **37**(146), 202–205 (1949)
15. Belsey, R.: Resection and reconstruction of the intrathoracic trachea. Br. J. Surg. **38**, 200–205 (1951)
16. Grillo, H.C.: Tracheal replacement: a critical review. Ann. Thorac. Surg. **73**, 1995–2004 (2002)
17. Wahidi, M.M., Ernst, A.: The Montgomery T-tube tracheal stent. Clinics in Chest Med. **24**(3), 437–443 (2003)
18. Xavier, R.G., Stefani Sanches, P.R., VieradeMacedo Neto, A., Kuhl, G., Bianchi Vearick, S., Dall'Onder Michelon, M.: Development of a modified Dumon stent for tracheal applicatons: an experimental study in dogs. Jornal Brasileiro de Pneumologia **34**, 21–26 (2008)
19. Noppen, M., Meysman, M., Claes, I., D'Haese, J., Vincken, W.: Screw-thread vs Dumon endoprothesis in the management of tracheal stenosis. Chest **115**, 532–535 (1999)
20. Madden, B.P., Datta, S., Charokopos, N.: Experience with ultraflex expandable metallic stents in the management of endobronchial pathology. Ann. Thorac. Surg. **73**, 938–944 (2002)
21. Noppen, M., Stratakos, G., D'Haese, J., Meysman, M., Vincken, W.: Removal of covered self-expandable matallic airway stents in benign disorders. Chest **127**, 482–487 (2005)
22. Freitag, L., Eicker, R., Linz, B., Greschuchna, D.: Theoretical and experimental basis for the development of a dynamic airway stent. Eur. Respirory J. **7**, 2038–2045 (1994)

23. Mitsuoka, M., Sakuragi, T., Itoh, T.: Clinical benefits and complications of Dumon stent insertion for the treatment of severe central stenosis or airway fistula. Jpn. Assoc. Thorac. Surg. **55**, 275–280 (2007)
24. Dumon, J.F., Cavaliere, S., Diaz-Gimenez, P., Vergnon, J.M.: Seven-year experience with the Dumon prosthesis. J. Bronchol. Interv. Pulmonol. **3**, 1–82 (1996)
25. Dumon, J.F.: A dedicated tracheobronchial stent. Chest **97**, 328–332 (1990)
26. Pereszlenyi, A., Igaz, M., Majer, I., Harustiak, S.: Role of endotracheal stenting in tracheal reconstruction surgery-retrospective analysis. Eur. J. Cardio-Thorac. Surg. **25**, 1059–1064 (2004)
27. Holzhauser, U., Lambert, R.K.: Analysis of tracheal mechanics and applications. J. Appl. Physiol. **91**, 290–297 (2001)
28. Costantino, M.L., Bagnoli, P., Dini, G., Fiore, G.B., Soncini, M., Corno, C., Acocella, F., Colombi, R.: A Medical Engineering & Physics numerical and experimental study of compliance and collapsibility of preterm lamb tracheae. J. Biomech. **37**, 1837–1847 (2004)
29. Gustin, B., G'Sell, C., Cochelin, B., Wourms, P., Potier-Ferry, M.: Finite element determination of the forces exerted by endotracheal tubes on the upper airways. Biomaterials **17**, 1219–1225 (1996)
30. Wiggs, B.R., Hrousis, C.A., Drazen, J.M., Kamm, R.D.: On the mechanism of mucosal folding in normal and asthmatic airways. J. Appl. Physiol. **83**, 1814–1821 (1997)
31. Rains, J.K., Bert, J.L., Roberts, C.R., Pare, P.D.: Mechanical properties of human tracheal cartilage. J. Appl. Physiol. **72**, 219–225 (1992)
32. Roberts, C.R., Rains, J.K., Paré, P.D., Walker, D.C., Wiggs, B., Bert, J.L.: Ultrastructure and tensile properties of human tracheal cartilage. J. Biomech. **31**, 81–86 (1998)
33. Zhongzhao, T., Ochoa, I., Bea, J.A., Doblare, M.: Theoretical and experimental studies on the nonlinear mechanical property of the tracheal cartilage. In: Annual International Conference of the IEEE Engineering in Medicine and Biology Society, pp. 1058–1061 (2007)
34. Yamada, H.: Mechanical Properties of Respiratory and Digestive Organs and Tissues. Williams & Wilkins, Bultimera (1970)
35. Lambert, R.K., Baile, E.M., Moreno, R.H., Bert, J., Pare, P.D.: A method for estimating the Young's modulus of complete tracheal cartilage rings. J. Appl. Physiol. **70**, 1152–1159 (1991)
36. Rains, J.: Mechanical properties of tracheal cartilage. Chemical and biological engineering theses and dissertations. University of British Columbia (1989)
37. Kempson G.E.: Mechanical properties of articular cartilage. In: Freeman, M.A.R. (eds.) 2nd edn. Pitman, London (1979)
38. Wang, Y.Q., Lin, Y. H., Teng, Z.Z.: A thin shell modeling for collapsible trachea. J. Fudan Univ. Natural Science **45**, 141–150 (2006)
39. Begis, D., Delpuech, C.L., Tallec, P., Loth, L., Thiriet, M., Vidrascu, M.: A finite-element model of tracheal collapse. J. Appl. Physiol. **64**, 1359–1368 (1988)
40. Gunst, S.J., Wu, M.F.: Plasticity in skeletal, cardiac, and smooth muscle selected contribution: plasticity of airway smooth muscle stiffness and extensibility: role of length-adaptive mechanisms. J. Appl. Physiol. **90**, 741–749 (2001)
41. Comín M., Peris J.L., Prat J.M., Dejoz R.J., Vera P.M., Hoyos J.V.: Biomecánica de la Fractura Ósea y Técnicas de Reparación, 2nd edn. Instituto de Biomecánica de Valencia (1999)
42. Stephens, N.L., Cardinal, R., Simmons, B.: Mechanical properties of tracheal smooth muscle: effects of temperature. Am. J. Physiol. Cell Physiol. **233**, C92–C98 (1977)
43. Holzapfel, G.A., Gasser, T.C., Ogden, R.W.: A new constitutive framework for arterial wall mechanics and a comparative study of material models. J. Elast. **61**, 1–48 (2000)
44. Williams P.L., Bannister L.H., Berry M.M., Collins P., Dyson M., Dussek J.E., Ferguson M.W.J.: Anatomía de Gray. 2. Spanish Edition (1998)
45. Holzapfel, G.A., Gasser, T.C., Ogden, R.W.: A new constitutive framework for arterial wall mechanics and a comparative study of material models. J. Elast. **61**, 1–48 (2000)
46. Peña, E., Calvo, B., Martínez, M.A., Doblaré, M.: On finite-strain damage of viscoelastic-fibred materials. Application to soft biological tissues. Int. J. Numer. Methods Eng. **74**, 1198–1218 (2008)

47. Holzapfel, G.A., Sommer, G., Gasser, C.T., Regitnig, P.: Determination of layer-specific mechanical properties of human coronary arteries with nonatherosclerotic intimal thickening and related constitutive modeling. Am. J. Physiol. Heart Circulatory Physiol. **289**, H2048–H2058 (2005)
48. Holzapfel, G.A., Gasser, T.C.: A viscoelastic for fiber-renforced composites at finite strains: continuum bases, computational aspects and applications. J. Computer Methods Appl. Mech. Eng. **190**, 4379–4403 (2001)
49. Korhonen, R.K., Laasanen, M.S., Töyräs, J., Rieppo, J., Hirvonen, J., Helminen, H.J., Jurvelin, J.S.: Comparison of the equilibrium response of articular cartilage in unconfined compression, confined compression and identation. J. Biomech. **35**, 903–909 (2002)
50. Sarma, P.A., Pidaparti, R.M., Moulik, P.N., Meiss, R.A.: Non-linear material models for tracheal smooth muscle tissue. Bio-Med. Mater. Eng. **13**, 235–245 (2003)
51. Sarma, P.A., Pidaparti, R.M., Meiss, R.A.: Anisotropic properties of tracheal smooth muscle tissue. J. Biomed. Mater. Res. **65**(1), 1–8 (2003)
52. Schaefer, G., Hassanien, A., Jiang, J.: Computational Intelligence in Medical Imaging. CRC Press, Boca Raton, FL (2009)
53. MIMICS Guide. Materialise Technologielaan 15, 3001 Leuven. http://www.materialise.com/mimics
54. McNeel R. Rhinoceros 4.0. http://www.rhino3d.com (2007)
55. Guyton, A.C., Hall, J.E.: Tratado de Fisiología Médica. Edición 9. McGraw-Hill Interamericana, pp. 872–873 (1997)
56. Boston Medical Products. 117 Flander Rd., Westborough. http://www.trachs.com/home/trachbronch/pdf/ttubecat.pdf
57. Herrero Jover J. ALMA IT Systems. Barcelona. http://www.alma3d.com (2005)
58. Beekhof, F.: CVMLCPP Common Versatile Multi-purpose Library for C++. CUI University of Geneva (2010)
59. Trabelsi, O., Pérez Del Palomar, A., López-Villalobos, J.L., Ginel, A., Doblaré, M.: Experimental characterization and constitutive modeling of the mechanical behavior of the human trachea. Med. Eng. Phys. **32**, 76–82 (2010)
60. Wang, L., Pinder, K.L., Bert, J.L., Okazawa, M., Paré, P.D.: Mechanical properties of the tracheal mucosal membrane in the rabbit. II. Morphometric analysis. J. Appl. Physiol. **88**, 1022–1028 (2000)
61. Bourgeois, C., Steinsland, E., Blanc, N., de Rooij, N.F.: Design of resonators for the determination of the temperature coefficients of elastic constants of monocrystalline silicon. Proceedings of the IEEE International Frequency Control Symposium (1997)
62. MEMS and Nanotechnology Exchange. Corporation for National Research Initiatives® (CNRI) (1999)
63. Verkerke, G.J., de Vries, M.P., Schutte, H.K., Van den Hoogen, F.J.A, Rakhorst, G.: Analysis of mechanical behavior of the Nijdam Voice prosthesis. Laryngoscope **107**, 1656–1660 (1997)
64. Sera, T., Satoh, S., Horinouchi, H., Kabayashi, K., Tanishita, K.: Respiratory flow in a realistic tracheostenosis model. J. Biomech. Eng. **125**, 461–471 (2003)
65. Logemann, J.A., Kahrilas, P.J., Cheng, J., Pauloski, B.R., Gibbons, P.J., Rademaker, A.W., Lin, S.: Closure mechanisms of laryngeal vestibule during swallow. Am. J. Physiol. **262**, G338–G344 (1992)
66. Rademaker, A.W., Pauloski, B.R., Logemann, J.A., Shanahan, T.K.: Oropharyngeal swallow efficiency as a representative measure of swallowing function. J. Speech, Lang. Hear. Res. **37**, 314–325 (1994)
67. Logemann, J.A., Pauloski, B.R., Rademaker, A.W., Kahrilas, P.J.: Oropharyngeal swallow in younger and older woman: Videofluoroscopic analysis. J. Speech, Lang. Hear. Res. **45**, 434–445 (2002)
68. Dantas, R.O., Dodds, W.J.: Effect of bolus volume and consistency on swallow-induced submental and infrahyoid electromyographic activity. Braz. J. Med. Biol. Res. **23**, 37–44 (1990)

69. Mendell, D.A., Logemann, J.A.: Temporal sequence of swallow events during the oropharyngeal swallow. J. Speech, Lang. Hear. Res. **50**(5), 1256–1271 (2007)
70. Wilson, E.M., Green, J.R.: Coordinative organization of lingual propulsion during the normal adult swallow. Dysphagia **21**(4), 226–236 (2006)
71. Inagaki, D., Miyaoka, Y., Ashida, I., Ueda, K., Yamada, Y.: Influences of body posture on duration of oral swallowing in normal young adults. J. Oral Rehabil. **34**(6), 414–421 (2007)
72. Steele, C.M., Van Lieshout, P.H.: The dynamics of lingual-mandibular coordination during liquid swallowing. Dysphagia **23**(1), 33–46 (2007)
73. Terk, A.R., Leder, S.B., Burrell, M.I.: Hyoid bone and laryngeal movement dependent upon presence of a tracheotomy tube. Dysphagia **22**(2), 89–93 (2007)
74. Mendell, D.A., Logemann, J.A.: Temporal sequence of swallow events during the oropharyngeal swallow. J. Speech Lang. Hear. Res. **50**(5), 1256–1271 (2007)
75. Honda, Y., Hata, N.: Dynamic imaging of swallowing in a seated position using open-configuration MRI. J. Magn. Reson. Imaging **26**(1), 172–176 (2007)
76. Crary, M.A., Carnaby Mann, G.D., Groher, ME.: Identification of swallowing events from sEMG Signals Obtained from Healthy Adults. Dysphagia **22**(2), 94–9 (2007)
77. Aboofazeli, M., Moussavi, Z.: Analysis of temporal pattern of swallowing mechanism. Conf. Proc. IEEE Eng. Med. Biol. Soc. **1**, 5591–5594 (2006)
78. Afkari, S.: Measuring frequency of spontaneous swallowing. Australasian Phys. Eng. Sci. Med. **30**(4), 313–7 (2007)
79. Moreno, M.C., Gil-Loyzaga, P., Gómez, F.R., Carcedo, C.F., Fernández, J.G., Martínez, E.P., Grande, A.H., Broto, J.P.: Magnetoelastic sensors as a new tool for laryngeal research. Acta Otolaryngology **127**(11), 1182–1187 (2007)
80. Robert, D.: Swallowing disorders following endotracheal intubation and tracheotomy. Reanimation **13**, 417–430 (2004)
81. Philips Medical Systems Nederland B.V. Integris V with High-Speed DICOM Image Interface R1 (MCV 2971). DICOM Conformance Statement (1998)
82. Crow, K.: Robust Product Design through Design of Experiments. DRM Associates (1998)
83. Ashby, W.R.: Design for a Brain. 2nd edn. Chapman & Hall, London (1960)
84. Miranda V.J.C.: Product design: techniques for robustness, reliability and optimization. Instituto Tecnológico y de Estudios Superiores de Monterrey, Campus Toluca (2004)
85. Montgomery, D.C.: Design and Analysis of Experiments, 7th edn (2008)
86. Berger, P.D., Maurer, R.E.: Experimental design with applications in management, engineering and the sciences. Technometrics: J. Statist. Phys. Chem. Eng. Sci. **45**(1), 105 (2003)
87. Roy, R.: The primer on the Taguchi method. Edition Van Nostrand Reinhold (1990)
88. Tsui, K.L.: An overview of Taguchi method and newly developed statistical methods for robust design. IIE Trans. **24**(5), 44–57 (1992)
89. Griscom, N.T., Wohl, M.E.B.: Dimensions of the growing trachea related to age and gender. Am. Roentgen Ray Soc. **146**, 233–237 (1986)
90. Rocabado B.J.L., Roldan T.R., Derosas A.C., Zuleta S.R., Hurtado S.G.: Manejo de la estenosis traqueal. Revista Chilena de Cirugía. 59(6):408-416. (2007)
91. Ragavan, A.J., Evrensel, C.A., Krumpe, P.: Interactions of airway oscillation, tracheal inclination and mucus elasticity. Significantly improve Simulated Cough clearance. Am. Coll. Chest Physicians **137**, 355–361 (2010)
92. Perez del Palomar A., Trabelsi O., Mena A., Lopez-Villalobos J.L., Ginel A., Doblare M.: Patient-specific models of human trachea to predict mechanical consequences of endoprosthesis implantation. Phil. Trans. R. Soc. A Math. Phys. Eng. Sci. **368** (1921): 2881–2896 (2010)
93. Neter, J., Kutner, M., Wasserman, W., Nachtsheim, C.: Applied Linear Statistical Models. McGraw-Hill, New York (1996)
94. Ryan, B.: Minitab Software. State College, Pensilvania (1972)
95. Darlington, R.B.: Regression and Linear Models. Mc Graw-Hill, New York (1990)

96. Hocking, R.R.: A biometrics invited paper. The analysis and selection of variables in linear regression. Biometrics **32**(1), 1–49 (1976)
97. Lindeman, R.H., Merenda, P.F., Gold, R.Z.: Introduction to Bivariate and Multivariate Analysis. Scott, Foresman, Glenview (1980)
98. Sharma, M.K., Jain, C.K.: Fluoride modeling using best subset procedure. J. Appl. Hydrol. **XVIII**, 12–22 (2005)
99. Neter, J., Wasserman, W.: Applied Linear Statistical Models, 2nd edn. Professional Publishing, Irwin (1985)

Part V
Brain

Image-Based Computational Fluid Dynamics for Patient-Specific Therapy Design and Personalized Medicine

Andreas A. Linninger

Abstract Recent advances in quantitative imaging allow unprecedented views into organ function and cellular chemistry of whole organisms in vivo. Novel imaging modalities enable the quantitative investigation of spatio-temporal reaction and transport phenomena in the living animal or the human body. Quantitative imaging modalities enable quantitative analysis with image-based computational fluid dynamics (iCFD). This methodology integrates medical imaging modalities with rigorous computational fluid dynamics. The future potential of patient-specific therapy design with iCFD will be illustrated with three case studies. This chapter will highlight the growing opportunities of patient-specific therapy design made possible by the advancement of quantitative medical imaging modalities in combination with mathematical modeling. Conclusions will emphasize on future steps towards individualized medicine using iCFD methods.

Keywords Biomedical engineering · Quantitative medical imaging · Image-based computational fluid dynamics

1 Introduction and Motivation

Advanced medical imaging technologies provide unprecedented measurements about the living organism. Quantitative medical imaging enables precise reconstruction of anatomical details as well as in vivo measurements of biological functions such as metabolite concentrations, flow rates of fluids or blood.

A. A. Linninger (✉)
Department of Chemical Engineering and Bioengineering,
University of Illinois at Chicago, Chicago, USA
e-mail: linninge@uic.edu

Anatomical details. Geometrical details about an individual subject's anatomy include sharp delineation of organ boundaries, specific tissue types, distinction of functional substructures in specialized organs such as the intricate gray and white matter architecture of the human brain. Beyond geometry, diffusion tensor imaging allows the assessment of anisotropic tissue properties such as the directional axonal fiber alignment in the brain. The precision of geometrical information captured with medical imaging is accurate enough so that differences among individuals can easily be discerned and accurately quantified. The accurate quantification of inter-patient anatomical variations is crucial to personalized medicine.

Biological functions. Key physiological and biochemical states of cells, tissues or organs can also be measured by quantitative medical imaging. Positron Emission Tomography (PET) is especially promising in opening a view to the concentrations of metabolites in vivo [12]. PET scans of entire animals are increasingly used to track the fate of tagged molecules inside the living organisms as a function of space and time. Different magnetic resonance imaging (MRI) modalities can accurately measure fluid flow rates; for example Cine MRI is suitable for measuring blood flow in vivo; fMRI accurately tracks relative cerebral blood flow inside the brain.

The advancements of medical imaging capabilities propel the fundamental understanding about biochemical and transport phenomenon. The precision in acquiring individual differences both in anatomical detail as well as functional behavior enables the design of individualized treatments. Accurate imaging data to capture anatomical and physiological details for an individual subject is an essential prerequisite for patient-specific medicine.

This chapter will introduce applications that capitalize on individualized information acquired by quantitative medical imaging. We will first characterize *image-based Computational Fluid Dynamics* (iCFD), a new process integrating anatomical and quantitative functional data into rigorous fluid dynamics calculations. We will review current state and future image-reconstruction techniques. Section 3 will introduce types of applications amenable to iCFD analysis. These case studies will highlight current iCFD capabilities. The conclusion section will pose open questions for future research.

2 Image-Based Computational Fluid Dynamics

Before modern imaging techniques became widely available, biological systems analysis was severely hindered by difficulties in accurately representing their complex anatomical shapes on a computer. Early engineering models of biological systems were constructed with simplified geometries of idealized shapes. For more realistic modeling of flow and reaction phenomena, *image reconstruction of patient-specific medical image data* enables the creation of computational models

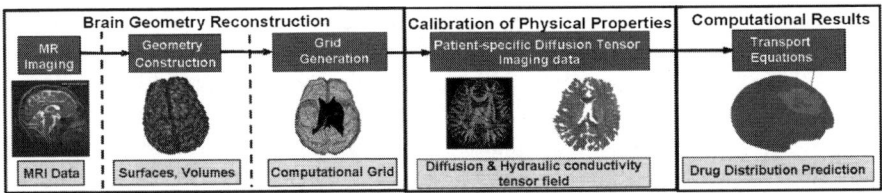

Fig. 1 Overview of computer-assisted brain analysis for the design of drug delivery therapies. The first step entails the collection of medical images from MRI. Geometry reconstruction detects sharp boundaries and functional regions inside the brain. Grid generation partitions surfaces and volumes into small tetrahedrons for finite volume discretization of the transport equations. Physical properties such as drug diffusion and hydraulic conductivity tensors are estimated from apparent water diffusion tensor measurement obtained by diffusion tensor imaging (DTI). Computational analysis solves the discretized transport equations to predict drug distribution

that precisely matches anatomical dimensions and shapes. The procedure of converting medical image data into realistic computational representations that preserve anatomical and physiological detail is termed *geometric image reconstruction*. The accurate mapping between biological and computational domain is also crucial for validating simulations with experimental data. It is a prerequisite for realistic computer simulations in an emerging field known as iCFD.

Our group has developed a workflow procedure that bridges the gap between image reconstruction and rigorous simulation of bio-transport problems. The proposed systems approach for image-based modeling of biological systems proceeds through three stages as depicted schematically in Fig. 1.

In stage 1, *brain geometry reconstruction* delineates the patients' individual brain geometry by converting medical image data into patient-specific three-dimensional surfaces and volume representation. Geometric image reconstruction involves the assembly of three-dimensional imaging data into computational meshes, usually unstructured grids. Computational *surface meshes* delineate system boundaries like an organ or the contrast between different zones such as fluid-filled spaces surrounded by soft tissues. Geometric image reconstruction also involves *smoothing, contrast enhancement* and *automatic edge detection*. The computational surface meshes are further divided into a finite number of volumes, typically tetrahedrons. Grid generation algorithms perform mesh regularization to optimally divide the domain into finite balance envelopes with suitable aspect ratios to form a *volumetric mesh*. Advanced semi-automated meshing software can be employed for rapid generation of multi-block structured or unstructured volume meshes.

In stage 2, *calibration of physical properties*, anisotropic tissue properties may be assigned to create patient-specific tissue models. Specifically, we estimate anisotropic transport properties using tensor fields acquired with DTI.

The combination of novel imaging techniques with modern systems methods is a crucial step for a rigorous quantification of *brain tissue anisotropy and heterogeneity*. Acquisition of advanced material properties from DTI reveals that porous brain tissue exhibits strong directionality of its material properties especially along the fibrous white matter tracts. For example, we previously estimated the elements of anisotropic diffusion and hydraulic conductivity tensors with advanced diffusion tensor imaging for a 32 year old volunteer [20]. A typical diffusion tensor near the putamen is given in (1):

$$\mathfrak{D}_e = 10^{-12} \begin{bmatrix} 28.61 & 0.580 & 1.040 \\ 0.580 & 23.44 & 1.230 \\ 1.040 & 1.230 & 21.04 \end{bmatrix} [\text{m}^2/\text{s}] \qquad (1)$$

Prior experiments determined the nerve growth factor (NGF), mean diffusivity as $D_e = 28 \pm 12 \cdot 10^{-12}$ m^2/s which is in reasonable agreement with the mean diffusivity of the calculated diffusion tensor [22, 30]. Computational meshes with anisotropic tensor fields mimic realistic tissue properties and constitute the physical domain for which the transport equations will be satisfied as outlined in the next step.

In stage 3, the equations of fluid motion defined over the discretized brain mesh obtained in step 1 are solved numerically. Volume is partitioned into a finite number of small non-overlapping tetrahedrons. These small tetrahedrons serve as balance envelopes over which the conservation laws of mass, species and momentum can be enforced.

A great aide in converting pixilated imaging data into physiologically consistent surfaces and domains is offered by automatic image reconstruction software. Several free tools available for spatial image reconstruction such as ImageJ, ITK-Snap, and VTK substantially accelerate the generation of computational maps [18, 33]. Commercial tools such as Mimics [24] have excellent features for contrast enhancement, automatic boundary detection and surface smoothing. Most software tools provide several file output formats for reconstructed surface or volume meshes. These file export formats are essential for importing reconstructed meshes obtained from imaging data into state-of-the-art CFD software such as ANSYS Fluent [3] or ADINA [1].

The information flow diagram for creating simulation case files from MRI images is summarized in Fig. 2. A series of MRI images in DICOM format are processed with Mimics, Materialise Incs, where the MRI images are stacked to create a 3D image of the object. Reconstructed surfaces can be exported as an STL file. The STL file can then be smoothed, repaired and exported for input in a commercial software. The optimized surface mesh is divided into a volume mesh by Delaunay triangulation (.msh). The volumetric mesh file is sent to iCFD simulations, which calculate flow and pressure fields as well as optionally species transport for applications with mass transfer and chemical reactions.

The next section will demonstrate current iCFD capabilities for designing patient-specific therapies and models for the cerebral blood flow.

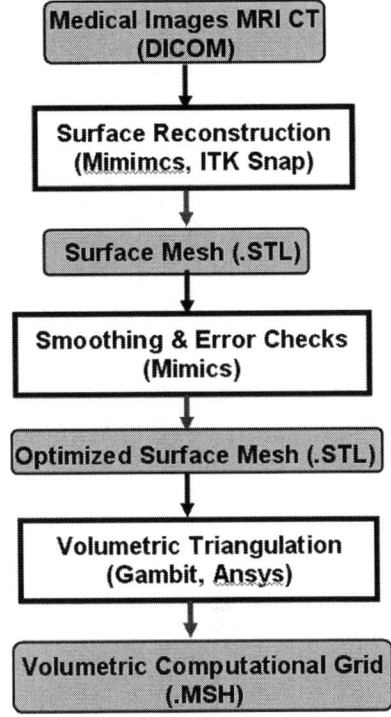

Fig. 2 Detailed flow chart of brain geometry reconstruction in the iCFD workflow. The flow chart describes the detailed steps of brain geometry reconstruction and the file formats needed for iCFD. Software applications for each step are included in parenthesis

3 Case Studies

3.1 Optimal Catheter Placement for Chemotherapy

This demonstration addresses the problem of optimal administration of chemotherapeutic agents for the patient-specific treatment of brain tumors by convection-enhanced drug delivery (CED). Optimal delivery of chemotherapeutics for treating malignant gliomas can be substantially improved with new computational techniques that maximize the effectiveness of therapy for the specific patient, while minimizing the risks associated with such therapies [13]. A glioma is a type of tumor that arises from *glial cells* in the brain or spine. In recent years, CED has received attention as means to deliver therapeutic agents directly to the target site inside the brain. This approach circumvents the problems posed by the impermeability of the blood brain barrier to macromolecules delivered systemically [5, 28]. However, successful chemotherapy requires effective drug dosage to destroy all cancerous cell at the tumor site as well as malignant cells spreading diffusely along white matter fiber tracts [13]. White matter anisotropy due to axonal bundle structure strongly impacts transport of drugs administered via CED [21, 22].

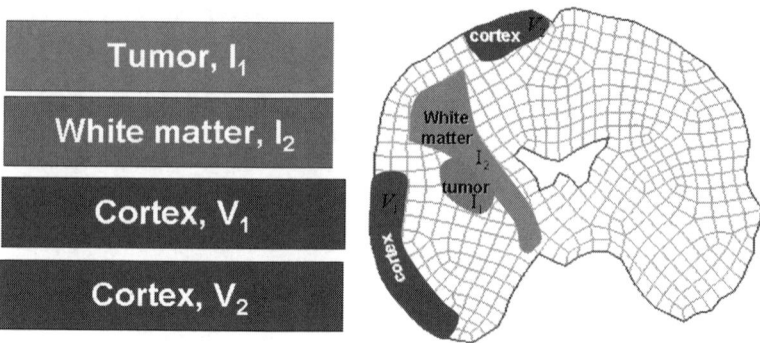

Fig. 3 Illustration of a simplified treatment plan in a coronal brain slice obtained from reconstructing a patient's MRI scan. The tumor (I_1) and adjacent white matter tracks (I_2) should be maximally dosed with therapeutic agent. Concentrations of chemotoxins must not exceed safe thresholds in the cortical regions V_1 and V_2

iCFD will be used to determine where the catheter infusion site should be placed. In addition, flow rate and concentrations of the chemotherapeutic agent can be optimized.

3.1.1 Treatment Plan

Chemotherapy with rigorous determination of the optimal catheter location is a distributed optimization problem with multiple objectives. On one hand, drug concentration should be highest in the vicinity of the tumor and the surrounding areas where recurrence may take place. The treatment should ensure final drug concentration above a therapeutically effective level in these regions. On the other hand, critical brain areas such as the sensory cortex should not be damaged by hazardous agent levels to reduce harmful side-effects. A simplified treatment plan to illustrate rigorous treatment design mandates chemotoxic drug concentration above the effective therapy concentration $Ct = 1.5$ in the tumor region (I_1) and the white matter region (I_2) is laid out in Fig. 3. To limit side-effects, cortex regions (V_1 and V_2) should be dosed below the toxicity limit, $Cp = 0.7$. The optimal catheter position is located by a novel optimization technique, which simultaneously maximizes drug concentration in the desired brain region, while ensuring that the drug concentration does not fall below a therapeutically effective level in all areas that need chemotherapy nor rise above the toxic threshold in functional brain areas that should be spared. Our hybrid algorithm has a genetic and a deterministic element to precisely locate optimal catheter and infusion parameters. A score function evaluates the match between the desired concentration profile formulated in the therapy goals and the drug distributions achieved by the current particular catheter placement. Genetic inheritance adjusts the catheter locations from among previous high scoring placements to identify the globally optimal solution. As a subroutine to the optimization, a multi-scale finite volume method with two reference systems was used for solving the transport equations.

3.1.2 Algorithm

Drug dispersion is typically solved over a coarse mesh of finite volumes. For a specific catheter location, the domain would have to be re-meshed to adjust for the interface between catheter outlet and porous brain tissue. Different computational meshes would have to be built for every iteration. To avoid this impractical approach, we have developed a method for optimal catheter placement without remeshing. To obtain continuous positional sensitivity of the objective function, the efflux from the catheter tip was projected to the cell walls with an analytical technique. With the face fluxes thus determined as an internal Neumann boundary, the convective diffusive species transport was solved over finite volumes. Combining these two computational levels, only a single mesh of the brain was necessary to locate the optimal catheter position. Using the novel multi-scale algorithm, it is possible to optimize catheter placement and design, as well as to precisely control drug distribution volume. The following sections show solutions to the distributed optimization problem coupled with the embedded transport problem.

3.1.3 Results

This study focused on the methodology of optimizing the catheter position based on the patient specific geometry. Coronal sections were reconstructed from MRI images of a human brain. The images were imported into Mimics reconstruction software to accurately delineate boundaries and surfaces of the specific individual brain [24]. This information was imported into Gambit to create a volumetric mesh [11]. A coronal slice from an MRI image of a human brain can be converted into a corresponding mesh [20, 22, 25, 26]. Locations I_1 are close to the tumor in which high therapeutic levels should be ensured. In addition, the white matter tracts I_2 should be treated. Areas, V_1 and V_2, are critical cortical areas in which dosage must not reach harmful levels. The optimal catheter position for CED chemotherapy was found by a novel optimization technique.

The score function measures the closeness of the candidate solution with respect to the desired treatment target. The concentration in the target area was maximized, while toxic concentrations were avoided in important cortical regions via a penalty approach. The genetic algorithm adjusted the catheter tip position via inheritance and mutation operators [19]. These steps were repeated for thirty generations until a minimum value for the score function was found. Conservation balances for mass, momentum and species transport over a mesh accurately representing the patient's specific brain geometry are given in:

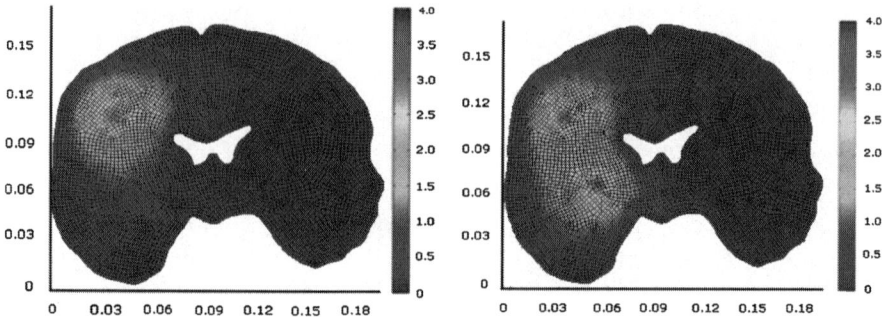

Fig. 4 Optimal catheter placement. The single catheter solution (*left*) only partially satisfies the treatment goals. The optimal dual port catheter achieves the therapeutic concentrations in the target areas (I^1 and I^2) without exceeding the toxicity thresholds in the cortical areas V^1 and V^2. (Reproduced from [19] with permission)

$$\vec{\nabla}\left(K\vec{\nabla}P\right) = 0 \qquad (2)$$

$$\vec{\nabla}\left(D\vec{\nabla}C\right) - \vec{U}\vec{\nabla}C = 0 \quad where \quad \vec{U} = K\vec{\nabla}P \qquad (3)$$

Here P represents pressure, $\vec{U}$ is the velocity vector, D is the drug diffusion tensor, and K is the hydraulic conductivity. The equations were discretized using the finite volume method.

Single Catheter Solution

For a single port catheter, the optimal catheter placement is presented in Fig. 4. With a single port, the optimal position is a compromise between reaching the tumor region (I_1) and also covering the white matter (I_2). Yet, even optimal placement is unsatisfactory, because only 70% of the target region can be reached with a therapeutically effective dosage. The damage to V_1 and V_2 is minimal with 99% of the area below toxic levels. The single catheter produces limited drug distribution which is not able to cover the tumor area and propagation path simultaneously. The overall score of 82% indicated that a one catheter solution cannot meet the chemotherapy specifications, and that the treatment will most likely not be effective in treating the brain tumor.

Dual-Port Catheter Solution

A better volume distribution can be achieved by using a two-port catheter as shown in Fig. 4. Dual-port catheter manages to completely cover the tumor region (I_1) and the white matter region (I_2). For the optimal solution to be safe, it should deliver the drug only at an effective therapy concentration of $C_t = 1$. At the same time, the optimal solution keeps drug concentrations below toxic levels in the cortex region, V_1 and V_2, below $C_p = 0.7$. The score function corresponds to 100% satisfaction of the treatment objectives. This new computational approach would combine advanced imaging technologies to realistically predict macromolecule distribution injected into the brain.

Even though this conceptual case study used drastic simplifications, it nevertheless demonstrated the feasibility of computer-aided patient-specific therapy design. For a realistic design and as pointer to future work, new tools need to be developed that capture the physician's intent in three-dimensions, optimization routines need to be integrated with automatic image reconstruction, finally the optimal solutions on the dual mesh approach need to be validated by a rigorous simulation with a finer mesh.

3.2 iCFD Predicts Intrathecal Drug Distribution

Intrathecal drug delivery, which is the infusion of therapeutic molecules into the cerebrospinal fluid (CSF), is an efficient infusion module to bypass the blood brain barrier and treat disorders of the central nervous system (CNS). The pulsations of the CSF due to the expansions of the cerebrovasculature cause rapid dispersion of drugs. High inter-patient variability is seen in drug distribution with identical infusions [8, 15, 23]. In this section, the quantitative iCFD methodology is used to explain the fundamental reasons for drug distribution variability and predict patient-specific drug distribution in vivo.

3.2.1 Construction of Patient-Specific Model

A patient-specific method is needed to better quantify actual drug concentrations in CSF and in spinal cord tissue. Drug pharmacokinetics such as binding, degradation, uptake and intracellular reactions that take place in vivo are incorporated in addition to drug transport. A patient-specific computational model is reconstructed from images of the patient's CNS obtained from MRI. An image reconstruction software, MIMICS [24], combines two dimensional MR images into a three-dimensional geometry. Automatic thresholding and manual segmentation were used to recreate the complex CSF-filled subarachnoid space in three dimensions. The reconstructed object shown in Fig. 5 was then discredized into a fine computational mesh of 600,000 tetrahedral volumes.

3.2.2 Computation of CSF Flow and Drug Transport

CSF pulsates inside the spine due to the periodic expansions of the cerebrovasculature [4]. Convective flow due to pulsations drastically enhances drug

Fig. 5 Building patient-specific model from MR images. The magnetic resonance image on the *left* captures the patient's CNS. The brain and spinal cord tissue as well as the CSF (narrow *white* spaces) were reconstructed to form the model seen on the *right*

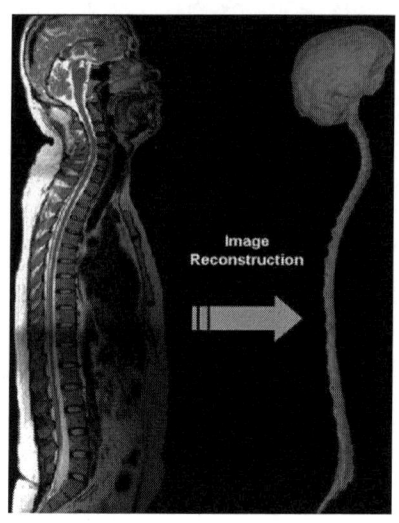

dispersion and micromixing. The patient's CSF pulsations were measured with Cine MRI and the pulsating flow field in the spine was recreated in the model. In addition, this model accounts for CSF production and reabsorption with physiologically consistent rates [17]. To compute drug biodistribution and kinetics, continuity (4), Navier–Stokes (5) and species transport (6) equations were applied in each control volume:

$$\vec{\nabla} \cdot (\rho \vec{u}) = 0 \qquad (4)$$

$$\rho \left(\frac{\partial \vec{u}}{\partial t} + \vec{u} \cdot \nabla \vec{u} \right) = -\nabla p + \nabla \cdot \bar{\bar{\tau}} + \rho \vec{g} + \vec{F} \qquad (5)$$

$$\frac{\partial C}{\partial t} + \vec{u} \cdot \vec{\nabla} C = \vec{\nabla} \cdot (D \vec{\nabla} C) \qquad (6)$$

where, ρ is the density, v is the kinematic viscosity, $\vec{g}$ is the gravity, F is the external force term, C is the concentration, D is the diffusivity, and u and p describe the periodic velocity and pressure fields, respectively. Drug mass transfer and perfusion into the spinal cord were computed with realistic tissues properties and resistances [21]. Drug diffuses inside the spinal cord according to Darcy's law and undergoes chemical reactions, R, as given in (7):

$$\varepsilon \frac{\partial C}{\partial t} + \vec{u} \cdot \vec{\nabla} C = \vec{\nabla} \cdot \left(D \vec{\nabla} C \right) + R \qquad (7)$$

where the porosity $\varepsilon = 0.3$.

Patient-specific computational analysis of drug distribution inside the CNS gives spatiotemporal concentration profiles of infused therapeutic molecules in target regions. Through quantitative analysis, regions of the tissue immersed in therapeutic concentrations and those immersed in toxic concentrations can be determined. If the therapy goal is not met, the original infusion protocol should be improved to enhance the efficiency of the delivery.

3.2.3 Gravity and Solution Baricity

The intrathecal distribution of drug solutions is influenced by gravity and patient posture [35]. The effect of gravity is a function of solution baricity, which is the ratio of drug solution density to patient CSF density. Anesthesiologists observed that hyperbaric and plain ropivacaine, a spinal anesthetic, achieved different levels of sensory block in supine patients [10]. Accounting for posture and solution baricity in predictions of drug dispersion gives insight into the different degrees of sensory block observed clinically.

Figure 6 shows simulation results for the infusion of hyperbaric and isobaric solutions into the subarachnoid space at the lumbar vertebrae for different patient postures after infusion. The iCFD simulations confirm that the distribution of drugs in hyperbaric solution is a strong function of patient posture due to gravity. In contrast, isobaric solution distributes evenly around the infusion site after infusion due to the lack of density difference between drug solution and CSF.

3.2.4 Patients' Physiological Parameters

Though not influenced by gravity, the infusion of isobaric solutions with identical dosage still produces significantly different distribution from patient to patient [31]. Our computational model predicts that drug dispersion is a strong function of CSF pulsations, and both frequency and magnitude of CSF pulsations impact drug transport speed. The effect of CSF pulsatile frequency and magnitude will be discussed in the following section. These simulations predict drug distribution for a 0.2 ml bolus injection of a model drug with a 17 gauge needle over 30 s at the second lumbar vertebra (L2).

CSF Pulsatile Frequency

The heart rate of the patient is the pulsatile frequency of CSF in the spine. While resting heart rate vary from 50 to 90 bpm in normal subjects, a wide range of heart rate, from as low as 15 bpm during emergency medical procedure to as high as 180 bpm during exercise, could result from a spectrum of activities [29]. Drug concentration profiles (normalized to the infusion concentration) along the spine was generated for three different heart rates: 43, 60 and 120 bpm, see Fig. 7 left. Immediately after the bolus injection, peak drug concentrations were found near the infusion site (L2) in the CSF. Concentration maxima at the infusion site after

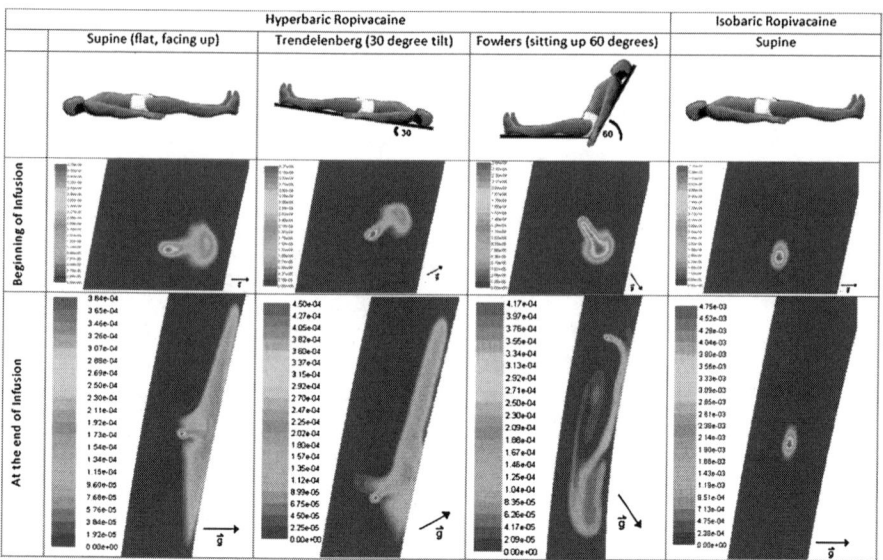

Fig. 6 Intrathecal infusion of hyperbaric and isobaric solutions as a function of posture—supine, Trendelenberg and Fowler's position. After infusion of isobaric drug solution, drugs distribute evenly around the infusion site. Distribution of drugs in hyperbaric solutions is strongly influenced by gravity. Drugs in hyperbaric solutions travel preferentially in dorsal subarachnoid space in supine patients, and accelerate toward the cervical spine in patients in Trendelenburg position. Hyperbaric drug solutions travel towards the sacral spine in patients in Fowler's position. The snapshots of the lower spine of the patient show drug distribution near the infusion site in the lumbar subarachnoid space, and the black line within the model is the boundary between the pulsating CSF and spinal cord. The injection site is between the second and third lumbar vertebrae. The vector indicates direction of gravity in relation to patient posture

injection for 43, 60, 120 bpm reached 1.1, 0.91, and 0.67% (normalized to infusion), respectively. The slowest heart rate (43 bpm) resulted in the highest drug concentration peak after injection.

CSF Pulse Magnitude

Natural variations in CSF pulse magnitudes in the spine can be found between individuals. CSF velocity magnitudes in the spine varies from 1.9 to 5.1 cm/s [14]. In order to determine the effect of CSF stroke volume on drug dispersion, stroke volumes of 1.0, 2.0, 3.0 ml at the cervical spine were investigated with iCFD simulations. Right after the bolus injection, peak concentrations were found near the infusion site (L2) in CSF. For stroke volumes of 1.0, 2.0, and 3.0 ml, peak concentrations in CSF were 1.26, 0.78, and 0.57% (normalized to infusion), respectively (see Fig. 7 right). Inter-patient variability in CSF flow dynamics in the spine such as stroke volume causes differences in drug distribution even for an identical injection protocol.

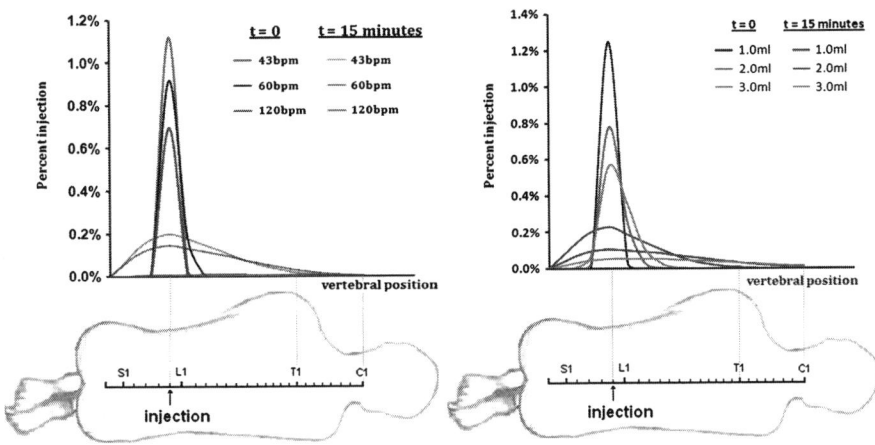

Fig. 7 Intrathecal drug spread as a function of (**a**) different heart rates (*left*) and (**b**) different CSF stroke volumes (*right*) in the cervical spine. **a** shows that right after injection, concentration peaks found near the injection site for CSF pulse frequency 43, 60, 120 bpm are 1.1, 0.91, and 0.67% of the injectate concentration, respectively. Fast dispersion caused by rapid CSF pulsations in the spine leads to lower concentration peaks, **b** shows that intrathecal drug distribution in the spine is a function of CSF stroke volume. Large pulse magnitude causes a broad distribution of drugs along the spine and a rapid drop in peak concentration

These iCFD simulations explain and quantify high inter-patient variability in drug dispersion for intrathecal infusions. The quantification of the effects of heart rate, CSF pulsations and solution baricity enables physicians to optimize IT infusion protocols and reduce risks for their patients.

3.3 Patient-Specific Model of the Cerebrovasculature

Cerebrovascular disorders rank among the leading causes of death in the United States [16]. Therefore, the dynamics of cerebral blood flow and its role in maintaining homeostasis of the CNS is of high clinical relevance. The case study will demonstrate that patient-specific computer models of cerebral vasculature can capture hemodynamic properties of an individual subject. Detailed CFD simulations based on patient-specific models will provide novel insights about normal and pathological cerebrovascular blood flow patterns and assist endovascologists and neurosurgeons in diagnosis and rational design of patient-specific treatments for stroke, atherosclerosis or arterial venous malformation. Computational studies permit a quantitative analysis of cerebral hemodynamics and may lead to fundamental understanding of complex dynamics like autoregulation, functional hyperemia, and pulsatile blood-soft tissue interaction in the brain. This case study describes computational methods to generate a comprehensive physiological model of the entire cerebral vasculature. Rather than limiting the analysis to a few vessel segments as in previous works, this project aims at predicting the hemodynamics of the entire human brain.

3.3.1 Flow Laws and Tree Optimization Determine Vasculature Topology

The main structures of the brain's arterial and venous tree can be reconstructed from medical imaging such as Magnetic Resonance Angiography (MRA) or Digital Subtraction Angiography (DAS). Typical resolution limit still detects successfully vessels with a diameter of 0.1–1 mm. Smaller vessels can be discerned by optical inspection, but are hard to reconstruct with automatic vessel reconstruction algorithms [7]. In the first step of our approach towards generating a complete vascular model for a specific patient, the major extracerebral arteries and veins were reproduced in 3D on a computer from high resolution MRI in combination with image processing tools [27]. We used MRI T1 and T2 images to reconstruct the brain geometry, and co-registered MRA data to segment the main arterial and venous trees [6]. The segmentation captures the subject's main vessels including the carotid, basilar artery, circle of Willis and anterior, middle, and posterior cerebral arteries as well as superior sagittal sinuses and jugular vein.

Small arterioles, micro-capillary bed and venules cannot be reconstructed from medical images for an individual patient. Therefore, step 2 uses automatic vessel generation techniques to complement the computer model with detailed cerebral microvasculature below the medical image resolution. We have deployed automatic growth and space filling algorithms to generate micro-vessels. Arterioles form binary trees which we build with a modified constrained constructive optimization algorithm proposed by Schreiner [32]. In our software entitled *Directed Interactive Growth Algorithm* (*DIGA*), numerous arteriole trees are grown by successively adding a large number of additional segments that spawn from the patient-specific arterial backbone. For each tree, new vessel segments are added so that the location of the bifurcation point minimizes the total blood volume of the entire vascular tree. Vessel diameters of the emerging tree are computed as a side condition of the non-linear optimization problem given in (8). These equality constraints enforce simplified blood flow dynamics and ensuring equal blood perfusion for each terminal segment. The resulting network is an acyclic binary graph with cylindrical arcs. The optimization of the blood volume subject to the flow equations is given in system (8).

$$\underset{\bar{x}}{Min} \quad V = \sum_{i=1}^{N} \pi r_i^2 l_i \tag{8}$$
$$\text{s.t.} \quad \Delta P = \alpha_i F_i \quad \forall \text{arcs on Network}$$

In (8), V is the total network volume, and r_i and l_i are the radius and length of vessel i, respectively. The pressure drop across the network and the outflow of all terminal segments are specified to match physiological perfusion rates.

Table 1 Summary of the components of the vasculature model

Type of vessel	Origin
Major arteries	Generated from medical MRA images: carotid arteries, basilar artery, circle of Willis, cerebral arteries
Arterioles	Synthesized by interactive binary tree space-filling algorithm
Pre-capillary vessels	Generated by fusing capillary mesh with arterioles
Capillaries	Built from patient-specific volumetric mesh using Voronoi tessellation
Post-capillary venules	Generated by fusing capillary mesh with venules
Venules	Synthesized by interactive binary tree space-filling algorithm
Major veins	Generated form MRA medical images: sagittal sinus and jugular vein

Hagen–Poiseuille's Law is used to calculate resistances, α_i, as a function of radii and length segment. Further, Murray's Law establishes anatomical diameter ratios between parent and bifurcating daughter branches. In our interactive tool, the user directs the generation algorithm to grow arteriole and venule trees in specified anatomical regions. These regions distinguish white and grey matter as well as functional sub-domains acquired from patient-specific medial images (Table 1). A user-defined number of terminal segments is grown inside the selected region by repeatedly solving the non-linear mathematical program in system (8).

Step 3 constructs realistic representations of the capillary mesh which cannot be created with the tree-generating algorithms, because the capillary bed does not form binary trees, but mesh-like structures [9]. For capillary meshes, we recommend the Voronoi tessellation with Bezier connectivity to achieve a desired degree of tortuosity. Our technique starts with a volumetric mesh of desired density. We used volumetric meshes created by Delaunay triangulation mentioned in step 1. Then, we construct its dual Voronoi mesh [34]. The capillaries are the arcs of this cyclic graph with a branching factor of three to four. The vessel density vessels per volume, average length, and diameters can easily be manipulated by adjusting the properties of the original tetrahedral volumetric mesh.

Figure 8 (bottom-left) shows an artificially generated network inside a surface mesh of the brain constructed from MRI data. A sample of a micro-vascular mesh using Voronoi tessellation is shown in bottom-right. The optical comparison confirms that these artificially generated structures are topologically similar to real vasculature. The large arteries near the surface of the brain qualitatively match the structure of the cerebral arteries running in the subarachnoidal space close to the cortical surface. The total cerebral vascular network has about 500,000 segments, and its major components are listed in Table 1.

3.3.2 Blood Flow and Pressure Simulations

The artificial cerebral networks generated by fusing patient-specific data with automatic space-filling microvasculature generation also enable the simulation of

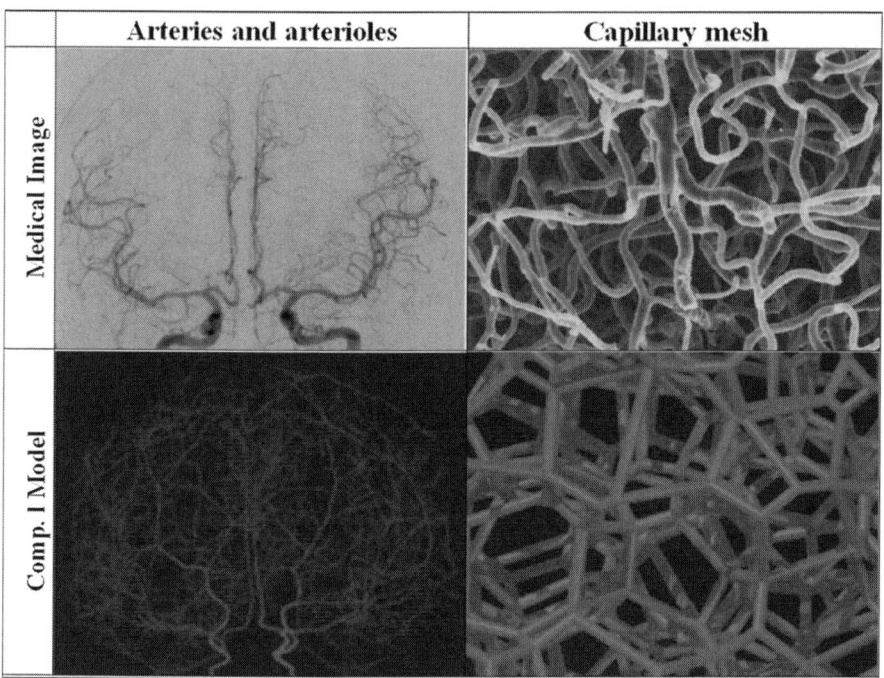

Fig. 8 Patient-specific topology of the cerebrovasculature model. *Top-left* image shows a magnetic resonance angiography picture of a patient's cerebral arteries. A patient-specific reconstructed arterial tree using DIGA is depicted in *bottom-left*. *Top-right* depicts scanning electron microscope image of cortical capillaries. A detailed view of a capillary mesh using three-dimensional Voronoi tessellation is shown at the *bottom right*

blood flow and pressure distribution throughout the cerebral vasculature bed. Flow rates and pressures in the distensible cerebral vasculature network can be computed by the mass and momentum balances given in Eqs. 9 and 10.

$$\frac{\partial A}{\partial t} + A\frac{\partial U}{\partial x} = 0 \qquad (9)$$

$$\frac{\partial U}{\partial t} + U\frac{\partial U}{\partial x} = -\frac{1}{\rho}\frac{\partial p_b}{\partial x} + \frac{8\pi\mu}{\rho}\frac{U}{A} \qquad (10)$$

x is the axial coordinate along a vessel, t is time, $A(t)$ is the cross-sectional area of the vessel, $U(x, t)$ is the average axial velocity, p_b is the average blood pressure over a vessel, ρ is the density, and μ is the blood viscosity [2, 36]. The continuity and momentum equations are coupled with the vessel distensibility equation, Eq. 11, which accounts for dynamic expansion and contraction of deformable blood vessels.

$$m\frac{\partial^2 A}{\partial t^2} + k\frac{\partial A}{\partial t} = Yf(A, A_0) - \Delta p \qquad (11)$$

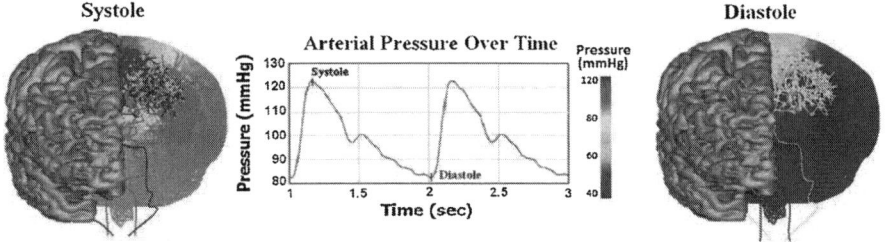

Fig. 9 A surface mesh of the brain was constructed from MRI data and a vascular network was grown using the automatic growth algorithm. Blood pressure simulations of the generated vasculature network are shown for systole (cardiac contraction) and diastole (cardiac recoil). Systolic input and diastolic input are indicated on the pressure waveform

Here, m is the wall mass of a segment, k is a dissipative dampening coefficient, Y is the Young's modulus, and A_0 is the vessel cross-sectional area at zero transmural pressure. Δp is the transmural pressure difference, $p_b - p^*$, where p^* is the intracranial pressure. The change in vessel lumen volume is calculated using $\Delta V = \pi l(r^2 - r_0^2)$, where l, r, and r_0 are the cylindrical vessel length, current radius, and radius at zero transmural pressure, respectively.

A dynamic simulation of the vasculature network was performed using measured pressure waveform of the basilar artery as input. The simulations displayed in Fig. 9 use an inlet pressure of 123 mmHg for the systolic phase of the cardiac cycle, and 85 mmHg for the diastolic phase. The venous pressure was assumed to be a constant at 2 mmHg.

The models we produce offer improvements over existing methodologies through the incorporation of microvasculature and discrete physiological domains. As a step towards understanding cerebral hemodynamics, simulations may be performed producing blood pressure distributions and flow rates throughout the vascular network. Future simulations will explore additional time variant properties of cerebrovasculature and the interaction of vasculature expansion with tissue deformation and CSF acceleration.

4 Conclusion

Disease state manifestation, drug transport speed, anatomical tissue and vascular structures, and blood flow rates in the CNS vary significantly from patient to patient. Tissue anisotropy in the brain is a function of a patient's brain size and anatomy. Identical infusion protocols produce variable results between individuals clinically. Cerebral vascularization is different in each subject. The iCFD methodology captures individual CNS anatomy, blood and CSF flow patterns for patient-specific prediction and therapy design. Three case studies on intraparenchymal and intrathecal drug delivery as well as blood flow predictions demonstrate the state-of-the-art in image-based modeling.

References

1. ADINA: ADINA. www.adina.com (2010)
2. Alastruey, J., Parker, K.H., Peiro, J., Byrd, S.M., Sherwin, S.J.: Modelling the circle of Willis to assess the effects of anatomical variations and occlusions on cerebral flows. J. Biomech. **40**(8), 1794–1805 (2007)
3. ANSYS: ANSYS fluent. http://www.ansys.com (2011)
4. Baledent, O., Henry-Feugeas, M.C., Idy-Peretti, I.: Cerebrospinal fluid dynamics and relation with blood flow: a magnetic resonance study with semiautomated cerebrospinal fluid segmentation. Invest. Radiol. **36**(7), 368–377 (2001)
5. Bobo, R.H., Laske, D.W., Akbasak, A., Morrison, P.F., Dedrick, R.L., Oldfield, E.H.: Convection-enhanced delivery of macromolecules in the brain. Proc. Natl. Acad. Sci. U. S. A. **91**(6), 2076–2080 (1994)
6. Bullitt, E., Aylward, L.: MIDAS—Community Designed Database of MR Brain Images of Healthy Volunteers. http://insight-journal.org/midas/community/view/21 (2011)
7. Bullitt, E., Aylward, S., Smith, K., Mukherji, S., Jiroutek, M., Muller, K.: Symbolic description of intracerebral vessels segmented from magnetic resonance angiograms and evaluation by comparison with X-ray angiograms. Med. Image Anal. **5**(2), 157–169 (2001)
8. Drasner, K.: Spinal anaesthesia: a century of refinement, and failure is still an option. Br. J. Anaesth. **102**(6), 729–730 (2009). doi:10.1093/Bja/Aep085
9. Duvernoy, H.M., Delon, S., Vannson, J.L.: Cortical blood vessels of the human brain. Brain Res. Bull. **7**(5), 519–579 (1981)
10. Fettes, P.D.W., Hocking, G., Peterson, M.K., Luck, J.F., Wildsmith, J.A.W.: Comparison of plain and hyperbaric solutions of ropivacaine for spinal anaesthesia. Br. J. Anaesth. **94**(1), 107–111 (2005). doi:10.1093/Bja/Aie008
11. Fluent Inc.: Fluent. www.fluent.com (2007)
12. Gjedde, A., Reith, J., Dyve, S., Leger, G., Guttman, M., Diksic, M., Evans, A., Kuwabara, H.: Dopa decarboxylase activity of the living human brain. Proc. Natl. Acad. Sci. U. S. A. **88**(7), 2721–2725 (1991)
13. Hall, W.A., Rustamzadeh, E., Asher, A.L.: Convection-enhanced delivery in clinical trials. Neurosurg. Focus **14**(2), e2 (2003)
14. Henry-Feugeas, M.C., Idy-Peretti, I., Blanchet, B., Hassine, D., Zannoli, G., Schouman-Claeys, E.: Temporal and spatial assessment of normal cerebrospinal fluid dynamics with MR imaging. Magn. Reson. Imaging **11**(8), 1107–1118 (1993)
15. Hocking, G., Wildsmith, J.A.: Intrathecal drug spread. Br. J. Anaesth. **93**(4), 568–578 (2004)
16. Hoyert, D., Kung, H.C., Smith, B.S.: Deaths: preliminary data for 2003. Natl. Vital Stat. Rep. **53**(15), 1–48 (2005)
17. Huang, T.Y., Chung, H.W., Chen, M.Y., Giiang, L.H., Chin, S.C., Lee, C.S., Chen, C.Y., Liu, Y.J.: Supratentorial cerebrospinal fluid production rate in healthy adults: quantification with two-dimensional cine phase-contrast MR imaging with high temporal and spatial resolution. Radiology **233**(2), 603–608 (2004)
18. Ibanez, L., Schroeder, W., Ng, L., Cates, J.: The ITK Software Guide: The Insight Segmentation and Registration Toolkit, version 2.4. Kitware, Clifton Park (2003)
19. Li, D., Ivanchenko, O., Sindhwani, N., Lueshen, E., Linninger, A.: Optimal catheter placement for chemotherapy. In: Proceedings of 20th European Symposium on Computer Aided Process Engineering (ESCAPE), pp. 223–228 (2010)
20. Linninger, A.A., Somayaji, M.R., Erickson, T., Guo, X., Penn, R.D.: Computational methods for predicting drug transport in anisotropic and heterogeneous brain tissue. J. Biomech. **41**(10), 2176–2187 (2008)
21. Linninger, A.A., Somayaji, M.R., Mekarski, M., Zhang, L.: Prediction of convection-enhanced drug delivery to the human brain. J. Theor. Biol. **250**(1), 125–138 (2008)

22. Linninger, A.A., Somayaji, M.R., Zhang, L., Hariharan, M.S., Penn, R.D.: Rigorous mathematical modeling techniques for optimal delivery of macromolecules to the brain. IEEE Trans. Biomed. Eng. **55**(9), 2303–2313 (2008)
23. Logan, M.R., McClure, J.H., Wildsmith, J.A.: Plain bupivacaine: an unpredictable spinal anaesthetic agent. Br. J. Anaesth. **58**(3), 292–296 (1986)
24. Materialise Inc.: Mimics. www.materialise.com/mimics (2011)
25. Morrison, P.F., Chen, M.Y., Chadwick, R.S., Lonser, R.R., Oldfield, E.H.: Focal delivery during direct infusion to brain: role of flow rate, catheter diameter, and tissue mechanics. Am. J. Physiol. Regul. Integr. Comp. Physiol. **277**(4), R1218–R1229 (1999)
26. Netti, P.A., Baxter, L.T., Boucher, Y., Skalak, R., Jain, R.K.: Macro- and microscopic fluid transport in living tissues: application to solid tumors. AIChE J. **43**(3), 818–834 (1997)
27. Nowinski, W.L., Volkau, I., Marchenko, Y., Thirunavuukarasuu, A., Ng, T.T., Runge, V.M.: A 3D model of human cerebrovasculature derived from 3T magnetic resonance angiography. Neuroinformatics **7**(1), 23–36 (2009)
28. Raghavan, R., Brady, M.L., RodrÃguez-Ponce, M.I., Hartlep, A., Pedain, C., Sampson, J.H.: Convection-enhanced delivery of therapeutics for brain disease, and its optimization. Neurosurg. Focus **20**(4), E12 (2006)
29. Rodeheffer, R.J., Gerstenblith, G., Becker, L.C., Fleg, J.L., Weisfeldt, M.L., Lakatta, E.G.: Exercise cardiac-output is maintained with advancing age in healthy-human subjects: cardiac dilatation and increased stroke volume compensate for a diminished heart-rate. Circulation **69**(2), 203–213 (1984)
30. Saltzman, W.M., Radomsky, M.L.: Drugs released from polymers: diffusion and elimination in brain tissue. Chem. Eng. Sci. **46**(10), 2429–2444 (1991)
31. Schiffer, E., Van Gessel, E., Fournier, R., Weber, A., Gamulin, Z.: Cerebrospinal fluid density influences extent of plain bupivacaine spinal anesthesia. Anesthesiology **96**(6), 1325–1330 (2002). doi:00000542-200206000-00010[pii]
32. Schreiner, W., Buxbaum, P.F.: Computer-optimization of vascular trees. IEEE Trans. Biomed. Eng. **40**(5), 482–491 (1993)
33. Schroeder, W., Martin, K., Lorensen, B.: The Visualization Toolkit: An Object-Oriented Approach to 3D Graphics, 4th edn. Kitware, Clifton Park (2006)
34. Tawhai, M.H., Burrowes, K.S.: Developing integrative computational models of pulmonary structure. Anat. Rec. B New. Anat. **275**(1), 207–218 (2003)
35. Wildsmith, J.A., McClure, J.H., Brown, D.T., Scott, D.B.: Effects of posture on the spread of isobaric and hyperbaric amethocaine. Br. J. Anaesth. **53**(3), 273–278 (1981)
36. Zagzoule, M., Marc-Vergnes, J.P.: A global mathematical model of the cerebral circulation in man. J. Biomech. **19**(12), 1015–1022 (1986)

Patient-Specific Modeling and Simulation of Deep Brain Stimulation

Karin Wårdell, Elin Diczfalusy and Mattias Åström

Abstract Deep brain stimulation (DBS) is widely used for reduction of symptoms caused by movement disorders. In this chapter a patient-specific finite element method for modeling and simulation of DBS electric parameters is presented. The individual's stereotactic preoperative MR-batch of images is used as input to the model in order to classify tissue type and allot electrical conductivity for cerebrospinal fluid, blood and grey as well as white matter. With patient-specific positioning of the DBS electrodes the method allows for investigation of the relative electric field changes in relation to anatomy and DBS-settings. Examples of visualization of the patient-specific electric entities together with the surrounding anatomy are given. The use of the method is exemplified on patients with Parkinson's disease. Future applications including multiphysics simulations and applicability for new DBS targets and symptoms are discussed.

1 Introduction

Deep brain stimulation (DBS) has become one of the most important brain stimulation techniques for clinical use [7]. The application of DBS for movement disorders such as Parkinson's disease, essential tremor and dystonia, is currently also expanding toward other diseases and symptoms such as epilepsy, Gilles de la Tourette's syndrome, obsessive compulsive disorders and other psychiatric illnesses. Research on DBS is performed in many centers and several thousand scientific papers have so far been published [16].

K. Wårdell (✉) · E. Diczfalusy · M. Åström
Department of Biomedical Engineering, Linköping University, Linköping, Sweden
e-mail: karin.wardell@liu.se

The clinical result from DBS is very dependent on the anatomical placement of the electrode in the brain, and thus the surgical implantation procedure as well as the stimulation parameter settings of the device. As DBS has become more commonly used, there are also an increasing number of reports of postoperative adverse events including speech disturbances, depression, mood changes and behavioral problems [14]. Such side disorders are often associated with electrode misplacement or the stimulation settings in relation to the pre-defined target area selected for electrode implantation. Since the mechanism of action of DBS is not fully understood, it is sometimes difficult to define the optimal target area in the brain which is related to a specific symptom. Increased knowledge of the mechanism of action and thus improved understanding of the therapeutic effect from DBS can be gained through different techniques. Examples are recording of the neural response to stimulation, anatomical and functional imaging, biochemical measurements of neurotransmitters with e.g. microdialysis and simulations of electrical modalities around the electrode.

Several research groups have used the finite element method (FEM) in order to develop computer-models of DBS electrodes and to set up simulations of the electric field around the electrode [3, 15, 21, 32]. The first generation of DBS-models were used to visualize the concept of electric parameters for different DBS stimulation settings, pre-selected target area and anatomical structures of the brain [3, 15, 21]. FEM has also been used to investigate the electrode-brain interface [31], the axonal tissue around deep brain structures directly activated by DBS [21] and the influence on the electric field from cerebrospinal (CSF) filled cystic cavities [3]. The latest modeling concepts are, however, patient- and treatment specific i.e. based on the individual's own anatomy and DBS electrode settings as input [6, 9, 28]. The activated tissue volume around the DBS electrode when positioned in the globus pallidus internus (GPi) has been studied by Vasques and colleagues [28]. Butson and co-workers used FEM for prediction of the volume of tissue activated based on diffusion tensor imaging (DTI) and an axon model [9]. This concept has been used in order to evaluate verbal influence from stimulation when the DBS electrode was positioned in the subthalamic nucles (STN) [22]. Patient-specific FEM simulations have also been used by our group in order to investigate the relationship between electrical field and both movement and speech intelligibility in patients with Parkinson's disease stimulated in the STN [5].

As the number of DBS implantations are constantly increasing and each implantation procedure requires accurate, precise and safe targeting of the brain structure for optimal clinical outcome, the technical aids for improving the DBS implantation and patient follow-up have become increasingly important [16]. A major challenge is therefore to develop tools that helps optimize the surgical procedure, the postoperative follow up, the patient safety, and also help reducing the total costs for health care. In this perspective, patient-specific modeling, simulation and visualization have the potential to become a useful tool for improvement of the surgical planning and the post-operative follow up.

In this chapter we present a patient-specific method for investigation of the DBS-electric distribution in relation to individual stimulation setting and brain anatomy by means of FEM modeling and simulations. Our approach [2, 6] uses the individual patient's preoperative batch of images in order to classify the electrical conductivity for different tissue types. With the help of treatment-specific positioning of the DBS electrode in relation to the postoperative MR images, the method allows for investigations of the relative electric field changes in relation to anatomy and DBS-settings. Visualization of the electrical entities together with anatomical structures can be done in numerous ways. Examples of visualization in 2D and 3D are given in this chapter. As it is of importance with knowledge of DBS and the surgical implantation procedure for setting up models and simulations, the chapter start with a review of these features.

2 Deep Brain Stimulation Systems

The most commonly used DBS electrodes (Fig. 1) have four contacts with a diameter of 1.27 mm and contact length of 1.5 mm and an intercontact distance of 0.5 mm or 1.5 mm (Medtronic's leads and electrode models 3389 and 3387, Medtronic Corporation). The stimulation waveforms are produced by an implantable pulse generator, for example, Soletra® is used for unilateral stimulation and Kinetra® for bilateral stimulation. The voltage is commonly set at a value between 1–5 V, the frequency between 120–185 Hz, the pulse width between 60–200 μs and the stimulation mode to mono- or bipolar. Each contact can be used as anode or cathode in bipolar electrode configuration or as cathode in the monopolar setting. If monopolar mode is used, the pulse generator case is used as anode. After placement of the electrode in the predefined target area the battery-operated pulse generator is implanted under the skin below the clavicle and connected to the electrode through a lead extension.

The Medtronic DBS-system was approved by the Food and Drug Administration as a treatment for tremor in 1997, for Parkinson's disease in 2002 and for dystonia in 2003. Currently several companies and research groups are developing new neurostimulation devices [25]. Among these, St. Jude Medical Incorporation received the European CE mark approval of the Libra® and LibraXP^{TM} DBS systems for treating symptoms of Parkinson's disease and the first clinical implantation was done in Europe in 2009. Boston Scientific Corporation recently initiated marketing the Vercise™ DBS System. The first implantations were done in Europe, in the end of 2010 on patients with Parkinson's disease. Another new approach is Steering Brain Stimulation (Sapiens Steering Brain Stimulation BV, The Netherlands). Their electrode uses 64 contacts that can be configured individually to provide tailored stimulation much beyond the capability of the probes that are commercially available today [20]. The electrode has so far been evaluated in animal studies. First implantations in patients are expected in the near future.

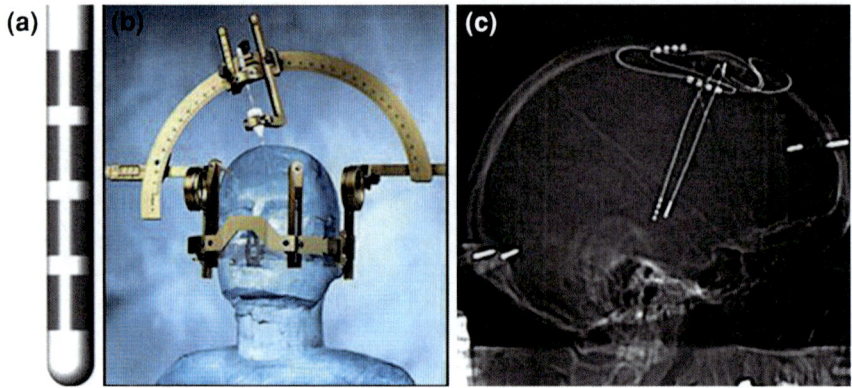

Fig. 1 a Medtronics DBS electrode 3389. b Leksell® stereotactic system (With permission from Elekta Instrument AB). c CT image with implanted DBS electrodes (With permission from Patric Blomstedt, Umeå University Hospital)

3 DBS Targets

The target area for implantation of the DBS electrode is preselected depending on the symptom that should be reduced, and is usually only slightly larger than the DBS electrode itself. This makes the positioning of the electrode crucial. The targeting, together with optimization of the stimulation parameters are therefore of utmost importance for an effective clinical outcome with minimal side-effects.

Structures deep within the brain such as the thalamus or the basal ganglia are commonly used as target areas (Fig. 2). One of the most common targets is the subthalamic nucleus (STN) [8]. It is used for reduction of symptoms elevating from Parkinson's disease and essential tremor. Targeting the GPi is preferred by many groups for symptoms caused by dystonia [29] but also frequently used for Parkinson's disease. Recent research shows that precise targeting of a subsection of these nuclei may be more efficient for reduction of well defined symptoms. For example, the STN being only approximately 240 mm^3 ($8 \times 6 \times 5$ mm^3) is believed to be organized into motor, limbic and associative functional portions [13].

DBS is currently also explored in a range of new target areas and symptoms, many of these related to various psychiatric disorders such as Gilles de la Tourette syndrome, obsessive–compulsive disorder, schizophrenia. Others are related to epilepsy, cluster headache and Alzheimers disease. Finding the best target related to a specific symptom or disorder is a delicate task which is a matter for intensive research at many clinics worldwide.

In order to explore new target areas for DBS and to perform postoperative follow-up on already implanted individuals, patient-specific modeling, simulation and visualization of DBS electric parameters has great potential to become a new tool that will help increase the understanding of the relation between DBS parameter settings and the clinical outcome.

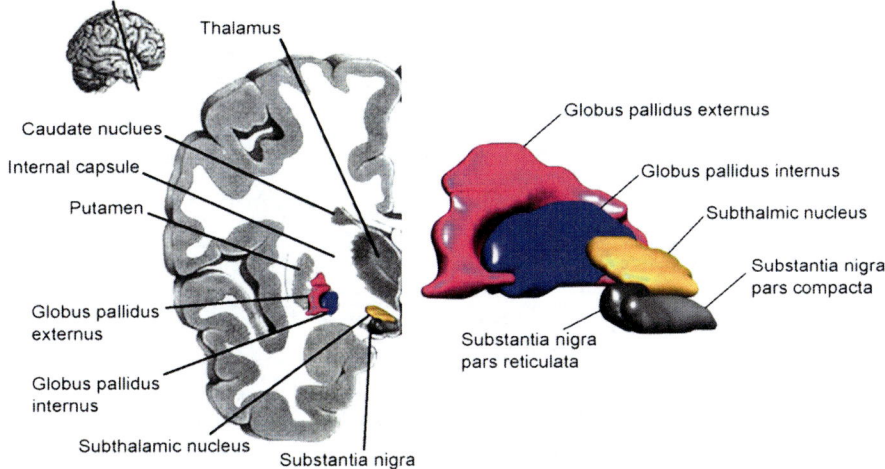

Fig. 2 Posterior view of the basal ganglia excluding the striatum. Image from [2]

4 Surgical Implantation of DBS Electrodes

Stereotactic technique is used in order to reach the target area in a safe way with high accuracy and precision during DBS surgery. The procedure can be distinguished according to preoperative planning, surgical implantation and postoperative follow-up (Fig. 3). A short review of the different steps with emphasizes to input parameters related to the modeling and simulation are presented in this section. A more detailed description of the surgical procedure and supporting techniques is presented in [16].

4.1 Preoperative Planning and Imaging

The preoperative phase is used for planning the trajectory and target coordinates and is highly dependent on high quality imaging of the brain with MRI and/or CT. The most common procedure is to use a frame based stereotactic system for the intervention. Before the imaging is taking place the stereotactic frame is firmly attached to the patient's head together with an indicator box. The indicator box produces reference points (fiducials) in the image batch which are used for the transformation of the target co-ordinates to the co-ordinates of the stereotactic system. The fiducials are also essential for setting up the models. Some centres acquire stereotactic MRI the day of implantation just before surgery while others perform the MRI some days before the implantation to be co-registered with a stereotactic CT of the day of surgery (Fig. 3). MRI sequences used can be T1, T2

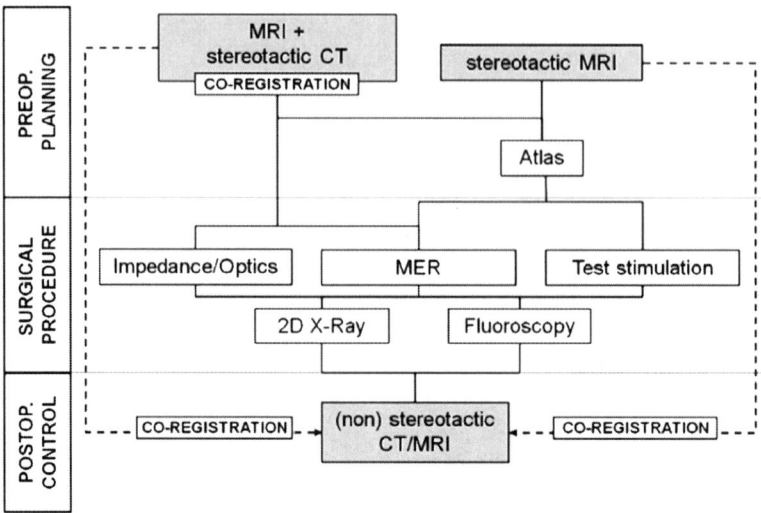

Fig. 3 Surgical procedure and current supporting techniques. Images nessecary as input for modeling and simulations are highlighted in grey. Modified from [16]

or proton density weighted and is depending on the target of interest. For setting up patient-specific models and simulations MRI is preferred before CT as it makes tissue segmentation easier. The surgical planning including calculation of target co-ordinates and trajectory is usually performed with commercially available stereotactic software such as SurgiPlan (Elekta Instrument AB, Stockholm, Sweden) iPlan (BrainLab AG, Munich, Germany) or Framelink (Medtronic Incorporation, Minneapolis, MN, USA).

Some clinics do the planning of entry point, trajectory and target area directly in the captured images while other users prefer to superimpose an anatomical brain atlas onto the MRI in order to improve the identification of the target aimed at. Modern versions of both the Schaltenbrand-Wahren and Talairach atlases are available for computer use [24] and sometimes integrated in the surgical planning systems. Åström, [2, 5] recently developed and used a 3D atlas based on Morels anatomical atlas [23] (Figs. 2 and 7b). Such 3D atlases help to increase the intuitive understanding of targets and their relation to the anatomy even more during the preoperative planning. They are also useful in the interpretation of the interaction between anatomy and simulated electrical parameters.

4.2 Surgical Implantation

As a next step the electrode implantation itself takes place. If the procedure requires the patient's feedback for specific testing, surgery is done with local anesthesia. After opening of the skull the probe can be inserted directly towards the pre-planned target by means of a guide. However, in order to reduce the effect from brain shift and thus fine tune the targeting, the electrode insertion procedure is usually companioned with intracerebral measurements. Microelectrode recording is the most commonly used technique and allows for neural activity studies in up to five trajectories surrounding the target area [12]. The final trajectory is then decided based on the results of the recordings. Both impedance measurement [33] and optical measurements [17, 30] are done along the pre-calculated trajectory using a guide with incorporated sensor for online feed-back. The feature of the optical technique includes both microvascular blood flow and tissue type discrimination along the trajectory. During surgery some clinics also perform intraoperative X-ray or fluoroscopy in order to check the electrode position. If the patient is awake the first testing of the DBS device is done before the probe is fixated. As a final step the pulse generator is implanted together with leads which connect it to the electrode.

4.3 Postoperative Follow-Up

Following the implantation the final electrode position and the absence of haemorrhage is controlled by means of CT or MRI, with or without the stereotactic frame. The electrode position will appear as electrode artefacts which in general hide parts of the anatomical structures. Therefore image co-registration is often part of the quality control as it allows a comparison between the planned and the final electrode position. The electrode artifacts in the postoperative images are also used for localizing the correct electrode position(s) when setting up patient-specific models. These positions are then transferred to the preoperative batch of MRI. A description of the procedure is presented under 7.5.2.

Post-operative follow-up also includes regular consultation with the patients. At these sessions the stimulation settings are fine tuned. For some disorders e.g. Parkinson's disease and essential tremor the response to stimulation is immediate and initial tests can be done already at surgery. For other disorders such as dystonia it can take months to find the best fit of stimulation parameters i.e. stimulation with optimal clinical effect and minimal adverse effect. During these session simulation of the electric field can be useful for helping optimising the electrode setting and thus the clinical effect.

5 Patient-Specific Models and Simulations of DBS Electric Field

Setting up patient-specific models and performing simulations include a number of steps. First, a *brain tissue model* of the brain region of interest is created based on the preoperative patient images. Each voxel in the tissue model is assigned an

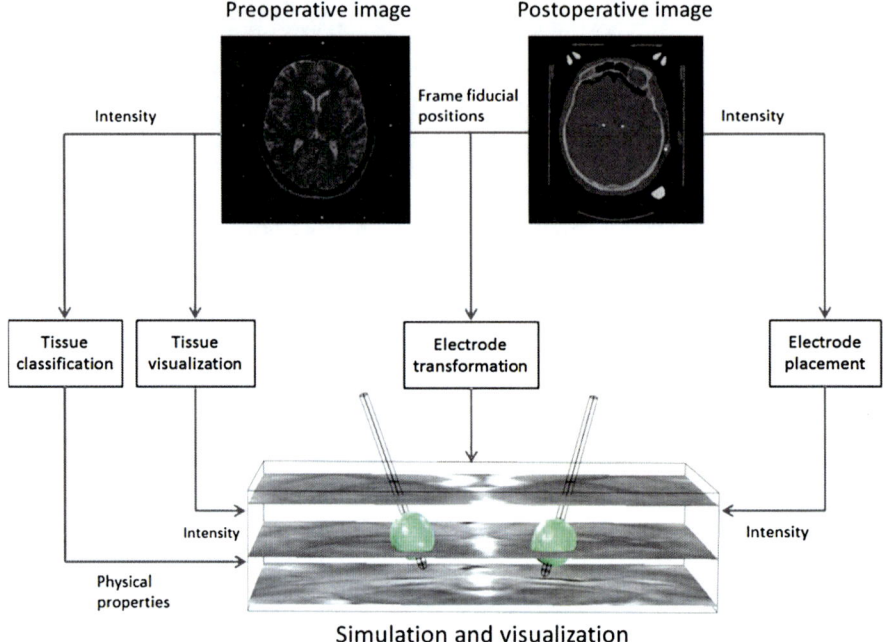

Fig. 4 Summary of the steps for setting up models and simulations of DBS-electric fields

electric conductivity value depending on the tissue type. Next, an *electrode model* of the DBS electrode used is set up based on its actual dimensions. The modelled electrode is positioned in the brain tissue model and assigned suitable electrical parameters. The patient-specific FEM simulation is then carried out based on the equation for steady currents in order to calculate the distribution of electrical parameters around the electrode. As a last step, the parameters of interest are visualized together with the preoperative MRI. An overview of the process is shown in Fig. 4.

5.1 Brain Tissue Model

The preoperative stereotactic images are used as the starting point for creating the patient-specific brain tissue model. Either MRI or DTI can be used as input to the brain tissue model.

When using MRI as input, the quality should be good enough to distinguish between grey matter, white matter, blood vessels and CSF. T1 or T2 weighted 1.5 Tesla MR images are preferred as input data, since these modalities allow intensity-based identification of all of the above mentioned tissue materials. Proton density weighted images can be used, but must be co-registered with either T1 or

Table 1 Electrical conductivity values for different brain matters at 130 Hz [1]

Brain matter	Electrical conductivity (S/m)
CSF	2.00
Grey matter	0.09
White matter	0.06
Blood	0.70

T2 weighted images for identification of CSF. The MRI-batch of preoperative images or co-registered images is used for intensity-based segmentation of the tissue in order to identify grey matter, white matter, CSF and blood vessels.

In the next step, a matrix is created where the intensity values are replaced with the corresponding tissue conductivity values for the DBS stimulation frequency used. For MRI voxels containing more than one tissue type, a linear interpolation function is used to allot an approximated electrical physical property value. The electrical conductivity property values are obtained from an online database by Andreuccetti [1]. Typical values for the stimulation frequency 130 Hz are presented in Table 1. As seen, the electrical conductivity is significantly higher in CSF than in other tissue types. This emphasizes the importance of segmentation of especially CSF-filled cystic cavities (Virchovs Robin spaces) in the deep brain structures before continuation with the simulation step [3]. An example of an MRI based brain tissue model is presented in Fig. 5b together with the original MR-image (Fig. 5a).

When DTI is used as input a different concept is nessecary in order to set up the brain tissue model. Brain tissue may be anisotropic in regions of myelinated fibre bundles. Tuch and colleagues [27] suggested that DTI can be used to non-invasively calculate anisotropic electrical conductivity tensors in patient-specific brain anatomy. Using this concept, the diffusion tensors can be represented by a symmetric positive definite 3×3 matrix, D:

$$D = \begin{bmatrix} D_{xx} & D_{xy} & D_{xz} \\ D_{xy} & D_{yy} & D_{yz} \\ D_{xz} & D_{yz} & D_{zz} \end{bmatrix} \quad [m^2 \, s^{-1}]$$

where the subscripts describes each direction. Tuch and collegues also showed that the diffusion and electrical conductivity tensors may be linearly related by:

$$\sigma = \frac{\sigma_e}{d_e} D \quad [S \, m^{-1}]$$

where σ is the electrical conductivity tensor, σ_e is the effective extracellular electrical conductivity, d_e is the effective extracellular diffusivity. The ratio of σ_e/d_e have been empirically derived to e.g. 0.844 S s mm^{-3} [27]. As a last step in the creation of the tissue model, each electrical conductivity value or tensor is transferred to the co-ordinate corresponding to the original location in the preoperative image.

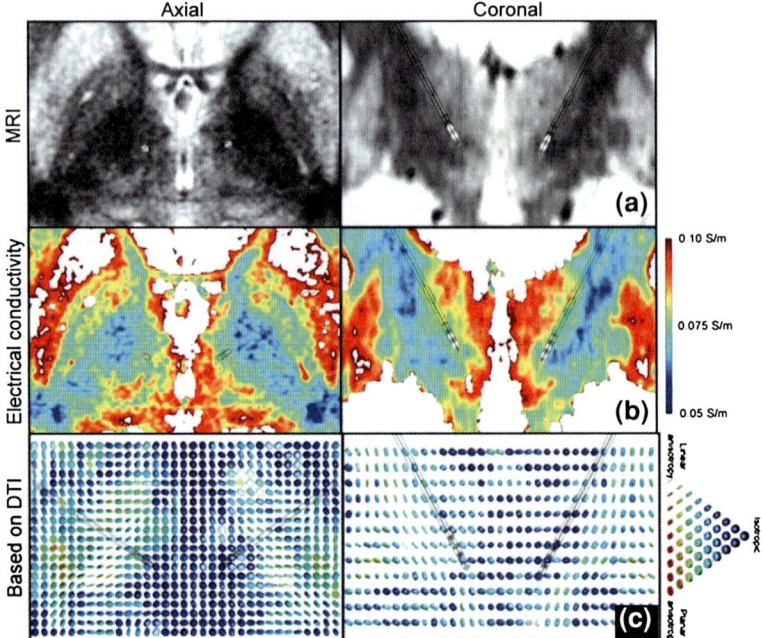

Fig. 5 Axial and coronal views of preoperative MRI (**a**) and with electric conductivity maps based on MRI displayed with colour-maps (**b**) and DTI displayed with superquadric glyph (**c**). The electrodes are positioned in the STN

The calculated anisotropic tissue conductivy can be visualized with superquadric glyphs [18]. A software tool was developed for visualization of diffusion tensors with superquadric glyphs together with axial and coronal MRI and DBS electrodes [2]. Figure 5c presents an example of a DTI brain tissue model.

5.2 Electrode Model

Electrode models, based on their actual geometrical dimensions, are predefined and set up in the FEM software. Examples of two different predefined electrode models (Medtronic 3389 and St Jude 6149) are presented in Fig. 6. In order to position the electrode models in the real target site, the electrode artefacts in the postoperative image batch are used. A second finite element model is set up, based on the post-operative images, where the electrode models are positioned at the centre of the artefacts. A transformation matrix is then created based on the positions of the fiducials in the pre- and postoperative images. The transformation matrix is applied to the electrode models, in order to place them at their accurate positions in the brain model.

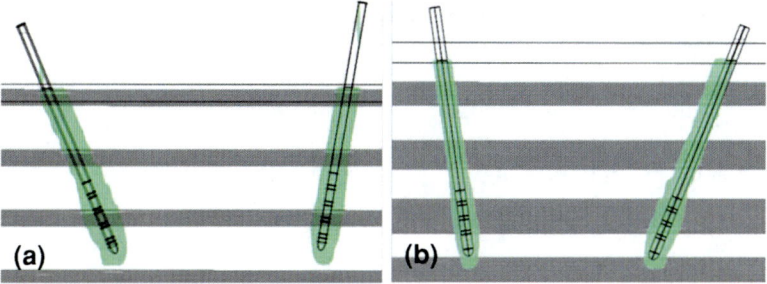

Fig. 6 Example of patient-specific positioning of electrode models with surrounding artefacts as seen in the postoperative images. **a** Medtronics electrode 3389. **b** St Jude's electrode 6149

5.3 Simulation

The equation for steady currents is used as the governing equation for calculation of the electric potential distribution in the vicinity of the electrodes [10]:

$$\nabla \cdot \bar{J} = -\nabla \cdot [\sigma \nabla V] = 0 \quad [\text{A m}^{-3}] \tag{1}$$

where $\bar{J}$ is the current density [A m^{-2}], σ the electrical conductivity [S m^{-1}], and V the electric potential [V]. Monopolar stimulation is simulated by setting the outer boundaries of the tissue model as anode, and the active electrode contact of each electrode as cathode. To mimic bipolar stimulation, one active contact of each electrode model is used as cathode, and the other as anode.

The model is solved using one of COMSOL Multiphysics pre-defined system solvers. The domains, including the brain model and the electrode models, are divided into a large number (usually ~2,000,000) of tetrahedral mesh elements with the highest mesh density close to the electrodes. The maximum element length is set to a suitable value, e.g. 1 mm, in order to retain the spatial resolution of the preoperative MRI. In a similar manner the simulation is performed when DTI is used as input, but due to the 3 × 3 tensor the calcuation time is increased.

5.4 Software and User Interface

A software tool (ELMA 1.0) developed in MatLab 7.6 (The MathWorks, USA) is used to create a patient-specific anatomical property matrices based on the pre- and postoperative MR-images. It makes all necessary preparations for setting up the simulations. A graphical user interface allows the user to choose which images to include, to define the anatomical region of interest, to perform image segmentation for allotation of the electrical conductivities, and to extract the transformation matrix for transfer of the electrode models to the simulations software. Finite element software (COMSOL Multiphysics, Comsol AB, Sweden) is then used for setting up the brain tissue models and electrode models, as well as running the FEM simulations. In this software electrode settings such as electrical potential, stimulations mode (mono- or bipolar) and choice of active contacts are done. The procedure is done using a conventional laptop and require for the moment approximatley 2–3 h depending on type of simulations performed.

6 Visualisation

Visualization is central for improving the interpretation of the results from patient-specific modeling and simulations of DBS. Various electrical entities as well as anatomy may be visualized for different rationales. The electric field, which is the first derivative of the electric potential, is a general property often visualized during simulations of DBS. Such visualizations are preferably done together with the anatomy in 2D or 3D.

An example of a patient-specific simulation of bilateral DBS implantation in STN is presented in Fig. 7 (left electrode: 4 V, contact 1; right electrode: 3.5 V, contact 7). In order to compare the electric field with the patient's own anatomy we have chosen to delineate the outer part of the electric field at the isolevel 0.2 V/mm, a border previously suggested by [15]. By using a fixed isolevel it is possible to make relative comparisons between simulations performed in the same patient but with different voltage or contact settings. This is often the case during evaluation of the clinical effect following various DBS system settings. The anatomical 2D structures as extracted from Morels atlas [23] are superimposed onto the MRI and electrical field images in Fig. 7a, c–d.

To further facilitate the interpretation of DBS simulations a 3D atlas has been created of the human thalamus and the basal ganglia based on axial slices of the stereotactic atlas by Morel [23]. The patient-specific anatomy can then be visualized in 3D together with patient MRI on both axial and coronal slides. The 3D atlas is created by segmentation of the axial atlas slices. Three dimensional surface objects of each structure and fibre path are generated by combining several aligned images. Each object is then filtered by 3D smoothing in order to remove rough edges and corners. In this way, depending on the positioning of the electrode, structures such as sub-nuclei of the thalamus and basal ganglia can be manually co-registered with patient MRI for improving the anatomic information in 3D. An example of a 3D atlas of a few relevant structures and fibre paths during DBS is presented in Fig. 7d. The DBS electrode and electric field is here animated and thus, not patient-specific.

When DTI is used as model input the visualization becomes more complex as the anisotropy (superquadatic glyphs) need to be presented together with the simulated field, and preferably also together with the preoperative MRI. An example is shown in Fig. 8. In this example two fixed isolevels are used 0.2 V/mm and 0.05 V/mm in order to make comparison with visualisation of the superquadratic glyphs easier.

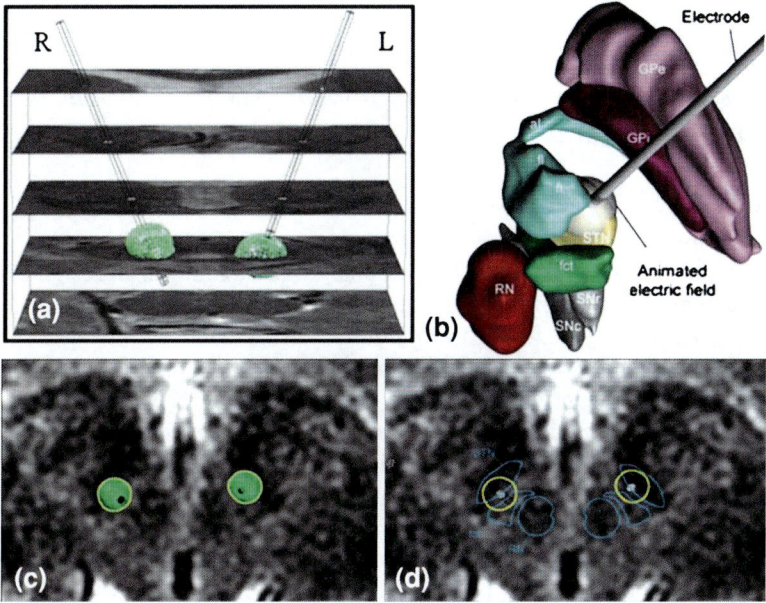

Fig. 7 a Electrodes positioned in the STN with simulated electric field with isolevel 0.2 V/mm presented with MRI. **b** Superior view of an anatomical structures together with animated DBS electrode and surrounding electric field. **c** Isolevels traced in yellow. **d** An anatomical atlas superimposed on the MRI together with traced isolevels

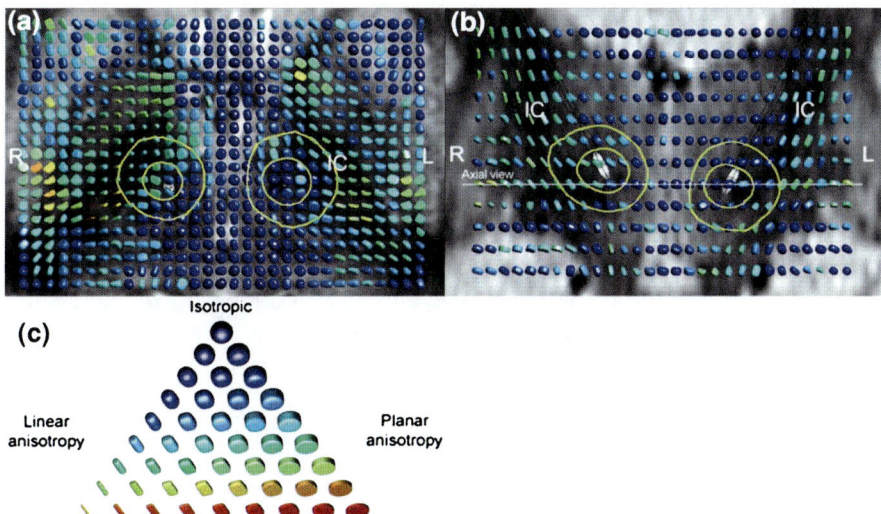

Fig. 8 Example of patient-specific visualization of electric field simulation using DTI as input. **a** axial and **b** cornal slice. The anisotropy in tissue is presented by the superquadratic glyps superimposed on the preoperative MRI. The electric fields are visualised with two isolevels 0.2 V/mm (*inner*) and 0.05 V/mm (*outer*) circles. Image from [2]

7 Clinical Examples of DBS Simulations

7.1 Parkinson's Disease: Simulations as Aid for Clinical Assessment

In the following an example of a patient with Parkinson's disease suffering from stimulation-induced speech impairments is presented [5, 26]. The patient underwent bilateral DBS surgery based on stereotactic T2-weighted 1.5 Tesla MRI with sequences enabling visualization of the STN. A postoperative fast spin echo T2-weigthed MRI was used for visualisation of the actual electrode position and consequently also used for setting up the models for the DBS-simulation. The electrode contacts located closest to the centre of the STN were used as active contacts. In order to identify these the FrameLink Planning Station^{TM} (Medtronic, Minneapolis, MN, USA) was used. Simulations were set up based on the pre- and postoperative MRI images and the DBS implanted electrode (Model 3389, Medtronics Inc. USA). In order to be able compare simulations of various stimulations settings, the result was presented as 0.2 V/mm isolevel superimposed on the axial and coronal preoperative MRI together with Morels 2D atlas.

Clinical assessments of speech intelligibility and movement were carried out during monopolar stimulation in the STN with an electric potential of 0, 2, and 4 V. Movement was evaluated using the Unified Parkinson's Disease Rating Scale part three (UPDRS-III) and speech by three sustained vowel phonation of "ah" and a 60 s monologue. The electric field was simulated for 2 and 4 V settings with frequency and pulse length set at 130 Hz and 60 μs respectively. This corresponded to the settings used during the clinical assessments (Table 2).

The patient suffered from acute stimulation-induced impairment of speech intelligibility during the high potential (4 V). Visualisation of the simulated electric field in relation to anatomy showed that the active electrode contacts were positioned slightly ventral, posterior and medial to the centre of the STN and that the electric field isolevel covered a major part of the fasciculus cerebello-thalamicus (fct) (Fig. 9). Movement as measured by the UPDRS-III was improved during both potential settings compared to off stimulation. In this particular patient the motor score was surprisingly the same during both high and low stimulation. A complete presentation of the study including ten patients is presented in [5].

7.2 Parkinson's Disease: Multiphysics Simulations

Modeling and simulations of DBS can also be done in a multiphysics approach. An example is given were patient-specific simulations of the electric field around DBS electrodes are combined with modeling and simulations of the tissue volume of influence around microdialysis probes. Equation (1) and a modified version of Fick's diffusion equation were implemented [11].

Table 2 Electrical settings and clinical effects

Left side contact (potential)	Right side contact (potential)	Speech intelligibility	UPDRS-III	Side-effects
0 (2 V)	5 (2 V)	70%	33	None
0 (4 V)	5 (4 V)	20%	33	Dysarthria

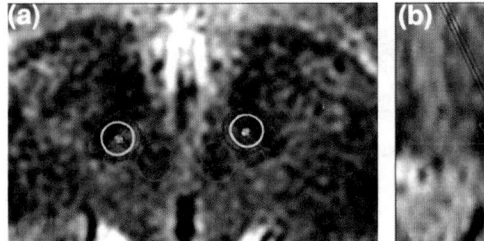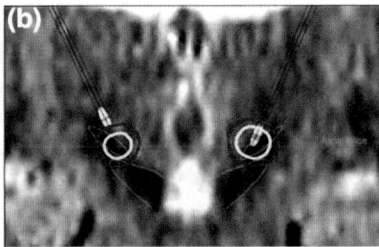

Fig. 9 a Axial and **b** coronal view of the electric field during electrical settings that induced dysarthria (*red*) and that did not induce dysarthria (*white*). The approximate boundaries of the red nucleus, the fct and the STN are traced with blue colour. Electric fields are presented with isolevel 0.2 V/mm

Stereotactic T2-weigthed 1.5 Tesla MRI was used for target identification. Bilateral implantation of DBS electrodes were done in the STN. In addition, microdialysis probes were implanted in the putamen (Put, right side) and the GPi (left and right side). This made monitoring of neurotransmitters such as dopamine, glutamate and serotonin in relation to changes of the DBS-voltage settings possible. Post-operative CT was used as identification of the electrode artifacts and thus positioning of the DBS electrodes for setting up the models. In a similar way, a small goldthread in the microdialysis probe tip was used to identify the anatomical position, and thus used for setting up patient-specific modeling and simulations of the tissue volume of influence around the membrane. Instead of identifying the electrode artifacts with help of the surgical planning system, the developed software (ELMA 1.0) was used.

An examples of a multiphysics simulation is presented in Fig. 10. An isolevel of 0.2 V/mm was used for presentation of the electric fields around the DBS electrodes positioned in the STN. The maximum tissue volume of influence (TVI_{max}) for dopamin was presented using a concentration level at $0.01*C_0$ nmol/L, where C_0 is the initial dopamin concentration in the tissue [11]. Patient-specific evaluation of microdialysis data in relation to the TVI_{max} and the electric field extensions is currenlty in progress.

8 Discussion and Conclusions

A method for patient-specific modeling and simulation of deep brain stimulation has been presented. By using MRI as input and by positioning the DBS electrodes at their real position it is possible to perform simulations on a patient-specific

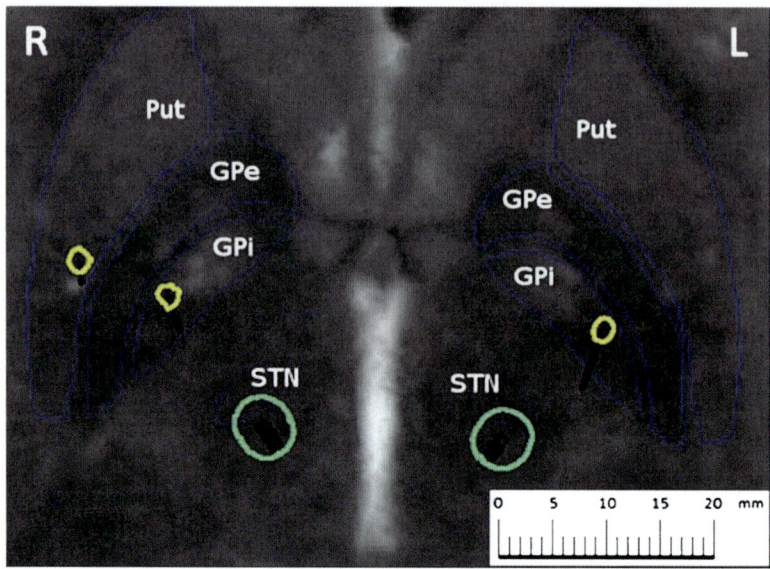

Fig. 10 Example of a patient-specific multiphysics simulation superimposed on an axial preoperative MRI slice. Electric field simulations are presented with isolevel 0.2 V/mm around the DBS electrodes positioned in the STN, and microdialysis TVI_{max} with a concentration isolevel of 0.01 C_0 around the three cathethers

basis. Simulations can be done for various frequency and voltage settings and also for various electrode models.

Our concept, based on electrical conductivity setting from MRI segmentation of grey and white matter, blood and CSF, makes the modeling and simulation easy to use and to adapt for new imaging concepts e.g. 3T MRI, now very quickly introduced at many neurosurgical clinics. 3T MRI will improve the simulation quality compared to 1.5T, due to more efficient image segmentation. Especially CSF-filled cysts common in structures such as the GPi, are expected to be visible. CSF, with its high conductivity has a great influence on the electric field shape and extension, a fact shown by [3]. An example was also presented were a DTI was used to set up the brain tissue model in the preparation for the simulations. DTI takes the tissues anisotropy into account, which may be important if the DBS electrode is positioned in or in the vicinity to white matter tracts. As long as DBS electrodes are positioned in grey matter, as is the case when aiming at the deep brain structures, the anisotropy is of less significance. As many electrodes are misplaced they may as well end up closely to white matter, and the anisotropy will then be important to take into account. The influence on simulation results from various image types (DTI, T2-MRI and pure homogenous) as model input has been studied by us [4]. These investigations show that the influence of heterogeneous and anisotropic tissue on the electric field may have a clinical relevance, especially in anatomic regions that are functionally subdivided and surrounded by multiple fibres of passage. However, at this point patient-specific models based on either MRI or DTI data appear suitable for presenting a rough estimation of the electric field.

In this chapter different exemples of visualisations have been given. It must be stressed that it is important to present the final simulation results together with the individual anatomy. One example commonly used by us is to superimpose the electric field on the preoperative axial or coronal MR-images. By using a fixed isolevel it is possible to do relative comparisons for instance if the DBS settings are changed during clinical assessments. Alternatively the electric potential or the activating function can be presented. Such visualisations are possible, and have been used by e.g. McIntyre et al., [9, 19, 21] and are also exemplified by Åström [2, 6]. In addition, it is very important to have in depth knowledge of the deep brain structures in the interpretation of the result. Therefore a 3D brain atlas was developed. This atlas is so far only applicable as a general guide, as the electric field simulation presented is animated (Fig. 7b). A future step in improvning the visualisation would be to integrate the presentation of patient-specific field simulations with the 3D atlas.

Along with the DBS electrode development also the number of pre-selected electrodes must be increased. Two types of Medtronics electrodes (3389 and 3387), as well as two type of St Jude electrode (6149, 6145) have already been implemented. A next step could be implementation of other suggested DBS-configurations proposed by e.g. Boston Scientific and the Sapiens. However, along with such introductions the software also needs to be modified to the specific simulation mode suggested by the manufactur i.e. to both include current and voltage control of the electrodes as well as feasable settings of stimulation parameters.

Already today, the simulation settings can relatively easy be adapted for other clinical applications and studies. So far we have used it on patients with Parkinson's disease where DBS electrodes have been positoned in the STN. We are currently adapting the model for use on patients with dystonia with DBS electrodes positioned in the GPi. Other interesting applications are Gilles de la Tourettes syndrome. A disorder for which the optimal target area is still under investigation. In order to explore new target areas for DBS and to perform postoperative follow-up on already implanted individuals, patient-specific modeling, simulation and visualization of DBS electric parameters has a great potential to become a new clinical tool that can both help increase the understanding of the relation between DBS parameter settings and the clinical outcome. A vision for the use in tomorrow's medicine would be a flexible, user friendly software that helps the clinician in both the preoperative planning phase and the postoperative follow up by simulating and presenting electrical fields and other entities in relation to the anatomy in a flexible way.

Acknowledgements The authors would like to thank the clinical collegues at the Unit of Functional Neurosurgery, London University Collegue and at the Division of Neurology and Neurosurgery at Linköping University Hospital for very valuable input and discussions during the development of the software. The work was financially supported as a group grant (311-2006-7661) by the Swedish Foundation for Strategic Research (SSF), the Swedish Research Council (VR) and the Swedish Governmental Agency for Innovation Systems (VINNOVA).

References

1. Andreuccetti, D., Fossi, R., Petrucci, C.: Dielectric Properties of Body Tissue. Italian National Research Council, Institute for Applied Physics, Florence, Italy (2005). http://niremf.ifac.cnr.it/tissprop/
2. Åström, M.: Modelling, simulation and visualisation of deep brain stimulation. Linköping studies in science and technology Dissertation, No. 1384, Department of Biomedical Engineering, Linköping University (2011). http://urn.kb.se/resolve?urn=urn:nbn:se:liu:diva-70090
3. Åström, M., Johansson, J.D., Hariz, M.I., Eriksson, O., Wårdell, K.: The effect of cystic cavities on deep brain stimulation in the basal ganglia: a simulation-based study. J. Neural. Eng. **3**, 132–138 (2006)
4. Åström, M., Lemaire, J., Wårdell, K.: Influence of heterogeneous and anisotropic tissue conductivity on electric field distribution in deep brain stimulation. *Submitted.* (2011)
5. Åström, M., Tripoliti, E., Hariz, M.I., Zrinzo, L.U., Martinez-Torres, I., Limousin, P., Wårdell, K.: Patient-specific model-based investigation of speech intelligibility and movement during deep brain stimulation. Stereotact. Funct. Neurosurg. **88**, 224–233 (2010)
6. Åström, M., Zrinzo, L.U., Tisch, S., Tripoliti, E., Hariz, M.I., Wårdell, K.: Method for patient-specific finite element modeling and simulation of deep brain stimulation. Med. Biol. Eng. Comput. **47**, 21–28 (2009) [Epub 2008 Oct 21]
7. Benabid, A.L.: Deep brain stimulation for Parkinson's disease. Curr. Opin. Neurobiol. **13**, 696–706 (2003)
8. Benabid, A.L., Chabardes, S., Mitrofanis, J., Pollak, P.: Deep brain stimulation of the subthalamic nucleus for the treatment of Parkinson's disease. Lancet Neurol. **8**, 67–81 (2009)
9. Butson, C.R., Cooper, S.E., Henderson, J.M., Mcintyre, C.C.: Patient-specific analysis of the volume of tissue activated during deep brain stimulation. Neuroimage **34**, 661–670 (2007)
10. Cheng, D.K.: Field and Wave Electromagnetics. Addison-Wesley Publishing Company Inc, New York (1989)
11. Diczfalusy, E., Zsigmond, P., Dizdar, N., Kullman, A., Loyd, D., Wårdell, K.: A patient-specific multiphysics model for prediction of analyte diffusion when using microdialysis in parallel to deep brain stimulation. *Submitted.* (2011)
12. Gross, R.E., Krack, P., Rodriguez-Oroz, M.C., Rezai, A.R., Benabid, A.L.: Electrophysiological mapping for the implantation of deep brain stimulators for Parkinson's disease and tremor. Mov. Disord. **21**, S259–S283 (2006)
13. Hardman, C.D., Henderson, J.M., Finkelstein, D.I., Horne, M.K., Paxinos, G., Halliday, G.M.: Comparison of the basal ganglia in rats, marmosets, macaques, baboons, and humans: volume and neuronal number for the output, internal relay, and striatal modulating nuclei. J. Comp. Neurol. **445**, 238–255 (2002)
14. Hariz, M.I., Rehncrona, S., Quinn, N.P., Speelman, J.D., Wensing, C.: Multicenter study on deep brain stimulation in Parkinson's disease: an independent assessment of reported adverse events at 4 years. Mov. Disord. **23**, 416–421 (2008)
15. Hemm, S., Mennessier, G., Vayssiere, N., Cif, L., El Fertit, H., Coubes, P.: Deep brain stimulation in movement disorders: stereotactic coregistration of two-dimensional electrical field modeling and magnetic resonance imaging. J. Neurosurg. **103**, 949–955 (2005)
16. Hemm, S., Wårdell, K.: Stereotactic implantation of deep brain stimulation electrodes: a review of technical systems, methods and emerging tools. Med. Biol. Eng. Comput. **48**, 611–624 (2010)
17. Johansson, J.D., Blomstedt, P., Haj-Hosseini, N., Bergenheim, A.T., Eriksson, O., Wårdell, K.: Combined diffuse light reflectance and electrical impedance measurements as a navigation aid in deep brain surgery. Stereotact. Funct. Neurosurg. **87**, 105–113 (2009)
18. Kindlmann, G.: Superquadric tensor glyphs. In: Proceedings IEEE TVCG/EG Symposium on Visualization, pp. 147–154. (2004)

19. Maks, C.B., Butson, C.R., Walter, B.L., Vitek, J.L., Mcintyre, C.C.: Deep brain stimulation activation volumes and their association with neurophysiological mapping and therapeutic outcomes. J. Neurol. Neurosurg. Psychiatry. **80**, 659–666 (2009) [Epub 2008 Apr 10]
20. Martens, H.C., Toader, E., Decre, M.M., Anderson, D.J., Vetter, R., Kipke, D.R., Baker, K.B., Johnson, M.D., Vitek, J.L.: Spatial steering of deep brain stimulation volumes using a novel lead design. Clin. Neurophysiol. **122**, 558–566 (2011)
21. Mcintyre, C.C., Mori, S., Sherman, D.L., Thakor, N.V., Vitek, J.L.: Electric field and stimulating influence generated by deep brain stimulation of the subthalamic nucleus. Clin. Neurophysiol. **115**, 589–595 (2004)
22. Mikos, A., Bowers, D., Noecker, A.M., Mcintyre, C.C., Won, M., Chaturvedi, A., Foote, K.D., Okun, M.S.: Patient-specific analysis of the relationship between the volume of tissue activated during DBS and verbal fluency. Neuroimage **54**(Suppl 1), S238–S246 (2011)
23. Morel, A.: Stereotactic Atlas of the Human Thalamus and Basal Ganglia. Informa Healthcare, New York (2007)
24. Nowinski, W.L., Thirunavuukarasuu, M., Benabid, A.L.: The cerefy clinical brainAtlas: enhanced edition with surgical planning and intraoperative support. CD-ROM (2005)
25. Panescu, D.: Emerging technologies. Implantable neurostimulation devices. IEEE Eng. Med. Biol. Mag. **27**, 100–105 (2008). 113
26. Tripoliti, E., Zrinzo, L., Martinez-Torres, I., Tisch, S., Frost, E., Borrell, E., Hariz, M.I., Limousin, P.: Effects of contact location and voltage amplitude on speech and movement in bilateral subthalamic nucleus deep brain stimulation. Mov. Disord. **23**, 2377–2383 (2008)
27. Tuch, D.S., Wedeen, V.J., Dale, A.M., George, J.S., Belliveau, J.W.: Conductivity tensor mapping of the human brain using diffusion tensor MRI. Proc. Natl. Acad. Sci. USA **98**, 11697–11701 (2001)
28. Vasques, X., Cif, L., Hess, O., Gavarini, S., Mennessier, G., Coubes, P.: Stereotactic model of the electrical distribution within the internal globus pallidus during deep brain stimulation. J. Comput. Neurosci. **26**, 109–118 (2008)
29. Vitek, J.L., Delong, M.R., Starr, P.A., Hariz, M.I., Metman, L.V.: Intraoperative neurophysiology in DBS for dystonia. Mov. Disord. **26**(Suppl 1), S31–S36 (2011)
30. Wårdell, K., Blomstedt, P., Richter, J., Antonsson, J., Eriksson, O., Zsigmond, P., Bergenheim, A.T., Hariz, M.I.: Intracerebral microvascular measurements during deep brain stimulation implantation using laser Doppler perfusion monitoring. Stereotact. Funct. Neurosurg. **85**, 279–286 (2007)
31. Yousif, N., Bayford, R., Bain, P.G., Liu, X.: The peri-electrode space is a significant element of the electrode-brain interface in deep brain stimulation: a computational study. Brain. Res. Bull. **74**, 361–368 (2007) [Epub 2007 Jul 26]
32. Yousif, N., Liu, X.: Modeling the current distribution across the depth electrode-brain interface in deep brain stimulation. Expert Rev. Med. Devices **4**, 623–631 (2007)
33. Zrinzo, L., Hariz, M.: Impedance recording in functional neurosurgery. In: Gildenberg, P.L., Lozano, A.M., Tasker, R. (eds.) Textbook of Stereotactic and Functional Neurosurgery. Springer, Berlin (2008)

Part VI
Patients-Specific Modeling in Diagnostics, Surgical Planning and Rehabilitation

Patient-Specific Modeling of Breast Biomechanics with Applications to Breast Cancer Detection and Treatment

Thiranja P. Babarenda Gamage, Vijayaraghavan Rajagopal,
Poul M. F. Nielsen and Martyn P. Nash

Abstract There are many challenges clinicians are faced with when diagnosing and treating breast cancer. Biomechanical modeling of the breast is a field of research that aims to assist clinicians by providing a physics based approach to addressing some of these challenges. This review describes the state of the art in the field, from aiding co-location of information between various medical imaging modalities used to identify tumours; to providing the ability to predict the location of these tumors during different biopsy or surgical procedures; to aiding temporal registration of follow-up medical images used to review the progress of suspicious lesions and therefore evaluate effectiveness of breast cancer treatments; to aiding implant selection for breast augmentation procedures and the subsequent prediction of the resulting appearance following such procedures. Significant technical challenges remain in terms of improving the accuracy of such biomechanical models. These include the precise determination and application of loading and

T. P. Babarenda Gamage · V. Rajagopal · P. M. F. Nielsen · M. P. Nash (✉)
Auckland Bioengineering Institute, The University of Auckland,
Level 6, 70 Symonds Street, Auckland, New Zealand
e-mail: martyn.nash@auckland.ac.nz

T. P. Babarenda Gamage
e-mail: psam012@aucklanduni.ac.nz

V. Rajagopal
e-mail: v.rajagopal@auckland.ac.nz

P. M. F. Nielsen
e-mail: p.nielsen@auckland.ac.nz

P. M. F. Nielsen · M. P. Nash
Department of Engineering Science, The University of Auckland,
Level 3, 70 Symonds Street, Auckland, New Zealand

boundary constraints applied during different clinical procedures, and accurate characterization of individual-specific mechanical properties of the different breast tissues. In addition to these more technical challenges, a number of practical challenges exist when translating biomechanical models from research based environments into clinical workflows, which demand general applicability, and ease and speed of use. This review outlines such challenges and provides an overview of the steps researchers are taking to address them. Once these challenges have been met, there is potential for extending the use of biomechanics to simulate more complex clinical procedures, from modeling needle insertions into breast tissue during real-time biopsy procedures, to simulating and predicting the outcome of different surgical procedures such as tumorectomies. Clinical adoption of such state-of-the-art modeling techniques has significant potential for reducing the number of misdiagnosed breast cancers while also helping improve clinical treatment of patients.

1 Introduction

Breast cancer is by far the most frequent form of cancer world wide in women [14], with some 1.2 million new cases being diagnosed each year. 400,000 women die from the disease each year [124]. Different imaging modalities are used to detect breast cancer with X-ray mammography, ultrasound imaging, Magnetic Resonance Imaging (MRI) and Positron Emission Tomography (PET), being the primary diagnostic tools. X-ray mammography in particular, is considered the gold standard and only feasible approach to the early detection of breast cancer for a number of reasons. These include high specificity, high throughput, with scans typically taking 10 s to image, and relatively low cost compared with other imaging modalities such as MRI, which can cost up to four times more than mammograms [49]. However, they produce harmful X-rays and only provide two dimensional (2D) images of the breast that are usually obtained through high levels of compression, resulting in difficulty in locating suspicious features in the uncompressed breast. Mammograms also have limited applicability to women with dense breasts due to decreased sensitivity [65]. MR, ultrasound and PET scans are typically acquired to overcome some of these issues however, they also suffer from various limitations. For example, in the case of MR images, while providing detailed three dimensional (3D) views of the breast in a benign manner, they have low specificity and usually require the use of contrast agents to be injected into the breast to better identify cancer (dynamic contrast enhanced MRI or DCE-MRI). Ultrasound imaging, while being portable, inexpensive and easy to use, is difficult to interpret due to its relatively low spatial resolution and contrast. PET scans, while being able to highlight the spread of cancer and therefore being useful in monitoring treatment responses, provide high specificity but poor image resolution (3–5 mm compared to 50 μm for mammograms) and sensitivity especially with smaller tumors, which are the most important to identify during screening [10]. Studies have shown that combining information from different medical imaging modalities can better detect breast cancer, as used together they mitigate the limitations of individual imaging modalities [68].

There are many existing challenges related to interpreting the information held within these different medical images, and include the co-location of tumors both within the same imaging modality (eg. temporal MR images) and across different modalities (eg. between MRI and X-ray mammograms) to aid diagnosis of breast cancer. Previous work has mainly focused on using registration algorithms to address these problems [8, 28, 35, 100, 112, 116, 130]. However, these can result in unrealistic deformations as these methods permit non-physical solutions, which do not satisfy the fundamental laws of mechanics. To partially address this, volume preserving algorithms have been applied to minimize tissue distortion during registration by applying more physically realistic constraints, for example to register pre- and post-contrast MR breast images [93, 94, 116]. However, standard intensity based techniques can still fail if the observed deformation is too large such as during large mammographic compressions. In addition, registration methods cannot provide any information regarding the location of tumors when target medical images have not been acquired.

Biomechanical models of the breast can be used to address these problems and help co-locate information between the various medical images to identify tumour locations, while also providing the ability to predict their location during different biopsy or surgical procedures. Such models have also been developed in tandem with newer registration algorithms which aim to apply more physics based constraints to the problem and therefore improve registration accuracy. A number of applications that are currently being investigated involve the use of biomechanical breast models to aid temporal registration of follow-up medical images used to review the progress of suspicious lesions or evaluate success of breast cancer treatments [98, 99], co-locate tumors between medical images [89, 91, 102, 122], guiding clinical biopsy procedures [5], track tumors for surgical assistance [19, 87] and for both aiding breast implant selection and predicting the outcome of breast augmentation procedures [74, 95–97]. Such applications typically require biomechanical models to be customized to an individual, and therefore the subject specific geometry and tissue distribution need to be modeled. Furthermore when simulating the mechanical behavior, the mechanical properties of an individual's tissues within the breast need to be identified in order to give reliable predictions of its mechanical behavior. The accuracy of these models also have to be assessed based on the application for which they were intended. These are some of the main technical challenges that need to be addressed when developing biomechanical models of the breast. From a practical standpoint, a number of challenges exist for translating these models into the clinic and be useful for the diagnosis and treatment of breast cancer. This review outlines some of these challenges and the steps researchers in the field of breast biomechanics are taking to solve these problems.

The review begins with an outline of breast anatomy in Sect. 2, followed by an overview of how this anatomy is represented using individual specific biomechanical models of the breast in Sect. 3. The theoretical aspects of simulating breast deformation under a variety of clinical procedures is then presented in Sect. 4, followed by a description of how the behavior of an individual's breasts are simulated under these procedures through the assignment of individual specific tissue mechanical properties in Sect. 5. An overview of the methods researchers have used to assess the performance of these models is then given in Sect. 6. Following on from these more technical aspects of modeling, Sect. 7 provides an overview of how breast models can be used in a number of applications and finally examples of how these models can be integrated into a clinical workflow are presented in Sect. 8.

2 Breast Anatomy

The breasts of an adult woman are tear-shaped glands situated on the chest and are responsible for lactation. They extend from the platysma myoides muscle on the second rib to the external oblique muscle on the seventh rib and sit on the pectoralis major and rib cartilage on the sternal margin (see Fig. 1). At the axillary margins the breasts sit on the serratus major, external oblique muscles and fascia of the thorax. The internal structure of the breast is organized in two layers of superficial and deeper connective tissue (fascia mammae), composed of collagen and elastin and are attached to the ligamentous tissue of the sternum. Between these two layers are the adipose, fibrous tissue and other functional components such as the lactiferous ducts. A fibrous cover is formed by the superficial layer which both passes between the glands and the skin and also enters the interior of the glands. The fibrous extensions of the suspensory ligaments of Cooper (also known as Cooper's ligaments, retinaculum fibrosa or fibrocollagenous septa), detach from the superficial fascia and reach the skin thereby supporting and holding the breasts in position on the chest wall allowing them to maintain their normal shape [26, 92]. It has also been recently hypothesized that Cooper's ligaments compartmentalize the adipose tissue within the breast [3] as shown in Fig. 1. A detailed review of the anatomy of the breast fascial system is provided by [92].

The skin is a heterogeneous structure containing four layers, which from outside to inside are the stratum corneum, the epidermis, the dermis and the subcutis or hypodermis. A detailed overview of the anatomy skin is provided by [1]. The mechanical behavior of the skin is largely determined by the dermis and the underlying subcutaneous adipose tissue, which can be considerably thick in certain parts of the body for healthy individuals. The average thickness of the skin on the breast is approximately 0.5–3 mm, which has previously been identified from film-screen mammography [78, 129] and breast CT scans [42]. Furthermore in vivo skin is generally in a state of pre-stress [61]. One of the contributors to this state of stress is the distribution of collagen fibers in skin leading to anisotropic mechanical behavior, limiting strain to a greater extent parallel to collagen fibers (along lines of maximal tension lines), compared with the transverse direction. These lines are called Kraissl's lines (oriented perpendicular to the action of the

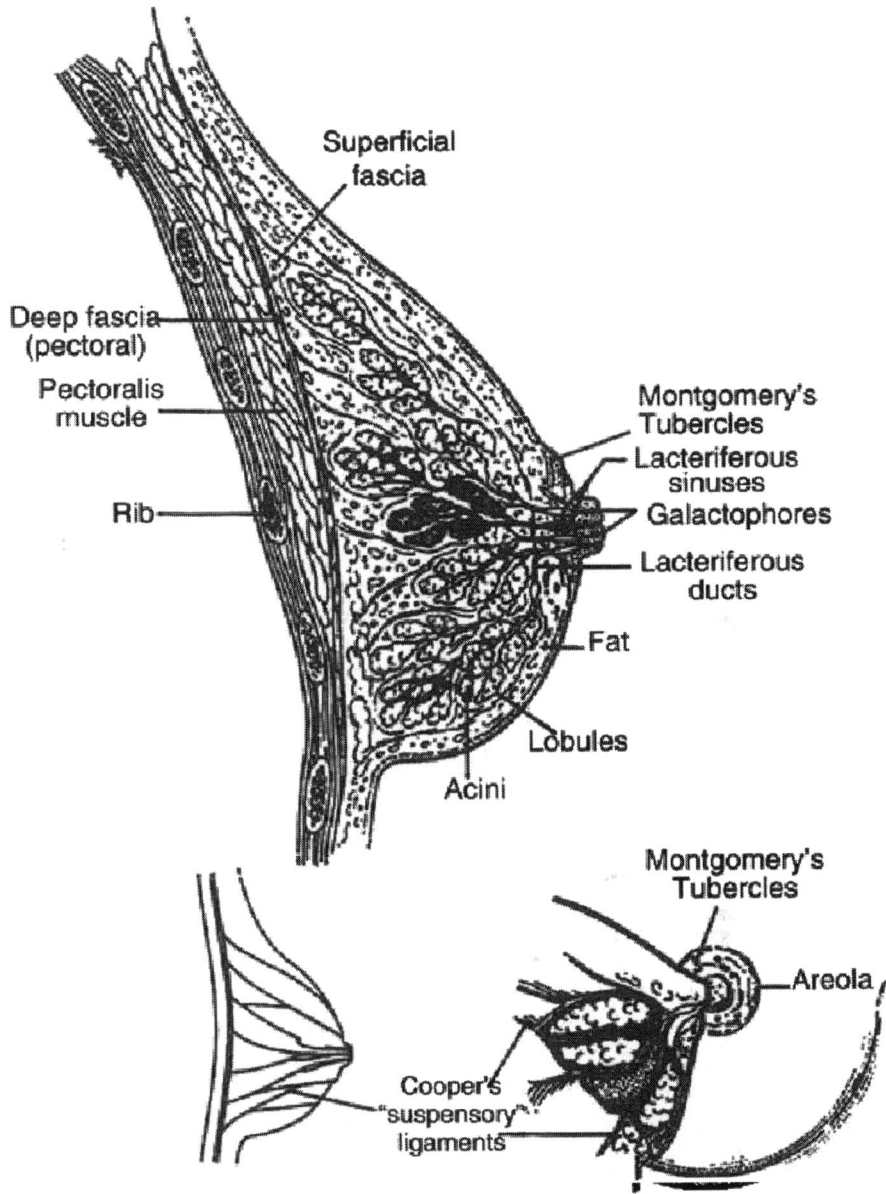

Fig. 1 Anatomy of the breast. With kind permission from springer science+business media: Pass [75], Fig. 2, (From Hindle [40])

underlying muscles), Borges's lines (follow furrows formed when the skin is relaxed) or Langer's lines (follow furrows formed in cadavers in a state of rigor mortis) [36, 128]. Other factors leading to this stress state include the degree of stretch or folding of skin around the boundaries of the breast, for example, in the cervical and axilla regions, where the arm connects to the shoulder.

A review of the normal and pathological breast from a histological perspective, and information regarding the prevalence of different types of tumors seen in the breast, are provided by [34]. In terms of their location, tumors are most commonly detected in the upper outer region (shoulder region) [11], as identified through lymphoscintigraphy where sentinel nodes of the breast lymphatic system closest to a tumor are identified and biopsied to determine the extent of the spread of breast cancer [59].

3 Creating Patient-Specific Anatomical Finite Element Models of the Breast

Like many other models created of the organ systems of the human body [45, 46], such as the heart [70], lungs and musculoskeletal system [31], biomechanical models of the breast are most commonly created using finite element (FE) models [133], as they have the ability to accurately and efficiently store detailed information about an organ's geometry and structure (described in Sect. 2), from clinical images. This is most commonly performed using clinical MR images acquired in the prone position, which provide 3D information of the boundaries (skin and air, breast tissue and muscle, muscle and rib) as well as information regarding the internal structure of the breast. While less common, computed tomography (CT) images in the prone or supine positions have also been used for model construction in breast augmentation applications [74, 95]. Some researchers have also acquired surface geometry of the breast in the prone position using 3D stereoscopy in near infrared (NIR) tomography applications [29], and in the supine or standing position for breast augmentation [74, 95, 96] or image-guided surgery applications [18–20]. Stereoscopic imaging is very important in the latter case where continuous real-time imaging of the breast surface is required during surgical interventions. However, a drawback is that they can only provide information relating to the skin-air boundary and therefore no information related to the internal boundaries of the breast. To address this, stereoscopic images can be accompanied by an MR scan prior to intervention, which along with markers placed on the breast surface, can be used to align the two image sets and therefore obtain the rib-muscle interface [19].

Once the boundaries of the breast have been identified and segmented, a number of meshing algorithms can be employed to construct anatomically realistic FE models for a specific individual. These include Delaunay triangulation, marching cube algorithms [74, 118] and non-linear geometric fitting [31, 71, 87]. A number of authors have also utilized meshing algorithms in number of software packages, such as MSC.Marc (MSC Software Corporation, Santa Ana, CA, USA) [110], ABAQUS FE (SIMULIA, *Dassault Systèmes*, Providence, Rhode Island, USA) [105, 109], ANSYS (Canonsburg, Pennsylvania, USA) [38, 120] and

Fig. 2 Segmented prone MRI skin data fitted to finite element model of the breast described using a cubic Hermite interpolation scheme

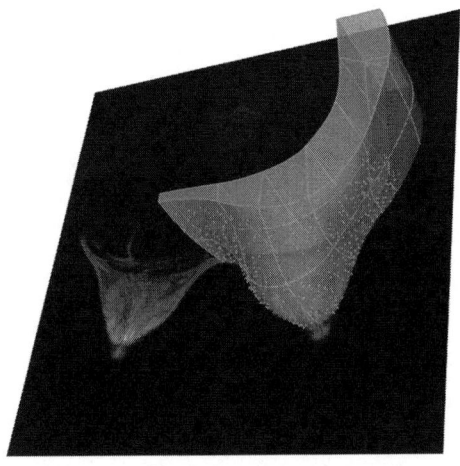

Harpoon V2 (Sharc Ltd, Manchester UK) [74]. Tetrahedral [74, 95, 120] and hexahedral [102] elements are the most widely used element types when creating FE meshes and are usually interpolated using standard Lagrange based interpolation schemes such as linear or quadratic Lagrange shape functions. Radial Basis Functions (RBF) which allow for smooth interpolation of incomplete surfaces [16, 17] have been used to approximate the pectoral muscle surface of the breast for augmentation applications [97]. Higher order C^1 continuous Hermite based interpolation schemes [15] have also been used to represent the geometry of the breast as they usually require fewer nodes, and therefore elements, to accurately describe the geometry than Lagrange based interpolation schemes. Higher order interpolation functions are also favored for their better rates of convergence when simulating large deformations [134], such as those seen in the breast, for example when tracking tumors between different orientations or in the case of large mammographic compressions. An example of using such a fitting procedure for creating an individual specific FE mesh from segmented MRI data in the prone orientation is shown in Fig. 2. For more information on the mathematical descriptions of these interpolation schemes as used in FE modeling, the reader is directed to [133] and [15].

Adipose, fibroglandular and pectoral muscle tissues are the most commonly modeled tissues within the breast and are usually segmented from the medical images in a number of ways including manual segmentation [3], standard thresholding [77, 105] and fuzzy C-means (FCM) based segmentation [110]. Many authors have also utilized segmentation algorithms in a number of software packages, including ANALYZE (Biomedical Imaging Resource, Mayo Foundation, Rochester, MN, USA) [115, 120] and Scion Image (Scion Corp., Maryland USA) [3]. Figure 3a shows an example of the segmented adipose and fibroglandular tissue data embedded within a FE mesh. In heterogeneous FE models constructed in previous studies, these tissues have been incorporated into the models by assigning a different tissue type to each element within the mesh based on the type of tissue each pixel or group of pixels represent in MR breast images.

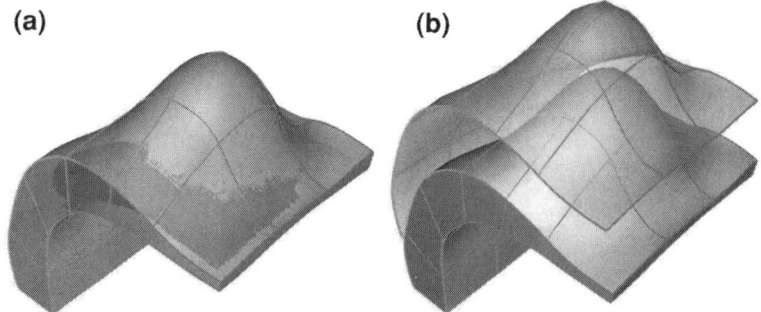

Fig. 3 a Fibroglandular (*yellow*), muscle (*red*) and adipose (*transparent*) tissue in FE models of the breast. **b** Explicitly modeling the skin tightly coupled to breast tissue a muscle (*dark grey*), fibroglandular and adipose (*light grey*)

Consequently such models of the breast can include many thousands of elements [41]. Tumors within the breast have also been modeled in ex vivo breast tissue samples in a similar manner [58, 72, 73, 106, 107].

The skin on the breast is usually modeled using 2D elements that are coupled to the surface of the 3D breast mesh [102, 74], and can be straightforwardly extracted from the 3D breast mesh as shown in Fig. 3b. Some studies have also used 3D elements closest to the breast surface in the 3D breast mesh to represent skin elements [120]. Due to the difficulties in identifying and segmenting Cooper's ligaments directly from medical images, these structures have only been modeled indirectly in breast biomechanical models by defining a preferred directional stiffness using anisotropic material descriptions or by incorporating the stiffness of these structures into the adipose tissue material descriptions (refer to Sect. 5 for more information on the types of material models used). The remaining structures and vessels (such as the nipple, blood vessels, and lymphatics) are usually not modeled discretely since their contribution to the overall mechanical deformation is assumed to be minor compared with the structures already discussed in this section.

4 Simulating Breast Biomechanics

Given an individual specific FE model of the breast, its mechanics can be simulated during different imaging and clinical procedures. This section provides an overview of the theoretical and computational techniques that have been used to model the mechanical behavior of the breast and its anatomical components subject to the various loading conditions these procedures apply on the breast.

The motivation for simulating breast biomechanics arose from the need to find correspondence between different medical images and thereby provide a map of the deformation occurring between and within different medical imaging procedures in order to better track regions of interest between these images. To this end, extensive research has focused on developing non-rigid registration algorithms to perform this task. Traditionally, these algorithms aimed to find a transformation map between the images that minimized the difference in image texture or other intensity-based similarity measurements, thus providing correspondence or deformation maps. However, due to their constraints being based entirely on image intensities they permitted physically implausible transformations, which limited their applicability. In light of such limitations, algorithms were developed to apply transformations to these medical images that were constrained partly by the laws of physics. One of the first attempts to directly incorporate the movement of the breast with image registration was performed by [53–55] for modeling X-ray mammographic compression. These models used purely geometrical constraints using a deformable cylinder to account for the kinematic transformations that occur during breast compression. However, a limitation of such models was that they neglected stress equilibrium within the breast. While solely modeling the kinematics of the breast does not strictly represent biomechanics, this work illustrated the utility of combining image registration techniques with geometrical models of the breast.

Since the work of Kita et al., the ability to represent realistic deformations of the breast has gained significant interest. During the past decade researchers have focused on using more physics based biomechanical models to predict breast deformation under different loading conditions during various imaging procedures and more recently during surgical interventions. These biomechanical models take into account both the kinematics and elasticity of the different breast tissues. A comprehensive survey of recent breast biomechanical models is provided in Table 1 for reference. The computational modeling theories currently used to simulate soft tissue biomechanics can be categorized into those that are based on heuristic approaches and those which use continuum mechanics and are listed below along with relevant references to their mathematical formulation.

- Heuristic
  - Mass-spring models [69, 135]
  - Mass-tensor model (MTM) [27]
- Continuum mechanics
  - Linear elasticity (small strain) [134]
  - Pseudo-non-linear elasticity [4, 5, 127]
  - Non-linear elasticity (finite strain) [12, 70]

Since heuristic approaches such as mass-spring models are quick and easy to implement while also being computationally inexpensive, they have been favored for solving many time dependent mechanical problems such as real-time surgical simulations [7, 60]. In this approach the tissue is discretized into a number of points (equivalent to mesh nodes) with each node being assigned a mass. The points are then connected using elastic springs and dampers with their motion being described by differential equations relating the stiffness of the springs to the elastic forces (both internally between springs and external body forces such as

Table 1 A survey of recent breast biomechanical models

Lead author	Modeling theories	Imaging modalities/ Degree of deformation	Parameters/ Anatomy	Boundary conditions	Performance*
Co-locating Tumors					
Ruiter [101], Ruiter et al. [102]]	FEM-non-linear, Skin (Membrane)	MR & X-ray mammography /21% compression	Literature/ Homogeneous Skin	Prescribe plate disp	Landmarks, 4.3 mm (X-ray mammo) & 3.9 mm (MR-mammo) error
Roose et al. [98, 99]	FEM-linear, Mass-tensor model	Temporal MR (prone)/ n/a	Literature/ Homogeneous,	Prescribe skin surface disp from registration	Node-node SSD, 70–80% decrease between follow-up MR images
Sarkar et al. [108]	FEM-linear	X-ray mammography (CC, MLO)/ compression % NR	Literature/ Homogeneous	Prescribe plate disp	Landmarks, 2.3 mm mean error between Visual
Shih et al. [110]	FEM-non-linear, Frictional contact	MR-mammography, pseudo CC,MLO mammograms/ 20, 40, 60% compression	Literature/ Adipose, Fibro	n/a	
Han et al. [38]	FEM-non-linear, GPU	MR-mammography/ 30% compression	In vivo/ Adipose, Fibro, muscle	Prescribe plate disp from registration	Landmarks, <4.2 mm Euclidean distance error
Lee et al. [63]	FEM-non-linear Frictionless contact	MR-mammography/ up to 38% compression	Literature/ Homogeneous	Rib-muscle tied contact	NCC, <4.4 mm RMS error
Reynolds et al. [91]	FEM-non-linear Unloaded state, Frictionless contact	X-ray mammography MR/up to 75% compression	Literature/ Homogeneous	Rib-muscle tied contact	Jaccard coefficient, 14–75% for tumors post-compression

(continued)

Table 1 (continued)

Lead author	Modeling theories	Imaging modalities/Degree of deformation	Parameters/Anatomy	Boundary conditions	Performance*
Tanner et al. [122]	FEM-linear, Skin (3D elements)	MR-mammography/up to 59% compression	In vivo/Muscle, skin Fibro, Tumor	Prescribe surface disp from registration	Landmarks, 4.1 mm Euclidean distance error
Predicting tumor location					
Rajagopal et al. [87, 88]	FEM-non-linear Unloaded state,	MR/ gravity (prone to supine)	In vivo/ Homogeneous	n/a	Landmarks, <4.8 mm Euclidean distance error
Carter et al. [19]	FEM-non-linear Unloaded state,	DCE-MR & stereoscopy/ gravity (supine to prone)	In vivo/Adipose, Fibro	Prescribe skin surface disp	Non-rigid-registration, 2.3 mm error
Augmentation/Implant modelling					
Roose et al. [95]	Mass-spring model (non-FEM)	CT, Stereoscopy/ Pre-post implant	Literature/Adipose, Fibro	NR	Skin surface, 8.5 mm max err pre-post implant
Roose et al. [96, 97]	FEM-linear Mass-tensor model	Stereoscopy/Pre-post implant	Literature/ Homogeneous	Prescribed implant disp, sliding contact	Skin surface, 4 mm mean error pre-post implant
del Palomar et al. [74]	FEM-non-linear Skin (Membrane)	CT, Stereoscopy/gravity (supine to standing)	In vivo/ Adipose,Fibro, Literature/ Skin	n/a	Skin surface, 2.4 mm error

All models assumed fixed chest wall (rib-muscle or muscle breast tissue) unless otherwise stated.
*Performance measures of the models were obtained by comparing model predicted locations/features/regions and their actual location which were independently identified on the medical images, for the different clinical procedures tested.
NR not reported, *SSD* sum of squared distances, *RMS* Root mean square

gravity). Such methods have previously been used to model breast deformation for simulating the outcome of augmentation procedures after placement of an implant [95]. While these methods are computationally efficient, they exhibit low accuracy and have also been known to produce non-physical solutions due to the discretization and arrangement of the springs which may, in certain arrangements, give rise to local minima, especially in the case of large deformations [95]. Heuristic approaches such as mass-tensor approximations have also been used, such as in the case of aiding registration of temporal mammograms [98, 99].

Continuum mechanics equations solved using the FE method represent the next and most commonly used class of mechanics theory for simulating breast deformation, with each of the breast tissues typically being constrained to be incompressible due to the large water content within the tissue [32]. They provide increased accuracy compared with mass-spring models but can be computationally expensive and thus difficult (in the case of linear or pseudo-linear elasticity) or impractical (in the case of non-linear elasticity) for use in dynamic real-time simulations. In most other situations where quasi-static behavior can be assumed (e.g. in applications other than surgical intervention where viscoelastic effects can be ignored), continuum mechanics based models are ideal for obtaining the desired accuracy.

For larger deformations, linear elasticity or pseudo-non-linear elasticity have often been used as a trade-off between biomechanical realism provided by fully nonlinear finite strain elasticity and computational efficiency. When using linear elasticity to model large deformations, excessive mesh distortion is a common problem. To address this problem many researchers have employed re-meshing or mesh moving algorithms that aim to correct mesh distortions and subsequently remap previously computed solutions onto these new meshes [114]. Such studies have also been performed in the context of breast modeling where re-meshing was shown to provide good numerical agreement between re-meshed linear elasticity and non-linear finite strain solutions for up to 30% cube compression [117]. However, with only relatively small compressions being applied, it is unclear whether these methods can provide accurate solutions for clinical compression (up to 75% compression), without introducing significant errors due to small strain approximations.

Pseudo-non-linear elasticity, which assumes the linear elastic modulus (or Young's modulus) is a function of strain, has also been used to model breast compression for biopsy procedures [4, 5, 132], and to simulate deformation due to gravity in the standing position for modeling static and dynamic thermography of the breast [48]. Fully non-linear large deformation models have also been used to model deformations due to gravity and compression and are discussed in Sects. 4.1 and 4.2 respectively. Numerical studies have been performed which highlight differences between linear, pseudo-non-linear and fully non-linear continuum mechanics formulations under different gravity loading conditions and concluded that both linear and pseudo-non-linear elasticity models were poor approximations to finite deformation models [127].

Skin is usually modeled using either 2D membrane elements [102] (neglecting bending stressing) or shell elements [109] (which account for bending stresses and torsion), assuming a finite skin thickness within each element. These 2D elements can then be coupled to the underlying 3D breast mesh using a number of methods including node-to-node, node-to-surface or surface-to-surface coupling, to model a variety of different scenarios including tight coupling or slip-stick type responses [134]. Some of these scenarios were investigated by [102], who showed improved accuracy over models neglecting skin, especially for the tightly coupled case when simulating relatively small (21%) mammographic compressions using an isotropic linear elastic skin model. However, it is unclear whether the use of linear elastic constitutive relations and tight coupling of skin to the underlying tissue would accurately model the behavior of skin under larger compressions or gravity loading. The state of pre-stress of skin on the breast has also not been investigated, and therefore its contribution under different clinical procedures is unknown.

An overview of how the breast is simulated under various clinical loading conditions is provided below.

4.1 Simulating Breast Deformation Under Gravity Loading

Simulating deformation of the breast under gravity loading is important for many clinical applications, such as tracking features of interest within the breast between the prone and supine positions. Large deformation theory used to simulate breast biomechanics during these procedures assumes that the breast is initially in a stress-free unloaded configuration. However, it is important to note that the breast is under the influence of gravity during the different imaging procedures. Thus, if the prone configuration is used as the reference state to predict the supine gravity-loaded configuration, errors in model predictions may arise because the prone configuration is not the true unloaded state of the breast. In addition, simply reversing the direction of gravity [74] does not produce the true unloaded state.

A numerically valid procedure for determining this unloaded state was described by [83–85], which included phantom validation experiments outlining the accuracy of this technique. It should be noted that since this unloaded configuration was identified using a biomechanical model, the accuracy of its prediction is largely influenced by other factors such as the boundary conditions and material parameters used in the model. Studies have been performed to assess how well the unloaded state can be approximated through neutral buoyancy studies [86–88] where a volunteer's breasts are immersed in water in the prone position. In this case, the density of the breast was assumed to be similar to that of water. Carter et al [21] also performed phantom studies under neutral buoyancy conditions. While such neutral buoyancy studies are impractical in a clinical setting, they do provide useful information to assess how well the unloaded configuration can be approximated. Figure 4 illustrates the numerical procedure used to determine the reference state of the breast and its comparison with neutral buoyancy experiments.

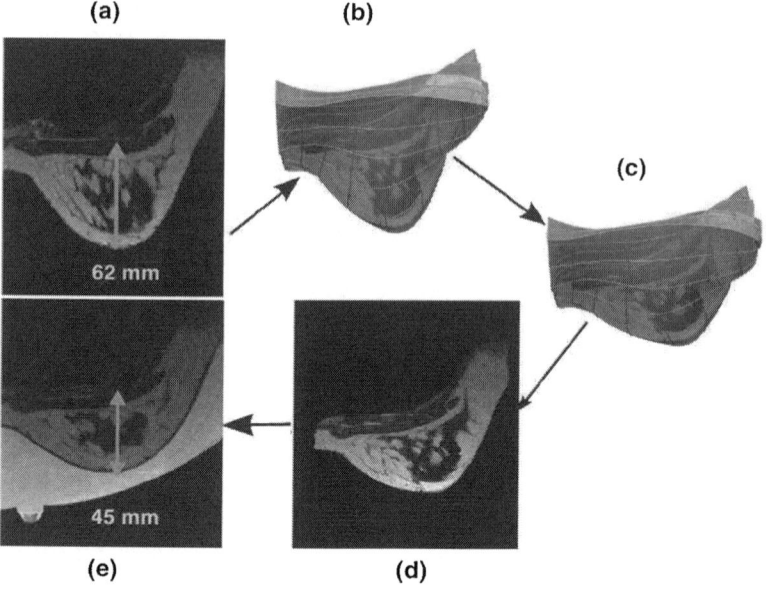

Fig. 4 Estimating the unloaded state of the breast. **a** MR image slice of the breast of a volunteer in the prone gravity-loaded state. **b** Prone MR image slice embedded in a finite element model that was fitted to the prone MR dataset. **c** Predicted unloaded shape of the breast with prone image warped to unloaded shape. **d** Synthetic MR image slice of the predicted unloaded configuration of the breast acquired using the biomechanical model. **e** MR image slice of the neutrally buoyant breast immersed in water (represented by the *white block* in the image). An indication of tissue displacement between this neutral buoyancy image and prone gravity-loaded image in (**a**) is given by the distance between nipple and pectoral muscle surface [90]

4.2 Modelling Mammographic Compression

The most common imaging modality used to screen the breast for cancer is X-ray mammography. During imaging the breast is compressed between two plates. This is performed predominantly in two directions; the cranio-caudal (CC) direction, which involves compression in the head-to-toe direction; and medio-lateral oblique (MLO) direction, which involves compression in the shoulder-to-opposite-hip direction. Clinical compressions usually range between 20% and 75% compression depending on the amount of breast tissue. The boundary loads on the surface of the breast during these compression procedures are difficult to measure, however, the orientation and separation of the compression plates are recorded (although in practice the plates may not be parallel). This information can be used together with FE models of the breast to simulate breast compression and the loads exerted by the compression plates. Breast compression has previously been modeled using two methods. The first involves the direct application of displacements to nodes on the surface of a FE model to mimic the effect of plate compression. These displacements are usually applied in the form of boundary conditions to the FE models and are most commonly

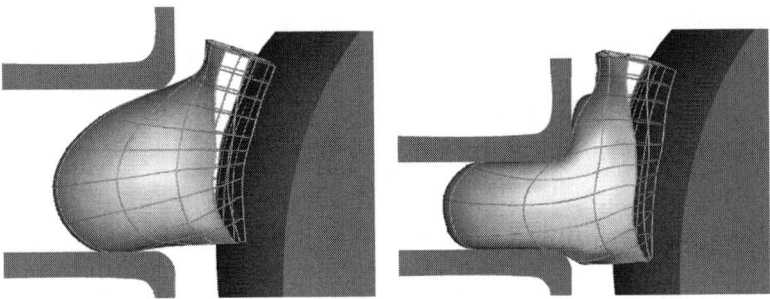

Fig. 5 Craniocaudal (CC) mammographic compression using frictionless contact mechanics and finite strain theory [23]

derived directly from medical images as discussed in Sect. 4.3. The second and more physically realistic approach to simulating compression involves applying constraints to the bodies in contact using contact mechanics theory. Readers are referred to [62] for more information regarding the theoretical formulation of the general contact mechanics for small and finite strain continuum mechanics problems. The theoretical aspects regarding the application of contact mechanics to breast compression has previously been studied in [22–24].

Patient-specific clinical compression studies have been performed by a number of authors as described in Table 1. Figure 5 illustrates an example of patient-specific CC compression using contact mechanics and finite strain theory. In most compression studies, gravity is rarely accounted for and therefore the unloaded state of the breast, as discussed in Sect. 4.1, is ignored (with the exception of [76, 77, 91]). Therefore this geometry is assumed to coincide directly with the segmented data used to construct the FE model. The largest compression simulations performed to date involved up to 75% compression on clinical datasets [91].

Controlled studies using phantoms have also been performed in a number of studies, aiming to validate the use of numerical techniques [22, 105] and registration algorithms with the aid of mechanical models [63, 64, 79]. The main purpose of these studies aimed to validate the applied theoretical and numerical methods, on simple geometries (e.g. cuboids), and to determine the accuracy with which these methods can be applied under ideal conditions. The conclusions from these experiments are valid only under the loading conditions considered in these studies. For example [5] stated that the use of a small strain approximation over finite strain mechanics did not introduce a significant error in 14% compression simulations of a phantom, for which only 1 mm difference in displacement errors between the two were observed. While this observation may hold for relatively small deformations in biopsy procedures ($<20\%$ compression), such approximations may not be valid with other modes of deformation such as those seen in larger breast compressions (up to 75%), and in tracking tumor locations between the prone and supine positions. This is also true for clinical studies, for example [109, 118] concluded that linear elasticity models provide similar results to those of finite elasticity and inaccurate assumptions of boundary conditions (discussed in Sect. 4.3) appear to have a much larger impact on the solutions than the chosen model parameters (discussed in Sect. 5), under 20% compression. It is uncertain whether these observations would apply to larger clinical compressions (up to 75%) or gravity loading.

4.3 Boundary Conditions for Breast Biomechanics

The reliability of any mechanical model is critically dependent upon both the boundary conditions and the material properties used to describe the mechanical response of the breast tissues. The former is discussed here while the latter is discussed in Sect. 5. An overview of boundary conditions applied in previous studies is provided in Table 1.

Typically, boundary conditions are applied to breast models to restrict the motion of the posterior surface of the breast (pectoral muscle/breast tissue or rib/pectoral muscle). In most cases this surface is assumed to be fixed for both mammographic compression and gravity loaded studies. On the other hand, Chung [23] assumed the breast tissue slides over the rib-pectoral muscle interface using frictionless contact mechanics constraints, described in Sect. 4.2. Registration algorithms have also been used to estimate boundary conditions at this interface by tracking the relative motion of the tissues between sets of images taken before and after an imaging procedure (mainly MR imaging) is performed. The estimated tissue displacement from the registration is then applied as boundary conditions on the FE models. The use of such techniques requires images to be acquired both before and after a clinical procedure has been performed which limits their applicability in clinical workflows.

A number of studies have also used similar techniques to apply surface displacements to free surfaces (typically the skin) in gravity loading and compression simulations as shown in Table 1. However, the use of such conditions do not reflect the physical situation. This is due the fact that the stress at a free surface is zero and there are no tethering points or reaction forces. Prescribing displacement at these surfaces violates this physical principle, resulting in non-zero reaction forces on free surfaces and can result in non-physical surface deformations in these regions. While boundary conditions obtained from registration can be useful in specific applications (such as co-locating tumors between MR images), biomechanical models which depend on such boundary conditions can no longer be used to predict deformation when medical images are unavailable for reference (e.g. when predicting the location of a tumor on the surgical table). Several researchers have also constructed fully physics-based biomechanical models of the breast (for example for MR-compression [25, 23, 91, 110] or for gravity loaded applications [74, 87]), which can operate independently of medical images once the initial patient-specific model has been created. These models aim to apply a minimal set of boundary conditions to the models and therefore have the ability to predict deformation under different loading conditions.

5 Patient-Specific Mechanical Properties of the Breast

In order for biomechanical models of the breast to obtain reliable predictions of the tissues mechanical behavior, it is important that the mechanical properties of the different tissues are identified. This task is made difficult due to the heterogeneous distribution of the mechanical properties, both structurally at the organ level and locally within each of its different tissue types. There is also significant variation in mechanical properties across the population depending on factors such as age and pathological state of the individual [57, 125]. Studies have also shown that different pathologies display different nonlinear elastic characteristics using MR elastography [111] and indentation tests on breast tissue [57, 125], as well as in other parts of the body such as the liver [131].

One method for estimating mechanical properties involves excising samples of tissue and performing predefined mechanical tests. For example, ex vivo studies relating to the breast have applied compression ($<15\%$ [125], $<25\%$ [57]), or indentation ($<40\%$ [51, 72, 73, 103, 104, 107]) on samples of breast tissue. These studies mainly involved fitting parameters to different hyperelastic constitutive relations. The mechanical properties of tumors have also been identified ex vivo using FE models [58, 72, 73, 106, 107, 126]. Linear, piecewise linear, exponential, Yeoh and Ogden phenomenological constitutive relations have mainly been used to describe the mechanical behavior of the different breast tissues in these studies. Statistical based material models such as the Arruda Boyce constitutive relation [2, 13] (initially developed to describe the behavior of rubber), have also been used to describe the behavior of breast tissue ex vivo [67]. More information on constitutive relations used in soft tissue parameter identification can be found in Humphrey [44]. The reader is directed to [123], for more information on constitutive relations used to describe tumor behavior. Gefen and Dilmoney [33] provide a more in-depth literature review of mechanical characteristics of different breast structures and the mechanical loads the breast is exposed to, during daily and sporting life.

While ex vivo experiments provide useful information about the orders of magnitudes of the mechanical properties, there are numerous drawbacks in using such parameters in patient-specific biomechanical models of the the breast. One of the main issues is the significant variation of these parameters across the population. Micro-structural and physiological changes can also occur in tissue over time leading to tissue deterioration [52]. Another drawback is the detachment of tissue from its surroundings during testing, which would normally contribute to its overall mechanical behavior and therefore influence the identification of parameters [9, 30, 66]. This is especially true in the case of the breast, where the supporting structures of Cooper's ligaments penetrate through the entire structure. Azar et al. [3, 4] hypothesized that these structures, compartmentalize adipose tissue, preventing it from being squeezed out of its location, and that the compression of adipose tissue increases the local pressure and the apparent stiffness of adipose tissue. The temperature at which ex vivo parameters are identified may also markedly change the mechanical properties of certain tissues within the breast, such as adipose tissue [3].

Due to such drawbacks, In vivo identification techniques can be more appropriate when modeling whole organ patient specific behavior. For example local In vivo breast tissue stiffness parameters have been identified in the past using a hand-held ultrasound indentation system and finite element models [37]. Elastography is another method of identifying in vivo mechanical properties of the breast, especially the tissues time dependent behavior. This involves application of relatively low frequency, small amplitude perturbations (e.g. indentation) to the breast while simultaneously imaging tissue deformation using ultrasound or MRI. Such techniques are typically useful for estimating the dynamic stiffness of the tissues, however, these are less important to clinical procedures that involve quasistatic loading, such as gravity or compression loading of the breast. The reader is directed to [6] for a more detailed overview on the use of elastography for identifying breast tissue parameters.

Patient-specific biomechanical models of the breast have also been used to identify in vivo tissue mechanical properties. This is typically achieved by applying a given loading condition to an individual's breasts. A patient-specific model can then be constructed to simulate this loading condition. The tissue parameters describing the behavior of for example, adipose, fibroglandular and muscle tissues are subsequently optimized to match the observed deformation (e.g skin surface) thereby identifying patient specific mechanical properties. These in vivo loading conditions typically correspond to clinical procedures such as MR-Mammography [38, 122] or prone to supine gravity loading (matching skin surface [74, 86, 87] or nipple to chest wall distance [19]). Cooper's Ligaments have also been modeled indirectly through the use of anisotropic constitutive relations [38, 120–122]. Examples of previously identified breast tissue parameters are presented in Table 2 , illustrating the wide variations of previously obtained parameters both ex vivo and in vivo. Accurate modeling of the different structures and tissues within the breast represents a trade off between incorporating more structural detail into the models and uniquely identifying their mechanical properties which can be difficult if sufficient data is unavailable. It is for this reason that robust identification of breast tissue parameters (which can also be easily determined in a clinical setting), is still an open question in the breast biomechanics field.

6 Assessing Performance of Biomechanical Models

The performance of breast biomechanical models is influenced by a number of factors including the choice of modeling theory used to define the mechanics of the breast, choice of boundary conditions for representing a particular loading state (as discussed in Sect. 4), and the estimation of the material parameters defining the mechanical behavior of the different tissues that constitute the breast (as discussed in Sect. 5) It is therefore important to be able to assess the performance of these

Table 2 Previously identified material parameters for the breast, illustrating the wide variations of parameters identified both ex vivo and in vivo

Lead author	Mode of deformation	Constitutive relation	Adipose Stiffness (kPa)	Fibroglandular tissue stiffness (kPa)	Tumor[a] Stiffness (kPa)
Krouskop et al. [57]	Ex vivo compression loading	Linear elastic relation	$E = 20 \pm 8$ at 20% compression	$E = 48 \pm 15$ at 20% compression	$E = 290 \pm 67$
Wellman et al. [125]	Ex vivo punch indentation	Linear elastic & exponential relations	$E = 17 \pm 8$ at 15% strain	$E = 272 \pm 168$ at 15% strain	$E = 2162$
Samani et al. [106, 107], O'Hagan et al. [72]	Ex vivo compression loading	Linear elastic & polynomial functions	$E = 3.25 \pm 0.9$ at 5% compression	$E = 3.24 \pm 0.61$ at 5% compression	$E = 16.38 \pm 1.55$
del Palomar et al. [74]	In vivo gravity loading	Neo-Hookean hyperelastic relation	$C_1 = 3$	$C_1 = 13$	n/a
Rajagopal et al. [87, 88]	In vivo gravity loading	Neo-Hookean hyperelastic relation	$C_1 = 0.08$ & 0.13	$C_1 = 0.08$ & 0.13	n/a

[a] ductal carcinoma insitu

models. A number of methods have been employed for this purpose and mainly fall into two categories, performance evaluation based on geometric data segmented from the medical images (such as skin surface or internal landmarks), and performance evaluation based on the medical images themselves through non-rigid registration algorithms. The medical images used to validate these models are typically MR images which provide structural detail in 3D. A survey of some of these existing methods is provided below along with references to their implementation and interpretation.

Geometric landmarks are commonly used to assess the performance of biomechanical models. These can be uniquely identifiable regions of tissue within the breast, as in the case of certain tumors, points on the surface of the skin (such as the nipple), or markers placed on the skin [such as small titanium spheres (CT scanning) [74] or vitamin E tablets [3]]. These landmarks can then be embedded into an FE model of the breast and tracked between different configurations, for example between the uncompressed and compressed breast during X-ray mammography. In this case the model's ability to predict landmark locations in the compressed breast can be determined by comparing the model predictions with the actual locations of landmarks from reference medical images of the compressed breast. Various methods have been used to make these comparisons such as Euclidean distance measures [87, 120], the Jaccard coefficient [47, 56, 91] (area or volume overlap) or the Dice coefficient (mean volume overlap) [56, 64]. Due to the difficulty in identifying viable landmarks within the breast, landmark based performance measures can be biased if landmarks are not uniformly distributed throughout the models, for example such as in the hypothetical case where landmarks are concentrated near the pectoral muscle where relatively little deformation is seen in most applications. Comparison of predicted and actual skin surfaces have also been used to assess the performance of breast models, for example, using closest segmented skin geometric point to model surface distances [31, 86] or symmetric mean absolute surface distances (SMAD) [64].

Image intensity based similarity measures provide an alternative method for evaluating the performance of models and have been frequently used as a performance measure for rigid and non-rigid registration (NRR) between different medical images. These usually involve embedding medical images, such as MR images, directly into the FE models. As the models deform during various procedures, the images are warped with the models. These model warped images can then be re-sampled and compared with actual MR images of the deformed breast using image intensity based similarity measures such as normalized cross-correlation (NCC) [19, 63] or normalized mutual information [109, 118, 120]. An example of such an approach is presented in Fig. 6 where model warped images are compared with clinical images for 38% breast compression of a volunteer using MR-mammography and NCC, where the arrows indicate the magnitude and direction of point-to-point prediction errors for a 2D image slice within the breast [63]. The performance of existing patient-specific biomechanical models using such metrics is presented in Table 1.

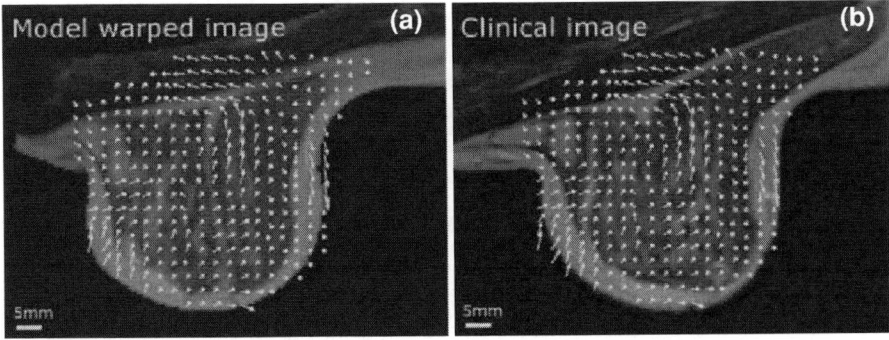

Fig. 6 An error vector field for comparing a model-warped image with a clinical image using NCC for 38% compression of the breast overlaid on a 2D slice of (**a**) the model-warped image, and (**b**) the clinical image [63]

7 Applications

Breast biomechanical models can be useful in a number of clinical applications to help temporal registration of follow-up medical images, co-location of tumors between medical images, for example between CC and MLO mammograms, guiding clinical biopsy procedures, tracking tumors for surgical assistance and for both aiding implant selection and predicting the outcome of breast augmentation procedures. A selection of these applications are discussed below.

7.1 Co-locating Tumors Between Medical Images

Registering different medical images and thereby allowing co-location of regions of interest within the breast, such as tumors, has been one of the main applications of breast biomechanical models. Previous work in this area includes the use of models for temporal registration of follow-up MRI images [98, 99] and registration of different pairs of mammogram images for both MR-Mammograhpy [4, 25, 38, 102, 105, 108, 118, 120, 132] and X-ray mammography, which is the main screening tool for diagnosing breast cancer [41, 50, 79–82, 89, 102, 110, 121]. In this section we highlight how such models can be used to co-locate features of interest between CC and MLO mammograms. Similar techniques can be used to aid tracking of features during various surgical procedures such as MR-guided biopsy [3].

Some of the main difficulties in interpreting X-ray mammograms are due, in-part, to the compression of the breast during imaging and the 2D nature of these images. It can therefore be unclear where features of interest are located between different X-ray views (such as CC or MLO mammograms), or where these features

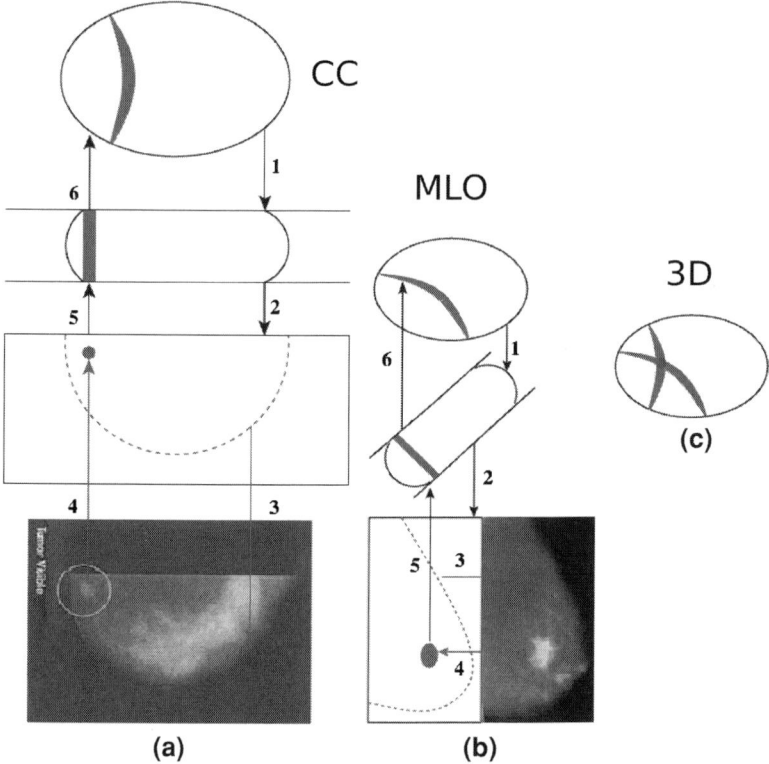

Fig. 7 Illustrates the procedure for co-localization of the 3D position of a lesion identified in a pair of (**a**) cranio-caudal (CC), (**b**) medio-lateral-oblique (MLO) mammograms with the 3D position of the lesion being identified by triangulating the 2D data (**c**). Refer to text for explanation of the numbered steps [22, 89]

lie in the uncompressed breast. The use of physics based biomechanical models can help in this respect as they can be used to realistically model the compression of the breast and allow tracking of tumors between the different X-ray views (see Sect. 4.2 for more information on how mammographic compression is modeled). This procedure is illustrated in Fig. 7 for co-localising the 3D positions of features of interest between a pair of CC and MLO mammograms [22, 89], with the process beginning with a patient-specific biomechanical model. With reference to Fig. 7, the steps are as follows:

1. 3D biomechanical models are used to simulate the compressed configurations of the breast for each 2D mammogram.
2. 2D projections of the compressed models are generated to provide skin contours for the two mammographic views.
3. The contours of the model projections are then matched to the mammograms using a rigid 2D alignment.

4. The locations of the features identified in the X-ray mammograms are then transfered to the projections of the compressed models.
5. For each feature, the associated 3D material points are marked parallel to the direction of the compression to form a straight tube of candidate points within the compressed breast model. A subset of these candidate points will be the 3D position of the feature.
6. For each compressed model, the tubes of candidate points are warped back to the common uncompressed breast model using the known compression deformation of the model. This enables the 3D locations of the features to be identified via the intersections of the unwarped tubes of points. This final step only involves reversing the geometric transformation calculated in Step 1 and all information necessary to do this is already carried by the biomechanical model.

7.2 Tracking Tumors for Surgical Assistance

Once suspicious tumors have been identified, treatment can involve the excision of the cancerous tissue, whilst preserving the remainder of the breast. These procedures are usually performed in the supine position making it difficult to identify their exact locations from diagnostic images, which are acquired in the prone position (DCE-MRI, MR-Mammography) or the standing positions (X-ray mammography). A large number of these tumors are identified on the surgical table by palpation, however, this procedure is difficult with smaller tumors (especially those situated close to the pectoral muscle) or ductal carcinoma in situ which may not be stiffer than the surrounding tissue. Furthermore it can also be difficult to distinguish the boundaries of these tumors for certain types of cancers (such as diffuse cancers), leading to uncertainty in determining which regions should be removed. Consequently, a large proportion of surgical operations need to be repeated due to the cancer not being completely excised [19]. Ultrasound imaging could be used to locate certain tumors in the supine position, however the images can be difficult to interpret due to their 2D nature and the deformation of the breast that occurs during imaging which can cause further difficulties in locating the tumors.

Biomechanical models of the breast can help in this respect, as they can be used to predict the location of tumors, for example, from prone DCE-MR images, to the supine position on the surgical table. An example of how a model can be used in this manner is shown in Fig. 8, with the process beginning from a patient-specific biomechanical model being constructed from prone DCE-MR images (8a). The techniques described in Sect. 4.1 can then be used to determine the unloaded state of the breast, which removes the effect of gravity (8b). This model can then be re-aligned with the patient on the surgical table and gravity can be applied (8c), producing the supine gravity deformed state (8d). The alignment of the model and

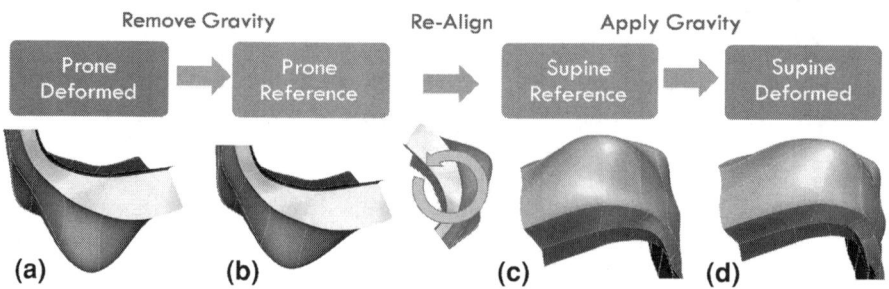

Fig. 8 Modeling procedure for obtaining the supine gravity loaded shape of the breast from the prone gravity loaded breast (See text for details)

the patient's breasts can be performed with the aid of stereoscopic camera systems. Such systems have previously been used to rigidly align models constructed in the supine position, with those in the standing position for predicting the outcome of breast augmentation procedures by tracking markers on the skin between the supine and standing positions [74].

Techniques have also been used to determine the locations of tumors on the surgical table using biomechanical models aided by stereoscopic camera systems [18–20]. Carter et al. [19] presented the first example of how such models could be used during surgery. In this approach DCE-MR images were acquired in the prone position to identify the location of the tumor prior to surgery. Non-contrast enhanced MR images in the supine position were also acquired prior to surgery and used to construct a patient specific model of the breast. The deformation from the supine to prone position was then simulated, allowing the tumor location identified in the DCE-MR prone images to be mapped back to supine model. However, during surgery the position of the patient on the operating table (in the supine position), could be different to their position in the previously acquired supine MR images used to create the original supine model. To account for this deformation and find the location of the tumor on the surgical table, a stereoscopic scan of the breast surface was obtained during surgery and used to provide boundary conditions for driving the original supine model to the configuration seen on the surgical table. The motivation behind taking this approach was mainly due to the relatively small deformation seen between the patient lying supine in the MR machine and on the surgical table, allowing linear elasticity theory to be employed to provide the location of the tumor during surgery in a very short amount of time (45 s). A detailed review on different applications of soft tissue modeling to image-guided surgery is provided by Carter et al. [20] and Hawkes et al. [39].

The different approaches outlined could be modified to begin directly from an MR-mammography procedure (prone position). However, these procedures rely on MR images being used to construct the initial 3D patient-specific model. Further research is required to apply these methods directly from X-ray mammography procedures (in the standing position) and determine the accuracy to which identified tumors can be predicted in the supine position when reference MR images are not available.

7.3 Breast Augmentation

Mastectomy in which part or all of a patient's breast is removed is a common treatment for breast cancer. In such situations, women often choose breast reconstruction following surgery. Aside from reconstructive surgery, there is a growing demand for breast augmentation for non-clinical, cosmetic or postural reasons, [for example in the case of congenital asymmetry of the beasts [95] or for breast reduction (augmentation mammoplasty)]. Many different implants are available to provide symmetrically shaped breasts. It is therefore difficult to ascertain pre-operatively, which implant would be best for a given individual [43]. More information on the different implants and the variety of reconstruction procedures are discussed in Spear and Spittler [113].

Biomechanical modeling of the breast can be used to aid the choice of implant and surgical method for breast augmentation procedures. These models can also be used to help predict surgical outcomes after such procedures are performed [74]. Previous biomechanical models of the breast have been created to help predict surgical outcome of augmenting procedures [95–97] where the implant was placed subglandular, above the pectoralis major muscle. Future applications of biomechanical models could potentially include the prediction of surface features on the breast and tissue density after mastectomy and potentially identify targets for intervention to improve cosmetic results [95].

8 Clinical Translation

In order for biomechanical models of the breast to be generally useful in a clinical setting, a number of practical challenges need to be addressed. These are mainly related to how research based workflows (where execution speed and ease of use are not essential), are translated into clinical workflows (where reliability, practicality and fast processing times are absolute requirements). One of the main challenges in applying breast biomechanical models in a clinical environment, is the availability of suitable images from which patient-specific anatomical models can be generated. Presently, MR images of the breast are routinely used in research applications to construct breast models, however acquiring these images is expensive and time consuming. Furthermore X-ray mammograms, are usually the starting point for breast cancer diagnosis. This presents a significant challenge in that the vast amount of information pertaining to the geometry and 3D structural arrangement of breast tissues is unavailable from the most commonly used form of

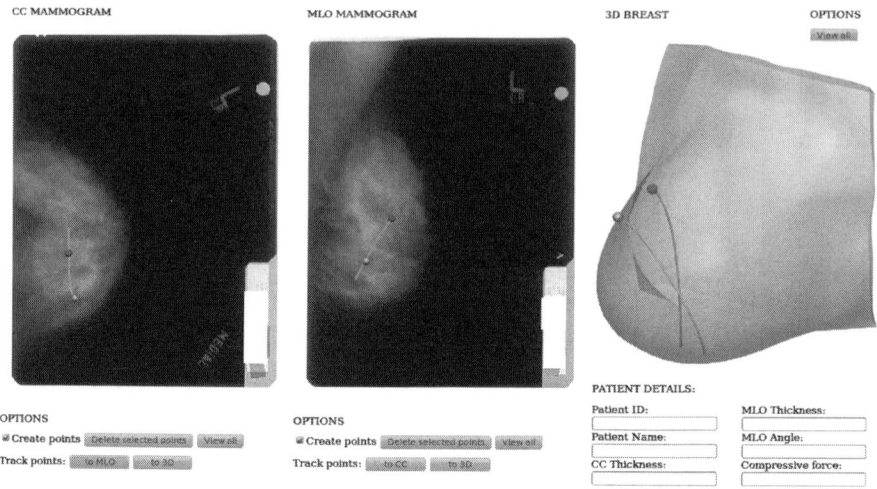

Fig. 9 Example of a graphical user interface that can be used to co-locate features between CC and MLO mammograms. A clinician can mark a region of interest on each mammogram. The resulting streamlines intersect on the 3D uncompressed breast model (shown on the *right*), defines the 3D location of the feature. Interface developed by Hayley Reynolds, Alan Wu, & Vijayaraghavan Rajagopal

clinical screening. Studies are being performed to address this issue by approximating the 3D shape of the uncompressed breast from 2D mammograms and a previously defined generic breast model [80]. Another approach is to construct statistical databases (using a similar approach to Tanner et al. [119]) relating mammograms and MR images, which can be used to better approximate the uncompressed breast using a few measurements (eg, age, weight and height). More information on how biomechanical models can be translated into the clinic, can be found in Rajagopal et al. [90].

Visualization and interpretation of results from biomechanical models through user-friendly interfaces is also important for clinical deployment, for example in the case of co-locating features between different X-ray mammogram images described in Sect. 7.1, where the ultimate goal is to pin-point the 3D location of features in the uncompressed breast. An example of a graphical user interface for solving such a problem using biomechanical models is presented in Fig. 9. Easy to use graphical interfaces could be integrated into clinical workflows by embedding them into picture archiving and communication systems (PACS) used for archiving medical images in the clinic. Pre-processing of the mechanics solutions to speed up the delivery of results could also be linked to these systems in a similar manner. The use of such graphical interfaces have a number of benefits including hiding of non-essential technical details from clinical operators while also providing a means to display the new information provided by biomechanical models in an interactive 3D environment.

9 Conclusion

We have seen how biomechanical models can be used to aid clinicians in a number of different applications, from helping co-locate tumors between different medical images and predicting their location (both during biopsy and surgical procedures), to helping determine the outcome of breast augmentation procedures. While substantial progress has been made in this field in recent years, a number of technical and practical challenges still exist. Current technical challenges include the precise determination and application of loading and boundary constraints applied during different clinical procedures, and accurate characterization of individual-specific mechanical properties of different breast tissues. Practical challenges include translating research based workflows into clinical workflows and providing clinicians with easy to use graphical interfaces for interpreting the new information provided by the biomechanical models. The application of biomechanical models also needs to be expanded to include applications related to ultrasound imaging, which is another important screening tool for breast cancer, that has been neglected in this field. Once these different aspects have been addressed, there is great potential for breast biomechanics to help reduce the number of misdiagnosed breast cancers. Addressing these challenges will also open the door to many new applications related to the treatment of breast cancer using biomechanical models, including aiding more advanced surgical and cosmetic procedures, from modeling biopsy needle insertion through the different tissues of the breast, to simulating breast tumorectomy, quadrantectomy, or mastectomy procedures.

Acknowledgements The financial support provided by the Foundation for Research, Science and Technology in New Zealand is gratefully acknowledged. We also thank Angela Lee, Richard Boyes, Hayley Reynolds, Jessica Jor and Tim Wu for their valuable contributions. M.P. Nash and P.M.F. Nielsen are supported by James Cook Fellowships administered by the Royal Society of New Zealand on behalf of the New Zealand Government.

References

1. Agache, P., Humbert, P.: Measuring the Skin. Editions Mdicales Internationales, Paris 2000, (2004)
2. Arruda, E.M., Boyce, M.C.: A three-dimensional constitutive model for the large stretch behavior of rubber elastic materials. J. Mech. Phys. Solids **41**(2), 389–412 (1993)
3. Azar, F.S., Metaxas, D.N., Schnall, M.D.: A finite element model of the breast for predicting mechanical deformations during biopsy procedures. In: Mathematical Methods in Biomedical Image Analysis, 2000. Proceedings. IEEE Workshop on, pp. 38–45 (2000)
4. Azar, F.S., Metaxas, D.N., Schnall, M.D.: A deformable finite element model of the breast for predicting mechanical deformations under external perturbations. Acad. Radiol. **8**(10), 965–975 (2001)
5. Azar, F.S., Metaxas, D.N., Schnall, M.D.: Methods for modeling and predicting mechanical deformations of the breast under external perturbations. Med. Image. Anal. **6**(1), 1–27 (2002)

6. Barbone, P.E., Oberai, A.A.: A review of the mathematical and computational foundations of biomechanical imaging. In: De, S., Guilak, F., Mofrad, R.K.M. (eds.) Computational Modeling in Biomechanics, pp. 375–408. Springer, The Netherlands (2010)
7. Baumann, R., Glauser, D., Tappy, D., Baur, C., Clavel, R.: Force feedback for virtual reality based minimally invasive surgery simulator. Stud. Health Technol. Inform. **29**, 564–579 (1996)
8. Behrenbruch, C.P., Marias, K., Armitage, P.A., Yam, M., Moore, N.R., English, R.E., Clarke, P.J., Leong, F.J., Brady, J.M.: Fusion of contrast-enhanced breast mr and mammographic imaging data. Br. J. Radiol. **77**(Spec No 2), S201–S208 (2004)
9. Berardesca, E., Elsner, P., Wilhelm, K., Maibach, H.: Bioengineering of the Skin: Methods and Instrumentation. CRC Press, Boca Raton (1995)
10. Berman, C.G.: Recent advances in breast-specific imaging. Cancer Control **14**(4), 338–349 (2007)
11. Blumgart, E.I., Uren, R.F., Nielsen, P.M.F., Nash, M.P., Reynolds, H.M.: Lymphatic drainage and tumour prevalence in the breast: a statistical analysis of symmetry, gender and node field independence. J. Anat. **218**(6), 652–659 (2011)
12. Bonet, J., Wood, R.: Nonlinear Continuum Mechanics for Finite Element Analysis. Cambridge University Press, Cambridge (1997)
13. Boyce, M.C., Arruda, E.M.: Constitutive models of rubber elasticity: a review. Rubber Chem. Technol. **73**, 504–523 (2000)
14. Boyle, P., Levin, B.: World Cancer Report 2008. International Agency for Research on Cancer (2008)
15. Bradley, C.P., Pullan, A.J., Hunter, P.J.: Geometric modeling of the human torso using cubic hermite elements. Ann. Biomed. Eng. **25**(1), 96–111 (1997)
16. Carr, J., Fright, W., Beatson, R.: Surface interpolation with radial basis functions for medical imaging. IEEE Trans. Med. Imaging **16**(1), 96–107 (1997)
17. Carr, J.C., Beatson, R.K., Cherrie, J.B., Mitchell, T.J., Fright, W.R., McCallum, B.C., Evans, T.R.: Reconstruction and representation of 3d objects with radial basis functions. In: Proceedings of the 28th Annual Conference on Computer graphics and Interactive Techniques, ACM, New York, NY, USA, SIGGRAPH '01, pp. 67–76 (2001)
18. Carter, T., Tanner, C., Crum, W., Beechey-Newman, N., Hawkes, D.: Medical imaging and augmented reality 3rd International Workshop, Shanghai, China, 17–18 August, 2006 Proceedings, Springer Berlin, Heidelberg, Chap A Framework for Image-Guided Breast Surgery, pp. 203–210 (2006)
19. Carter, T., Tanner, C., Beechey-Newman, N., Barratt, D., Hawkes, D.: Mr navigated breast surgery: method and initial clinical experience. Med. Image Comput. Comput. Assist. Interv. **11**, 356–363 (2008)
20. Carter, T.J., Sermesant, M., Cash, D.M., Barratt, D.C., Tanner, C., Hawkes, D.J.: Application of soft tissue modelling to image-guided surgery. Med. Eng. Phys. **27**(10), 893–909 (2005)
21. Carter T.J., Tanner C., Hawkes D.J.: Determining material properties of the breast for image-guided surgery. In: Miga M.I., Wong K.H. (eds.) Medical Imaging 2009: Visualization, Image-Guided Procedures, and Modeling. SPIE 7261:726124 (2009)
22. Chung, J., Rajagopal, V., Nielsen, P.M.F., Nash, M.P.: A biomechanical model of mammographic compressions. Biomech. Model. Mechanobiol. **7**, 43–52 (2008)
23. Chung, J.H.: Modelling mammographic mechanics. PhD thesis, The University of Auckland (2008)
24. Chung, J.H., Rajagopal, V., Laursen A Tod., Nielsen, P.M.F., Nash, M.P.: Frictional contact mechanics methods for soft materials: Application to tracking breast cancers. J. Biomech. **41**, 69–77 (2008)
25. Chung, J.H., Rajagopal, V., Nielsen, P.M.F., Nash, M.P.: Modelling mammographic compression of the breast, Medical Image Computing and Computer-assisted Intervention MICCAI 2008, Lecture Notes in Computer Science, Chap Modelling Mammographic Compression of the Breast, vol. 5242/2008, pp. 758–765. Springer, Heidelberg (2008)

26. Cooper, A.P.: On the Anatomy of the Breast. Longman Publishing, London (1840)
27. Cotin, S., Delingette, H., Ayache, N.: A hybrid elastic model for real-time cutting, deformations, and force feedback for surgery training and simulation. Visual Comput. **16**, 437–452 (2000)
28. Dawant, B.: Non-rigid registration of medical images: purpose and methods, a short survey. In: IEEE, International Symposium on Biomedical Imaging, pp. 465–468 (2002)
29. Dehghani, H., Doyley, M.M., Pogue, B.W., Jiang, S., Geng, J., Paulsen, K.D.: Breast deformation modelling for image reconstruction in near infrared optical tomography. Phys. Med. Biol. **49**(7), 1131 (2004)
30. Elsner, P., Berardesca, E., Wilhelm, K., Maibach, H.: Bioengineering of the Skin. CRC Press, Boca Raton (2002)
31. Fernandez, J.W., Mithraratne, P., Thrupp, S.F., Tawhai, M.H., Hunter, P.J.: Anatomically based geometric modelling of the musculo-skeletal system and other organs. Biomech. Model. Mechanobiol. **2**(3), 139–155 (2004)
32. Fung, Y.C.: Biomechanics Mechanical Properties of Living Tissue. Springer, Berlin (1993)
33. Gefen, A., Dilmoney, B.: Mechanics of the normal woman's breast. Technol. Health Care **15**(4), 259–271 (2007)
34. Guinebretiere, J., Menet, E., Tardivon, A., Cherel, P., Vanel, D.: Normal and pathological breast, the histological basis. Eur. J. Radiol. **54**(1), 6–14 (2005)
35. Guo, Y., Sivaramakrishna, R., Lu, C.C., Suri, J., Laxminarayan, S.: Breast image registration techniques: a survey. Med. Biol. Eng. Comput. **44**, 15–26 (2006)
36. Haerle, F., Champy, M., Terry, B.: Atlas of Craniomaxillofacial Osteosynthesis: Microplates, Miniplates and Screws, edn. 2. Thieme, Stuttgart (2009)
37. Han, L., Burcher, M., Noble, J.: Non-invasive measurement of biomechanical properties of in vivo soft tissues. In: Dohi, T., Kikinis, R. (eds.) Medical Image Computing and Computer-Assisted Intervention MICCAI 2002, Lecture Notes in Computer Science, vol. 2488, pp. 208–215. Springer, Heidelberg (2002)
38. Han, L., Hipwell, J.H., Taylor, Z.A., Tanner, C., Ourselin, S., Hawkes, D.J.: 10th International Workshop, IWDM 2010, Girona, Catalonia, Spain, June 16–18, 2010. Proceedings, Springer Berlin, Heidelberg, Chap Fast Deformation Simulation of Breasts Using GPU-Based Dynamic Explicit Finite Element Method. Lecture Notes in Computer Science, pp. 728–735 (2010)
39. Hawkes, D., Barratt, D., Blackall, J., Chan, C., Edwards, P., Rhode, K., Penney, G., McClelland, J., Hill, D.: Tissue deformation and shape models in image-guided interventions: a discussion paper. Med. Image Anal. **9**(2), 163–175 (2005)
40. Hindle, W.H. (ed.): Breast Care: A Clinical Guidebook for Women's Primary Healthcare Providers. Springer, New York (1999)
41. Hipwell, J.H., Tanner, C., Crum, W.R., Schnabel, J.A., Hawkes, D.J.: A new validation method for X-ray mammogram registration algorithms using a projection model of breast X-ray compression. IEEE Trans. Med. Imaging **26**(9), 1190–1200 (2007)
42. Huang, S.Y., Boone, J.M., Yang, K., Kwan, A.L.C., Packard, N.J.: The effect of skin thickness determined using breast ct on mammographic dosimetry. Med. Phys. **35**(4), 1199–1206 (2008)
43. Hudson, D.A.: Factors determining shape and symmetry in immediate breast reconstruction. Ann. Plast. Surg. **52**(1), 15–21 (2004)
44. Humphrey, J.: Review paper: continuum biomechanics of soft biological tissues. Proceedings of the Royal Society of London Series A: Mathematical, Physical and Engineering Sciences 459(2029), 3–46 (2003)
45. Hunter, P., Nielsen, P.: A strategy for integrative computational physiology. Physiology **20**(5), 316–325 (2005)
46. Hunter, P.J.: The IUPS physiome project: a framework for computational physiology. Prog. Biophys. Mol. Biol. **85**, 551–569 (2004)
47. Jaccard, P.: The distribution of the flora in the alpine zone. 1. New Phytologist **11**(2), 37–50 (1912)

48. Jiang, L., Zhan, W., Loew, M.H.: Modeling static and dynamic thermography of the human breast under elastic deformation. Phys. Med. Biol. **56**(1), 187 (2011)
49. Kahan, Z.: Breast Cancer, a Heterogeneous Disease Entity The Very Early Stages. 1st edn. Springer, Berlin (2011)
50. Kellner, A.L., Nelson, T.R., Cervino, L.I., Boone, J.M.: Simulation of mechanical compression of breast tissue. IEEE Trans. Biomed. Eng. **54**(10), 1885–1891 (2007)
51. Kerdok, A.E., Jordan, P., Liu, Y., Wellman, P.S., Socrate, S., Howe, R.D.: Identification of nonlinear constitutive law parameters of breast tissue. In: Proceeding of 2005 summer Bioengineering Conference, June 22–26, Vail Cascade Resort and Spa, Vail, Colorado (2005)
52. Kerdok, A.E., Ottensmeyer, M.P., Howe, R.D.: Effects of perfusion on the viscoelastic characteristics of liver. J. Biomech. **39**(12), 2221–2231 (2006)
53. Kita, Y., Highnam, R., Brady, M.: Correspondence between different view breast X-rays using a simulation of breast deformation. In: Computer Vision and Pattern Recognition, 1998. Proceedings. 1998 IEEE Computer Society Conference on, pp. 700–707 (1998)
54. Kita, Y., Highnam, R., Brady, M.: Correspondence between different view breast x rays using curved epipolar lines. Comput. Vision Image Understanding **83**(1), 38–56 (2001)
55. Kita, Y., Tohno, E., Highnam, R.P., Brady, M.: A cad system for the 3d location of lesions in mammograms. Med. Image Anal. **6**(3), 267–273 (2002)
56. Klein, A., Andersson, J., Ardekani, B.A., Ashburner, J., Avants, B., Chiang, M.C., Christensen, G.E., Collins, D.L., Gee, J., Hellier, P., Song, J.H., Jenkinson, M., Lepage, C., Rueckert, D., Thompson, P., Vercauteren, T., Woods, R.P., Mann, J.J., Parsey, R.V.: Evaluation of 14 nonlinear deformation algorithms applied to human brain mri registration. NeuroImage **46**(3), 786–802 (2009)
57. Krouskop, T.A., Wheeler, T.M., Kallel, F., Garra, B.S., Hall, T.: Elastic moduli of breast and prostate tissues under compression. Ultrason. Imaging **20**(4), 260–274 (1998)
58. Krouskop, T.A., Younes, P.S., Srinivasan, S., Wheeler, T., Ophir, J.: Differences in the compressive stress-strain response of infiltrating ductal carcinomas with and without lobular features–implications for mammography and elastography. Ultrason. Imaging **25**(3), 162–170 (2003)
59. Krynyckyi, B.R., Kim, C.K., Goyenechea, M.R., Chan, P.T., Zhang, Z.Y., Machac, J.: Clinical breast lymphoscintigraphy: optimal techniques for performing studies, image atlas, and analysis of images. Radiographics **24**(1), 121–145 (2004)
60. Kuhnapfel, U., Cakmak, H.K., Maass, H.: Endoscopic surgery training using virtual reality and deformable tissue simulation. Comput. Graphics **24**(5), 671–682 (2000)
61. Langer, K.: On the anatomy and physiology of the skin: I the cleavability of the cutis. Br. J. Plast. Surg. **31**(1), 3–8 (1978)
62. Laursen, T.A.: Computational Contact and Impact Mechanics, Fundamentals of Modeling Interfacial Phenomena in Nonlinear Finite Element Analysis. 1st edn. Springer, Berlin (2002)
63. Lee, A.W., Rajagopal, V., Chung, J.H., Nielsen, P.M., Nash, M.P.: Method for validating breast compression models using normalised cross-correlation. In: Miller, K., Nielsen, P.M. (eds.) Computational Biomechanics for Medicine, pp. 63–71. Springer, New York (2010)
64. Lee, A.W.C., Schnabel, J.A., Rajagopal, V., Nielsen, P.M.F., Nash, M.P.: Breast image registration by combining finite elements and free-form deformations. In: Mart, J., Oliver, A., Freixenet, J., Mart, R. (eds.) Digital Mammography, Lecture Notes in Computer Science, vol. 6136, pp. 736–743. Springer, Heidelberg (2010)
65. Lehman, C.D.: Role of mri in screening women at high risk for breast cancer. J. Magn. Reson. Imaging **24**(5), 964–970 (2006)
66. Lim, K., Chew, C., Chen, P., Jeyapalina, S., Ho, H., Rappel, J., Lim, B.: New extensometer to measure in vivo uniaxial mechanical properties of human skin. J. Biomech. **41**(5), 931–936 (2008)
67. Liu, Y., Kerdok, A.E., Howe, R.D.: International Symposium, ISMS 2004, Cambridge, MA, USA, 17–18 June 2004. Proceedings Chap A Nonlinear Finite Element Model of Soft

Tissue Indentation, Lecture Notes in Computer Science, pp. 67–76. Springer, Heidelberg (2004)
68. Malur, S., Wurdinger, S., Moritz, A., Michels, W., Schneider, A.: Comparison of written reports of mammography, sonography and magnetic resonance mammography for preoperative evaluation of breast lesions, with special emphasis on magnetic resonance mammography. Breast Cancer Res. **3**, 1–6 (2000)
69. Mollemans, W., Schutyser, F., Van Cleynenbreugel, J., Suetens, P.: Tetrahedral mass spring model for fast soft tissue deformation. In: Ayache, N., Delingette, H. (eds.) Surgery Simulation and Soft Tissue Modeling, Lecture Notes in Computer Science, vol. 2673, pp. 1002–1003. Springer Berlin, Heidelberg (2003)
70. Nash, M.P., Hunter, P.J.: Computational mechanics of the heart. J. Elast. **61**(1), 113–141 (2000)
71. Nielsen, P.M.F.: The anatomy of the heart: a finite element model. PhD thesis, University of Auckland (1987)
72. O'Hagan, J.J., Samani, A.: Measurement of the hyperelastic properties of tissue slices with tumour inclusion. Phys. Med. Biol. **53**(24), 7087 (2008)
73. O'Hagan, J.J., Samani, A.: Measurement of the hyperelastic properties of 44 pathological ex vivo breast tissue samples. Phys. Med. Biol. **54**(8), 2557 (2009)
74. del Palomar, A.P., Calvo, B., Herrero, J., Lopez, J., Doblar, M.: A finite element model to accurately predict real deformations of the breast. Med. Eng. Phys. **30**(9), 1089–1097 (2008)
75. Pass, H.A.: Benign and malignant diseases of the breast. In: Norton, J.A., Barie, P.S., Bollinger, R.R., Chang, A.E., Lowry, S.F., Mulvihill, S.J., Pass, H.I., Thompson, R.W. (eds.) Surgery, pp. 2005–2035 Springer, New York (2008)
76. Pathmanathan, P., Gavaghan, D.J., Whiteley, J.P., Rajagopal, V., Nielsen, P.M.F., Nash, M.P.: Predicting tumour location by simulating large deformations of the breast using a 3d finite element model and nonlinear elasticity. Medical Image Computing and Computer-Assisted Intervention MICCAI 2004, Lecture Notes in Computer Science, vol. 3217 pp. 217–224. Springer-Verlag Berlin Heidelberg (2004)
77. Pathmanathan, P., Gavaghan, D.J., Whiteley, J.P., Chapman, S.J., Brady, J.M.: Predicting tumor location by modeling the deformation of the breast. IEEE Trans. Biomed. Eng. **55**(10), 2471–2480 (2008)
78. Pope, T.L., Read, M.E., Medsker, T., Buschi, A.J., Brenbridge, A.N.: Breast skin thickness: normal range and causes of thickening shown on film-screen mammography. J. Can. Assoc. Radiol. **35**(4), 365–368 (1984)
79. Qiu, Y.: Three dimensional finite element model for lesion correspondence in breast imaging. Master's thesis, University of South Florida (2003)
80. Qiu, Y.: Temporal registration of mammograms by finite element simulation of mr breast volume deformation. PhD thesis, University of South Florida (2009)
81. Qiu, Y., Goldgof, D.B., Li, L., Sarkar, S., Zhang, Y., Anton, S.: Correspondence recovery in 2-view mammography. In: Proceedings of the 2004 IEEE International Symposium on Biomedical Imaging: From Nano to Macro, Arlington, VA, USA, 15–18 April 2004, IEEE, pp. 197–200 (2004)
82. Qiu, Y., Sun, X., Manohar, V., Goldgof, D.: Towards registration of temporal mammograms by finite element simulation of mr breast volumes. In: Miga MI, Cleary KR (eds.) Medical Imaging 2008: Visualization, Image-guided Procedures, and Modeling, SPIE, vol. 6918, p. 69182F (2008)
83. Rajagopal, V., Chung, J., Nielsen, P.M.F., Nash, M.P.: Finite element modelling of breast biomechanics: Finding a reference state. Conf. Proc. IEEE Eng. Med. Biol. Soc. **3**, 3268–3271 (2005)
84. Rajagopal, V., Chung, J., Nielsen, P.M.F., Nash, M.P.: Finite element modelling of breast biomechanics: directly calculating the reference state. Conf. Proc. IEEE Eng. Med. Biol. Soc. **1**, 420–423 (2006)
85. Rajagopal, V., Chung, J., Nielsen, P.M.F., Nash, M.P.: Determining the finite elasticity reference state from a loaded configuration. Int. J. Numer. Methods Eng. **72**, 1434–1451 (2007)

86. Rajagopal, V., Lee, A., Chung, J.H., Warren, R., Highnam, R.P., Nielsen, P.M.F., Nash, M.P.: Towards tracking breast cancer across medical images using subject-specific biomechanical models. Med. Image Comput. Comput. Assist. Interv. **10**, 651–658 (2007)
87. Rajagopal, V., Lee, A., Chung, J.H., Warren, R., Highnam, R.P., Nash, M.P., Nielsen, P.M.: Creating individual-specific biomechanical models of the breast for medical image analysis. Acad. Radiol. **15**(11), 1425–1436 (2008)
88. Rajagopal, V., Nash, M.P., Highnam P, Ralph., Nielsen, P.M.F.: The breast biomechanics reference state for multi-modal image analysis. International Workshop on Digital Mammography IWDM 2008, Lecture Notes in Computer Science, vol. 5116, pp. 385–392. Springer-Verlag Berlin Heidelberg (2008)
89. Rajagopal, V., Chung, J.H., Highnam, R.P., Warren, R., Nielsen, P.M., Nash, M.P.: Mapping microcalcifications between 2d mammograms and 3d mri using a biomechanical model of the breast. In: Miller, K, Nielsen P.M. (eds.) Computational Biomechanics for Medicine, pp.17–28. Springer, New York (2010)
90. Rajagopal, V., Nielsen, P.M.F., Nash, M.P.: Modeling breast biomechanics for multi-modal image analysis-successes and challenges. Wiley Interdisciplinary Rev.: Syst. Biol. Med. **2**(3), 293–304 (2010)
91. Reynolds, H.M., Puthran, J., Doyle, A., Jones, W., Nielsen, P.M., Nash, M.P., Rajagopal, V.: Mapping breast cancer between clinical X-ray and mr images. In: Computational Biomechanics for Medicine V, MICCAI 2010 Workshop pp.78-88, 24 September (2010)
92. Riggio, E., Quattrone, P., Nava, M.: Anatomical study of the breast superficial fascial system: the inframammary fold unit. Eur. J. Plast. Surg. **23**(6), 310–315 (2000)
93. Rohlfing, T., Maurer, C.R., Bluemke, D.A., Jacobs, M.A.: An alternating-constraints algorithm for volume-preserving non-rigid registration of contrast-enhanced MR breast images. In: Gee, J.C., Maintz, J.B.A., Vannier M.W. (eds.) Biomedical Image Registration – 2nd International Workshop, WBIR 2003, Philadelphia, PA, USA, June 23–24, 2003. Lecture Notes in Computer Science, vol. 2717, pp. 291–300. Springer, Heidelberg (2003a)
94. Rohlfing, T., Maurer J, C.R., Bluemke, D., Jacobs, M.: Volume-preserving nonrigid registration of mr breast images using free-form deformation with an incompressibility constraint. IEEE Trans. Med. Imaging **22**(6), 730–741 (2003b)
95. Roose, L., Maerteleire, W.D., Mollemans, W., Suetens, P.: Validation of different soft tissue simulation methods for breast augmentation. Int. Congr. Ser. **1281**, 485–490 (2005)
96. Roose, L., De Maerteleire, W., Mollemans, W., Maes, F., Suetens, P.: Simulation of soft-tissue deformations for breast augmentation planning. In: Harders, M., Szkely, G. (eds.) Biomedical Simulation, Lecture Notes in Computer Science, vol. 4072, pp. 197–205. Springer, Heidelberg (2006)
97. Roose, L., Maerteleire, W.D., Mollemans, W., Maes, F., Suetens, P.: Pre-operative simulation and post-operative validation of soft-tissue deformations for breast implantation planning. In: Cleary, K.R., Galloway, R.L., Jr (eds.) Medical Imaging 2006: Visualization, Image-Guided Procedures, and Display, SPIE, vol. 6141, p. 61410Z (2006)
98. Roose, L., Mollemans, W., Loeckx, D., Maes, F., Suetens, P.: Biomechanically based elastic breast registration using mass tensor simulation. In: Larsen, R., Nielsen, M., Sporring, J. (eds.) Medical Image Computing and Computer-Assisted Intervention MICCAI 2006, Lecture Notes in Computer Science, vol. 4191, pp. 718–725. Springer, Heidelberg (2006)
99. Roose, L., Loeckx, D., Mollemans, W., Maes, F., Suetens, P.: Adaptive boundary conditions for physically based follow-up breast mr image registration. In: Metaxas, D., Axel, L., Fichtinger, G., Szkely, G. (eds.) Medical Image Computing and Computer-Assisted Intervention MICCAI 2008, Lecture Notes in Computer Science, vol. 5242, pp. 839–846. Springer, Heidelberg (2008)
100. Rueckert, D., Sonoda, L.I., Hayes, C., Hill, D.L., Leach, M.O., Hawkes, D.J.: Nonrigid registration using free-form deformations: application to breast mr images. IEEE Trans. Med. Imaging **18**(8), 712–721 (1999)

101. Ruiter, N.V.: Registration of X-ray mammograms and MR-volumes of the female breast based on simulated mammographic deformation PhD thesis, der Universitat Mannheim (2003)
102. Ruiter, N.V., Stotzka, R., Muller, T.O., Gemmeke, H., Reichenbach, J.R., Kaiser, W.A.: Model-based registration of X-ray mammograms and MR images of the female breast. In: IEEE Transactions on Nuclear Science, Institute of Electrical and Electronics Engineers, New York, NY, ETATS-UNIS, vol. 53, p. 8, anglais (2006)
103. Samani, A., Plewes, D.: A method to measure the hyperelastic parameters of ex vivo breast tissue samples. Phys. Med. Biol. **49**(18), 4395 (2004)
104. Samani, A., Plewes, D.: An inverse problem solution for measuring the elastic modulus of intact ex vivo breast tissue tumours. Phys. Med. Biol. **52**(5), 1247 (2007)
105. Samani, A., Bishop, J., Yaffe, M.J., Plewes, D.B.: Biomechanical 3-d finite element modeling of the human breast using mri data. IEEE Trans. Med. Imaging **20**(4), 271–279 (2001)
106. Samani, A., Bishop, J., Luginbuhl, C., Plewes, D.B.: Measuring the elastic modulus of ex vivo small tissue samples. Phys. Med. Biol. **48**(14), 2183 (2003)
107. Samani, A., Zubovits, J., Plewes, D.B.: Elastic moduli of normal and pathological human breast tissues: an inversion-technique-based investigation of 169 samples. Phys. Med. Biol. **52**(6), 1565–1576 (2007)
108. Sarkar, S., Zhang, Y., Qiu, Y., Goldgof, D., Li, L.: 3d finite element modeling of nonrigid breast deformation for feature registration in -ray and mr images. In: Applications of Computer Vision, 2007. WACV '07. IEEE Workshop on, p. 38 (2007)
109. Schnabel, J.A., Tanner, C., Castellano-Smith, A.D., Degenhard, A., Leach, M.O., Hose, D.R., Hill, D.L.G., Hawkes, D.J.: Validation of nonrigid image registration using finite-element methods: application to breast mr images. IEEE Trans. Med. Imaging **22**(2), 238–247 (2003)
110. Shih, T.C., Chen, J.H., Liu, D., Nie, K., Sun, L., Lin, M., Chang, D., Nalcioglu, O., Su, M.Y.: Computational simulation of breast compression based on segmented breast and fibroglandular tissues on magnetic resonance images. Phys. Med. Biol. **55**(14), 4153–4168 (2010)
111. Sinkus, R., Weiss, S., Wigger, E., Lorenzen, J., Dargatz, M., Kuhl, C.: Nonlinear elastic tissue properties of the breast measured by mr-elastography: initial in vitro and In vivo results. In: Proceedings ISMRM 10th Annual Meeting, p. 33 (2002)
112. Sivaramakrishna, R.: 3d breast image registration–a review. Technol. Cancer Res. Treat. **4**(1), 39–48 (2005)
113. Spear, S.L., Spittler, C.J.: Breast reconstruction with implants and expanders. Plast. Reconstr. Surg. **107**(1), 177–87 (2001)
114. Stein, K., Tezduyar, T.E., Benney, R.: Automatic mesh update with the solid-extension mesh moving technique. Comput. Methods Appl. Mech. Eng. **193**, 2019–2032 (2004)
115. Tanner, C., Degenhard, A., Schnabel, J., Smith, A., Hayes, C., Sonoda, L., Leach, M., Hose, D., Hill, D., Hawkes, D.: A method for the comparison of biomechanical breast models. In: Mathematical Methods in Biomedical Image Analysis, 2001. MMBIA 2001. IEEE Workshop on, pp. 11–18 (2001)
116. Tanner, C., Schnabel, J., Degenhard, A., Castellano-Smith, A., Hayes, C., Leach, M., Hose, D., Hill, D., Hawkes, D.: Validation of volume-preserving non-rigid registration: application to contrast-enhanced mr-mammography. In: Dohi, T., Kikinis, R. (eds.) Medical Image Computing and Computer-Assisted Intervention MICCAI, Lecture Notes in Computer Science, vol. 2488, pp. 307–314. Springer, Heidelberg (2002)
117. Tanner, C., Carter, T., Hawkes, D.J.: 3d rezoning for finite element modelling of large breast deformations. In: European Modelling Symposium, pp. 51–53 (2006a)
118. Tanner, C., Schnabel, J.A., Hill, D.L.G., Hawkes, D.J., Leach, M.O., Hose, D.R.: Factors influencing the accuracy of biomechanical breast models. Med. Phys. **33**(6), 1758–1769 (2006b)
119. Tanner, C., Hipwell, J., Hawkes, D.: Statistical deformation models of breast compressions from biomechanical simulations. In: Krupinski, E. (ed.) Digital Mammography, Lecture Notes in Computer Science, vol. 5116, pp. 426–432. Springer, Heidelberg (2008)

120. Tanner C., White M., Guarino S., Hall-Craggs M.A., Douek M., Hawkes D.J. (2009) Anisotropic behaviour of breast tissue for large compressions. In: ISBI'09: Proceedings of the 6th IEEE international conference on Symposium on Biomedical Imaging. IEEE Press, Piscataway, NJ, pp. 1223–1226
121. Tanner, C., Hipwell, J.H., Hawkes, D.J.: Digital Mammography, Breast Shapes on Real and Simulated Mammograms, Chap Breast Shapes on Real and Simulated Mammograms, Lecture Notes in Computer Science, pp. 540–547. Springer, Heidelberg (2010)
122. Tanner, C., White, M., Guarino, S., Hall-Craggs, M.A., Douek, M., Hawkes, D.J.: Large breast compressions: observations and evaluation of simulations. Med. Phys. **38**(2), 682–690 (2011)
123. Unnikrishnan, G.U., Unnikrishnan, V.U., Reddy, J.N., Lim, C.T.: Review on the constitutive models of tumor tissue for computational analysis. Appl. Mech. Rev. **63**(4), 040801 (2010)
124. Veronesi, U., Boyle, P., Goldhirsch, A., Orecchia, R., Viale, G.: Breast cancer. The Lancet **365**(9472), 1727–1741 (2005)
125. Wellman, P.S., Howe, R.D., Dalton, E., Kern, K.A.: Breast Tissue Stiffness in Compression is Correlated to Histological Diagnosis, Technical Report. Harvard University, Cambridge (1999)
126. Wessel, C., Schnabel, J., Brady, S.: Towards more realistic biomechanical modelling of tumours under mammographic compressions. In: Mart, J., Oliver, A., Freixenet, J., Mart, R. (eds.) Digital Mammography, Lecture Notes in Computer Science, vol. 6136, pp. 481–489. Springer, Heidelberg (2010)
127. Whiteley, J.P., Gavaghan, D.J., Chapman, S.J., Brady, J.M.: Non-linear modelling of breast tissue. Math. Med. Biol. **24**(3), 327–345 (2007)
128. Wilhelmi, B.J., Blackwell, S.J., Phillips, L.G.: Langer's lines: to use or not to use. Plast. Reconstr. Surg. **104**(1), 208–214 (1999)
129. Willson, S.A., Adam, E.J., Tucker, A.K.: Patterns of breast skin thickness in normal mammograms. Clin. Radiol. **33**(6), 691–693 (1982)
130. Wirth, M.: A nonrigid approach to medical image registration: matching images of the breast. PhD thesis, RMIT University (2000)
131. Yeh, W.C., Li, P.C., Jeng, Y.M., Hsu, H.C., Kuo, P.L., Li, M.L., Yang, P.M., Lee, P.H.: Elastic modulus measurements of human liver and correlation with pathology. Ultrasound Med. Biol. **28**(4), 467–474 (2002)
132. Yin, H.M., Sun, L.Z., Wang, G., Yamada, T., Wang, J., Vannier, M.W.: Imageparser: a tool for finite element generation from three-dimensional medical images. Biomed. Eng. Online **3**, 31 (2004)
133. Zienkiewicz, O., Taylor, R.: The Finite Element Method: The Basis. vol. 1, 5th edn. Butterworth-Heinemann, London (2000)
134. Zienkiewicz, O., Taylor, R.: The finite element method for solid and structural mechanics. 6th edn. Elsevier Butterworth-Heinemann, London (2005)
135. Zyganitidis, C., Bliznakova, K., Pallikarakis, N.: A novel simulation algorithm for soft tissue compression. Med. Biol. Eng. Comput. **45**, 661–669 (2007)

Soft-Tissue Simulation for Cranio-Maxillofacial Surgery: Clinical Needs and Technical Aspects

Hyungmin Kim, Philipp Jürgens and Mauricio Reyes

Abstract Computerized soft-tissue simulation can provide unprecedented means for predicting facial outlook pre-operatively. Surgeons can virtually perform several surgical plans to have the best surgical results for their patients while considering corresponding soft-tissue outcome. It could be used as an interactive communication tool with their patients as well. There has been comprehensive amount of works for simulating soft-tissue for cranio-maxillofacial surgery. Although some of them have been realized as commercial products, none of them has been fully integrated into clinical practice due to the lack of accuracy and excessive amount of processing time. In this chapter, state-of-the-art and general workflow in facial soft-tissue simulation will be presented, along with an example of patient-specific facial soft-tissue simulation method.

1 Introduction

1.1 Facial Aesthetics

The human face is the most complex anatomical, physiological and functional region of the body. The senses of vision, hearing, smell and taste are concentrated in the facial region. And the skin of the face has the highest number of sensibility receptors. The oral cavity serves for ingestion and creation of speech and the upper airway allows respiration. Twenty-eight mimic muscles help to express emotions and thoughts. The face is the most important part of the body that helps us to

---

H. Kim · P. Jürgens · M. Reyes (✉)
University of Bern, Bern, Switzerland
e-mail: mauricio.reyes@istb.unibe.ch

Fig. 1 Golden rectangle (Source: Wikipedia)

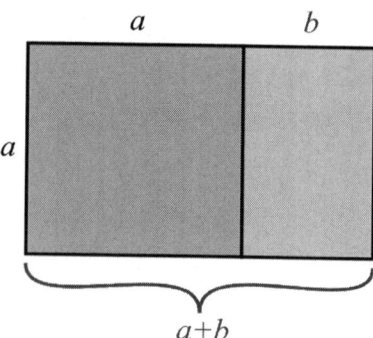

identify other people and recognize in unknown person's gender, approximate age and even constitution. While a disfigured face impresses and causes certain reactions on a beautiful face fascinates and attracts them.

In general beauty is associated with other positive characteristics: we think that a beautiful person is more likable, communicative, successful and popular than others. But how can we define beauty? David Hume's statement "Beauty exists merely in the mind which contemplates them" [18] is a contradiction to the attempt of defining objective standards for beauty. But if an intercultural comparison is applied, there is a high accordance about the perception of an aesthetic facial appearance among ethnical groups and different age classes [9, 30]. Even little children are more attracted to attractive adults, than to not-attractive individuals [29].

Over the past centuries various disciplines in science and arts have tried to answer the question which criteria make a face attractive. The face of a human being is considered to be attractive if it presents the following properties:

- Symmetry [18]
- Average [29]
- Individual characteristics [8]

An average face means a facial appearance close to the mean of the population. If an unknown person presents an average face the person seems familiar to the opponent, because it corresponds to the previously learned mean of the category facial appearance [28]. The average face is subconsciously associated with a good phenotype meaning that the stabilizing effect of natural selection of certain facial appearances is recognized [53].

Comparable factors can be detected when analyzing symmetry: symmetric facial proportions indicate healthy genes. It is even supposed that this is a characteristic for a higher resistance against parasites [53].

In contrary to symmetry and average, certain unique characteristics allow us to recognize an individual face, which is essential for the perception of an individual and the development of a personal identity [8].

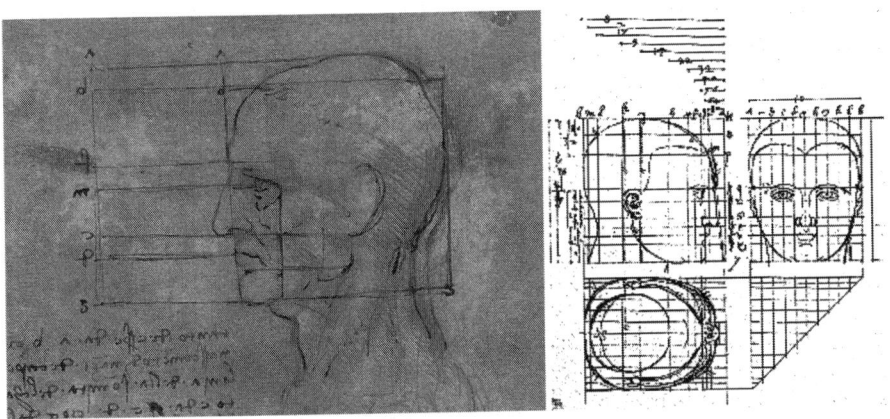

Fig. 2 Proportions of the human head: Leonardo da Vinci (*left*) (Source: Wikipedia), Albrecht Dürer (*right*) [12]

Besides symmetry and average an extremely important factor are proportions. The embodiment of harmonic and aesthetic proportions and even mathematical beauty is the golden ratio mainly applied in architecture and arts.

$$\frac{a+b}{a} = \frac{a}{b} \equiv \varphi = \frac{1+\sqrt{5}}{2} = 1.6180339887\ldots \qquad (1)$$

When analyzing facial proportions, we also find proportions close to the golden ratio (Fig. 1): most famous are the proportion analysis of Albrecht Dürer and Leonardo da Vinci as in Fig. 2.

1.2 Cranio-Maxillofacial Surgery

Cranio-maxillofacial (CMF) surgery is a surgical specialty that deals with the treatment of inborn or acquired facial disfigurements. These conditions can be such as cleft lip- and palate, craniofacial malformations, aftermath of facial trauma or of ablative tumor surgery. Surgical interventions in the CMF area and even their planning make high demands on the spatial sense of the surgeons. This is on one the hand due to the close proximity of highly vulnerable anatomical structures and on the other hand due to the complex morphology of the region. Modern image-guided techniques are the basis for diagnostics, therapy and documentation. These technologies enable us to produce patient-specific models of the clinical situation. They give us the possibility to perform accurate planning and transfer the planning to the operation theatre. These technologies have made their way into the clinical routine of highly advanced treatment centers [6, 13, 20, 21, 22, 40]. One of the most evident indications for the use of virtual planning tools in CMF Surgery is the planning of surgical intervention for patients suffering of malocclusion. Malocclusion can either be caused by a malposition of teeth in the level of the alveolar crest or by an incorrect positioning of the upper and lower jaw relative to each other. For the former, an orthodontic treatment will deliver satisfactory results. For the latter, only a surgical procedure will provide a causal therapy. These interventions are called orthognathic surgeries and their aim is to change the position of the maxillary and mandibular bone, relative to each other and to the skull base [1, 10, 17]. As these interventions are highly elective, an accurate and extensive preoperative planning has to be conducted. The technique for planning and performing the simultaneous relocation of maxilla and mandible was first described in 1970 by Obwegeser [43]. To update the planning procedure several systems for virtual three-dimensional visualization and procedure planning based on volume datasets have been recently introduced in some clinical centers, routinely substituting the conventional two-dimensional cephalogram based planning-approach, and especially improving the prediction of soft tissue deformations [3, 42, 54, 57, 58]. In order to ensure an optimal pre-operative skeletal planning of the patient with his postoperative facial appearance, a highly reliable and accurate prediction system is required [25, 26].

In order to realize the pre-operative surgical plan in the operation theatre, the planning and prediction software should be linked to navigation system for the intra-operative control of the relocation of the upper and lower jaw [6, 20, 22, 61]. As a consequence, a planning console for CMF surgery should contain the following modules to cover all the clinical requirements:

- Import module of DICOM data
- User-friendly segmentation and mesh modeling
- Module for virtual osteotomy and relocation of bone segments
- Three-dimensional cephalometry
- Fast and accurate soft-tissue simulation
- Connection to intra-operative navigation system

1.3 Soft-Tissue Simulation in CMF Surgery: State-of-the-Art

There have been already extensive amount of studies in order to overcome the difficulty in soft-tissue modeling. Vannier et al. [56] originally introduced the idea of predicting post-operative facial appearance for CMF surgery using computer-assisted approach. Later, face modeling was intensively studied by the computer graphics community, whose main interest was in simulating facial expressions in real-time. Terzopoulos et al. [50] and Lee et al. [32] suggested multi-layered facial tissue model consisting of mass-spring elements. On the other hand, Koch et al. [27] suggested a non-linear finite-element approach using C^1-continuous surface element and linear springs connected to bony structures. Keeve et al. [24]

presented a Mass-Spring Model (MSM) with prismatic element, and compared the result with Finite-Element Model (FEM) in terms of accuracy and computational cost. Sarti et al. [48] proposed a voxel-based FEM approach based on a super computer, which can be directly applied to voxel grid of the Computed Tomography (CT) data at the cost of large memory space. Zachow et al. [59] suggested a fast tetrahedral volumetric FEM, which can be applicable to clinical practice. Later, Gladilin et al. [16] extended this to functional simulation with facial muscles. Alternatively, Cotin et al. [7] proposed a hybrid method using Mass-Tensor Model (MTM) for enhanced local deformations in simulation of surgical sites. Later, MTM was extended to non-linear, anisotropic elasticity by Picinbono et al. [46]. Chabanas et al. [5] proposed a mesh morphing algorithm to minimize the laborious efforts on preparing finite-element mesh. Mollemans et al. [39] first applied MTM to CMF soft-tissue simulation, and the method was qualitatively and quantitatively evaluated on ten clinical cases. Some of the previously developed methods have been realized as commercial products, e.g. SurgiCase CMF (Materialise, Leuven, Belgium), Maxilim (Medicim NV, Mechelen, Belgium), Dolphin 3D Surgery (Dolphin Imaging & Management Systems, Chatsworth, CA), 3dMDvultus (3dMD, Atlanta, Ga), In Vivo (Anatomage Inc, San Jose, CA).

Even though considerable amount of research has been conducted for resolving soft-tissue simulation, up to our knowledge none of it has been fully integrated into the clinical workflow, due to insufficient accuracy or excessive processing time required for simulation. The state-of-the-art simulation results around error-sensitive regions, such as nose and lips area, are still far from clinical reality, e.g. nose tip stays in stationary position under the effect of considerable movements of the underlying skeletal segments. This results in the need for patient-specific simulation strategy, that considers individual facial muscular structures for soft-tissue simulation. At the same time, the excessive amount of processing time for patient-specific modeling should be avoided, in order to render the developed method useful for daily clinical practice. Accordingly, a simulation framework enabling reduced modeling efforts, and enhanced computational performance is anticipated. Fortunately, rapid development in computing hardware, e.g. multi-core central processing unit (CPU) or graphics processing unit (GPU), makes this situation favorable.

1.4 Computational Models for Soft-Tissue Simulation

The computational methods, developed for simulating soft-tissues behavior so far, can be categorized into the following three types: MSM, FEM, and MTM. MSM is known to be the most computationally efficient method due to its simplicity in physics. However, the coarse approximation of true physics is not suitable for surgical planning. On the other hand, the mathematical background of FEM is based on continuum mechanics, and its utmost accuracy has been proved in other

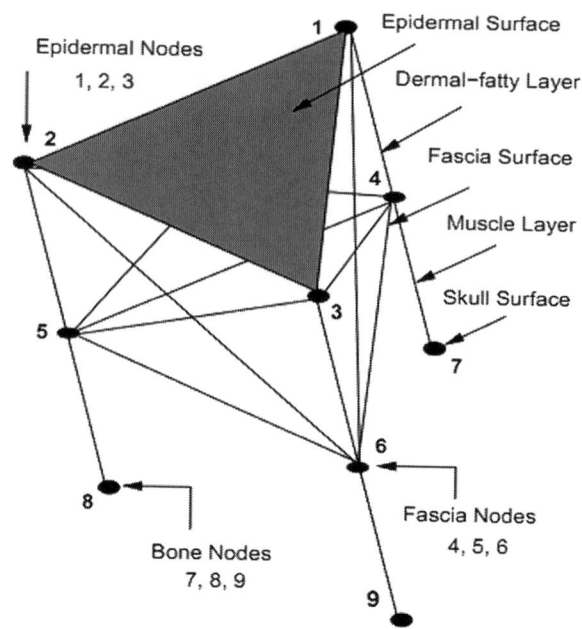

Fig. 3 Multi-layered mass-spring model (Source: Lee et al. [32])

scientific applications. However, it usually requires considerable amount of time for preparation of mesh and computation of soft-tissue deformations. This makes FEM hard to be used in daily clinical applications without using specialized hardware-specific implementation. MTM was developed as a golden mean between MSM and FEM, which is known to be bio-mechanically relevant as FEM, and computationally efficient as MSM. Those models have been actively used in various applications with its unique features, although recent developments in computation hardware has bridged the gap between these strategies.

1.4.1 Mass-Spring Model

In MSM, the object is discretized into mass points and inter-connections between each of these mass points. As shown in Fig. 3, layer-based MSM has been widely used for characterizing multiple soft-tissue layers, e.g. epidermis, dermis, SMAS (Superficial Muscular Aponeurotic System), fascia and muscles.

The governing equation for MSM can be described by Eq. 2 and is known as Hooke's law, where the elastic forces $\mathbf{f}_s$ is proportional to the changes in length of the spring element ΔL between the actual and the rest status at the rate of spring constant k_s.

$$\mathbf{f}_s = k_s \Delta L \qquad (2)$$

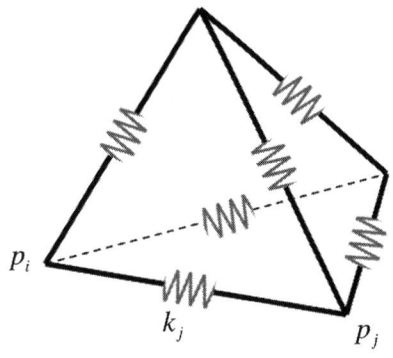

Fig. 4 Tetrahedral mass-spring model: 3D coordinate of point i and j ($\mathbf{p}_i, \mathbf{p}_j$), spring constant of connecting spring k_j

Alternatively, tetrahedral-based MSM was proposed by Mollemans et al. [37], where automatic mesh generation method can be used for constructing volumetric meshes from surface meshes. The governing equation can be expressed by Eq. 3, which correlates the total internal elastic force of point i ($\mathbf{f}_i^{int}$) with the set of vertices connected to the mass point i (Ψ_i), 3D coordinate of point i and j ($\mathbf{p}_i, \mathbf{p}_j$), the length of the spring at rest L_j^0, and spring constant of connecting spring k_j as shown in Fig. 4.

$$\mathbf{f}_i^{int} = \sum_{\forall \mathbf{p}_j \in \Psi_i} k_j \left(\| \mathbf{p}_j - \mathbf{p}_i \| - L_j^0 \right) \frac{\mathbf{p}_j - \mathbf{p}_i}{\| \mathbf{p}_j - \mathbf{p}_i \|} \quad (3)$$

According to Van Gelder et al. [15], the spring constant of spring can be expressed by

$$k_j = \frac{E \sum_{e \in \Lambda_j} V_e}{(L_j^0)^2} \quad (4)$$

where Λ_j is the collection of all neighboring tetrahedra containing spring j, V_e is the initial volume of the neighboring tetrahedra, and L_j^0 is the initial length of the spring.

Since there is little interest in the transient phase of the deformation for CMF soft-tissue simulation, the direct computation approach proposed by Teschner et al. [52] can be used, which enables the deformation to be computed in iterative ways by setting the internal force in equilibrium state to be zero.

1.4.2 Finite-Element Model

Soft-tissue simulation can be regarded as the task of finding the displacement vector field which maps all points in initial shape to deformed shape, where the deformation caused by the displacements or forces on the boundary, described by the deformation function $\Phi(\mathbf{x})$ as shown in Fig. 5.

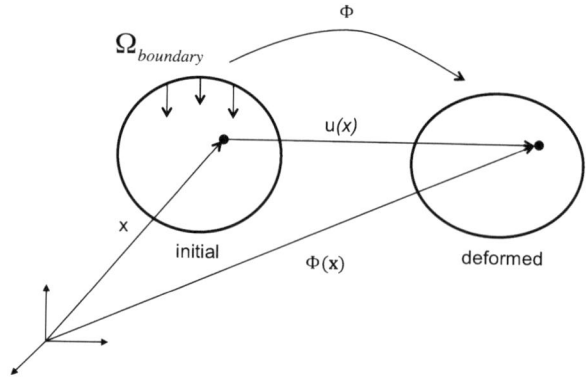

Fig. 5 Problem description of soft-tissue simulation: the deformation function $\Phi(\mathbf{x})$, displacement vector **u**, spatial position vector **x**, domain for boundary condition $\Omega_{boundary}$

The displacement vector **u** is defined by

$$\mathbf{u} = \Phi(\mathbf{x}) - \mathbf{x} \quad (5)$$

The change of metric in the deformed body can be measured by the right Cauchy-Green deformation tensor **C**, which are expressed by the deformation gradient.

$$\mathbf{C} = \nabla\Phi^T \nabla\Phi \quad (6)$$

The amount of deformation can be expressed by the 3×3 Green-Lagrange strain tensor $\mathbf{E}(\mathbf{x})$, whose diagonal terms represent the length variation and off-diagonal terms state the shear effect along the three axes.

$$\mathbf{E} = \frac{1}{2}(\mathbf{C} - \mathbf{I}) = \frac{1}{2}(\nabla\Phi^T \nabla\Phi - I) \quad (7)$$

According to Eq. 5, $\nabla\Phi = I + \nabla\mathbf{u}$ can be deduced, and Eq. 7 can be expressed as

$$\mathbf{E} = \frac{1}{2}((I + \nabla\mathbf{u})^T(I + \nabla\mathbf{u}) - I) = \frac{1}{2}(\nabla\mathbf{u} + \nabla\mathbf{u}^T + \nabla\mathbf{u}^T \nabla\mathbf{u}) \quad (8)$$

Assuming small displacements, linearized Green-Lagrange strain tensor $\mathbf{E}_L$ can be used:

$$\mathbf{E}_L = \frac{1}{2}(\nabla\mathbf{u} + \nabla\mathbf{u}^T) = \begin{bmatrix} \varepsilon_{xx} & \frac{\gamma_{xy}}{2} & \frac{\gamma_{xz}}{2} \\ \frac{\gamma_{xy}}{2} & \varepsilon_{yy} & \frac{\gamma_{yz}}{2} \\ \frac{\gamma_{xz}}{2} & \frac{\gamma_{yz}}{2} & \varepsilon_{zz} \end{bmatrix} \quad (9)$$

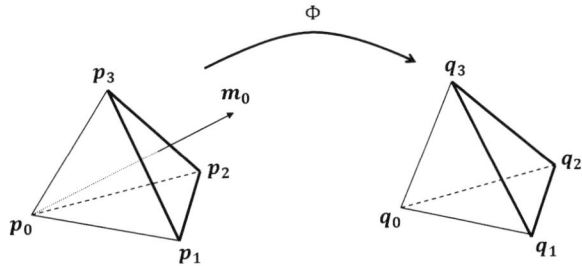

Fig. 6 Deformation of tetrahedral element

The constitutive equation for a linear isotropic material is given by Hooke's law [13]:

$$\begin{bmatrix} \sigma_{xx} \\ \sigma_{yy} \\ \sigma_{zz} \\ \sigma_{xy} \\ \sigma_{yz} \\ \sigma_{zx} \end{bmatrix} = \begin{bmatrix} \lambda+2\mu & \lambda & \lambda & 0 & 0 & 0 \\ \lambda & \lambda+2\mu & \lambda & 0 & 0 & 0 \\ \lambda & \lambda & \lambda+2\mu & 0 & 0 & 0 \\ 0 & 0 & 0 & \mu & 0 & 0 \\ 0 & 0 & 0 & 0 & \mu & 0 \\ 0 & 0 & 0 & 0 & 0 & \mu \end{bmatrix} \begin{bmatrix} \varepsilon_{xx} \\ \varepsilon_{yy} \\ \varepsilon_{zz} \\ \gamma_{xy} \\ \gamma_{yz} \\ \gamma_{zx} \end{bmatrix} \Leftrightarrow \sigma = \mathbf{D}\varepsilon \quad (10)$$

where Lamé coefficients λ and μ are expressed in relationship with Young's modulus E and Poisson's ratio v:

$$\lambda = \frac{Ev}{(1+v)(1-2v)} \quad (11)$$

$$\mu = \frac{E}{2(1+v)} \quad (12)$$

FEM theory explains that the elastic deformation of the entire object can be described by the elastic behavior of each element descritizing the object. Therefore, the displacement **u** inside the linear tetrahedral element shown in Fig. 6 can be expressed as a linear combination of the displacements of the four vertices of the tetrahedron:

$$\mathbf{u} = \sum_{k=0}^{3} \mathbf{N}_k^e \mathbf{u}_k^e = \mathbf{N}^e \mathbf{u}^e \quad (13)$$

where

$$N_k^e = \frac{a_k + b_k x + c_k y + d_k z}{6V^e},$$

here V^e is the volume of the tetrahedron, and a_k, b_k, c_k, d_k are defined by

$$a_k = \begin{vmatrix} x_{k+1} & y_{k+1} & z_{k+1} \\ x_{k+2} & y_{k+2} & z_{k+2} \\ x_{k+3} & y_{k+3} & z_{k+3} \end{vmatrix}$$

$$b_k = \begin{vmatrix} 1 & y_{k+1} & z_{k+1} \\ 1 & y_{k+2} & z_{k+2} \\ 1 & y_{k+3} & z_{k+3} \end{vmatrix}$$

$$c_k = -\begin{vmatrix} x_{k+1} & 1 & z_{k+1} \\ x_{k+2} & 1 & z_{k+2} \\ x_{k+3} & 1 & z_{k+3} \end{vmatrix} \quad (14)$$

$$d_k = -\begin{vmatrix} x_{k+1} & y_{k+1} & 1 \\ x_{k+2} & y_{k+2} & 1 \\ x_{k+3} & y_{k+3} & 1 \end{vmatrix}$$

The strain ε in Eq. 10 can be expressed as:

$$\varepsilon = \begin{bmatrix} \varepsilon_{xx} \\ \varepsilon_{yy} \\ \varepsilon_{zz} \\ \gamma_{xy} \\ \gamma_{yz} \\ \gamma_{zx} \end{bmatrix} = \begin{bmatrix} \frac{\partial u}{\partial x} \\ \frac{\partial v}{\partial y} \\ \frac{\partial w}{\partial z} \\ \frac{\partial u}{\partial y} + \frac{\partial v}{\partial x} \\ \frac{\partial v}{\partial z} + \frac{\partial w}{\partial y} \\ \frac{\partial u}{\partial z} + \frac{\partial w}{\partial x} \end{bmatrix} = \mathbf{B}\mathbf{u}^e = [\mathbf{B}_k \mathbf{B}_{k+1} \mathbf{B}_{k+2} \mathbf{B}_{k+3}]\mathbf{u}^e \quad (15)$$

where $\mathbf{B}_k$ is expressed by

$$\mathbf{B}_k = \begin{bmatrix} \frac{\partial N_k}{\partial x} & 0 & 0 \\ 0 & \frac{\partial N_k}{\partial y} & 0 \\ 0 & 0 & \frac{\partial N_k}{\partial z} \\ \frac{\partial N_k}{\partial y} & \frac{\partial N_k}{\partial x} & 0 \\ 0 & \frac{\partial N_k}{\partial z} & \frac{\partial N_k}{\partial y} \\ \frac{\partial N_k}{\partial z} & 0 & \frac{\partial N_k}{\partial x} \end{bmatrix} = \frac{1}{6V^e}\begin{bmatrix} b_k & 0 & 0 \\ 0 & c_k & 0 \\ 0 & 0 & d_k \\ c_k & b_k & 0 \\ 0 & d_k & c_k \\ d_k & 0 & b_k \end{bmatrix} \quad (16)$$

The principle of virtual work gives the relationship between the work done by the internal stresses and external work performed by the elastic force in equilibrium state:

$$\delta \mathbf{u}^{e^T} \mathbf{q}^e = \int_{V^e} \delta \varepsilon^T \sigma dV \quad (17)$$

where $\mathbf{q}^e$ is the elastic force vector, $\delta \mathbf{u}^e$ is the virtual displacement vector.
Combined with Eq. 16, the following equation is deduced.

$$\delta \mathbf{u}^{e^T} \mathbf{q}^e = \int_{V^e} (\mathbf{B}\delta \mathbf{u}^e)^T \sigma dV = \delta \mathbf{u}^{e^T} \int_{V^e} \mathbf{B}^T \sigma dV \quad (18)$$

Since Eq. 18 must be valid for any virtual displacement $\delta\mathbf{u}^e$,

$$\mathbf{q}^e = \int_{V^e} \mathbf{B}^T \sigma dV = \mathbf{K}^e \mathbf{u}^e \tag{19}$$

where $\mathbf{K}^e = \int_{V^e} \mathbf{B}^T \mathbf{D} \mathbf{B} dV$

The total elastic force $\mathbf{F}_k$ on vertex k can be obtained by summing the elastic force contributions of neighboring tetrahedra containing vertex k:

$$\mathbf{F}_k = \sum_{\forall j \in \Omega_k} \mathbf{q}_k^j = \sum_{\forall j \in \Omega_k} \mathbf{K}^j \mathbf{u}^j \tag{20}$$

Combining Eq. 20 in matrix form for m vertices:

$$\mathbf{F} = \mathbf{K}\mathbf{u} \tag{21}$$

where elastic force vector $\mathbf{F}$ and displacement vector $\mathbf{u}$ have dimension of $(m \times 3) \times 1$, and $\mathbf{K}$ is called the global stiffness matrix.

For all joined vertices which we know the displacements, the boundary condition is given as,

$$\mathbf{u}_k = \mathbf{u}_k^0 \tag{22}$$

For all free vertices where displacements need to be updated,

$$\mathbf{F}_k = 0 \tag{23}$$

Since the global matrix $\mathbf{K}$ is very sparse whose element $\mathbf{K}_{ij}$ is only non-zero if model point i and j are vertices of a common tetrahedron, special algorithms can be used for solving these linear matrix system to speed up the calculation.

1.4.3 Mass-Tensor Model

In MTM, the object is discretized with tetrahedral meshes as shown in Fig. 6.

Similar to the finite element theory, the displacement inside each tetrahedron e can be expressed by a linear interpolation of the displacement of the four vertices u_j

$$\mathbf{u}(\mathbf{x}) = -\sum_{j=0}^{3} \frac{\mathbf{m}_j \cdot (\mathbf{x} - \mathbf{p}_{j+1})}{6V^e} \mathbf{u}_j \tag{24}$$

where V^e is the initial volume of tetrahedron e, and $\mathbf{m}_j$ is the area vector defined by

$$\mathbf{m}_j = (-1)^{j+1} \left(\mathbf{p}_{j+1} \times \mathbf{p}_{j+2} + \mathbf{p}_{j+2} \times \mathbf{p}_{j+3} + \mathbf{p}_{j+3} \times \mathbf{p}_{j+1} \right) \tag{25}$$

The nodal force exerted at jth vertex of tetrahedron e is defined as the derivative of the elastic energy:

Table 1 Hyperelastic material models

St-Venant Kirchoff	$w = \frac{\lambda}{2}(tr\mathbf{E})^2 + \mu tr\mathbf{E}^2$
Neo-Hookean	$w = \frac{\mu}{2} tr\mathbf{E} + f(I_3)$
Fung isotropic	$w = \frac{\mu}{2} e^{tr\mathbf{E}} + f(I_3)$
Fung anisotropic	$w = \frac{\mu}{2} e^{tr\mathbf{E}} + \frac{k_1}{k_2}\left(e^{k_2(I_4-1)} - 1\right) + f(I_3)$
Veronda–Westman	$w = c_1 e^{\gamma tr\mathbf{E}} + c_2 tr\mathbf{E}^2 + f(I_3)$
Mooney–Rivlin	$w = c_{10} tr\mathbf{E} + c_{01} tr\mathbf{E}^2 + f(I_3)$

$$\mathbf{f}_j^e = -\frac{\partial W}{\partial \mathbf{p}_j} \qquad (26)$$

where elastic energy required to deform a body can be expressed as a function of the invariants ($I_1 = tr\mathbf{E}, I_2 = tr\mathbf{E}^2, I_3 = det\mathbf{E}$) of strain tensor $\mathbf{E}$:

$$W = \int w(I_1, I_2, I_3) dX \qquad (27)$$

There are several hyperelastic material models are available as shown in Table 1.

By using St-Venant Kirchoff elastic enery model in Table 1 and linearized strain tensor in Eq. 9, the following equation can be deduced:

$$\mathbf{f}_j^e = \sum_{k=0}^{3} \mathbf{K}_{jk}^e \mathbf{u}_k \qquad (28)$$

where the 3×3 stiffness tensors $\mathbf{K}_{jk}^e$ is expressed as

$$\mathbf{K}_{jk}^e = \frac{1}{36 V^e}\left(\lambda(\mathbf{m}_j \otimes \mathbf{m}_k) + \mu(\mathbf{m}_k \otimes \mathbf{m}_j) + \mu(\mathbf{m}_j \cdot \mathbf{m}_k)\mathbf{I}\right) \qquad (29)$$

Here the area vector is defined by Eq. 25, its tensor product is defined by $\mathbf{a} \otimes \mathbf{b} = \mathbf{a}\mathbf{b}^T$, and V^e is the initial volume of tetrahedron e.

The total elastic force exerted on vertex j is the sum of the local force across all tetrahedra containing vertex j:

$$\mathbf{f}_j = \sum_{\forall e \in \Omega_j} \mathbf{f}_j^e = \sum_{\forall e \in \Omega_j} \sum_{k=0}^{3} \mathbf{K}_{jk}^e \mathbf{u}_k \qquad (30)$$

where Ω_j is the collection of all tetrahedra connected to vertex j.

By rewriting Eq. 30 into vertex and edge components, one can obtain the following final expressions for forces at each node j

$$\mathbf{f}_j = \mathbf{K}_{jj}\mathbf{u}_j + \sum_{\forall k \in \Psi_j} \mathbf{K}_{jk} \mathbf{u}_k \qquad (31)$$

where $\mathbf{K}_{jj} = \sum_{e \in \Omega_j} \mathbf{K}_{jj}^e$, $\mathbf{K}_{jk} = \sum_{e \in \Omega_j} \mathbf{K}_{jk}^e$, and Ψ_j is the collection of all the neighboring vertices connected to vertex j.

Since the stiffness tensors are only dependent on initial mesh configuration and material properties, they can be pre-computed and applied to various boundary conditions without re-computation.

By introducing direct computation [52], the solution can be obtained in iterative manners.

$$\mathbf{u}_j^t = -\mathbf{K}_{jj}^{-1} \left(\sum_{\forall k \in \Psi_j} \mathbf{K}_{jk} \mathbf{u}_k^{t-1} \right) \qquad (32)$$

2 Essential Procedures for Facial Soft-Tissue Simulation

Facial soft-tissue simulation consists of mainly five sub-tasks as shown in Fig. 7: image acquisition, tissue modeling, boundary condition, and tissue property assignment, actual simulation. The detailed description on each step will be presented in the following sub-sections.

2.1 Image Acquisition

CT scan is commonly used as a standard protocol for computer-assisted 3D planning. Since Magnetic Resonance Imaging (MRI) scan is needed to extract sub-anatomical soft structures. However, it is still required to have laborious efforts to segment anatomical structures from MRI scans. Additionally, 3D scan data can provide up-to-date outer skin profile with texture in high-precision. During image acquisition, it is highly recommended not to have any unnatural soft-tissue deformation caused by external forces (e.g. fixation bandage), and to ensure an appropriate imaging field-of-view (FOV) covering the whole frontal soft-tissue region, as shown in Fig. 8.

2.2 Tissue Modeling

2.2.1 Segmentation of Hard and Soft-Tissue

Generally, extraction of bony structures can be automated on CT scans by using simple intensity-based thresholding. On the other hand, segmentation of soft-tissue requires certain amount of manual adjustment. Image-based segmentation tools, such as lasso or magic wand in imaging software, such as Amira (Mercury Computer Systems, Germany), can be effectively used to accelerate manual segmentation

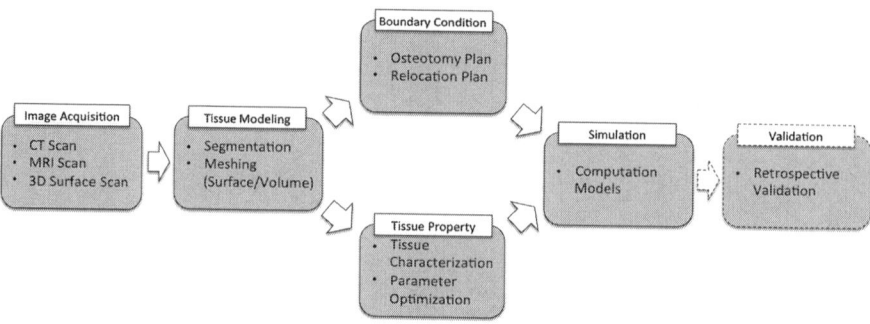

Fig. 7 General steps to solve soft-tissue simulation problem

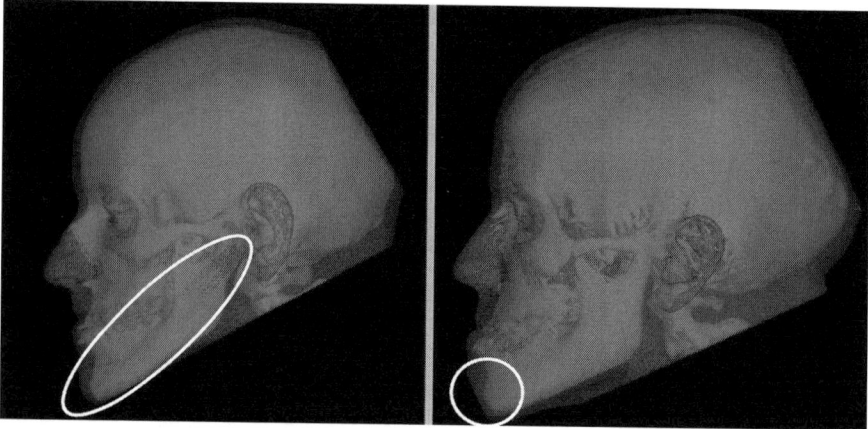

Fig. 8 Examples of undesirable scans: unnatural deformation caused by fixation bandage (*left*), FOV does not cover the whole frontal soft-tissue region (*right*)

procedures. As for advanced segmentation method, level-set based segmentation has been applied for segmentation of extra-cranial soft-tissue layer [36], and has proved to be effective in defining weak boundaries between intra- and extra-cranial soft-tissue. Additional efforts might be needed if it is required to separate upper and lower jaw around temporomandibular joint (TMJ). Furthermore, it is quite common to have streak artifacts caused by metal filling on patient's teeth, which makes segmentation task difficult for both hard and soft-tissues, as shown in Fig. 9 [34].

2.2.2 Surface and Volumetric Meshing

Based on the segmentation result, surface mesh can be generated using conventional marching cube algorithm [33], followed by automatized tetrahedral volumetric mesh generation will be followed. Due to strict limitation in converting

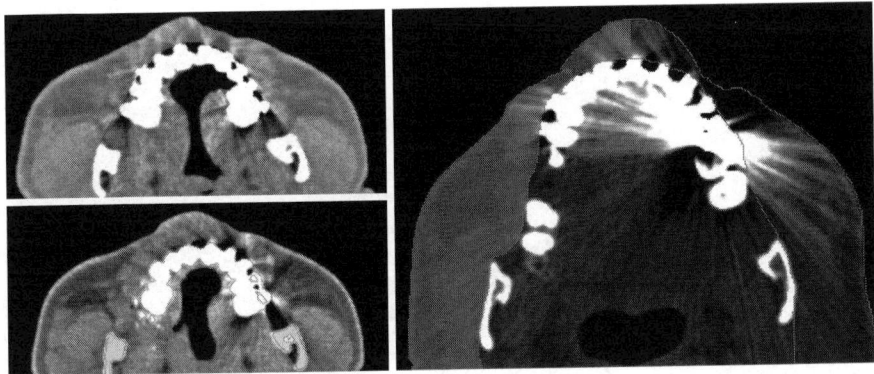

Fig. 9 Segmentation of extra-cranial soft-tissue and bony structures (*left*), metal streak artifacts are commonly present in clinical CT scans (*right*)

surface mesh to volumetric mesh, certain amount of iterative refinement in segmentation result is commonly required.

2.3 Boundary Condition Assignment

Once both entire hard- and soft-tissue models are created, bony segments can be prepared by the osteotomy planning software. Each movable part will define the corresponding contact regions with soft-tissue model. Then, a relocation planning module will define the planned movement of the parts, which will provide the corresponding movement of the affected regions on the soft-tissue.

As shown in Fig. 10, volumetric nodes are classified into four types: fixed, joined, sliding and free nodes. Fixed nodes are defined along the most posterior plane of volumetric soft-tissue model, which can be assumed to be fixed during the simulation. Joined nodes are defined on the interface between relocated bone segments and soft-tissue, which are defined by the osteotomy and relocation plan. Additionally, sliding nodes are defined on the interface around the teeth and mucosa area.

Sliding nodes has been modeled by putting additional springs between soft- and hard-tissue layers in MSM [51], and considering only tangential component of local force in bone–tissue interface using MTM [25, 26].

2.4 Tissue Property Assignment

The characterization of soft-tissue properties can be performed in the following ways, as shown in Fig. 11.

- In vitro rheology

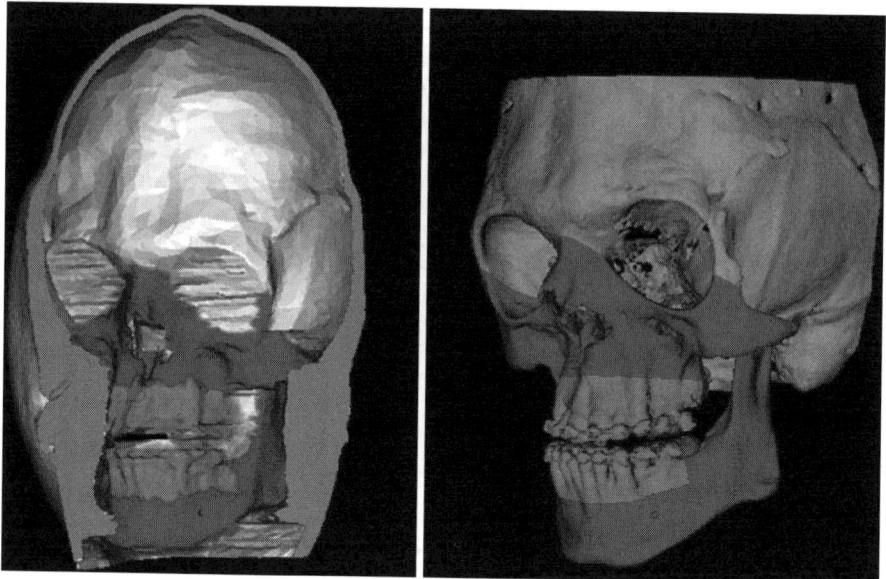

Fig. 10 Classification of volumetric nodes: fixed (*pink*), joined (*orange*), sliding (*blue*) and free (*white*)

- In vivo rhelogy
- In silico rhelogy
- Elastometry

In vitro rheology can be performed in a laboratory using loading devices with matured technique. However, it is not suitable for soft-tissue due to perfusion. In vivo rheology can provide stress–strain relationships at several locations. However, influence of boundary condition in real biological tissues is not well understood. In silico rheology is suitable for surgery simulation, which is based on computational approaches. The physical parameters can be deduced by the optimization process [38, 60]. Elastometry based on MRI or ultrasound imaging can provide non-invasive ways to measure internal tissue properties. However, it is only validated for linear elastic materials.

2.5 Simulation

The calculation of deformation has to be performed based on the computational methods presented in Sect. 1.4. The computational models could be selected depending on the limiting factors of the application, e.g. pre-processing difficulty, computation time or level of required accuracy.

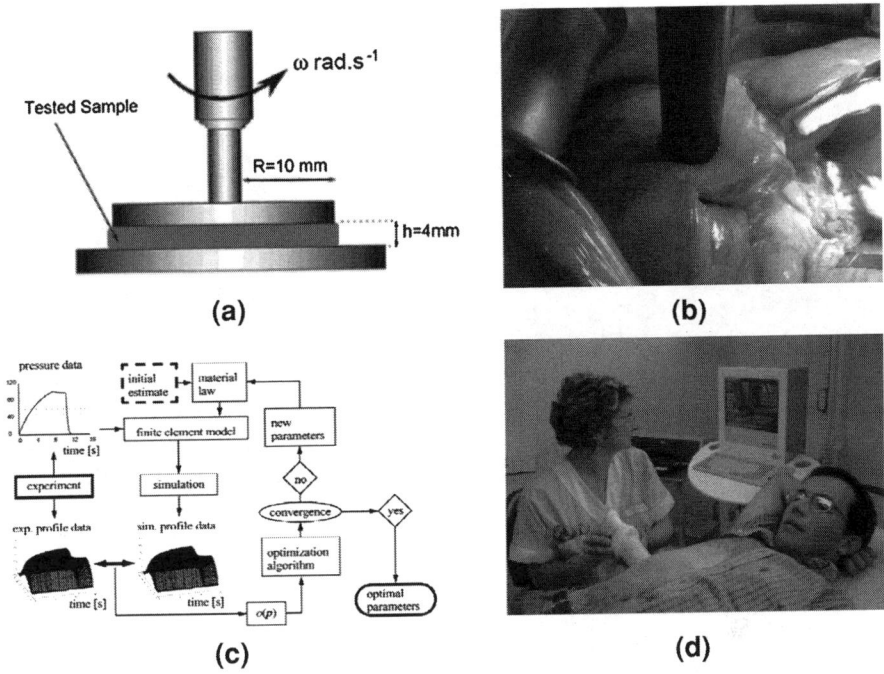

Fig. 11 Various tissue characterization methods: (**a**) In vitro rheology [35] (**b**) In-vivo rheology [41] (**c**) In silico rheology [23] (**d**) Elastometry [31]

2.6 Validation

In order to assess the simulation accuracy, normally retrospective validation scheme is used, which compares the surface to surface distance between the simulation result and actual post-operative result. Since there is no guarantee that actual operation has been performed as it is planned, the actual displacement of bone segments during the operation has to be reproduced by matching the pre-operative and post-operative images. First, the post-operative data needs to be aligned with pre-operative data by using volumetric image registration method on the unaltered skull part, such as skull-base, as shown in Fig. 12 (*left*). Then, the actual bone-related planning is reproduced by performing ICP-based surface registration between pre- and post-operative bone segments, see Fig. 12 (*right*) [2].

For simplicity, the distance between simulation and post-operative result has been measured based on closest matched points. However, this could result in underestimation of true discrepancy between simulation result and actual outcome. In order to produce a more reliable assessment, the surface-to-surface distance should be measured between anatomically corresponding points. Several methods based on Thin-Plate-Spine + Rapid Point Matching (TPS+RPM) [39], Spherical

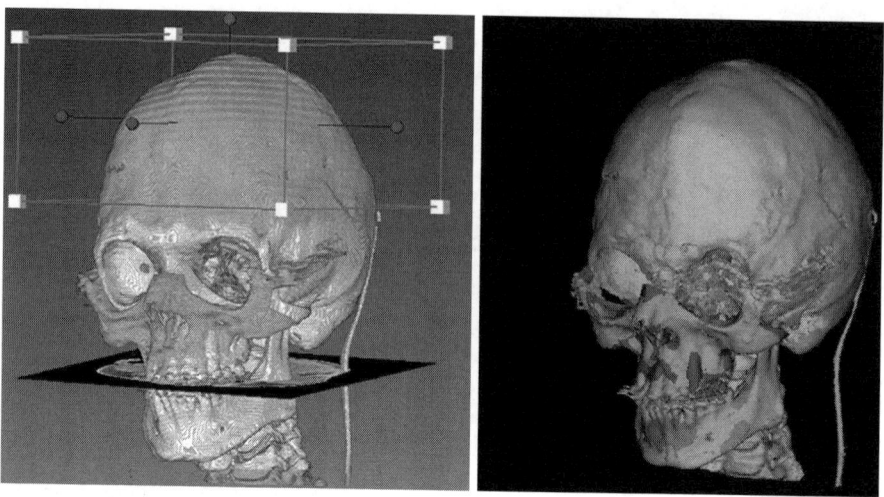

Fig. 12 Reproduction of bone-related planning for retrospective validation study: registration of pre- and post-operative images (*left*), registration of pre-operative bone segments to post-operative yellow skull model (*right*)

Harmonic Point Distribution Model (SPHARM-PDM) [44], Thin-Plate-Spine + Closest Point matching (TPS+CP) [25, 26] have been proposed so far.

In TPS+CP distance metric, one surface is transformed into the target surface based on manually defined corresponding anatomical landmarks, followed by anatomical corresponding points defined by a closest point matching on the deformed shape, see Fig. 13. Finally, surface-to-surface distance is measured on undeformed shape based on this revised anatomical corresponding points. Therefore, this metric can be useful for measuring surface to surface distance on deforming anatomical parts such as facial soft-tissues [25, 26].

3 Patient-Specific Facial Soft-Tissue Simulation

An example of patient-specific simulation pipeline is depicted in Fig. 14 [25, 26]. In addition to the general procedure depicted in Fig. 7, muscle modeling part is introduced in order to incorporate patient-specific facial muscles. The detailed description on patient-specific consideration on this study will be presented in the following sub-sections.

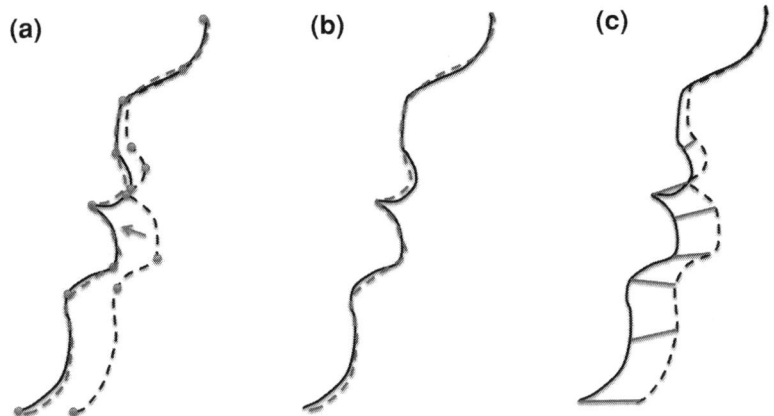

Fig. 13 TPS+CP based surface-to-surface distance measurement: (**a**) landmark-based implicit TPS transform of original geometry, (**b**) closest point matching on transformed shape, (**c**) anatomically adjusted correspondences

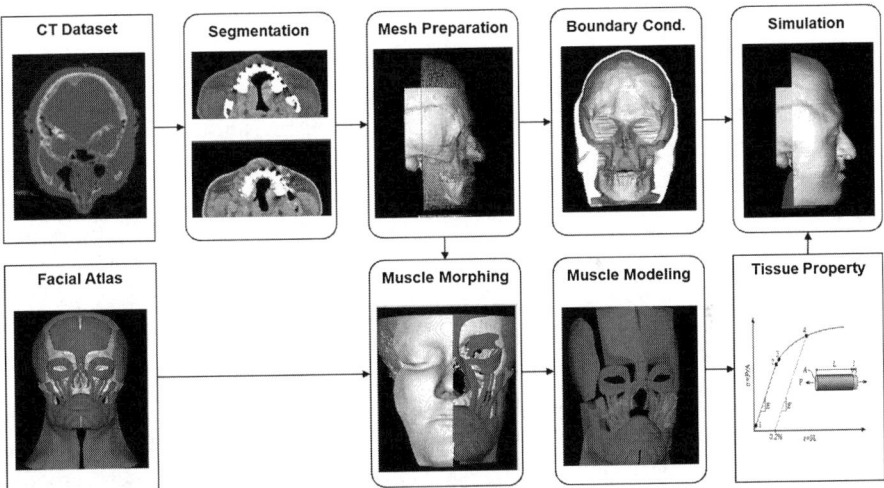

Fig. 14 The pipeline of the anatomically considered facial soft-tissue simulation method

3.1 Facial Muscle Modeling

It is almost impossible to identify individual muscles from clinical CT scan. Even with MR scan, which is not included in a typical clinical workflow, segmentation of individual muscles is still a challenging and time-consuming task. In order to

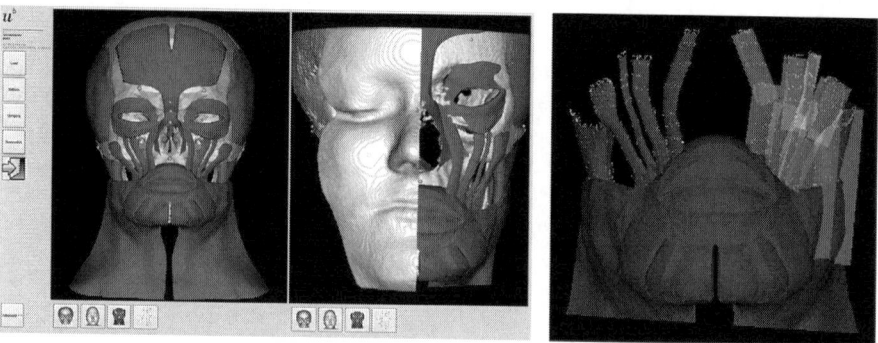

Fig. 15 Morphing of template muscles to patient-specific anatomy (*left*), and extraction of morphed muscle direction using oriented-bounding boxes (*right*)

minimize the efforts on facial muscles segmentation, construction of patient-specific facial muscles was considered through a template-to-patient deformation process. Publicly available facial template model [49] was used. The morphing procedure was driven by landmark-based thin-plate-spline (TPS) deformation [4] based on 32 anatomical landmarks, which are commonly used for measuring skin depth in forensic science [45].

In order to obtain the direction of muscles, oriented-bounding boxes (OBB) were extracted for linear-type muscles. As shown in Fig. 15, the number of OBBs for individual muscle can be adjusted according to the curvature of the muscle. For example, the masseter muscle was represented by single OBB, while the other muscles were represented by double OBBs. The longitudinal direction of each bounding box was regarded as the direction of muscle for each segment.

3.2 Equivalent Tissue Property Assignment

Up to our knowledge, there was no consensus in the literature on mechanical properties of facial soft-tissue. Different material parameters have been proposed in previous works depending on the characterization method and material model. Consequently, Young's modulus of muscle tissue along and across fiber direction were adopted from a previous study [11]. Previous works on estimating optimal facial tissue properties by comparing their simulation models with real postoperative data [38, 60] were referred to as well. The parameters that we selected for our simulation are shown in Table 2. Since there was no previous study found on anisotropic Poisson ratio of muscle, isotropic Poisson ratio was adopted.

Since it is difficult to remesh volumetric facial tissue according to the intersection with muscle surfaces, the equivalent material property was calculated by considering the volumetric proportion of muscle in each tetrahedron. The approach

Table 2 Material properties used for simulation

	Young's modulus (MPa)	Poisson ratio
Fat	0.003	0.46
Muscle across fiber	0.79	0.43
Muscle along fiber	0.5	0.43

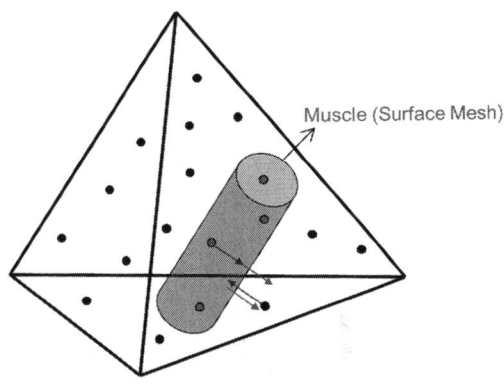

Fig. 16 Calculation of intersecting portion of muscle in each tetrahedron. Random points (*black dots*) are generated inside a tetrahedron and geometrical tests, described by the *arrows*, allow definition of material properties

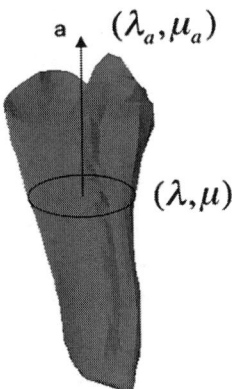

Fig. 17 Definition of Lamé coefficients along the direction of muscle

proposed by Uesu et al. [55] was adopted, which employs point sampling in tetrahedron. Then, geometrical test, based on directional comparison of surface normal and closest distance vector, as shown in Fig. 16, was performed to calculate equivalent material properties.

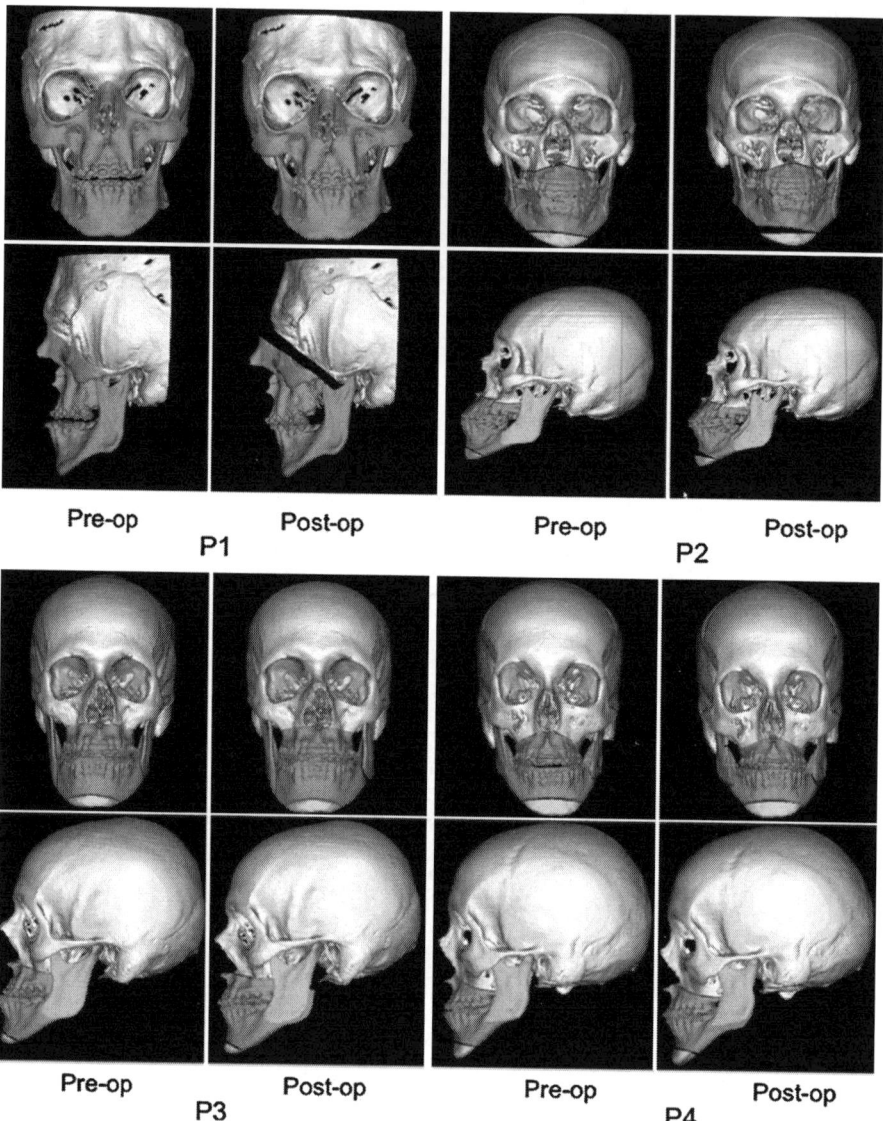

Fig. 18 Reproduced relocation planning of four clinical cases. On each case pre- and postoperative situation is displayed, with different colors representing the main structures of interest for the surgical procedures

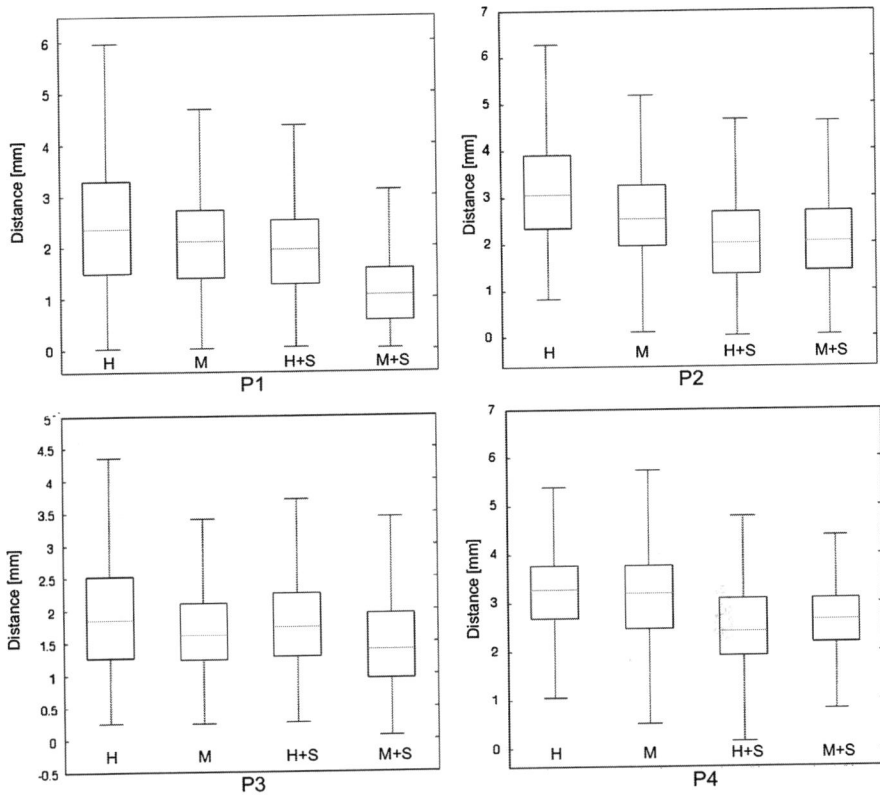

Fig. 19 Boxplot of distance errors between simulations and post-operative result for four clinical cases (P1, P2, P3, P4); Each graph represents the results with homogeneous (H), muscle template (M), homogeneous + sliding contact (H+S), and muscle template + sliding contact (M+S)

3.3 Transversely-Isotropic Mass-Tensor Modeling

The extension to transversely isotropic MTM can be achieved by adding 3×3 anisotropic stiffness tensor term to isotropic stiffness tensor in Eq. 29 [47], where the differences in Lamé coefficients along and across the muscle direction are defined by $\triangle \lambda = \lambda_a - \lambda$, $\triangle \mu = \mu_a - \mu$ (Fig. 17).

$$\begin{aligned}\mathbf{A}^e_{jk} = \mathbf{K}^e_{jk} &+ \frac{1}{144 V^e}(\triangle\lambda(\mathbf{a}\cdot\mathbf{m}_k)(\mathbf{m}_j \otimes \mathbf{a}) \\ &- (\triangle\lambda + 2\triangle\mu)(\mathbf{a}\cdot\mathbf{m}_j)(\mathbf{a}\cdot\mathbf{m}_k)(\mathbf{a}\otimes\mathbf{a}) \\ &+ \triangle\mu(\mathbf{a}\cdot\mathbf{m}_k)(\mathbf{a}\otimes\mathbf{m}_j) + \triangle\mu(\mathbf{a}\cdot\mathbf{m}_j)(\mathbf{m}_k\otimes\mathbf{a}) \\ &+ \triangle\mu(\mathbf{m}_j\cdot\mathbf{m}_k)(\mathbf{a}\otimes\mathbf{a}) + \triangle\mu(\mathbf{a}\cdot\mathbf{m}_j)(\mathbf{a}\cdot\mathbf{m}_k)\mathbf{I})\end{aligned} \qquad (33)$$

Table 3 Statistical results for four patient data

		Average error	Standard deviation	90th Percentile	95th Percentile
P1	H	2.50	1.21	4.10	4.40
	M	2.21	1.23	3.39	4.16
	H+S	2.07	1.24	3.68	4.39
	M+S	1.21	0.92	2.15	2.90
P2	H	3.25	1.27	5.13	5.85
	M	2.68	1.07	4.00	4.65
	H+S	2.11	1.06	3.42	4.06
	M+S	2.14	0.98	3.42	4.04
P3	H	2.03	2.41	3.01	3.38
	M	1.80	2.36	2.61	2.87
	H+S	1.75	0.71	2.61	2.78
	M+S	1.58	1.90	2.58	2.84
P4	H	3.23	0.83	4.25	4.48
	M	3.10	0.99	4.27	4.51
	H+S	2.43	0.84	3.46	3.67
	M+S	2.56	0.74	3.39	3.62

The further extension to non-linear MTM can be achieved by introducing the complete Green–St.Venant strain tensor in Eq. 8 into St.Venant–Kirchhoff model [46].

3.4 Validation with Clinical Cases

In order to assess the accuracy of the proposed patient-specific simulation strategy, the retrospective validation of four clinical cases was performed. The actual bone-related plannings were reproduced by the procedures presented in Sect. 2.6, and are shown in Fig. 18.

The simulation of homogeneous and transversely isotropic elasticity tissue models was conducted, and accuracy in different tissue models was compared. In addition, sliding contact was also considered as part of the simulation. For the calculation of surface-to-surface distance, the TPS+CP distance measurement scheme, presented in Sect. 2.6, was used.

As shown in Fig. 19 and Table 3, there was a statistically significant decrease in mean surface-to-surface distance by incorporating transversely isotropy of muscles and sliding contact. The statistical significance was confirmed by Wilcoxon test ($p < 0.05$). The qualitative improvements can be visualized by color-mapped distance errors, as shown in Fig. 20. The negative value means that predicted skin surface lies inside the post-operative skin surface.

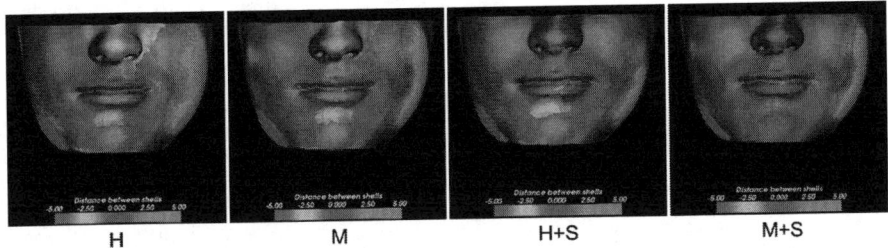

Fig. 20 Example of color-mapped distance map between simulations for patient three; Colormap starts from red to blue in the range of [−5 mm, 5 mm]

4 Conclusions

There have been various computational methods developed for simulating facial soft-tissue. Each computational model can be selected depending on its own characteristics. In order to enhance simulation accuracy, patient-specific soft-tissue simulation for CMF surgery has been developed and retrospectively validated with post-operative CT data. As for minimizing efforts for building patient-specific models, a subject-specific template-based morphing technique has been proposed. The directional behavior of facial muscles has been modeled with computationally efficient transversely-isotropic MTM. By introducing sliding boundary condition on error-sensitive regions, the simulation accuracy has been improved in statistically significant amount.

Acknowledgements This work was supported by the Swiss National Center of Competence in Research "Computer Aided and Image Guided Medical Interventions (Co-Me)", and the AO/ASIF Foundation, Davos, Switzerland.

References

1. Barich, F.: Class II, division 2 (angle) malocclusion: report of a case. Am. J. Orthod. **37**(4), 286–293 (1951)
2. Besl, P., McKay, H.: A method for registration of 3-d shapes. Pattern Anal. Mach. Intell. IEEE Trans. **14**(2), 294–299 (1992)
3. Bianchi, A., Muyldermans, L., Martino, M.D., Lancellotti, L., Amadori, S., Sarti, A., Marchetti, C.: Facial soft tissue esthetic predictions: validation in craniomaxillofacial surgery with cone beam computed tomography data. J. Oral. Maxillofac. Surg. **68**(7), 1471–1479 (2010)
4. Bookstein, F.: Principal warps: thin-plate splines and the decomposition of deformations. Pattern Anal. Mach. Intell. IEEE Trans. **11**(6), 567–585 (1989)
5. Chabanas, M., Luboz, V., Payan, Y.: Patient specific finite element model of the face soft tissues for computer-assisted maxillofacial surgery. Med. Image. Anal. **7**(2), 131–151 (2003)
6. Chapuis, J., Schramm, A., Pappas, I., Hallermann, W., Schwenzer-Zimmerer, K., Langlotz, F., Caversaccio, M.: A new system for computer-aided preoperative planning and intraoperative

navigation during corrective jaw surgery. Inf. Technol. Biomed. IEEE Trans. **11**(3), 274–287 (2007)
7. Cotin, S., Delingette, H., Ayache, N.: A hybrid elastic model for real-time cutting, deformations, and force feedback for surgery training and simulation. Vis. Comput. **16**(8), 437–452 (2000)
8. Cunningham, M.: Measuring the physical in physical attractiveness: quasi-experiments on the sociobiology of female facial beauty. J. Pers. Soc. Psychol. **50**(5), 925–935 (2000)
9. Cunningham, M., Roberts, A.: "Their ideas of beauty are, on the whole, the same as ours": consistency and variability in the cross-cultural perception of female physical attractiveness. J. Pers. Soc. Psychol. **68**(2), 261–279 (1995)
10. Downs, W.B.: Variations in facial relationships; their significance in treatment and prognosis. Am. J. Orthod. **34**(10), 812–840 (1948)
11. Duck, F.A.: Physical Properties of Tissue: A Comprehensive Reference Book, Academic Press, London (1990)
12. Dürer A.:Vier Bücher von menschlicher Proportion. Nürnberg (1528)
13. Ewers, R., Schicho, K., Undt, G., Wanschitz, F., Truppe, M., Seemann, R., Wagner, A.: Basic research and 12 years of clinical experience in computer-assisted navigation technology: a review. Int. J. Oral. Maxillofac. Surg. **34**(1), 1–8 (2005)
14. Fung, Y.C.: Biomechanics: Mechanical Properties of Living Tissues. Springer, New York (1993)
15. Gelder, A.V.: Approximate simulation of elastic membranes by triangulated spring meshes. J. Graph. Tools **3**(2), 21–42 (1998)
16. Gladilin, E.: Biomechanical modeling of soft tissue and facial expressions for craniofacial surgery planning. Ph.D. thesis, Free University Berlin (2003)
17. Hopkin, G.B.: The growth factor in the prognosis of treated cases of angle class 3 malocclusion. Rep. Congr. Eur. Orthod. Soc. **41**, 353–363 (1965)
18. Hume, D.: A treatise of human nature. http://www.books.google.com (2003)
19. Johnston, V.S., Solomon, C.J., Gibson, S.J., Pallares-Bejarano, A.: Human facial beauty: current theories and methodologies. Arch. Facial. Plast. Surg. **5**(5), 371–377 (2003)
20. Juergens, P., et al.: A computer-assisted diagnostic and treatment concept to increase accuracy and safety in the extracranial correction of cranial vault asymmetries. J. Oral. Maxillofac. Surg. (2011, in press)
21. Juergens, P., Klug, C., Krol, Z., Beinemann, J., Kim, H., Reyes, M., Guevara-Rojas, G., Zeilhofer, H.F., Ewers, R., Schicho, K.: Navigation-guided harvesting of autologous iliac crest graft for mandibular reconstruction. J. Oral. Maxillofacial. Surg. (2011, in press)
22. Juergens, P., Ratia, J., Beinemann, J., Krol, Z., Schicho, K., Kunz, C., Zeilhofer, H.F., Zimmerer, S.: Enabling an unimpeded surgical approach to the skull base in patients with cranial hyperostosis, exemplarily demonstrated for craniometaphyseal dysplasia. J. Neurosurg. **115**, 528–535 (2011)
23. Kauer, M., Vuskovic, V., Dual, J., Szekely, G., Bajka, M.: Inverse finite element characterization of soft tissues. Med. Image. Anal. **6**(3), 275–287 (2002)
24. Keeve, E., Girod, S., Kikinis, R., Girod, B.: Deformable modeling of facial tissue for craniofacial surgery simulation. Comput. Aided Surg. **3**(5), 228–238 (1998)
25. Kim, H., Juergens, P., Weber, S., Nolte, L.: A new soft-tissue simulation strategy for craniomaxillofacial surgery using facial muscle template model. Prog. Biophys. Mol. Biol. **103**(2–3), 284–291 (2010)
26. Kim, H., Jürgens, P., Nolte, L., Reyes, M.: Anatomically-driven soft-tissue simulation strategy for cranio-maxillofacial surgery using facial muscle template model. Med. Image Comput. Comput.-Assist. Interv. **6361**, 61–68 (2010)
27. Koch, R.M., Gross, M.H., Carls, F.R., von Büren, D.F., Fankhauser, G., Parish, Y.I.H.: Simulating facial surgery using finite element models. In: Proceedings of ACM SIGGRAPH, pp. 421–428. ACM Press, New York (1996)
28. Langlois, J., Roggman, L.: What is average and what is not average about attractive faces? Psychol. Sci. **5**, 214–220 (1994)

29. Langlois, J., Roggman, L., Casey, R.: Infant preferences for attractive faces: rudiments of a stereotype? Dev. Psychol. **23**(3), 363–369 (1987)
30. Langlois, J.H., Roggman, L.A.: Attractive faces are only average. Psychol. Sci. **1**, 115–121 (2008)
31. Ledinghen, V.D., Vergniol, J.: Transient elastography (fibroscan). Gastroenterol. Clin. Biol. **32**(6 Suppl 1), 58–67 (2008)
32. Lee, Y., Terzopoulos, D., Waters, K.: Realistic modeling for facial animation. Proceedings of the 22nd annual conference on computer graphics and interactive techniques, pp. 55–62 (1995)
33. Lorensen, W., Cline, H.: Marching cubes: a high resolution 3d surface construction algorithm. Proceedings of the 14th annual conference on computer graphics and interactive techniques, p. 169 (1987)
34. Man, B.D., Nuyts, J., Dupont, P., Marchal, G., Suetens, P.: Metal streak artifacts in X-ray computed tomography: a simulation study. Nucl. Sci. IEEE Trans. **46**(3), 691–696 (1999)
35. Marchesseau, S., Heimann, T., Chatelin, S., Willinger, R., Delingette, H.: Fast porous viscohyperelastic soft tissue model for surgery simulation: Application to liver surgery. Prog. Biophys. Mol. Biol. **103**(2–3), 185–196 (2010)
36. Mollemans, W.: Facial modelling for surgery systems. Ph.D. thesis, Katholieke Universiteit Leuven (2007)
37. Mollemans, W., Schutyser, F., Cleynenbreugel, J.V., Suetens, P.: Tetrahedral mass spring model for fast soft tissue deformation. Simulation and Soft Tissue Modeling **2673**, 145–154(2003)
38. Mollemans, W., Schutyser, F., Nadjmi, N., Maes, F., Suetens, P.: Parameter optimisation of a linear tetrahedral mass tensor model for a maxillofacial soft tissue simulator. Biomed. Simul. **4072**, 159–168 (2006)
39. Mollemans, W., Schutyser, F., Nadjmi, N., Maes, F., Suetens, P.: Predicting soft tissue deformations for a maxillofacial surgery planning system: from computational strategies to a complete clinical validation. Med. Image Anal. **11**(3), 282–301 (2007)
40. Mueller, A.A., Paysan, P., Schumacher, R., Zeilhofer, H.F., Berg-Boerner, B.I., Maurer, J., Vetter, T., Schkommodau, E., Juergens, P., Schwenzer-Zimmerer, K.: Missing facial parts computed by a morphable model and transferred directly to a polyamide laser-sintered prosthesis: an innovation study. Br. J. Oral Maxillofac. Surg. (2011, in press)
41. Nava, A., Mazza, E., Furrer, M., Villiger, P., Reinhart, W.H.: In vivo mechanical characterization of human liver. Med. Image Anal. **12**(2), 203–216 (2008)
42. Nkenke, E., Zachow, S., Benz, M., Maier, T., Veit, K., Kramer, M., Benz, S., Häusler, G., Neukam, F.W., Lell, M.: Fusion of computed tomography data and optical 3d images of the dentition for streak artefact correction in the simulation of orthognathic surgery. Dentomaxillofac. Radiol. **33**(4), 226–232 (2004)
43. Obwegeser, H.: [the one time forward movement of the maxilla and backward movement of the mandible for the correction of extreme prognathism]. SSO Schweiz Monatsschr Zahnheilkd **80**(5), 547–556 (1970)
44. Paniagua, B., Cevidanes, L., Zhu, H., Styner, M.: Outcome quantification using spharm-pdm toolbox in orthognathic surgery. Int. J. CARS **6**(5), 617–626 (2011)
45. Phillips, V., Smuts, N.: Facial reconstruction: utilization of computerized tomography to measure facial tissue thickness in a mixed racial population. Forensic Sci. Int. **83**(1), 51–59 (1996)
46. Picinbono, G., Delingette, H., Ayache, N.: Non-linear anisotropic elasticity for real-time surgery simulation. Graph. Model. **65**(5), 305–321 (2003)
47. Picinbono, G., Lombardo, J.C., Delingette, H., Ayache, N.: Anisotropic elasticity and force extrapolation to improve realism of surgery simulation. Robotics and Automation, 2000. Proceedings ICRA '00. IEEE International Conference on, vol. 1, pp. 596–602 (2000)
48. Sarti, A., Gori, R., Lamberti, C.: A physically based model to simulate maxillo-facial surgery from 3d ct images. Future Gener. Comput. Syst. **15**(2), 217–222 (1999)

49. Smith, D., Oliker, A., Carter, C., Kirov, M., McCarthy, J.: A virtual reality atlas of craniofacial anatomy. Plast. Reconstr. Surg. **120**(6), 1641 (2007)
50. Terzopoulos, D., Waters, K.: Physically-based facial modeling, analysis, and animation. J. Vis. Comput. Animat. **1**(2), 73–80 (1990)
51. Teschner, M., Girod, S., Girod, B.: Optimization approaches for soft-tissue prediction in craniofacial surgery simulation. Med. Image Comput. Comput.-Assist. Interv. **1679**, 1183–1190 (1999)
52. Teschner, M., Girod, S., Girod, B.: Direct computation of nonlinear soft-tissue deformation. Proc. Vision, Modeling, Visualization VM'00, pp. 383–390 (2000)
53. Thornhill, R.G.: Human facial beauty: average, symmetry and parasite resistance. Hum. Nat. **4**, 239–269 (1993)
54. Troulis, M.J., Everett, P., Seldin, E.B., Kikinis, R., Kaban, L.B.: Development of a three-dimensional treatment planning system based on computed tomographic data. Int. J. Oral. Maxillofac. Surg. **31**(4), 349–357 (2002)
55. Uesu, D., Bavoil, L., Fleishman, S., Shepherd, J., Silva, C.: Simplification of unstructured tetrahedral meshes by point sampling. Volume Graphics, 2005. Fourth International Workshop on, pp. 157–238 (2005)
56. Vannier, M.W., Gado, M.H., Marsh, J.L.: Three-dimensional display of intracranial soft-tissue structures. AJNR Am. J. Neuroradiol. **4**(3), 520–521 (1983)
57. Westermark, A., Zachow, S., Eppley, B.L.: Three-dimensional osteotomy planning in maxillofacial surgery including soft tissue prediction. J. Craniofac. Surg. **16**(1), 100–104 (2005)
58. Xia, J., Ip, H.H., Samman, N., Wong, H.T., Gateno, J., Wang, D., Yeung, R.W., Kot, C.S., Tideman, H.: Three-dimensional virtual-reality surgical planning and soft-tissue prediction for orthognathic surgery. IEEE Trans. Inf. Technol. Biomed. **5**(2), 97–107 (2001)
59. Zachow, S., Gladiline, E., Hege, H., Deuflhard, P.: Finite-element simulation of soft tissue deformation. Proc. CARS 2000, pp. 23–28 (2000)
60. Zachow, S., Hierl, T., Erdmann, B.: A quantitative evaluation of 3d soft tissue prediction in maxillofacial surgery planning. Proc. 3. Jahrestagung der Deutschen Gesellschaft fur Computer-und Roboter-assistierte Chirurgie eV, Munchen (2004)
61. Zeilhofer, H.F., Kliegis, U., Sader, R., Horch, H.H.: Video matching as intraoperative navigation aid in operations to improve the facial profile. Mund Kiefer Gesichtschir **1**(Suppl 1), S68–70 (1997)

Patient-Specific Modeling of Subjects with a Lower Limb Amputation

Sigal Portnoy and Amit Gefen

Abstract The rehabilitation outcomes following lower limb amputation depend on decision-making in the surgery room and in optimal fitting of the prosthetic components. Presently, the surgical and rehabilitation processes are performed according to general guidelines and on the experience of the surgeon/prosthetist. Patient-specific models of the residual limb and its interaction with the prosthetic socket have been created for the last two decades for research purposes. However, no modeling technique has yet to be integrated with the clinical community, as a tool for surgical and rehabilitative decision-making. In this chapter, we will review the main advancements in patient-specific modeling of the lower limb residuum over the last decades and discuss its potential use as both a tool for clinicians and as a patient-specific monitor aimed to prevent injury to the residuum.

1 Introduction

Lower limb amputations (LLA) are performed at a ratio of almost 5–1, when compared to upper limbs amputations [8]. In 2005 there were approximately 623,000 transtibial amputation (TTA) and transfemoral amputation (TFA) patients in the United States alone [57]. The main cause for limb amputation is vascular

S. Portnoy (✉)
Department of Rehabilitation, Hadassah Medical Center,
Jerusalem, Israel
e-mail: msbm@eng.tau.ac.il

A. Gefen
Department of Biomedical Engineering, Tel Aviv University,
Tel Aviv, Israel

disease, usually with comorbidity of diabetes. Other causes for limb amputation are trauma, congenital abnormalities, tumors and physical deformations [6].

Reacquiring mobility following an LLA involves a long and complicated rehabilitation process. Its progress and success depend on the physical and mental health condition of the patient, as well as on the skills and expertise of the caretakers and the rehabilitation professional staff. Slocum [45], an American surgeon, mentioned four interdependent factors which are important to the survival of the amputee: a durable residual limb, a functional well-fitted prosthesis, proper training in the use of the prosthesis and psychological guidance to a healthy mental attitude. Environmental factors, e.g. family support and life situation before the amputation, were also found to influence the outcomes of losing a limb [44].

The first factor mentioned by [45], the durability of the residual limb, depends mostly on the comorbidities of the patient and on the surgical planning and execution. An example of a comorbidity that may affect the durability of the residuum is neuropathy, i.e. impaired bio-sensory feedback. LLA patients may suffer from neuropathy due to diabetes, vascular disease complications, or nerve severance caused by the LLA itself [19]. This condition may lead to over-loading of the residuum, as pain is not registered by the patient, thus endangering the integrity of the limb. Another comorbidity that may endanger the residuum is diabetes. Diabetes increases the elastic modulus of skin [34, 49], thereby increasing its susceptibility to breakdown.

The surgical planning and execution may also influence the durability of the residuum. While the aetiology of the amputation usually dictates the height of the amputation, each surgeon may use different techniques, thereby affecting the durability of the residuum and the outcomes of the rehabilitation process. Criteria for amputation level are investigated using invasive or non-invasive measurements, such as fluorescein dye angiography, thermography, and laser Doppler flowmetry [42]. Surgeons try to amputate at the distal most likely site to recover, since a more proximal amputation level increases energy costs during load-bearing [10]. Other surgical decisions that may affect the durability of the residuum are the bevelment at the distal end of the truncated bones, the tibia and fibula in TTA or the femur in TFA. A distal tibiofibular bone-bridging in TTA is a controversial surgical technique that increases the surface area of the bones and allows distal end weight-bearing [29]. All the above mentioned parameters, i.e. patients' comorbidities and different amputation techniques, create significant inter-subject variability that makes it difficult to impose general guidelines for the surgical and rehabilitation processes, for producing the optimal results.

The second factor mentioned by Slocum, a functional well-fitted prosthesis, is one of the great challenges for the prosthetists, clinicians and the engineering communities. The majority of LLA patients uses a prosthetic socket and wears silicone or gel liners to improve interface fit. The stiffness of the liner influences the peak interface stresses [21]. The most common prosthetic socket is the patellar tendon-bearing (PTB) prosthesis, which is designed to distribute the stresses in the soft tissues of the residual limb, mainly toward the patellar tendon and also towards the medial flair and tibia condyles [41]. The prosthetic socket is not a replica of the residuum, but it is shaped to channel the stresses from pressure sensitive areas to pressure tolerant areas [43]. Poor prosthetic fitting may lead to pain, oedema, dermatitis, blisters, skin irritation, verrucous hyperplasia, and sometimes flap necrosis and osteomyelitis [7, 24, 25, 47].

A minority of amputees, usually TFA patients who experienced difficulties using the traditional socket, avoid complications resulted from the limb/socket interface by transferring the loads directly to the truncated bone using osseointegration, also termed direct skeletal attachment. This method, developed by Brånemark in 1952 [3], eliminates the need for a prosthetic socket. This compelling solution provides the LLA patient with a feeling of reduced weight on the residuum, eliminates the need for new socket design and replacement due to limb occasional volume changes or socket wear, allows better control over the prosthesis, and is more easily donned and doffed [15]. However, this new interface is a source for a different set of complications. Mainly, it requires two surgeries [15]: the first for implantation of a titanium fixture, inserted intramedullary into the femur. During a healing period of six months, the patient may use a prosthetic socket. The second surgery involved insertion of an abutment into the distal end of the fixture. The implant holds a risk for superficial infections at the skin penetration area, pain, deep infections, and risk of implant loosening due to stress shielding.

In the last decades, the rapid technology advancements have greatly improved the well-being of the LLA residuum by providing better suspension systems, socket fabrication materials, shock absorbers, computer-controlled prosthetic joints and more [8]. These improved features allow an enhanced control over the socket or implant design, the user's stability, balance, and reduce the risk of falling. The concept of better stability during locomotion of an LLA patient implies to a more symmetrical gait pattern, as the patient feels more comfortable transferring more load to the amputated limb. This is important in order to produce an energy-conserving and aesthetic ambulation, as well as to prevent excessive loads on the sound limb, which may accelerate cartilage degradation and cause joint pain.

As mentioned, the rehabilitation outcomes following LLA depend on decision-making in the surgery room and in optimal fitting of the prosthetic components. Presently, the surgical and rehabilitation processes are performed according to general guidelines and rely on the experience of the surgeon/prosthetist. There are some prosthetists who are assisted by current technology in the design and manufacturing of a patient-specific prosthetic socket. These prosthetists use surface scanning of the geometry of the residuum. Once the geometry of the limb is captured, a 3D model of the limb topology is created, allowing the prosthetists to use computer aided design (CAD) controlling a virtual prosthesis which is then manufactured with computer aided manufacture (CAM). However, many prosthetists still use the traditional method and create the prosthetic socket using a plaster cast mold of the residuum, which prolongs the fitting process and may not be the optimal design for the patient. The quality evaluation of a non-CAD socket design and its ability to disperse pressures evenly along the limb relies only on the

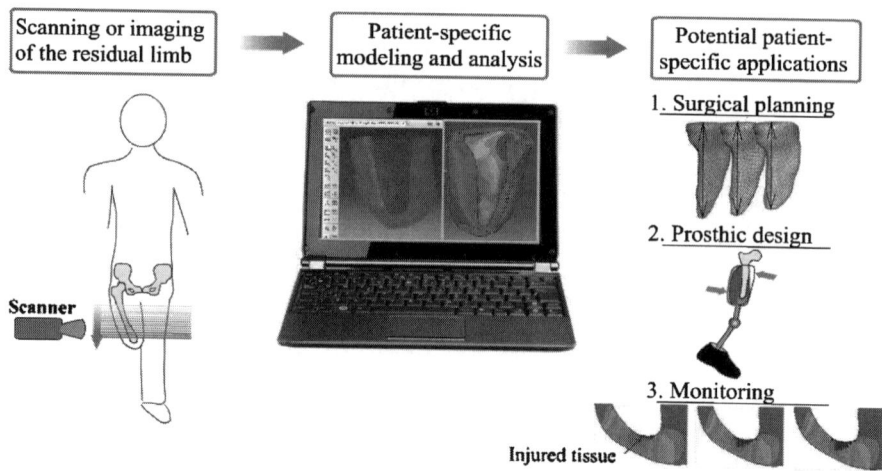

Fig. 1 Digitizing either the external surface or the internal geometry of the residuum is performed using touch probes, laser scanners or imaging systems. After the patient-specific model of the residuum is created, several applications may be practiced: surgical planning, socket design or monitoring for the prevention of soft tissue injury

patient's feedback and examination of the skin condition once the new prosthesis is removed. However, even with the current advances in computational algorithms, where a CAD/CAM system is utilized for the design and fabrication of the prosthetic socket, there are still no standardization of quantitative measures that point to the optimal socket shape that will most likely preserve the integrity of the soft tissues and provide comfort and stability to the use. Even with the most cutting-edge technology used by the prosthetists, the design is still based solely on the subjective experience of the prosthetist [55].

Elaborated 3D patient-specific models of the residual limb and mechanical analyzes of its interaction with the prosthetic socket have been created for the last two decades for research purposes alone. As depicted in Fig. 1, the process of creating the patient-specific model requires digitizing either the external surface or the internal geometry of the residuum, performed using touch probes, laser scanners, or imaging systems, e.g. computerized tomography (CT), ultrasound, and magnetic resonance imaging (MRI). After the patient-specific model of the residuum is created, several applications may be practiced: surgical planning, prosthetic design or monitoring for the prevention of soft tissue injury (Fig. 1).

No modeling technique has yet to be integrated with the clinical community, as a tool for surgical and rehabilitative decision-making or for patient-specific monitoring of the mechanical state of the residuum. In this chapter we will review the main advancements in patient-specific modeling of the LLA residuum and discuss its potential use as both a tool for surgeons, clinicians and as a monitor for the amputees themselves, aimed to prevent injury to the residuum.

2 Patient-Specific Surface Modeling of the Residuum

Currently, the prevailing method for realizing a prosthetic socket in most of the small to medium orthopaedic laboratories is by creating a negative plaster cast of the residuum, which is then used to create a positive plaster cast. The prosthetist then performs manual modifications on the positive cast and a temporary socket is manufactured to be subjectively evaluated by the amputee. This process is repeated until the final prosthesis is manufactured [9]. The drawbacks of this commonly-exercised method are that it is time-consuming and greatly relies on the skills and expertise of the prosthetist for the creation of the negative mold and performing the modifications that would produce an optimal prosthetic fit to the individual.

Several CAD/CAM systems for socket design and fabrication were created in the 80s. This unique method was termed computer aided socket design (CASD; [5]). The first prototype was presented at the fourth World Congress of the International Society for Prosthetics and Orthotics (ISPO) in 1983 [17]. Currently, several commercial CASD systems exist, e.g. OMEGA (The Ohio Willow Wood Company Inc., Ohio, USA), BioSculptor (Florida, USA), Infinity CAD Systems (Michigan, USA), Canfit^{TM} (VORUM Research Cooperation, Vancouver, Canada), Rodin4D (Le Haillan, France), Fourroux Prosthetics (Alabama, USA), and more. CASD systems obtain the patient-specific external topology of the residuum using a shape digitizer or a laser scanner [2, 16, 27, 40]. The scanning hardware, which once required a large designated room space [4, 27], nowadays can be replaced with a simple and portable handheld scanning device. Today, it takes seconds–minutes for the CASD software to digitize the patient-specific residuum topology after the scanning. The virtual socket is easily created and virtual rectifications that produce either pressure relief or load-bearing areas are produced manually or with specially-designed templates [2, 27]. Then, once the virtual socket is finalized, it takes up to 30 min for the CAM automated carver to fabricate the designed socket [14]. The accuracy of two different shape digitizers (Ohio Willow Wood, Mount Sterling, Ohio), used by a group of practitioners and a group of students to measure common anthropometric dimensions of a TTA residuum, was found to be smaller than 1 cm [14]. The measurements were consistent across subjects and accurate, when compared to hand calipers and tape measurements. These finding lead the author to the conclusion that these capturing methods are an efficient and reliable tool for 3D digitization of the patient-specific residual limb's surface [14]. Supporting Geil's conclusion, [17] performed clinical trials (n = 90) to evaluate the effectiveness of CASD/CAM systems and reported that the produced sockets were characterized by the subjects as more comfortable or better fitted than manually-manipulated socket designs. Some researchers and practitioners, however, do not believe that casted, scanned, or digitized exact anatomic data of the LLA residuum necessarily benefit the design of the prosthetic socket [46], claiming that simple anatomical measurements may suffice for this application. However, no studies have yet shown to prove or disprove their hypothesis.

The CAM system itself was found to be a simple and user-friendly device that does not require special facilities to ensure operator safety [16]. In addition to the aforementioned benefits, current CAM systems, e.g. selective laser sintering (SLS), are able to incorporate features to the prosthetic socket that cannot be produced manually [37]. For example, [38] used a CASD/CAM system and created five TTA patient-specific sockets with a special feature. Namely, they enhanced the flexibility of the socket's inner wall over the pressure sensitive regions of the fibular head and distal tibial end using ahexagonal set of triangular antilever beams, composed of radial slots. The amputees ambulated with the SLS socket for two weeks and their gait pattern was analyzed using gait analysis, however, no significant differences between SLS and conventional sockets were observed [38].

The benefits of CASD/CAM system in producing patient-specific socket design by acquiring the surface geometry of the residuum are mainly the reduction in the examination time and the total cost of socket design, as several trial sockets are rarely required. It allows for a fast and comfortable examination of the residuum and produce precise record, by quantifying the state of the residuum at the time of the examination, and the detailed description of the designed socket. Additionally, this technology may allow remote fabrication of the prosthetic socket according to on-site evaluation of the LLA patient in facilities that do not have access to equipment for socket fabrication, e.g. in developed countries [50].

3 Patient-Specific Internal Modeling of the Residuum

In the late 1980s, several research groups pioneered the field of finite element analysis (FEA) of the mechanical state of the TTA [35, 48] and TFA [20] residual limbs. FEA is a numerical technique which is widely used for a variety of biomechanical engineering applications [28], ranging from cardiovascular stent or dental implant designs to complex simulations of childbirth.

The patient-specific modeling process comprises of three steps:

1. *Acquiring the patient-specific geometry* using an imaging technique, e.g. CT or MRI.
2. *Creating the 3D patient-specific geometrical model* of the residuum using a CAD engineering software, e.g. SolidWorks (Dassault Systemes, Massachusetts, USA), Simpleware (Simpleware Ltd, United Kingdom), and Mimics (Materialise, Belgium).
3. *Configuring the model properties and boundary conditions*: Meshing the geometric model, assigning material properties, setting the boundary conditions and interactions of the instances.

Performing the mechanical analysis can be achieved using FEA packages such as ABAQUS (Simulia, Providence, RI, USA), Nastran (MSC Software, CA, USA), ANSYS (NASDAQ, PA, USA), and more. FEA calculates mechanical parameters across the model, e.g. motion (displacements, velocities, and accelerations), strains, stresses, reaction forces, energy, and volume. These parameters might be experimentally measured on the surface of the residuum but cannot be measured internally. Furthermore, the effect of different factors that may influence the mechanical state of the residuum may be studied without the need to build an actual prosthesis. The quantification of these parameters is important when studying the pressure-related soft tissues injury of the residuum, caused by its interaction with the prosthetic socket during load-bearing, as well as the implant loosening and stress shielding caused by osseointegration. These extensive analyses cannot be accomplished by surface modeling alone, and require the geometry of the truncated bones and the surrounding soft tissues.

A number of challenges in LLA FEA are faced by researchers, attempting to study the mechanical state of the residuum under different loading patterns or to harness the capabilities of this technique to be easily utilized in the clinics. The first challenge is creating the patient-specific external and internal geometry. For current methods to be of practical use in the clinic, the imaging and modeling techniques must be fast, user-friendly, and preferably absent of exposure to radiation. The ideal modeling tool should manage rapid generation of the finite element model, with optimized mesh and minimal loss of geometrical information. The second challenge arises from the complexities of non-linearity of the non-homogenous anisotropic viscoelastic materials and due to the nonlinear boundary conditions, i.e. slip/friction contact problem. The last challenge is in determining the exact loading conditions.

In this subchapter, we present several studies where patient-specific FEA of the LLA residuum were performed either for studying the socket/residuum interaction (Sect. 3.1) or for quantifing the internal mechanical condition at the bone/titanium implant interface (Sect. 3.2).

3.1 Socket/Residuum Interaction

Patient-specific FEA of a TTA residuum was accomplished by Childress and his group approximately 25 years ago. They created the first FEA of a residuum [48], which geometry was based on transverse CT scans of a TTA residuum. The model included a prosthetic socket and was aimed at predicting the interfacial pressures. Similarly, the first FEA of a TFA residuum was created by [20]. The authors aimed to emulate the process that the prosthetist performs in order to rectify the plaster cast. The locations of the femur of two TFA patients were measured using an ultrasonic apparatus. The surface topology of each residuum was measured using two diametrically opposed contracting/retracting probes. The model was loaded with an interface pressure loading function and the resulted outer shape of the residuum was considered.

The results provided by FEA were mostly validated by comparing its calculation of interface stresses with the measured interface pressures [36] or measured interface stresses (shear and pressure; [39, 56]). These studies showed that the interface loading conditions is greatly affected by various patient-specific factors, e.g. the shape of the residuum and the mechanical properties of the tissues. The individual gait patterns of each patient also influence the residuum loading state. Beil et al. [1] measured interface pressures during ambulation with a normal total-surface weight-bearing suction socket and a vacuum-assisted socket. The later has been shown to eliminate daily volume loss. The authors commented that the muscle hypertrophy and volume loss, attributed to LLA limb shape change over time, is a direct cause for higher loads at bony prominences, elevating the risk of injury to the skin and underlying soft tissues.

As detailed in the Introduction, poor prosthetic fitting may lead to various complications, e.g. pain and tissue bruising or breakdown. One of the less reported complications that may arise from the compression of the soft tissues of the residual limb be the truncated bones against the hard prosthetic socket is deep tissue injury (DTI; [11, 26]). DTI is a dangerous pressure-related lesion that initiates at the bone proximity. Contrary to the more familiar pressure ulcers or decubitus ulcers that first appear as a non-blanchable redness that does not subside after the load is relieved, DTI cannot be detected on the skin's surface. Similarly to bed-ridden or spinal cord injury patients that cannot detect pain at the site of injury, prosthetic-users may also suffer from neuropathy caused by diabetes, vascular disease or nerve severance that occurred during the surgical LLA procedure. Since the nervous system of these patients does not alert them against the danger of DTI, they are more likely to be injured.

All of the aforementioned studies that used FEA to investigate the mechanical state of the residuum, focused on interface stresses, which cannot predict the internal mechanical state at the bones' proximity and assess the danger to DTI. Consequently, a more complex analysis, concerning the internal strain and stress distribution in the residuum, was warranted. Accordingly, we performed FEA to quantify the internal strains in the soft tissues of the residual limbs of TTA patients during load-bearing in an attempt to determine inter-subject variability, identify patients who are more susceptible to DTI and quantify the effect of surgical and morphological factors on the mechanical state of the LLA residuum [31, 32].

For this purpose, we recruited five traumatic unilateral TTA patients, who use their prosthesis daily in an active lifestyle. The study group comprised of one 32 years old female and four male, aged 44–63 years old. Two of the subjects suffered from diabetes. The exact geometry of the residuum of each subject was obtained in an open-MRI ("SignaSP" model, General Electric Co., CT, USA). The residuum, contained in a plaster cast, was scanned twice: once while the subject was standing on top of the MRI table in a position where the contained residuum was unloaded and situated in the scanning volume. The second scan was performed in the same manner after the subject was requested to place the limb on the MRI table and apply comfortable static load during the scan. The load apply by each subject during the second scanning procedure was recorded using 0.2 mm-thick force sensors (FlexiForce, Tekscan Co, MA, USA), placed between the plaster cast and the MRI table. The parameters for the MRI sequences

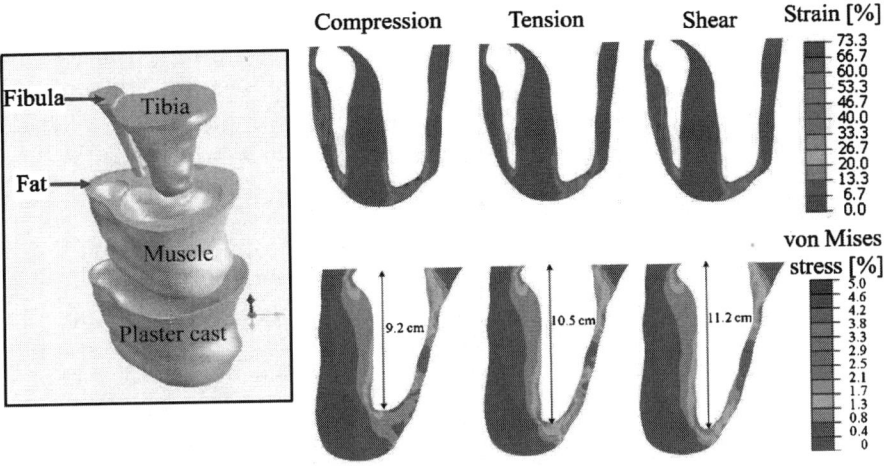

Fig. 2 The MRI-based 3D patient-specific Solid model of the residuum (*left frame*) comprises of the truncated bones, the tibia and the fibula, the muscle and fat tissues, enveloped by a skin tissue and confined in a plaster cast. The internal compression, tension and shear strains are depicted on the *upper right frame*. The von Mises stresses produced under the same loading conditions of the model, following virtual manipulation of the tibial length, are depicted on the *lower right frame*

were: field intensity 0.5 T, spatial resolution of 0.1 mm, slice thickness 4 mm, and the acquired axial and coronal images were T1-weighted.

The axial scans acquired during the first sequence, where no load was applied to the residuum, were uploaded into a commercial Solid modeling package (SolidWorks 2009, SolidWorks, MA, USA). The contours of both truncated bones, the tibia and the fibula, were visually detected from the MRI scans, as well as the contours of the muscle and fat tissues, surrounding the bones. The four instances, i.e. tibia, fibula, muscle and fat, were then transformed into 3D solid models. The models of the bones were subtracted from the fat and muscle models to create a cavity without gaps. The model of the plaster cast was created as an envelope to the residual limb model. An example of a patient-specific 3D Solid model of a TTA residuum is depicted on the left frame of Fig. 2.

Once the exact geometry of the unloaded residuum was registered into a virtual model, we uploaded it into a commercial FEA package (ABAQUS v6.8, SIMULIA, RI, USA). After configuring a thin layer around the residuum, representing the skin, we performed non-linear large-deformations FEA. In each model, all of the interfaces between the bones, muscle and fat tissues were tied together, since there was no available data of tissue-on-tissue friction characteristics. A friction coefficient of 0.7 was set at the skin/cast interface. We constrained the outer layer of the cast for all translations and rotations and imposed downward displacement of the bones, pressing the soft tissue towards the distal end of the cast, as performed by the subjects during the second scanning sequence. The displacement of the bones of each subject was calculated as the difference between the minimal soft tissue thickness under the tibia, found in the coronal scans, between the unloaded and loaded residuum sequences. The averaged measured displacement was 1.4 ± 0.3 cm [31].

Since the bones are at least one order of magnitude stiffer than the soft tissues, we assumed that the bones are rigid. Therefore meshing the interior volume of the bones did not contribute to our study; contrarily it would have elongated the time of analysis so we used only the external surfaces of the two bones in our FEA. We meshed the bones with 4-node quadrilateral surface elements ("SFM3D4" in ABAQUS). The skin tissue was meshed with 6-node quadratic triangular membrane elements ("M3D6"). We assumed that the plaster cast is a homogeneous, isotropic and linear-elastic material, with elastic modulus of 1 GPa and Poisson's ratio of 0.3 [31]. We meshed the plaster cast, as well as the muscle and fat tissues with second-order 10-node modified quadratic tetrahedron elements ("C3D10 M"). The soft tissues were assumed to be isotropic, hyperelastic, and homogeneous and were modeled using the Generalized Mooney-Rivlin Solid strain energy function:

$$W = C_{10}(I_1 - 3) + C_{11}(I_1 - 3)(I_2 - 3) + \frac{1}{D_1}(J - 1)^2 \quad (1)$$

where the invariants of the principal stretch ratios λ_i are $I_1 = \lambda_1^2 + \lambda_2^2 + \lambda_3^2$ and $I_2 = \lambda_1^{-2}\lambda_2^{-2}\lambda_3^{-2}$ the relative volume change is $J = \lambda_1^2\lambda_2^2\lambda_3^2$ and C_{10}, C_{11}, D_1 are constitutive parameters taken from the literature.

In order to predict DTI in the residuum during prolonged standing we used an injury threshold, obtained by [12], who applied large compressive strains to tissue-engineered muscle constructs. The findings suggested that exposure to 77% strain causes immediate cell death, but following long-term exposures for a duration that exceeds 3 h, damaging strains were 52%.

One of the main outcomes of our FEA study was that internal strains and stresses in the TTA residuum were higher in the bones proximity than in the muscle periphery (right upper frame in Fig. 2). Anatomically, the geometrical characteristics of the residuum of the TTA participants (n = 5) varied substantially between subjects. We found no muscle strains above the immediate injury threshold (77%) in the simulations of the standing posture, indicating that all subjects were not at immediate risk for DTI. Two subjects with minimal residuum fat padding were the only ones identified as theoretically prone to DTI at long-term weight-bearing periods, but since it is unreasonable for a prosthetic-user to continuously load the residuum for more than 3 h, we concluded that the subjects were under no risk for DTI development [31].

It is likely, however, that the limb would bear continuous load during sitting in common scenarios like during a long flight, or at the cinemas. In these cases, the soft tissues of the residuum must withstand prolonged strains and stresses which may ultimately lead to DTI. Accordingly we repeated the aforementioned method for the creation of a patient-specific 3D finite element model of a TTA patient. Except that in this FEA we did not impose bone displacement but surface pressures on the virtual skin surface. Additionally, when simulating sitting postures, the proximal soft tissues at the knee joint were constrained for mediolateral and posterior-anterior movements. The surface pressures were measured using two pressure-sensing pads, designed in association with Sensor Products Co. (NJ, USA). The two thin and flexible piezo-resistive pressure sensing mats comprised of a total of 325 pressure sensors and were able to completely envelop the residuum, placed inside the prosthetic socket. Data were sampled at 19 Hz while the subject was instructed to sit with a straight back posture, keep the hips mostly on the chair and parallel to the ground and both knees at a 90° flexion and place both feet on the floor. Then the subject was asked to remain seated and extend his amputated limb forward, to a knee-flexion angle of 30° [30, 33].

We utilized a well-established algorithm for DTI simulations [23] that combines FEA with empirical injury threshold and damage law for muscle tissue subjected to prolonged continuous loading. The muscle injury threshold is the compression stress, σ, which causes permanent muscular damage. Considering the time of exposure, t, an animal-model-based relationship can be formulated as [22]:

$$\sigma = \frac{K}{1 + e^{\alpha(t-t_0)}} + C \qquad (2)$$

where $K = 23$ kPa, $\alpha = 0.15$ min^{-1}, $t_0 = 95$ min and $C = 8$ kPa [22].

The stiffness of the necrotic tissue is increased, thereby affecting the strain and stress distribution in its proximity. The altered mechanical properties of injured muscle tissue were measured by Gefen et al. [13]:

$$\frac{G_{injured}}{G_{uninjured}} = A_0 + A_1\sigma + A_2 t \qquad (3)$$

where $G_{injured}$ and $G_{uninjured}$ are the long-term shear moduli of the injured and uninjured muscle tissues, respectively, and the empirical constants are $A_0 = 0.742$, $A_1 = 0.022$ kPa^{-1}, and $A_2 = 0.012$ min^{-1}.

We performed two FEAs, for two sitting postures (30 and 90° knee flexions where we advanced the time step in 15 min intervals to simulate the evolution of internal stresses in the soft tissues of the residuum over a total period of 75 min.

We wrote a designated code (LabView 8, National Instruments, TX, USA) that reads the output report file along with the last created ABAQUS input and then updates the assignment of mechanical properties to the "injured" muscle elements, detected by Eq. 2, using the muscle damage law in Eq. 3. The FEA was resubmitted with the updated muscle stiffness.

Our results showed that maintaining a sitting posture of 90°-knee-flexion, may cause internal injuries in muscle tissue near the tibial tuberosity and distal tibial end (predicted injured muscle tissue volume >1,000 mm^3). Contrarily, the 30°-knee-flexion simulations produced minor injury around the distal tibial end (predicted injured muscle tissue volume <14 mm^3). The predicted injury rates at 90°-knee-flexion were over one order of magnitude higher than those during 30°-knee-flexion [30, 33].

The aforementioned technique of patient-specific FEA of the LLA residuum can be utilized for prosthetics design and liner evaluation and, if numerically-simplified, it can also be used for patient's continuous monitoring and alert, as will be described later on. A third application of patient-specific FEA, mentioned in the Introduction (Fig. 1) is as a tool for the surgeon. The analyses could supply answers to questions that arise while planning an LLA surgical procedure. For example, what would be the effect of a shorter tibia or a blunter tibial end on the stress distribution in the soft tissues of the residuum of a specific patient? What effect would the location of the surgical scar have on internal stresses in the residuum?

In a sequel study [31] we attempted to supply answers to these questions and answer two additional questions concerning the effect of stiffness of the muscle or the presence of a fibular osteophyte on the internal stresses in the TTA residuum. Namely, we studied the effect of five biomechanical factors: bone lengths, tibial bevelment, existence of a fibular osteophyte, stiffness of the muscle flap, and location of a surgical scar, on the internal tissue loads. For this purpose, we randomly chose one of the aforementioned patient-specific models as the reference case and then applied various manipulations. We studied each of the five aforementioned factors separately during static loading. The effect of these factors on the internal loads in the LLA residuum during dynamic activities, e.g. gait or stair climbing, has yet to be studied.

A total set of 12 variant model configurations were built and analyzed. The geometrical manipulations, i.e. shorter tibial lengths, different tibial end bevelments and an osteophyte at the distal fibular end, were conducted on the Solid model, whereas analyses where mechanical properties of the soft tissues of the TTA residuum were modified, i.e. changes in stiffness of the muscle tissue and scarring in different locations and depths, were conducted by modifying the mechanical properties of elements directly in the finite element model.

We found that when shortening the tibia from its MRI-measured length of 11.2–9.2 cm, and thereby creating a thicker muscle flap (see bottom frame on the right of Fig. 2), peak loads at the distal edge of the tibia were relieved considerably. Specifically, von Mises stresses under the tibia decreased by up to $\sim 80\%$. This finding is attributed to the fact that the shorter bone is supported by a thicker muscular layer, dispersing the loads more uniformly. Also, there is an increase in the cross-sectional area of the distal tibial end for the more proximal amputation levels, as the tibia broadens towards the knee. An interesting finding following the bone-shortening simulations is the decrease of stresses near the skin's surface (bottom frame on the right in Fig. 2). This finding may suggest that although high stresses remain at the vicinity of the distal tibial end, the surface stresses considerably decrease to a level where surface pressure measurement would read zero loads. As discussed before, in this condition under the simulated loads, where off-shelf pressure sensing devices would not alert to the risk injury, DTI may occur under prolonged loading conditions.

We found that when decreasing the tibial bevelment, i.e. the distal inclination of the bone, surgically shaped during the LLA, caused propagation of internal stresses from the bone proximity towards the skin's surface, thereby theoretically reducing the risk for DTI.

The addition of a distal fibular osteophyte mostly increased the strain and stress concentrations directly under the fibula, which may endanger the muscle tissue in that region.

We elevated the stiffness of the muscle tissue for simulation of biological and pathobiological variation across LLA patients, as well as muscle spasticity. This was found to be the most influential factor to affect the rise of von Mises stresses in the soft tissues of the virtual residuum on its skin's surface. This finding suggests that, theoretically, a stiffer muscle is more prone to developing DTI as well as superficial pressure ulcers.

Last, we simulated surface surgical scar at different locations as well as one deep scar that includes the fat tissue below the scan and is clinically termed adherent scar. The inclusion of a surgical in modeling the residuum of LLA patients has never been considered before in the literature. We found the internal stresses shifted toward the location of the virtual scar. This finding suggests that patient-specific FEA may be useful for surgical planning of LLA, to determine where the optimal location of a scar should be.

One of the limitations of patient-specific 3D FEA is the time needed for model construction and time until the completion of the analysis and visualization of the results. Due to this limitation, patient-specific 3D FEA are not yet a part the clinical setting for the purposes illustrated in Fig. 1, namely, patient-specific socket design, surgical manipulation and monitoring internal and superficial stresses for the prevention of soft tissue breakdown.

Since current technology does not yet allow for FEA to be quickly and easily utilized for these purposes, other analytical algorithms might be considered. In a recent study [30, 33] we report integrating a solution of a 2D axisymmetric indentation problem [54] into a portable patient-specific holter-like internal stress monitor, built in our laboratory (Fig. 3a). This device samples real-time interface pressure, measured by thin sensors (FlexiForce, Tekscan Co, MA, USA), and calculates internal stresses under the tibia, using anatomical data of the tibial radius and the thickness of the soft tissues under the tibia, measured using Sagittal X-rays of the patient before the device is activated (Fig. 3a).

Yang [54] calculated the indentation of a flat-ended cylinder (tibia) into a compressible elastic layer (the soft tissues) during load-bearing at a time t, which is defined as $\delta(t)$ as:

$$\delta(t) = \frac{F(t) \cdot h}{\pi a^2 \cdot (\lambda + 2G)} \quad (4)$$

where a is the radius of the tibia, h is the thickness of the soft tissues under the distal tibial end, $F(t)$ is the time dependent limb-socket force, and G and λ are the effective shear modulus and Lame constant of the soft tissues, respectively [54].

The von Mises stress, $\sigma_v(t)$ is expressed by [54]:

$$\sigma_v(t) = \left(A \cdot \frac{a}{h} + B\right) \sqrt{\frac{[\sigma_{rr}(t) - \sigma_{zz}(t)]^2 + \sigma_{zz}(t)^2 + \sigma_{rr}(t)^2 + 6\sigma_{rz}(t)^2}{2}} \quad (5)$$

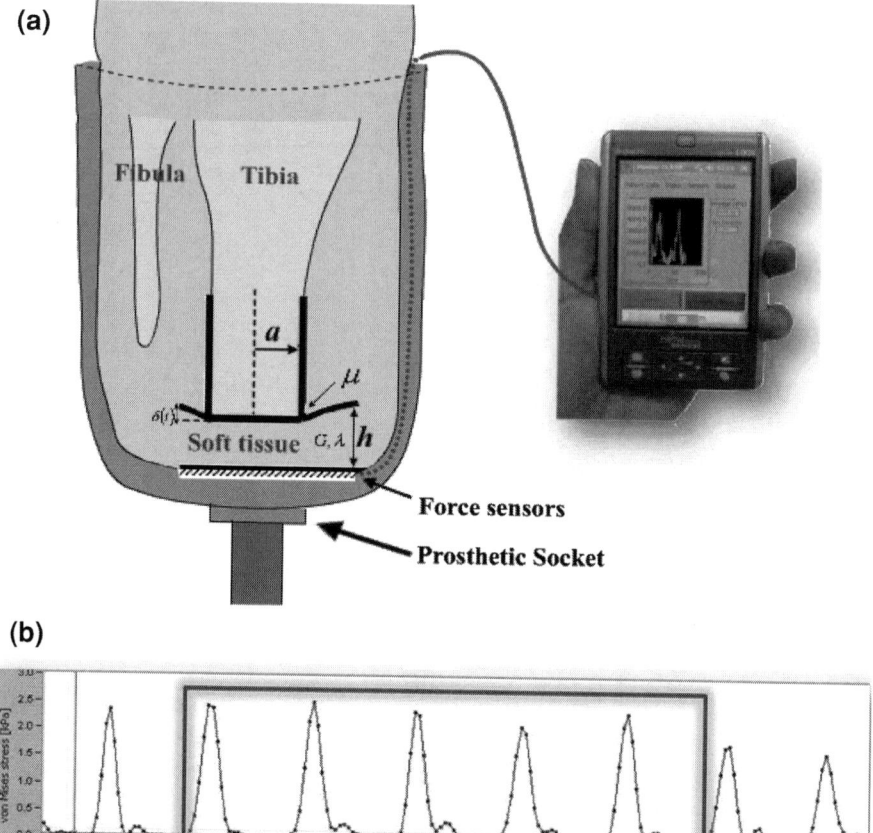

Fig. 3 **a** The tibia is represented by a flat-ended cylinder (with a radius a), thrust by a force $F(t)$ into a compressible elastic layer (with mechanical properties G and λ), which represents the overlying soft tissues (with initial thickness h). Using this model, an axisymmetric indentation problem is solved. The soft tissue layer is assumed to be bonded to a rigid substrate that represents the socket of the prosthesis. The geometric parameters, a and h, are measured on anterior-posterior X-rays of each subject in order to create the patient-specific model. Real-time contact forces between the residuum and prosthetic socket are measured by the stress monitor device using a set of four paper-thin and flexible force sensors. The sensors are attached to the residuum at its distal end, directly under the truncated tibia. Analogue force data from the sensors are converted to digital data via a drive circuit and data-acquisition card, combined together to one holter-like unit. Digital force data are then transmitted to the device. **b** Internal stresses under the tibia are calculated, presented and recorded

where the stress components of the von Mises stress are:

$$\sigma_{zz}(t) = -(\lambda + 2G)\frac{\delta(t)}{h}, \quad \sigma_{rr}(t) = -\frac{\lambda\delta(t)}{h}, \quad \sigma_{rz}(t) = -\mu(\lambda + 2G)\frac{\delta(t)}{h}$$

and μ is the coefficient of friction between the tibia and the soft tissues. A numerical correction term $A \cdot (a/h) + B$ adjusts for variations of the tibial radius to soft tissue thickness ratio (a/h) over a relatively large range across TTA patients. The values of A and B were obtained by means of large deformation FEAs, as detailed in [30, 33].

We created two duplicate real-time portable internal stress monitor devices, to allow for a two-center study (in the Netherlands and Israel). Real-time interface forces between the residuum and the prosthetic socket, $F(t)$, were measured at a frequency of 24 Hz. The analogue data was converted to digital data via a drive circuit and data-acquisition card (USB-6008 OEM, National Instruments), and transmitted to a handheld computer (Fujitsu-Siemens Pocket LOOX N560, Fujitsu Co., Japan; Fig. 3a). The stress monitor calculates, in real-time, the instantaneous average patient-specific internal von Mises stress in the soft tissues under the truncated tibia. The stresses are calculated and presented on the screen of the handheld computer using a LabView eight module for handheld PC (National Instruments Co., TX, USA; Fig. 3a, b).

We evaluated patient-specific internal stresses in the residual limbs of 18 prosthetic-users. Each subject ambulated on plane floor, grass, up-stairs, up-slope, down-stairs, and down-slope.

Our results showed that the internal stresses were the highest while subjects descended a slope. Peak and root mean square (RMS) stresses calculated while descending a slope were approximately 2-times higher for vascular subjects (n = 7) compared to traumatic patients (n = 11), but were similar between the two sub-groups for the other ambulation tasks.

Overall, the described patient-specific portable internal stress monitor was shown to be an efficient, practical tool for real-time evaluation of internal stresses in the TTA residuum. Pending integration of appropriate dynamic tissue injury thresholds, the device can be adopted in the future for alerting to the danger of DTI [30, 33].

3.2 Bone/Implant Interaction

As detailed in the Introduction, the problematic limb/socket interface may be avoided by an alternative technique of transferring the loads directly to the truncated bone using osseointegration [3]. The direct skeletal attachment provides a feeling of less weight on the residuum, eliminates the need for new socket design and replacement, allows better control over the prosthesis, and is more easily donned and doffed [15]. However, aside from the long heeling period and the risk for superficial infections at the skin penetration area, the implant is also at risk of loosening due to stress shielding.

Xu et al. [53] investigated the strain and stress distribution in the direct skeletal implant of a TFA patient by performing 3D patient-specific FEA. The patient-specific model was created using pre-surgery CT of the residuum. Implants of different diameters were then virtually inserted into the femoral model and the mechanical conditions of the bones were calculated under typical walking loads. Local stress variations were found near the contact discontinuity areas.

Isaacson et al. [18] used the exposed exoprosthetic attachment to the femur of TFA patients as a cathode for applying external electrical stimulation, directly to the truncated bone. Controlled electrical stimulation is believed to improve the fixation of the skeletal implant by recruitment of growth factors, assisting with calcium deposition and osteoblast migration and secretion of additional extracellular matrix [18]. The research team performed CT-based patient-specific FEA (n = 11) using a module for bioelectric field problems. One of the authors' finding was that the presence of heterotopic ossification (HO) significantly affect the bioelectricity in the residuum. This was attributed to the fact that larger volumes of HO requires a higher potential difference to satisfy the electric field and current density criterion needed to accelerate bone healing using simulations.

Although cases of femoral bone loss associated with stress shielding that may lead to periprosthetic fracture do exist [52], computational studies of the bone remodeling process under cyclic loading conditions of the LLA gait have yet to be published. This is surprising in light of the relatively high number of publications dealing with subject-specific FEA of stress shielding patterns following hip or knee replacement. Although these studies are not in the scope of this chapter, [51] who created patient-specific models of stress shielding following hip replacement concluded that subject-specific FEA may be useful for explaining the variation in bone adaptation responsiveness between different subjects.

4 Discussion

The integration of CASD/CAM system with 3D FEA, displaying the internal mechanical condition of the LLA residuum is the next expected leap in applied technology innovations for the design of patient-specific sockets or direct skeletal attachments. Transforming the subjective art of socket design into intelligence precise science will require multidisciplinary effort of prosthetists, biomedical engineers, clinicians and programmers. The integration of CASD/CAM systems in the daily clinical establishments will be beneficial in terms of (1) reducing the examination time, (2) reducing the total cost of socket design, as several trial sockets would not be necessary, (3) allowing a fast and comfortable examination of the residuum, (4) producing precise record quantifying the state of the residuum at the time of the examination, and (5) the detailed description of the designed socket maybe reviewed for future use and then a precise replica of the socket may be fabricated. While initial work towards realization of this goal is underway [9], the development of sophisticated tools for LLA surgical planning and control or for monitoring and protection of the LLA residuum from injury is unattended.

References

1. Beil, T.L., Street, G.M., Covey, S.J.: Interface pressures during ambulation using suction and vacuum-assisted prosthetic sockets. J. Rehabil. Res. Dev. **39**(6), 693–700 (2000)
2. Boone, D., Burgess, E.: Automated fabrication of mobility aids: clinical demonstration of the UCL computer aided socket design system. J. Prosthet. Orthot. **1**(3), 187–190 (1989)
3. Brånemark, R., Brånemark, P.I., Rydevik, B., Myers, R.R.: Osseointegration in skeletal reconstruction and rehabilitation: a review. J. Rehabil. Res. Dev. **38**, 175–181 (2001)
4. Commean, P.K., Smith, K.E., Vannier, M.W.: Design of a 3-D surface scanner for lower limb prosthetics: a technical note. J. Rehabil. Res. Dev. **33**(3), 267–278 (1996)
5. Dean, D., Saunders, C.G.: A software package for design and manufacture of prosthetic sockets for transtibial amputees. IEEE Trans. Biomed. Eng. **32**(4), 257–262 (1985)
6. Dillingham, T.R., Pezzin, L.E.: Postacute care services use for dysvascular amputees: a population-based study of Massachusetts. Am. J. Phys. Med. Rehabil. **84**, 147–152 (2005)
7. Dudek, N.L., Marks, M.B., Marshall, S.C.: Skin problems in an amputee clinic. Am. J. Phys. Med. Rehabil. **85**, 424–429 (2006)
8. Esquenazi, A.: Amputation rehabilitation and prosthetic restoration. From surgery to community reintegration. Disabil. Rehabil. **26**(14–15), 831–836 (2004)
9. Facoetti, G., Gabbiadini, S., Colombo, G., Rizzi, C.: Knowledge-based system for guided modeling of sockets for lower limb prostheses. Comput. Aided Design Appl. **7**(5), 723–737 (2010)
10. Gailey, R.S., Wenger, M.A., Raya, M., Kirk, N., Erbs, K., Spyropoulos, P., Nash, M.S.: Energy expenditure of trans-tibial amputees during ambulation at self-selected pace. Prosthet. Orthot. Int. **18**(2), 84–91 (1994)
11. Gefen, A.: Deep tissue injury from a bioengineering point of view. Ostomy Wound Manage. **55**(4), 26–36 (2009)
12. Gefen, A.: How much time does it take to get a pressure ulcer? Integrated evidence from human, animal, and in vitro studies. Ostomy Wound Manage. **54**, 26–35 (2008)
13. Gefen, A., Gefen, N., Linder-Ganz, E., Margulies, S.S.: In vivo muscle stiffening under bone compression promotes deep pressure sores. J. Biomech. Eng. **127**, 512–524 (2005)
14. Geil, M.D.: Consistency, precision, and accuracy of optical and electromagnetic shape-capturing systems for digital measurement of residual-limb anthropometrics of persons with transtibial amputation. J. Rehabil. Res. Dev. **44**(4), 515–524 (2007)
15. Hagberg, K., Brånemark, R.: One hundred patients treated with osseointegrated transfemoral amputation prostheses–rehabilitation perspective. J. Rehabil. Res. Dev. **46**(3), 331–344 (2009)
16. Herbert, N., Simpson, D., Spence, W.D., Ion, W.: A preliminary investigation into the development of 3-D printing of prosthetic sockets. J. Rehabil. Res. Dev. **42**(2), 141–146 (2005)
17. Houston, V.L., Burgess, E.M., Childress, D.S., Lehneis, H.R., Mason, C.P., Garbarini, M.A., LaBlanc, K.P., Boone, D.A., Chan, R.B., Harlan, J.H., Brncick, M.D.: Automated fabrication of mobility aids (AFMA): below-knee CASD/CAM testing and evaluation program results. J. Rehabil. Res. Dev. **29**(4), 78–124 (1992)
18. Isaacson, B.M., Stinstra, J.G., MacLeod, R.S., Pasquina, P.F., Bloebaum, R.D.: Developing a quantitative measurement system for assessing heterotopic ossification and monitoring the

bioelectric metrics from electrically induced osseointegration in the residual limb of service members. Ann. Biomed. Eng. **38**(9), 2968–2978 (2010)
19. Kosasih, J.B., Silver-Thorn, M.B.: Sensory changes in adults with unilateral transtibial amputation. J. Rehabil. Res. Dev. **35**, 85–90 (1998)
20. Krouskop, T.A., Muilenberg, A.L., Doughtery, D.R., Winningham, D.J.: Computer-aided design of a prosthetic socket for an above-knee amputee. J. Rehabil. Res. Dev. **24**(2), 31–38 (1987)
21. Lin, C.C., Chang, C.H., Wu, C.L., Chung, K.C., Liao, I.C.: Effects of liner stiffness for transtibial prosthesis: a finite element contact model. Med. Eng. Phys. **26**, 1–9 (2004)
22. Linder-Ganz, E., Engelberg, S., Scheinowitz, M., Gefen, A.: Pressure-time cell death threshold for albino rat skeletal muscles as related to pressure sore biomechanics. J. Biomech. **39**, 2725–2732 (2005)
23. Linder-Ganz, E., Gefen, A.: Stress analyses coupled with damage laws to determine biomechanical risk factors for deep tissue injury during sitting. J. Biomech. Eng. **131**, 011033-1–011033-13 (2009)
24. Lyon, C.C., Kulkarni, J., Zimerson, E., VanRoss, E., Beck, M.H.: Skin disorders in amputees. J. Am. Acad. Dermatol. **42**, 501–507 (2000)
25. Mak, A.F.T., Zhang, M., Boone, D.A.: State-of-the-art research in lower-limb prosthetic biomechanics-socket interface: a review. J. Rehabil. Res. Dev. **38**, 161–174 (2001)
26. Mak, A.F., Zhang, M., Tam, E.W.: Biomechanics of pressure ulcer in body tissues interacting with external forces during locomotion. Annu. Rev. Biomed. Eng. **12**, 29–53 (2010)
27. Oberg, K., Kofman, J., Karisson, A., Lindstrom, B., Sigblad, G.: The CAPOD system-A scandinavian CAD/CAM system for prosthetic sockets. J. Prosthet. Orthot. **1**(3), 139–148 (1989)
28. Panagiotopoulou, O.: Finite element analysis (FEA): applying an engineering method to functional morphology in anthropology and human biology. Ann. Hum. Biol. **36**(5), 609–623 (2009)
29. Pinzur, M.S., Beck, J., Himes, R., Callaci, J.: Distal tibiofibular bone-bridging in transtibial amputation. J. Bone Joint Surg. Am. **90**(12), 2682–2687 (2008)
30. Portnoy, S., Siev-Ner, I., Shabshin, N., Gefen, A.: Effects of sitting postures on risks for deep tissue injury in the residuum of a transtibial prosthetic-user: a biomechanical case study. Comput. Methods Biomech. Biomed. Eng. **5**, 1 (2010)
31. Portnoy, S., Siev-Ner, I., Shabshin, N., Kristal, A., Yizhar, Z., Gefen, A.: Patient-specific analyses of deep tissue loads post transtibial amputation in residual limbs of multiple prosthetic users. J. Biomech. **42**, 2686–2693 (2009)
32. Portnoy, S., Siev-Ner, I., Yizhar, Z., Kristal, A., Shabshin, N., Gefen, A.: Surgical and morphological factors that affect internal mechanical loads in soft tissues of the transtibial residuum. Ann. Biomed. Eng. **37**, 2583–2605 (2009)
33. Portnoy, S., van Haare, J., Geers, R.P.J., Kristal, A., Siev-Ner, I., Seelen, H.A.M., Oomens, C.W.J., Gefen, A.: Real-time subject-specific analyses of dynamic internal tissue loads in the residual limb of transtibial amputees. Med. Eng. Phys. **32**, 312–323 (2010)
34. Reihsner, R., Melling, M., Pfeiler, W., Menzel, E.J.: Alterations of biochemical and two-dimensional biomechanical properties of human skin in diabetes mellitus as compared to effects of in vitro non-enzymatic glycation. Clin. Biomech. (Bristol, Avon) **15**(5), 379–386 (2000)
35. Reynolds, D.P.: Shape design and interface load analysis for below-knee prosthetic sockets. Ph.D. Dissertation, University College, University of London (1998)
36. Reynolds, D.P., Lord, M.: Interface load analysis for computer-aided design of below-knee prosthetic sockets. Med. Biol. Eng. Comput. **30**(4), 419–426 (1992)
37. Rogers, B., Bosker, G.W., Crawford, R.H., Faustini, M.C., Neptune, R.R., Walden, G., Gitter, A.J.: Advanced trans-tibial socket fabrication using selective laser sintering. Prosthet. Orthot. Int. **31**(1), 88–100 (2009)

38. Rogers, B., Bosker, G., Faustini, M., Walden, G., Neptune, R.R., Crawford, R.: Case report: variably compliant transtibial prosthetic socket fabricated using solid freeform fabrication. J. Prosthet. Orthot. **20**(1), 1–7 (2008)
39. Sanders, J.E., Daly, C.H.: Normal and shear stresses on a residual limb in a prosthetic socket during ambulation: comparison of finite element results with experimental measurements. J. Rehabil. Res. Dev. **30**(2), 191–204 (1993)
40. Saunders, C.G., Bannon, M., Sabiston, R.M., Panych, L., Jenks, S.L., Wood, I.R., Raschke, S.: The CANFIT system: shape management technology for prosthetic and orthotic applications. J. Prosthet. Orthot. **1**(3), 122–130 (1989)
41. Selles, R.W., Janssens, P.J., Jongenengel, C.D., Bussmann, J.B.: A randomized controlled trial comparing functional outcome and cost efficiency of a total surface-bearing socket versus a conventional patellar tendon-bearing socket in trans-tibial amputees. Arch. Phys. Med. Rehab. **86**, 154–161 (2005)
42. Silver-Thorn, M.B.: Investigation of lower limb tissue perfusion during loading. J. Rehab. Res. Develop. **39**, 597–608 (2002)
43. Silver-Thorn, M.B., Steege, J.W., Childress, D.S.: A review of prosthetic interface stress investigations. J. Rehab. Res. Develop. **33**, 253–266 (1996)
44. Sinha, R., van den Heuvel, W.J., Arokiasamy, P.: Factors affecting quality of life in lower limb amputees. Prosthet. Orthot. Int. **35**(1), 90–96 (2011)
45. Slocum, D.B.: An Atlas of Amputation. The C.V. Mosby Company, St Louis (1949)
46. Smith, D.G., Burgess, E.M.: The use of CAD/CAM technology in prosthetics and orthotics–current clinical models and a view to the future. J. Rehabil. Res. Dev. **38**(3), 327–334 (2001)
47. Smith, E., Ryall, N.: Residual limb osteomyelitis: a case series from a national prosthetic centre. Disabil. Rehabil. **31**, 1785–1789 (2009)
48. Steege, J.W., Schnur. D.S., Vorhis, R.L., Rovick, J.S.: Finite element analysis as a method of pressure prediction at the below-knee socket interface. In: Proceedings of RESNA 10th Annual Conference, California, pp. 814–816 (1987)
49. Van-Schie, C.H.M.: A review of the biomechanics of the diabetic foot. Lower Extrem. Wounds **4**, 160–170 (2005)
50. Walsh, N.E., Lancaster, J.L., Faulkner, V.W., Rogers, W.E.: A computerized system to manufacture prostheses for amputees in developing countries. J. Prosthet. Orthot. **1**(3), 165–181 (1989)
51. Weinans, H., Sumner, D.R., Igloria, R., Natarajan, R.N.: Sensitivity of periprosthetic stress-shielding to load and the bone density–modulus relationship in subject-specific finite element models. J. Biomech. **33**(7), 809–817 (2000)
52. Wik, T.S., Foss, O.A., Havik, S., Persen, L., Aamodt, A., Witsø, E.: Periprosthetic fracture caused by stress shielding after implantation of a femoral condyle endoprosthesis in a transfemoral amputee-a case report. Acta Orthop. **81**(6), 765–767 (2010)
53. Xu, W., Xu, D.H., Crocombe, A.D.: Three-dimensional finite element stress and strain analysis of a transfemoral osseointegration implant. Proc. Inst. Mech. Eng. H **220**(6), 661–670 (2006)
54. Yang, F.: Asymptotic solution to axisymmetric indentation of a compressible elastic thin film. Thin Solid Films **515**, 2274–2283 (2006)
55. Zhang, M., Mak, A.F., Roberts, V.C.: Finite element modeling of a residual lower-limb in a prosthetic socket: a survey of the development in the first decade. Med. Eng. Phys. **20**(5), 360–373 (1998)
56. Zhang, M., Roberts, C.: Comparison of computational analysis with clinical measurement of stresses on below-knee residual limb in a prosthetic socket. Med. Eng. Phys. **22**(9), 607–612 (2000)
57. Ziegler-Graham, K., MacKenzie, E.J., Ephraim, P.L., Travison, T.G., Brookmeyer, R.: Estimating the prevalence of limb loss in the United States 2005–2050. Arch. Phys. Med. Rehabil. **89**(3), 422–429 (2008)

Patient-Specific Modeling of the Cornea

Roy Asher, Amit Gefen and David Varssano

Abstract This chapter reviews the up-to-date literature in regard to patient-specific modeling of the human cornea. First, the relevant anatomical, morphological and physiological aspects are summarized. Then, examples of non-patient-specific seldom are addressed before existing patient-specific models are discussed. Finally, an overview is presented, of the development of a patient-specific model that can aid the clinician to prognose response to different intraocular pressure controlling medications. This feature can be very useful in cases of keratoconus—a disease of the cornea which involves changes to the corneal microstructure as well as to its macrostructure, gradually causing a conic-like corneal shape that distorts vision.

1 Anatomy and Physiology of the Cornea

The human eyeball is an imperfect globe that lies recessed in a bony socket called "the orbit". Only a small portion of the anterior surface of the eyeball is exposed. The cornea forms the transparent outer covering of the visible colored portion of the eyeball and has a smaller radius of curvature than the remaining portions of

R. Asher (✉) · A. Gefen
Department of Biomedical Engineering,
Faculty of Engineering, Tel Aviv University,
69978 Tel Aviv, Israel
e-mail: asherel@zahav.net.il

D. Varssano
Cornea and External Disease, Department of Ophthalmology,
Tel Aviv Medical Center, Sackler Faculty of Medicine,
Tel-Aviv University, 69978 Tel Aviv, Israel

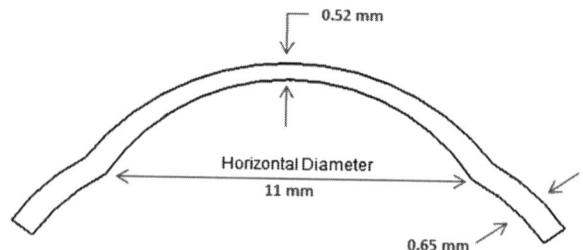

Fig. 1 The corneal thickness in the central region is 0.52 mm. The cornea thickens toward its periphery where its value is about 0.65 mm. The horizontal diameter is in the range of 11–12 mm in 95% of the cases

the eyeball. The visible white opaque portion of the eyeball is called the sclera, fashioned by tough white fibrous tissue and the visible colored portion of the eyeball is due to the underlying iris. The interface between the air and cornea forms most of the optical power of the eye. The curvature of the cornea is responsible for roughly two-thirds of the refraction of light in the eye and the slightest imperfection in its shape results in astigmatism and refractive error. In its central region, the radius of curvature of the human cornea is 7.86 mm with a standard deviation of 0.26 mm. The horizontal diameter (Fig. 1) is in the range of 11–12 mm in 95% of the cases. The central area that lies directly in front of the pupil is the main optical zone and is about 3–4 mm in diameter. The corneal thickness in the central region is 0.52 mm with a standard deviation of 0.04 mm. The cornea thickens toward its periphery, where its value is about 0.65 mm (Fig. 1) [32].

The cornea performs three major functions:

(a) Protects the inner contents of the eye.
(b) Maintains the shape of the eye.
(c) Refracts light.

Disease and injury can alter the shape, transparency and thickness of the cornea, leading to serious changes in the visual performance of the eye. For this reason, understanding the biomechanical response of the cornea, particularly when a disease is present, is of great importance. This task is made difficult by the layered structure of the cornea and its lack of homogeneity. Both of these factors mean that losing corneal tissue due to disease affects the corneal tissue's mechanical properties and makes predicting the effects of pathological factors more difficult [2].

2 Microstructure and Morphology of the Cornea

The human cornea is divided into five layers lying parallel to its surface. The layers from outside to inside are: the epithelium, Bowman's layer, the stroma, Descemet's membrane and the endothelium [26]. At the micro level, the stroma, which forms about 90% of the corneal thickness, is made up of interlacing layers (300–500 lamella) of collagen fibrils lying parallel to the corneal surface and embedded within a matrix of proteoglycans. This construction not only raises questions on whether the stroma can be considered a homogeneous structure from a macro viewpoint; it also leads to difficulties in understanding the material behavior [2]. In accordance with this microstructure, the response of the cornea is not weakly or mildly anisotropic, it is usually highly anisotropic. There has been strong evidence that for highly anisotropic composite materials isotropic theory would not provide even a rough approximation to their behavior under most types of loading conditions. It is therefore important to know how the orientation of collagen fibrils influences the overall corneal mechanical behavior [23].

The transparency of the cornea is a direct consequence of the structural organizations of the macromolecular components of the stroma, the predominant layer in the cornea. This includes the requirement of a regular distribution of the collagen fibrils in addition to their being of equal diameter [13]. It is relatively straightforward to show that if the diameter is the same for all fibrils, and if the fibrils sit on a hexagonal lattice whose sides are smaller than half the wavelength of the incoming light, all secondary light will cancel out in all directions and will only be able to propagate in the forward direction, making the cornea transparent. The distance between adjacent fibrils must be more or less the same but still less than half the incoming wavelengths. This, in fact, is an accurate description of the positional constraints obeyed by the collagen fibrils in the cornea, which is likely to be a direct consequence of the way fibrils are kept in position by proteoglycans. It is likely that if either diseases or adverse physiological conditions impair the uniformity of the distances between adjacent fibrils, or the uniformity of the fibril diameters, corneas can lose their transparency [22].

Collagen fibrils reinforce biological materials and, because they are strongest axially, knowledge of their orientation within a particular tissue may be used to model its mechanical performance [39]. In the cornea, the relationship between fibril orientation and tissue mechanics is of considerable interest. According to Boot et al. [4], in 1938, Kokott was the first to suggest that collagen fibrils in the deeper layers of the stroma in the central cornea are not isotropically arranged, but rather adopt a preferential orientation along the superior-inferior (vertical direction) and nasal-temporal (horizontal direction) corneal meridians. X-ray scattering studies later confirmed this, and indicated that the preferred orientation is more prevalent in the posterior half of the stroma. Notably, some have proposed that this preferential fibril orientation exists in order to take up the stress of the ocular rectus muscles along the cornea. In a recent study [4], wide-angle X-ray scattering was used in order to quantify the relative number of stromal collagen fibrils directed along the two preferred corneal lamellar directions. The data suggest that, on average, the two directions are populated in equal proportion at the corneal center. Approximately one-third of the fibrils throughout the stromal depth tend to lie within a 45° sector of the superior-inferior meridian, and similarly for the nasal-temporal direction. Perhaps most interesting of all is the observation that an annulus of collagen fibers encircle the limbus. Meek et al. [27, 1] suggested that the change in orientation of the fibrils as they move toward the limbus accounts for the flattening of curvature of the cornea in this region, and their finding of the circumferential ring of fibrils at the limbus also provides a very elegant explanation as to how the transition in curvature between the cornea and sclera is achieved.

The majority of corneal diseases result from disruption to the fibril arrangement in the corneal stroma. Recently Grytz and Meschke [17] introduced a novel remodeling algorithm characterized by the reorganization of the collagen fibril architecture. This is achieved by allowing for a continuous re-orientation of preferred fibril orientations and a continuous adaptation of the fibril dispersion at the meso-scale, stimulated by the stress environment at the macro-scale. The proposed remodeling algorithm was applied to an incompressible finite element model of the corneo-scleral shell subjected to physiological intraocular pressure (IOP). The remodeling process showed that the corneal fibrils near the limbus experience a substantial change in their morphology. Due to the different curvatures of the scleral sphere and the corneal sphere the tensile stress in circumferential direction is much higher than in meridional direction at the limbus. Consequently, the mean limbus evolve toward the circumferential direction. The concentration of the collagen fibril orientations towards the circumferential direction at the limbus predicted by the remodeling algorithm can be interpreted as the development of a fibrillar annulus such as described by Newton and Meek [27].

It is straight forward that the direction of collagen in different parts of the corneal tissue has implications for tissue shape. The cornea is approximately spherical in the central 4 mm or so, but then flattens. Moreover, the cornea is more curved than the rest of the eyeball, so flattening allows a smoother transition at the limbus. This change of curvature moving radially outwards across the limbus has its own mechanical implications.

3 Mechanical Properties of the Cornea

Some previous finite element (FE) models of the cornea assumed homogeneity and isotropy of the tissue. In the central cornea there is no overall fiber bundle orientation in the stroma, the major load bearing layer of the cornea. However, there is evidence that this apparent randomness gives way to a preferential circumferential orientation when moving from the central cornea toward the limbus [37, 17]. Fibers in the limbus have long been known to be circumferentially oriented [25]. This knowledge of preferential fiber orientation that was recently backed up by Grytz and Meschke [17] in their computational analysis is inconsistent with the usual assumption of an isotropic constitutive law for the cornea. The determination of the appropriate constitutive assumptions is critical to mechanical models.

When subjecting soft biological tissues that exhibit preferred directions in their microstructure (e.g. cornea, arterial walls and heart tissue) to small strains (less than 2–5%), their mechanical behavior can generally be modeled using conventional anisotropic linear elasticity. Under large deformations, however, these materials exhibit highly anisotropic and nonlinear-elastic behavior due to rearrangements in the microstructure, such as reorientation of the fiber directions with deformation. The accurate simulation of these nonlinear large-strain effects requires the use of constitutive models formulated within the framework of anisotropic hyperelasticity. An example of an invariant-based energy function is the form proposed by Holzapfel et al. [14, 18] for arterial walls (Eq. 1). The model assumes that the directions of the collagen fibers within each family are dispersed (with rotational symmetry) about a mean preferred direction. The structure parameter k ($0 \leq k \leq 1/3$) characterizes the level of dispersion of the collagen orientations. When $k = 0$, the fibers are perfectly aligned (no dispersion). When $k = 1/3$, the fibers are randomly distributed and the material becomes isotropic. C_{10}, K_1 and K_2 are material parameters [16]. The parameter k, characterizing the level of dispersion of the collagen orientation, could be derived, for example, from X-ray scattering findings such as those that were reported by Boot and Meek on [5].

$$U = C_{10}(\bar{I}_1 - 3) + \frac{k_1}{2k_2} \sum_{\alpha=1}^{N} \{\exp(k_2 \langle \bar{E}_\alpha \rangle_2)\}$$

$$\bar{E}_\alpha = K(\bar{I}_1 - 3) + (1 - 3k)(\bar{I}_{4(\alpha\alpha)} - 1)$$

(1)

Equation 1. An invariant-based energy function proposed by Holzapfel et al. [14, 18] for arterial walls. C_{10}, K_1 and K_2 are material parameters. The structure parameter k ($0 \leq k \leq 1/3$) characterizes the level of dispersion of the collagen orientations.

Based on their X-ray findings, Boot and Meek [5] discuss a possible expression for determining the elastic modulus of the cornea in any particular direction (Eq. 2). The expression is defined by three main factors:

(1) E_f—the fibril elastic modulus
(2) ß—the fibril volume fraction
(3) η—the fibril orientations with respect to that direction

$$E_c = E_f \beta \eta + E_g(1 - \beta)$$

(2)

Equation 2. E_c is the elastic modulus of the cornea in any particular direction. E_f is the fibril elastic modulus. ß is the fibril volume fraction. η is the fibril orientations with respect to that direction [5].

Given that the fibril diameter does not change significantly over the cornea until the limbus is reached, and that the intermolecular collagen spacing within fibrils appears likewise constant, there is no reason to suppose that the elastic modulus of the fibrils (E_f) will vary for different meridians. The author also refers to the fibril volume fraction (ß) expecting it to be on average similar along different corneal meridians. As a result, the elastic modulus discussed will be, to a first approximation, simply proportional to the reinforcement efficiency (η). This quantitative measure expresses the ability of a pattern of collagen fibrils to reinforce a tissue in tension. For a laminated structure, in which the fibril direction is restricted to the lamellae planes, η takes the form of Eq. 3. Where g(θ) is the probability of finding a fibril oriented at an angle θ to an applied load [39].

$$\eta = \int_0^\pi \frac{g(\theta)\cos^4\theta d\theta}{g(\theta)d\theta} \tag{3}$$

Equation 3. g(θ) is the probability of finding a fibril oriented at an angle θ to an applied load [39].

Taking into account that the X-ray findings show two preferred corneal meridians that on average are populated in equal proportions and are distributed in a Gaussian spread about these directions, a normalized elastic modulus will simply take the form of η (Eq. 3). The implication of the elastic modulus as a function of angle is that the normal human cornea will be equipped to withstand tensile stress more effectively along its superior-inferior and nasal-temporal trajectories than along the oblique directions [5].

4 Keratoconus

The main characteristic in keratoconus is the deterioration of the cornea's structure mainly in the form of localized loss of up to 75% of the corneal tissue [6]. As a result, under IOP the cornea changes shape, with serious implications on its refractive power [3, 11]. In keratoconus, corneal topography is characteristically altered; Mandell and Polse [24] stated that compared to normal corneas, many keratoconic corneas become more hyperbolic in shape and furthermore that most keratoconic corneas have characteristically large differences between their central and peripheral thickness. Keratoconus may develop from a presumably normal cornea by an increased distensibility of keratoconic tissue. Thus the progressive alteration of the keratoconic corneal shape may be based on elastic deformation. Theoretically, prolonged increased IOP, bursts of intense force (such as in eye rubbing), decreased corneal tissue strength, decreased corneal tissue mass or a combination may be pathogenetic factors.

In the literature, topics concerning tissue mass and strength have attracted the most attention. The magnitude of the corneal tissue mass depends on the corneal thickness. In general, keratoconus is characterized as a corneal thinning disorder. The variability of the corneal thinning has been shown in different in vivo studies comparing normal and keratoconic eyes. By histopathological studies Pouliquen [34] found normal sized collagen fibers with a decrease in the number of collagen lamella. Biomechanical studies have demonstrated a decreased amount of normal stromal collagen in keratoconus. In accordance with Polack [33], thinning of the cornea without actual collagenolysis occurs when collagen lamellas are released from their attachment to other lamella or to Bowman's layer and slide.

Thus, a thinning of the keratoconic cornea does not necessarily imply a decrease in the total corneal tissue mass [10].

In any stage of the disease, every layer and tissue of the cornea may become involved in the pathological process. Many theories have been proposed regarding the pathogenesis of keratoconus. Among these are theories which either implicate alteration in the basal Bowman's layer and the anterior stroma as the site of the first change, or the possible existence of primary pathology of keratocytes. However the primary lesion seems to be in the anterior part of the cornea with secondary destruction toward a posterior direction [38].

A possible hypothesis for the pathogenesis of keratoconus is that keratoconus may involve a primary lesion in the anterior part of the central cornea which affects Bowman's layer and the basal epithelial cells. This may result in reduced attachment of the anterior collagen lamella to Bowman's layer and increased sliding of the collagen bundles which form the lamella due to altered synthesis of matrix substance. The reduced attachment and the increased sliding of the collagen structures correspond to increased distensibility of the corneal tissue. The process of attachment and sliding of the corneal collagen structures may be influenced by external factors and also by constitutional genetics. Therefore, the reduced ocular rigidity and the association between keratoconus and connective tissue diseases such as Osteogenesis imperfecta, Marfan's syndrome and Ehlers-Danlos syndrome may reflect a common constitution for decreased attachment and increased sliding of collagen structure [20]. The association of keratoconus to atopy and such external factors as contact lens wear and eye rubbing may be due to a damage of the basal corneal epithelium cells and perhaps Bowman's membrane, thus initiating the development of keratoconus.

Intra ocular pressure has an important role in the mechanical behavior of the normal and keratoconic corneas. Measures that were taken using the Goldmann applanation tonometry in a group of untreated corneas of 120 patients (203 eyes) by Emara et al. [12] revealed that the mean IOP is 15.6 ± 2.7 mmHg (range 10–24 mmHg). The corneal tissue was observed to behave linearly to a range of IOPs between 2 and 4 kPa (15–30 mmHg). Beyond this pressure, the modulus of elasticity grew suddenly [7]. According to Edmund [10], no significant difference was found in the IOPs between normal and keratoconic eyes.

5 Laser-Assisted In Situ Keratomileusis

Laser-assisted in situ keratomileusis (LASIK), the most commonly performed refractive surgery, involves the creation of a lamellar flap followed by patterned photoablation of the underlying stroma. In a recent large-scale review of LASIK outcomes over a decade-long study period, most patients (95%) were highly satisfied with their outcome. However, 5% expressed dissatisfaction with the surgical outcome based on the low quality of life scores that were attributed in part to refractive regression, poor night vision, and residual refractive error [36].

In LASIK (flap creation) without tissue removal, central flattening and peripheral steepening were shown to occur [35]. Attaining a level of reliability and predictability is a principal aim of any refractive treatment regardless of the surgical technique that is used. One of the better possible approaches is to combine the current technology of wavefront analysis with a biomechanical model. The latter can be used on a patient-specific basis to predict with great accuracy how the components of the eye, especially the cornea, react to the surgical plan. One of the cornerstones of this biomechanical approach to improve visual outcome is the finite element method.

6 Modeling the Human Cornea

Numerical modeling based on the finite element method is used to accurately predict the performance of normal and keratoconic human corneas and to provide a detailed account of their response to various mechanical changes. Previous studies attempted this approach with varying degrees of success. Buzard, [8] and Bryant and McDonnell, [7] developed a finite element model in which corneas were modeled using two-dimensional axisymmetric elements. Their work confirmed the effectiveness of numerical modeling in corneal biomechanics, but was unable to model asymmetrical effects that could be found in injuries or diseases such as keratoconus. A more detailed three-dimensional model was used by Pinsky and Dayte, [32] to predict the immediate change in corneal topography following refractive surgery. This model was based on a linear-elastic behavior pattern.

In a study conducted by Gefen [15] a more accurate biomechanical model of normal and keratoconic corneas was developed by employing a more realistic representation of the corneal geometry and anisotropic mechanical properties. Cases of keratokonous were simulated by global or localized tissue "weakening" as well as localized asymmetric thinning distortions of the cornea, considered separately or together in the models, in order to determine their relative individual/combined influences on the shape of the cornea. Computational modeling of the cornea based on the FE method was used in order to predict the mechanical performance of normal and keratoconic human corneas and to provide a detailed account of their response to various changes in tissue mechanical properties, geometry and IOP. The following general assumptions were made:

(1) The normal unloaded cornea was assumed to be symmetric, half-sphere shaped at the central region, and clamped at the sclera.
(2) Corneal tissue deformations under the IOP values considered in the paper (15–25 mmHg) were relatively small compared with the cornea's radius of curvature.
(3) Corneal tissue was treated as a linear elastic, nearly-incompressible material with three-dimensional orthotropic mechanical properties.

The assumption of orthotropic material behavior is a progress over previous work, and was taken to explore the possibility that softening of the cornea at certain material directions contributes to keratoconus. A preliminary sensitivity analysis to determine the effects of the value assigned to Err (the perpendicular elastic moduli of the corneal tissue) on model predictions showed that a physiologically reasonable variation of ±40% had negligible influence on corneal displacements and the respective dioptric maps. Contrarily, the value assigned to $E\varphi\varphi$ (the meridional elastic moduli of the corneal tissue) did show considerable influence on the model outcomes, and so, $E\varphi\varphi$ was assumed to be involved in the pathogenesis of keratoconus. Accordingly, in some keratoconus simulation cases $E\varphi\varphi$ was reduced separately or simultaneously with $Gr\varphi$.

The geometry of a normal cornea as described in the literature was used as a basis for building an appropriate finite element model (FEM). This FEM was then used as a reference case for the computational model which was fundamental for understanding the mechanical behavior of the cornea, just before applying the factors involved in the pathogenesis of keratoconus (tissue thinning and degradation of mechanical properties). Two cases of tissue thinning, which may characterize two severity levels of keratoconus were simulated by building appropriate FE models for them.

Stress distribution has an importance in the research of the pathology of keratoconus. Thus, the stress distribution through the corneal center was calculated for all corneal conditions where an average IOP of 15 mmHg was assumed. First attention should be paid to the location of the maximal stress point. In all cases it was obtained at the tissue thinnest point. In the normal cornea model the thinnest point and maximal stress region are located at the cornea apex, but in some simulated cases the thinnest point was located in a small distance from the cornea apex. The maximal stress region like the cone center was obtained at the cornea thinnest point. The minimum thickness of the cornea tissue has a significant influence on the stress level. The maximal stress value changes exponentially to the tissue thickness. This means that in an advanced condition of keratoconus, every small IOP increase bears a significant increase in the stress level. Based on these findings the authors offer an optional explanation to how the IOP value may have a major role in the pathogenesis of keratoconus. It may include the following scenario: IOP raise bears a high stress level in the thinnest point of the cornea tissue. Some of the tissue fibers cannot withstand this stress level and start to fail. As a result tissue degradation and a decrease in the mechanical properties occur. Then a high stress level is obtained even under lower IOP and so on and so forth. Other scenario may support the possibility that the whole process begins with tissue degradation from any reason and the IOP actually stays stable.

Based on their findings Gefen et al. [15] report that tissue thinning alone appears as insufficient to cause the marked degradation of vision seen in severe keratoconus patients. This was based on the cases that simulated regional thinning or a decrease in mechanical properties and did not yield a keratoconus like topography when they were applied as an individual factor in the FE model. But when integrating them together into one FE Model a maximal displacement of 230 μm and dioptric power of 54.3 D under a normal IOP of 15 mmHg was observed. Those are numbers correspond to the mean values that were observed in the literature (56.2 ± 8.5 diopters). If it is assumed that regional thinning and a decrease in mechanical properties are conditional to the pathogenesis of keratoconus, it is reasonable to also assume that it is not a coincidence that both are inherent in keratoconus. It is possible that a decrease in one of them may cause a decrease in the other, yet the question of which of them starts the whole process and what is the relationship between them is not fully understood. Furthermore, Gefen et al. [15] results demonstrate that the corneal topography and dioptric power in the diseased cornea are substantially more sensitive to changes in the level of IOP compared with the normal cornea, when subjected to the same extent of change in IOP. This is consistent with clinical experience and papers reporting that lowering the IOP by means of anti-glaucoma drugs reduces dioptric powers of weaker corneas, e.g. post-LASIK (laser in situ keratomileusis), especially in ectatic corneas in which there is a marked dilation or distention due to post-operative biomechanical changes [21].

Gefen et al. [15] used FE models that allowed characterization of the biomechanical interactions in keratoconus, toward understanding the etiology of this poorly studied malady. These models were based on a linear-elastic anisotropic constitutive law. An alternative material modeling approach, allowed by commercial FE solvers, is to employ a hyperelastic (rubber-like non-linear) material model for the corneal tissue. Common strain energy density functions characterizing hyperelastic materials usually depend on principal stretch ratios, and thus, are unable to account for anisotropy.

Pandolfi et al. [29] introduced a finite deformation, hyperelastic, orthotropic material model which accounts for the high orientation of the collagen fibrils in the microstructure of the cornea. A large deformation constitutive model, as proposed by Holzapfel et al. [18], was adopted to capture the mechanical response of the cornea. The special choice of the anisotropic strain-energy function greatly simplifies the derivation of the related tangent moduli. Additionally it was assumed that no contributions derive from the anisotropic part of the energy in the case of compressive stretches. The material is characterized by six material constants that are related to the stiffness of the different components of the tissue, namely the underlying matrix and a set of two families of reinforcing collagen fibers with prescribed orientation. Both the external and internal surfaces of the cornea were well approximated by segments of ellipsoids. Pandolfi et al. [29] documented a parametric solid model and a finite element generator able to describe the external and internal surfaces of the cornea by using geometrical data such as curvatures, conic constants, and thicknesses available from standard in vivo medical measurements. The finite element implementation of the anisotropic material model was used to reproduce uniaxial tests on corneal samples and to simulate the mechanical behavior of the whole cornea. The simulations showed good correspondence with the uniaxial tests. The material model was able to capture the uniaxial response with good accuracy and the pressure–displacement curve obtained with the finite element model was remarkably close to a pair of experimental curves obtained by Bryant and McDonnell [7]. The main application of the corneal model described by Pandolfi et al. is the development of a consistent numerical model for the simulation of laser refractive surgery (e.g., LASIK). The author states that the numerical model does not account for some important features of living corneas, which should be considered for realistic simulations, such as the stress-free configuration of the cornea that is not known a priori. Typically, the geometry of the cornea as is registered by standard measurements is the one deformed by the IOP. This was addressed by Pandolfi et al. in a subsequent paper [30]. The developed computer code was equipped with an interpolation procedure able to provide the geometrical parameters in the unloaded configuration and, consequently, to compute the refractive power of the simulated eye under the effect of various IOPs. The finite element model was then used to evaluate the refractive power of the cornea undergoing refractive surgery intervention. The main goal of the paper was to apply the distributed collagen fibril model to the analysis of the response of the human cornea in terms of stress and refractive powers. Four analyses were performed; the first analysis refers to the baseline and is used as a reference (defined as "baseline" case). The second analysis was performed by considering equal distributed collagen fibril orientations (defined as "isotropic" case). This was done by setting the k parameter, earlier described in this chapter as part of the Holzapfel et al. [14, 18] energy function, to be equal to 1/3 for both fibril families. The third analysis refers to a situation in which both sets of collagen fibrils have ideal alignments (defined as "fibers" case). The fourth analysis was performed by assuming the highest degree of anisotropy ($k = 0$) for the nasal-temporal set of fibrils while keeping the baseline distribution of k for the superior-inferior set of fibrils (defined as "mixed" case). The finite element computations performed were able to provide the three-dimensional stress distributions under various levels of IOPs. Referring to the physiological pressure value of 16 mmHg, stress maps were compared for different values of the material cases. In all performed analyses, the von Mises stress σ_M fell within the range 0–20 kPa. An IOP of 40 mmHg was also tested with the different cases. The output values were in agreement with the results reported by Hjortdal [19]. In general, an increase in the fibril stiffness (k parameter) of the nasal-temporal fibril set induces an increase in the von Mises stress along the same direction. The variation of the fibril stiffness induces also a modification of the stress distributions across the corneal thickness. The maximum principal Cauchy stress in the nasal-temporal meridional section of the cornea increases with the stiffness parameters, and the gradient of the stress across the thickness increases as well. Interestingly, as far as the optical performance is concerned, the k parameters related to the fibrils in the nasal-temporal direction remarkably affects the refractive power along the vertical superior-inferior meridian. The author offers a possible explanation of this behavior noting that under the effect of the IOP the softer meridian is more prone to flatten than the stiffer meridian. The anisotropic response of the cornea predicted by the numerical analysis for uneven distribution of the fibril stiffness can be used to justify astigmatic refraction of a cornea.

Models such as those developed by Gefen [15] and Pandolfi [29], emphasizes the great importance of the microstructure of the cornea, in particular, the highly oriented collagen fibrillar structure, which provides the cornea with the unique mechanical and optical properties. The use of an energy function such as the Gerhard et al. [14, 18] that was adapted for the cornea by Pandolfi [31] can allow the simulation of high IOP which can then in turn produce large deformations in the cornea. Nowadays, corneal topography could be easily recovered by using corneal topographers. This can allow the development of patient-specific models of the human cornea.

7 Methods for Acquiring Patient-Specific Data

In the following part two different corneal topographers "Orbscan" and "Galile" are introduced. These can be used for acquiring patient-specific corneal topography data for the development of patient-specific finite element models.

7.1 Orbscan

The Orbscan is a three-dimensional (3D) slit-scan topography system designed for analysis of the corneal and anterior chamber surfaces. It uses a calibrated video and scanning slit-beam system to independently measure the x, y, and z locations of several thousand points on each surface of the cornea and anterior chamber. These points are used to construct topographic maps [40]. Auffarth et al. [4] used the Orbscan Topography System to evaluate corneal topography in a series of 38 keratoconus patients. The Orbscan Topography System was used to evaluate a series of keratoconus patients and analyze their topographic maps. Analysis of the Orbscan color maps was focused on quantitative topographic parameters at three points: the central point of the cornea; the apex, the point with maximum reading on the anterior elevation best-fit sphere map; and the thinnest point, the spot with minimum value on the pachymetry map. Thirty-three patients (86.8%) had bilateral keratoconus and five (13.2%) had unilateral keratoconus. On the anterior elevation best-fit-sphere map, 68 cone apexes were located in the inferior temporal quadrant; three cone apexes were located above the horizontal meridian. The mean radius of the apex was 1.0 ± 0.5 mm. The mean elevation of the apex was 0.117 ± 0.076 mm. The mean tangential and composite curvatures were identical, 56.2 ± 8.5 diopters (D). The mean pachymetry of the thinnest point was 0.457 ± 0.094 mm (range 0.237–0.593 mm). The mean distance between the apex and the thinnest point was 0.917 ± 0.729 mm (range 3.364–0.068 mm) ($P < 0.001$). The correlation between apex elevation and apex composite curvature was high ($r = 0.938$ $P < 0.0001$).

7.2 Galilei

The Galilei™ Dual Scheimpflug Analyzer (Ziemer Group; Port, Switzerland) is a new device for corneal and anterior segment analysis that combines Placido-based corneal topography and dual Scheimpflug anterior segment imaging [35]. The integrated system exploits the advantages of both technologies in a single exam, along a common reference axis. This type of imaging allows assessment of anterior and posterior corneal topography, anterior chamber depth, as well as anterior and posterior topography of the lens. The principal advantage of Dual Scheimpflug imaging is that corresponding corneal thickness data from each view can simply be averaged to compensate for unintentional misalignment, which results in a corrected measurement value at the corresponding location [35]. Simple averaging of the thicknesses in the two corresponding Scheimpflug views reduces this error by a factor of 10, without the need for correcting the misalignment. The dual Scheimpflug system has only to take into account decentration and allocate each averaged thickness and posterior height value to its proper location, whereas single Scheimpflug systems additionally have to make estimations on the variable surface inclination for calculating correct thicknesses or posterior heights [35]. The Galilei™ integrates an infrared Placido topographer with the visible light of the slit system. The performance of Placido topography has been well documented in the literature, and the technology is well understood. The principle advantage of integrating Placido topography with Scheimpflug imaging is to improve the accuracy of central anterior corneal curvature calculation [35]. In summary, the advantages of the Galilei™ System's Dual Scheimpflug Imaging with integrated Placido topography include:

1. Direct measurement of anterior corneal surface curvature.
2. Direct measurement of elevation of all anterior segment structures.
3. Pachymetry calculation that is insensitive to decentration.
4. Motion correction from top view camera, in place for Placido capture.
5. Greater coverage area in combining both technologies.
6. Same reference axis for both technologies.

8 Existing Patient Specific Finite Element Models for the Human Cornea

Most biomechanical models of the cornea have been simplified geometric models of sphere or an ellipsoid or an axisymmetric model. They have not been based on actual patient data. The following are models that are based on patient-specific data.

Deenadayalu et al. [9] studied the effect of varying parameters on the refractive change induced by the LASIK flap. Three-dimensional finite element analysis modeling of actual patient data (topography, pachymetry and axial length) was used in a finite element analysis to determine the response and change in optical power of the cornea as a function of the material property of the cornea (corneal elasticity), flap diameter and thickness, and IOP. In order to acquire the data, a commercially available corneal topographer (Humphrey Atlas Eclipse Corneal Topography System Model 995) was used in the form of 25 rings, each divided into 180 evenly spaced data points. The data were then transformed into spatial Cartesian (x, y, z) coordinates. Some amount of data is usually missing in the superior area of the cornea due to the shadow of the upper lid and lashes. Along each ray, a cubic polynomial was represented as $z = g(r)$ with r as the independent variable. The cubic spline interpolation was used to estimate the values of z for the missing data points. In addition, in order to describe the sclera, points were mapped by extrapolating beyond the last corneal point on each ray. The transition from the cornea to the sclera was modeled biomechanically with a change in elasticity at the transition, but the change in curvature at the limbus was unaccounted for in the model. The finite element nodes were generated to coincide with the topographic data points. The structure was modeled with two layers of elements across the thickness. The thickness of the cornea was obtained clinically from ultrasonic pachymetry data at 33 points on the cornea, with the thickness at the remaining points estimated by linear interpolation. The finite element mesh consisted of 26,402 nodes (79,206 degrees-of-freedom) and 16,200 elements, of which 9,000 were used to model the cornea. All of the degrees-of-freedom were assumed to be fixed for the nodes at the periphery of the finite element model. The elements that were used to mesh the cornea were linear triangular prism and hexahedral, and for the meshing of the sclera linear hexahedral elements were used. The cornea was treated as orthotropic. The modulus of elasticity in the direction perpendicular to the corneal plane, $E11$, was assumed to be 0.125% of the in-plane elastic modulus $E22$ or $E33$. The in-plane moduli values were assumed to be equal, $E22 = E33$. The in-plane Poisson's ratio was assumed to be 0.49 (nearly incompressible). The other two Poisson's ratios were each assigned a value of 0.01. The corneal stiffness is primarily in-planar and the stiffness in the direction perpendicular to the corneal plane is negligible. The shear moduli were assumed to be small relative to the in-plane modulus of elasticity. The sclera was treated as linear isotropic and the ratio of the modulus of elasticity of the sclera to the in-plane modulus of elasticity of the cornea was assumed to be 2.5 throughout the study [9]. The commercial finite element program ABAQUS (ABAQUS Inc, Providence, RI) was used for subjecting the model to a nonlinear static structural analysis. The response of the structure at each node was used to update the coordinates of the reference configuration and thereby map the postoperative topography. The postoperative (x, y, z) locations of the nodes on the anterior surface of the cornea were computed as the sum of the reference configuration and the nodal displacements (x, y, z).

Two patients were selected for the study and the data was used to model four eyes. In all four eyes, a hyperopic shift occurred with the creation of a LASIK flap. The amount of hyperopic shift noted in each modeled eye increased in relation to increased elasticity, flap thickness and IOP. It did not change appreciably in relation to flap diameter. The author states that from the model, it is not possible to determine whether the hyperopic shifts occurred as a result of relative thickening of the peripheral cornea, changes in the stress/strain relationship of the corneal collagen fibrils, or some combination of both.

Two parameters from those that were changed in the study are:

(a) The in-planar modulus of elasticity of the cornea—varying from 2 MPa to 12 MPa.
(b) IOP—varying from 15 to 25 mmHg.

A series of finite element simulations was carried out with a variety of corneal elasticity from very high ($E11 = E22 = 2$ MPa), gradually changing to a relatively stiff cornea (low elasticity) ($E11 = E22 = 12$ MPa), holding the other parameters to constant values. Peripheral steepening and central flattening were observed from the flap formation in each case (hyperopic shift). The amount of corneal deformation, or hyperopic shift, was proportional to the elasticity and decreased as the stiffness increased. As for the IOP, the magnitudes of the peripheral corneal steepening increased as the pressures increased from 15 to 25 mmHg. The central hyperopic shift increased from 0.75 D at 15 mmHg to 1.25 D at 25 mmHg [9]. As could be expected, the parameter that had the most effect on the refractive change in power distribution was the modulus of elasticity. Patient eyes (simulated) with a highly elastic cornea (modulus of elasticity = 2 MPa) could undergo as much as a 2.0 D hyperopic shift with just the introduction of the flap. Also, higher IOP induced a greater hyperopic shift than lower IOP. These outcomes contribute to the argument that a better understanding of the corneal biomechanical response to LASIK may be essential to achieving optimal visual performance. There is a growing body of evidence that the combination of wavefront technology and corneal topography provides the surgeon with the optimal information on the vision characteristics of the eye [9].

In another study, a patient-specific FE model of corneal refractive surgery was reported by Roy et al. [36]. The model allows an analysis of the biomechanical changes associated with a specific LASIK treatment plan and their impact on the accuracy of the postoperative optical result. The model was also used for the estimation of the magnitude of corneal elastic property reductions based on clinical topographic changes between the preoperative and one-week postoperative examinations. 3D models of corneas, of the right and left eyes of a 35 year old female patient who underwent LASIK for myopia with astigmatism, were constructed using tomographic data from a commercial anterior segment imaging system (Pentacam, Oculus Optikgeräte GmbH, Germany). The x, y, and z coordinates from the elevation maps of the anterior and posterior surfaces were interpolated using orthogonal Zernike polynomials up to the sixth order, having 28 terms and with a 5 mm normalization radius. The interpolated elevation data (x, y and z) were used to obtain a 3D surface using a commercial computer aided drafting (CAD) package (PROENGINEER WILDFIRE, PTC, Needham, MD). Each point along a radius was connected by a cubic spline to form a curvilinear edge in 3D. Multiple edges that form the shape of a corneal surface were then blended to obtain a 3D surface for the anterior and posterior cornea. These surfaces (anterior and posterior) were then joined at the limbus to form a 3D solid, representing the in vivo pre-LASIK cornea. A scleral shell was extended from the cornea to an axial length of 3.5 mm in order to simulate an unrestrained limbus. The posterior borders of the sclera were restrained completely. The posterior surfaces of the cornea and sclera were loaded with IOPcc obtained from the ocular response analyzer. IOPcc is the cornea compensated IOP and is considered to be less sensitive to changes in the corneal biomechanical properties and thickness after LASIK compared with the Goldmann applanation [36]. In the clinical setting, corneal topography is measured at a specific IOP and is distinct from the unloaded shape that would be obtained at an IOP of 0 mm Hg. To solve for the undeformed state, the authors developed a custom inverse model by using the commercial FE analysis package ABAQUS (Simulia Inc.) and PYTHON scripting language. In the inverse model, initial estimates of the unloaded shapes of the cornea and sclera were loaded to clinical IOP and then compared with the in vivo shape. The coordinates of the unloaded shape were then corrected based on the difference between the coordinates of the in vivo geometry and the FEM prediction. The resulting unloaded shape was then loaded again to the same IOP. This procedure was repeated until a user-specified tolerance of 0.1% (the difference between the coordinates of two FEM results from successive simulations of loading the corneoscleral model from zero to in vivo IOP) was achieved. The mesh for each model consisted of linear, eight node 3D hexahedral elements [36].

Experimentally derived stress versus strain curves that could be found in literature were fitted to a reduced polynomial material model (Eq. 4).

$$W = C_{10}(I_1 - 3) + C_{20}(I_1 - 3)^2 \tag{4}$$

Equation 4. A reduced polynomial material model that was fit by Roy et al. [36]. W is the strain energy potential and I_1 is the strain invariant. C_{10} and C_{20} are hyperelastic constants.

Where W is the strain energy potential and I_1 is the strain invariant. C_{10} and C_{20} are hyperelastic constants obtained from the fitting of the experimental stress versus strain data and were determined for each of the three meridia. The magnitudes of C10 and C20 along the other meridia were then interpolated. In order to ensure a smooth gradient in C_{10} and C_{20} across all meridians of the cornea a sinusoidal function was chosen for the interpolation (Eq. 5).

$$C = a \cdot \sin^2(\theta) + b \cdot \sin(\theta) + c \tag{5}$$

Equation 5. A sinusoidal function that was chosen for the interpolation. θ is the meridian and a, b, and c are the constants obtained from the regression.

Where θ is the meridian and a, b, and c are the constants obtained from the regression. The reduced polynomial form and the sinusoidal function were also implemented using user-defined subroutines in ABAQUS. The sclera was also modeled using the reduced polynomial form as an isotropic and hyperelastic material. Scleral elastic properties were assumed to be three times the stiffness of the vertical meridian of the cornea [36].

In both eyes, a reduction in corneal elastic properties or weakening of the cornea from the preoperative state resulted in more accurate axial power estimates by the FEM compared with in vivo. These results suggest that the anterior corneal flattening associated with a myopic correction is overestimated if elastic properties are assumed to be as high after surgery as before surgery; conversely, weakening of the cornea, which favors less biomechanical flattening by allowing more forward directed central corneal displacement after LASIK, is required to produce the best match to the clinical response. The tendency toward overcorrection due to excessive biomechanical flattening with stiff corneas and undercorrection in corneas with lower elastic properties supports the hypothesis that a myopic undercorrection may be a clinical marker of weaker corneal elastic properties [36].

Some reasons among those that make the later model better than previous FEM-based studies are the fact that most previous models have not incorporated patient-specific clinical measurements of IOP and corneal geometry, have not included the corneo-scleral limbus and have not modeled the effects of corneal meridional elastic property variation. Using such a model, in which a specific surgical algorithm can be simulated to predict the postoperative outcome and to determine the magnitude of elastic property change within the treatment zone, could assist in achieving the best fit to actual clinical outcome. This is made possible due to the relevant patient-specific clinical information which is the basis for the model. It should also be noted that in addition to material properties and ablation profile, the IOP may influence the shape of the cornea before and especially after corneal surgery and is a potential limitation in patient-specific modeling [36].

The presented method for patient-specific 3D computational modeling of corneal refractive surgery provides representations of clinical corneal shape changes in LASIK and allows inverse estimation of the surgical impact on corneal elastic properties. FEMs such as the one generated in this study do not take a long time to solve (a few minutes) and do not require significant computational resources. The accuracy of such models will depend on the quality of the clinical geometry data supplied, and the results could be regarded as device specific. It is therefore only reasonable to use the most accurate and reliable device available on the market in order to reduce sampling errors.

9 A Patient-Specific FEM of the Cornea to Obtain Prognosis of Keratoconus.

It is our interest to acquire a method for obtaining a simple patient specific model that can aid the clinician to prognose response to different IOP controlling medications in cases of keratoconus. These cases are more vulnerable to an increase in IOP, and this may bear clinical implications, such as related to pharmaceutical

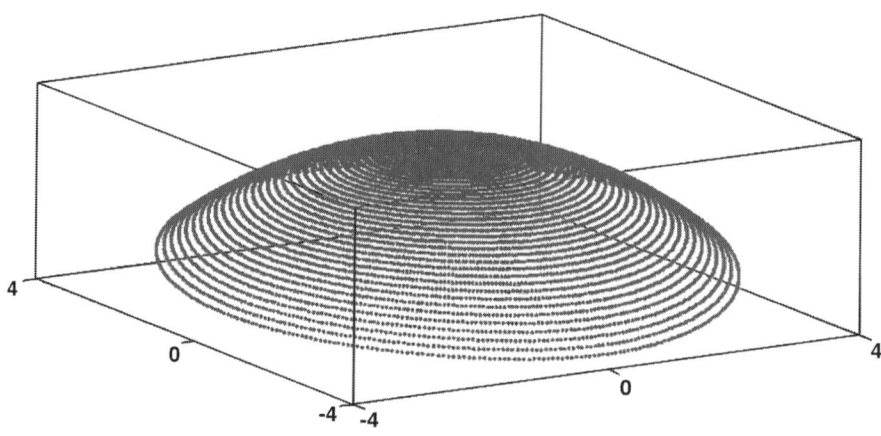

Fig. 2 An example of a sampled anterior corneal surface in a scattered presentation. Plotted with Matlab (Mathworks, Matlab 2009). The 2D support is an 8 mm × 8 mm square

regulation of IOP [15]. For that matter, a method is presented for creating patient-specific FE models based on the patient's geometry of the cornea, true IOP (IOPT) and modulus of elasticity.

The ability of the GALILEI™ Dual Scheimpflug Analyzer (Ziemer Group; Port, Switzerland) to extract highly accurate and densely sampled data from any given cornea was a key factor in the selection of the device. Moreover, for modeling the geometry of a keratoconic cornea it is important for the topographer to be able to sample non-symmetrical surfaces in a reliable method. The GALILEI, as was previously explained in this chapter, meets this required demand. The data of patients with a known condition of keratoconus was acquired and exported using comma delimited files in a cylindrical coordinates system. These files include information about the corneal anterior and posterior height, curvature, pachymetrey and power maps. Data containing the corneal anterior surface height were loaded into Matlab (Mathworks, Matlab 2009) where they were analyzed. First, the data was reorganized and transferred from cylindrical to Cartesian coordinates, allowing a scattered presentation of the surface (Fig. 2).

Then, a best fit sphere (BFS) was fitted to the data and used in order to calculate the surface elevation BFS maps (Fig. 3a). These maps were also obtained directly from the GALILEI (Fig. 3b). The calculated maps were compared to the maps obtained by the GALILEI in order to verify the accuracy of the imported data.

Finally, after successful verification, the data containing in average 28,000 sampled points was transferred in the format of a cloud file into CAD software (Dassault Systèmes, SolidWorks Corp., Solidworks 2010). The same process was done for the posterior surface as well. On the basis of the anterior and posterior cloud files, meshes and surfaces of the anterior and posterior surfaces

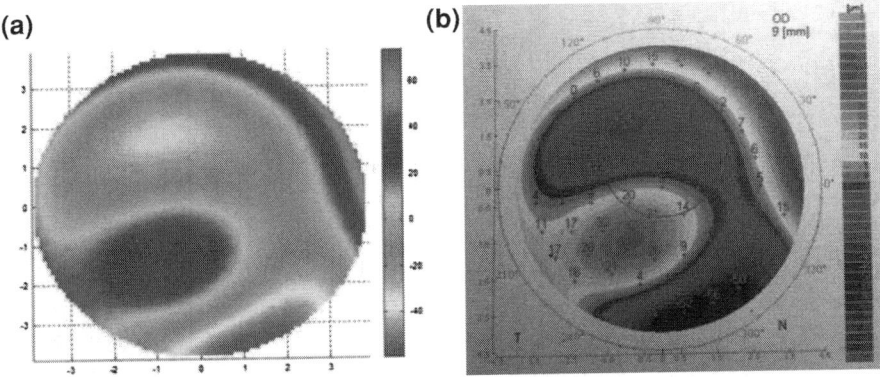

Fig. 3 Surface elevation BFS maps. **a** Calculated using the exported data, **b** obtained directly from the GALILEI. Note that the scales are different but represent the same values

Fig. 4 A *solid* patient-specific cornea part. Also shown in the figure are the sampled data points that were used for the creation of the solid part

were created. By assembling both the anterior and the posterior surfaces together a solid cornea part was obtained (Fig. 4).

The solid cornea part was then imported into a commercial FE analysis package (Dassault Systèmes, Simulia, Abaqus 6.10) for further analysis. Material properties were set to linear elasticity with a nearly incompressible Poisson's ratio (0.45). Boundary conditions were defined as "clamped" throughout the entire circumference and a load of 10 mmHg was applied on the posterior surface, according to the physiological IOP. The mesh for the models consists of linear, eight-node 3D hexahedral elements.

Assuming that the clinician has access to the equipment needed to determine the patient's corneal topography, IOPT and modulus of elasticity, a simple patient-specific model such as the one aforementioned can aid him when prognosing a response to different IOP controlling medications. Output fields such as "elastic strain" (Fig. 5) and "maximum von Mises equivalent stress" can allow access to information such as the magnitude of the local elevation and curvature which directly affects the dioptric power of the eye.

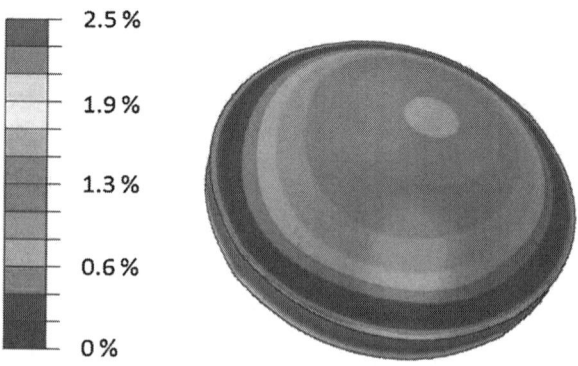

Fig. 5 The patient-specific cornea's maximum principal elastic strain output. Material properties were set to linear elasticity and Poisson's ratio to 0.45. Boundary conditions were defined as "clamped" throughout the entire circumference and a load of 10 mmHg was applied on the posterior surface. The mesh for the models consists of linear, eight node 3D hexahedral elements

In a study that was conducted by Orssengo et al. [28], a mathematical model was developed for estimating the mechanical properties of the cornea in vivo. The cornea was modeled as a shell, and for the purpose of modeling the behavior of the cornea during applanation tonometry, a combination of the equations for the deformations of a shell due to applanation pressure and IOP was used. Thanks to a relationship between the two pressures at certain corneal dimensions called the "calibration dimensions", it is possible to determine the modulus of elasticity of the cornea (Eq. 6). Based on previous experiments, a calibration corneal thickness of 0.52 mm was used. The cornea is known to be thicker when traveling from the center towards the limbal region; therefore, it could be argued that the assumption of constant corneal thickness throughout the entire cornea might produce large deviations from the true value. Nonetheless, the calculated modulus of elasticity was found to agree with published experimental results. Moreover, in its central region—the main optical zone—the cornea has a thickness of approximately 0.52 mm [32], namely, deviations from the true modulus value will be larger closer to the limbal region, and therefore, less significant.

$$E = 0.0229 \cdot IOPT \tag{6}$$

Equation 6. The theoretical modulus of elasticity of the cornea in MPa, based on a calibration corneal thickness of 0.52 mm [28] (IOPT in mmHg).

For a patient-specific FE model that relies on true geometrical data of the cornea, such as the model presented above, inputs such as the patient-specific modulus of elasticity and IOPT complete a simple yet full patient-specific defined FE model. The task of acquiring the geometric topography of the cornea, IOPT and an estimate of the cornea's modulus of elasticity is a relatively simple process and is not time consuming—a feature that is of utmost importance in the clinical environment. Thanks to this, simulating an increase or decrease of IOP for patients with cases of keratoconus using the aforementioned patient-specific model could be easily done.

10 Concluding Remarks

The ability of simple as well as complex FE models to serve as a means for better understanding the outcomes of LASIK refractive surgeries or for gaining more knowledge about the pathogenesis of keratoconus has been thoroughly discussed. Under large deformations, soft biological tissues that exhibit preferred directions in their microstructure are highly nonlinear and anisotropic. Therefore, accurate simulation of these nonlinear large-strain effects requires the use of constitutive models that are able to account for the anisotropic and hyperelastic behavior. Examples of such models were brought forward and discussed in this chapter. Nevertheless, when subjected to small strains, the mechanical behavior of these materials can generally be modeled using conventional anisotropic linear elasticity. The importance of integrating patient-specific data into such FE models is straight-forward and lies in the ability to give information that considers the specific case at hand. It is common nowadays to find in eye clinics advanced equipment such as topographers and analyzers that can easily produce patient-specific data. Also, average computers are able to solve FE models in a relatively short amount of time. This attributes to the desirability of developing simple methods for acquiring patient-specific models such as the one aforementioned. There is no significant difference in the IOP between normal and katoconic eyes, but the latter are more vulnerable to an increase in IOP, and this may bear clinical implications, such as related to pharmaceutical regulation of IOP. The use of patient-specific FE models to prognose responses to pharmaceutical regulation of IOP are a useful and simple solution. The implication of a high increase in IOP on the cornea, particularly in the conditions characterizing keratoconus, are yet to be explored and may account for the pathogenesis of keratoconus.

Acknowledgment The authors wish to acknowledge the help of Sharon Sherry for the technical editing of this chapter.

References

1. Aghamohammadzadeh H, Newton RH, Meek KM.: X-ray scattering used to map the preferred collagen orientation in the human cornea and limbus. Structure **12**(2), 249–256 (2004)
2. Anderson, K., El-Sheikh, A., Newson, T.: Application of structural analysis to the mechanical behaviour of the cornea. J. R. Soc. Interface **1**(1), 3–15 (2004)
3. Andreassen, T.T., Simonsen, A.H., Oxlund, H.: Biomechanical properties of keratoconus and normal corneas. Exp. Eye Res. **31**(4), 435–441 (1980)
4. Auffarth, G.U., Wang, L., Völcker, H.E.: Keratoconus evaluation using the orbscan topography system. J. Cataract Refract. Surg. **26**(2), 222–228 (2000)
5. Boote, C., Dennis, S., Meek, K.: Spatial mapping of collagen fibril organisation in primate cornea-an X-ray diffraction investigation. J. Struct. Biol. **146**(3), 359–367 (2004)
6. Bron, A.J.: Keratoconus. Cornea **7**(3), 163–169 (1988)
7. Bryant, M.R., McDonnell, P.J.: Constitutive laws for biomechanical modeling of refractive surgery. J. Biomech. Eng. **118**(4), 473–481 (1996)

8. Buzard, K.A.: Introduction to biomechanics of the cornea. Refract. Corneal Surg. **8**(2), 127–138 (1992)
9. Deenadayalu, C., Mobasher, B., Rajan, S.D., Hall, G.W.: Refractive change induced by the LASIK flap in a biomechanical finite element model. J. Refract. Surg. **22**(3), 286–292 (2006)
10. Edmund, C.: Corneal tissue mass in normal and keratoconic eyes: in vivo estimation based on area of horizontal optical sections. Acta Ophthalmol. (Copenh.) **66**(3), 305–308 (1988)
11. Edmund, C.: Corneal topography and elasticity in normal and keratoconic eyes: a methodological study concerning the pathogenesis of keratoconus. Acta Ophthalmol. Suppl. **193**, 1–36 (1989)
12. Emara, B., Probst, L.E., Tingey, D.P., Kennedy, D.W., Willms, L.J., Machat, J.: Correlation of intraocular pressure and central corneal thickness in normal myopic eyes and after laser in situ keratomileusis. J. Cataract Refract. Surg. **24**(10), 1320–1325 (1998)
13. Gallagher, B., Maurice, D.: Striations of light scattering in the corneal stroma. J. Ultrastruct. Res. **61**(1), 100–114 (1977)
14. Gasser, T.C., Ogden, R.W., Holzapfel, G.A.: Hyperelastic modelling of arterial layers with distributed collagen fibre orientations. J. R. Soc. Interface **3**(6), 15–35 (2006)
15. Gefen, A., Shalom, R., Elad, D., Mandel, Y.: Biomechanical analysis of the keratoconic cornea. J. Mech. Behav. Biomed. Mater. **2**(3), 224–236 (2009)
16. Govindarajan, S.M., Hurtado, J.A.: Anisotropic Hyperelastic Models in Abaqus. In: Heinrich, G., Kaliske, M., Lion, A., Reese, S. (eds.) Constitutive Models for Rubber VI, vol. 59, pp. 365–369. CRC Press, Oxford (2010)
17. Grytz, R., Meschke, G.: A computational remodeling approach to predict the physiological architecture of the collagen fibril network in corneo-scleral shells. Biomech. Model. Mechanobiol. **9**(2), 225–235 (2010)
18. Holzapfel, G.A., Gasser, T.C., Ogden, R.W.: A new constitutive framework for arterial wall mechanics and a comparative study of material models. J. Elasticity **61**(1–3), 1–48 (2000)
19. Hjortdal, J.O.: Regional elastic performance of the human cornea. J. Biomech. **29**(7), 931–942 (1996)
20. Ihalainen, A.: Clinical and epidemiological features of keratoconus genetic and external factors in the pathogenesis of the disease. Acta Ophthalmol. Suppl. **178**, 1–64 (1986)
21. Kamiya, K., Aizawa, D., Igarashi, A., Komatsu, M., Shimizu, K.: Effects of antiglaucoma drugs on refractive outcomes in eyes with myopic regression after laser in situ keratomileusis. Am. J. Ophthalmol. **145**(2), 233–238 (2008)
22. Knupp, C., Pinali, C., Lewis, P.N., Parfitt, G.J., Young, R.D., Meek, K.M., Quantock, A.J.: The architecture of the cornea and structural basis of its transparency. Adv. Protein Chem. Struct. Biol. **78**, 25–49 (2009)
23. Li, L.Y., Tighe, B.: The anisotropic material constitutive models for the human cornea. J. Struct. Biol. **153**(3), 223–230 (2006)
24. Mandell, R.B., Polse, K.A.: Keratoconus: spatial variation of corneal thickenss as a diagnostic test. Arch. Ophthalmol. **82**(2), 182–188 (1969)
25. Maurice, D.M.: The cornea and sclera. In: Davson, H. (ed.) The Eye, pp. 489–600. Academic Press, New York (1969)
26. Maurice, D.M., Davson, E.: The Cornea and Sclera. In: Davson, H. (ed.) The Eye, pp. 1–158. Academic Press, Orlando (1984)
27. Newton, R.H., Meek, K.M.: The integration of the corneal and limbal fibrils in the human eye. Biophys. J. **75**(5), 2508–2512 (1998)
28. Orssengo, G.J., Pye, D.C.: Determination of the true intraocular pressure and modulus of elasticity of the human cornea in vivo. Bull. Math. Biol. **61**(3), 551–572 (1999)
29. Pandolfi, A., Manganiello, F.: A model for the human cornea: constitutive formulation and numerical analysis. Biomech. Model. Mechanobiol. **5**(4), 237–246 (2006)
30. Pandolfi, A., Fotia, G., Manganiello, F.: Finite element simulations of laser refractive corneal surgery. Eng. Comput. **25**(1), 15–24 (2008)
31. Pandolfi, A., Holzapfel, G.A.: Three-dimensional modeling and computational analysis of the human cornea considering distributed collagen fibril orientations. J. Biomech. Eng. **130**(6), 061006 (2008)

32. Pinsky, P.M., Datye, D.V.: A microstructurally-based finite element model of the incised human cornea. J. Biomech. **24**(10), 907–922 (1991)
33. Polack, F.M.: Contributions of electron microscopy to the study of corneal pathology. Surv. Ophthalmol. **20**(6), 375–414 (1976)
34. Pouliquen, Y.B., Graf, Y., Kozak, J., Bisson, J., Faure, F., Bourles, F.: Etude morphologique et biochimique de kératocone. I. Etude morphologique, Arch. Ophtalmol. **30**, 497–532 (1970)
35. Roberts, C.J., Zuger, B.J.: The advantage and principle of dual scheimpflug imaging for analyzing the anterior segment of the human eye. www.ziemergroup.com (2006)
36. Roy, A.S., Dupps Jr, W.J.: Patient-specific modeling of corneal refractive surgery outcomes and inverse estimation of elastic property changes. J. Biomech. Eng. **133**(1), 011002 (2011)
37. Sayers, Z., Koch, M.H., Whitburn, S.B., Meek, K.M., Elliott, G.F., Harmsen, A.: Synchrotron X-ray diffraction study of corneal stroma. J. Mol. Biol. **160**(4), 593–607 (1982)
38. CC, T.E.N.G.: Electron microscope study of the pathology of keratoconus: I. Am. J. Ophthalmol. **55**, 18–47 (1963)
39. Yarker, Y.E., Hukins, D.W., Nave, C.: X-ray diffraction studies of tibial plateau cartilage using synchrotron radiation. Connect. Tissue Res. **12**(3–4), 337–343 (1984)
40. Yaylali, V., Kaufman, S.C., Thompson, H.W.: Corneal thickness measurements with the Orbscan Topography System and ultrasonic pachymetry. J. Cataract Refract. Surg. **23**(9), 1345–1350 (1997)

Part VII
Systems Biology Approaches in Patients-Specific Modeling

Computational Modeling of Gene Translation and its Potential Applications in Individualized Medicine

Tamir Tuller and Hadas Zur

Abstract Gene translation is a central cellular process in all living organisms including humans. Thus, predictive modeling of this process, based on various patient specific genomic and cellular features, should have important applications in the diagnosis and understanding of various diseases. In this chapter we survey the recent advances in modeling gene translation and its potential clinical applications.

1 Introduction

Gene translation is the second major stage of gene expression. It is a complex process through which an mRNA sequence is decoded by the ribosome to produce a specific protein. The elongation step of this process is an iterative procedure in which each codon in the mRNA sequence is recognized by a specific tRNA molecule, which adds one additional amino acid to the growing peptide [61]. As gene translation is a central process in all living organisms, its understanding has ramifications for human health [3, 13, 32], biotechnology [2, 19, 33, 51, 57, 62, 67, 71] evolution and systems biology [13, 35, 57, 67].

T. Tuller (✉)
Faculty of Mathematics and Computer Science,
Department of Molecular Genetics,
Weizmann Institute of Science, Rehovot, Israel
e-mail: tamir.tuller@weizmann.ac.il

H. Zur
School of Computer Science, Tel Aviv University,
Tel Aviv, Israel

In recent years there has been a sharp increase in large scale technologies for measuring different features related to gene translation including ribosomal densities [2, 24], protein abundance of endogenous [17, 37, 43, 56, 63] and heterologous genes [33, 70], tRNA levels [8–10, 34, 57], and mRNA folding [30].

These data should be useful for developing computational predictive modeling of translation, which will advance analyzing and understanding human diseases. In this chapter we will survey the recent approaches for modeling this process and their potential applications for human health and in particular individualized medicine.

2 Gene Expression and Gene Translation

Proteins are biochemical compounds consisting of one or more polypeptides, and after they fold into their three dimensional form, they perform most of the biological functions in the cell. A polypeptide is a single linear polymer chain of amino acids. The sequence of amino acids in a protein is usually defined by the sequence of nucleotides that composes a gene, which is encoded in the genetic code, and usually specifies 20 amino acids that are redundantly encoded in 61 codons.

Gene expression is the process by which information from a gene is used in the synthesis of a functional gene product (usually a protein). The process of gene expression is used by all known life forms, eukaryotes (including multicellular organisms), prokaryotes (bacteria and archaea) and viruses, to generate the macromolecular machinery for life. Several steps in the gene expression process may be regulated, including the transcription, post transcriptional modifications such as RNA splicing, translation, and post-translational modifications of a protein. Gene regulation gives the cell control over structure and function, and is the basis for cellular differentiation, morphogenesis and the versatility and adaptability of any organism. Gene regulation may also serve as a substrate for evolutionary change, since control of the timing, location, and the expression levels of a gene can have a profound effect on the functions (actions) of the gene in a cell or in a multicellular organism. Here we generally mention two important stages of this process: the transcription and the translation (See Fig. 1; more details regarding gene expression, transcription, and translation can be found, for example, in [1]).

Transcription: The gene itself is typically a long stretch of DNA which carries genetic information encoded by genetic code. Every molecule of DNA consists of two strands, each of them having two ends named $5'$ and $3'$ and an orientation: the two strands are oriented in an anti-parallel direction. For a certain gene, the coding strand contains the genetic information while the template strand (non-coding strand) serves as a blueprint for the production of RNA. Transcription is the process by which RNA copies (mRNA) are produced from DNA, and is performed by RNA polymerase, which adds one RNA nucleotide at a time to a growing RNA strand (based on the coding DNA strand). This RNA is complementary to the template $3' \rightarrow 5'$ DNA strand, which is itself complementary to the coding $5' \rightarrow 3'$ DNA strand. Therefore, the resulting $5' \rightarrow 3'$ RNA strand is identical to

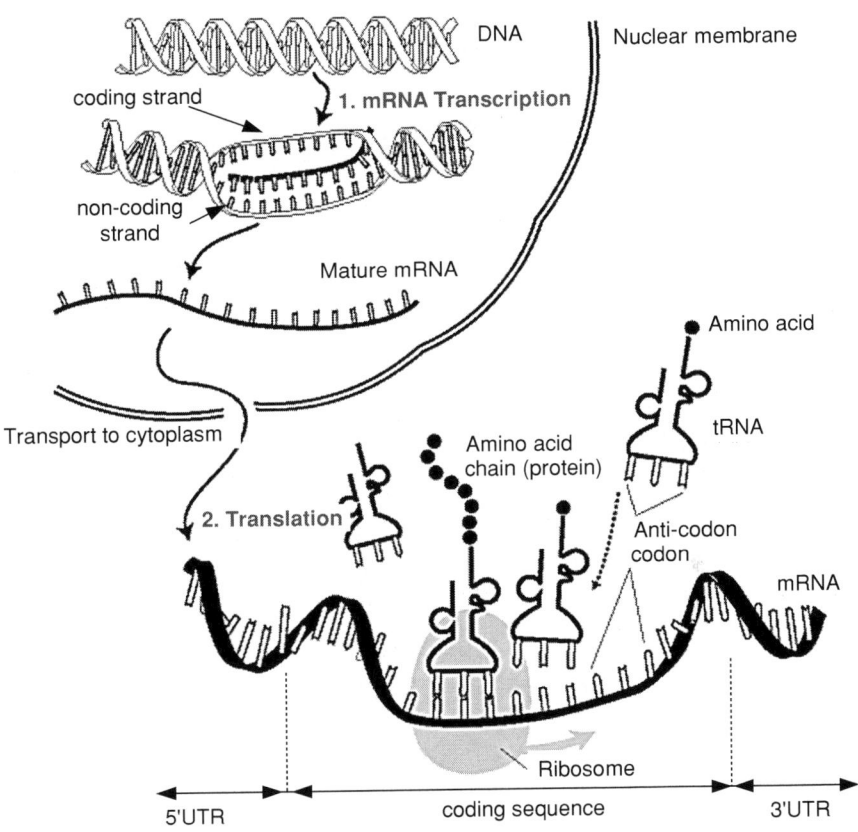

Fig. 1 a Gene expression involves two major stages: the transcription of DNA to mRNA, and the translation of mRNA to protein by the ribosome. **b** Gene translation consists of three major steps: *1* initiation—the binding of the ribosome to the mRNA, *2* elongation—an iterative phase of decoding the codon sequence and replacing it with an amino acid sequence, *3* termination—the disassociation of the small and large subunits of the ribosome from the mRNA

the coding DNA strand, with the exception that thymines (T) are replaced with uracils (U) in the RNA. A coding DNA strand reading "ATG" is transcribed as "AUG" in RNA.

Translation: For some RNA (non-coding RNA) the mature RNA is the final gene product. RNA genes have many important roles in gene translation, regulation of protein abundance, and are part of the machinery comprising the ribosome (see, for example, [41]). One type of non-coding gene is the transfer RNA (tRNA); as we explicate in the following sections, it has an imperative role in gene translation.

In the case of messenger RNA (mRNA) the RNA is an information carrier coding for the synthesis of one or more proteins. Every mRNA consists of three parts: 5′ untranslated region (5′UTR), protein-coding region or open reading frame

(ORF) and 3′ untranslated region (3′UTR). The coding region carries information for protein synthesis encoded by genetic code in the form of triplets of nucleotides. The translation from a coding sequence to an amino acid sequence is performed by the ribosome. The ribosome is a large complex of proteins and ribosomal RNA (rRNA) that is composed of two subunits, known as the large and small subunits. The small subunit binds to the mRNA sequence, while the large subunit binds to the amino acids and tRNA molecules [1].

There are 64 possible codons, of which 61 encode amino acids. The codon encoding the amino acid Methionine (AUG) also serves as a start codon—and in most of the genes should appear at the beginning of the coding sequence for proper translation of it. Three codons are stop codons and appear at the end of the coding sequence—when the ribosome approaches the stop codons translation terminates, and the mature protein is released.

Gene translation has three major stages:

Initiation: the binding of the ribosome to the 5′UTR with the aid of initiation factors and the initiator Methionine. The association of the ribosome's large and small subunites and the initiator Methionine, enables commencement of the next phase, elongation. See more details about this stage, for example, in [54].

Elongation: an iterative stage in which each triplet of nucleotides of the coding region (codon) is decoded by the ribosome to its amino acid. Each codon corresponds to a binding site complementary to an anticodon triplet in the transfer RNA molecule. However in practice, due to wobble interactions, a codon may be recognized by more than one tRNA molecule, with different binding affinities. For example, both the tRNA with the anti-codon TGA and tRNA with the anti-codon CGA recognize the codon TCG; the affinity of the second tRNA is higher as it is the exact complementary of this codon (See all the interactions between tRNAs and codons in Fig. 2). Transfer RNAs with the same anticodon sequence always carry an identical type of amino acid. In addition, all the tRNA molecules that recognize the same codon carry the same amino acid.

Amino acids are then chained together by the ribosome according to the order of triplets (codons) in the coding region. The ribosome facilitates transfer RNA in binding to messenger RNA, and utilizes the amino acid from each transfer RNA to produce a protein.

The last stage of translation is termination: as the ribosome approaches a stop codon, there is a dissociation of the ribosome's large and small subunites, and the protein is released from the ribosome.

Unlike initiation and termination, the machinery used during translation elongation (e.g. the ribosome, the tRNAs, and the genetic code) and hence mechanisms underlying this stage, have been highly conserved across the three kingdoms of life [28]. Thus, models of this stage that were inferred based on genomic and cellular measurements of unicellular organisms (e.g. *E. coli* and *S. cerevisiae*) can be employed for analyzing other (even mutli-cellular) organisms (e.g. human). Indeed, often translation elongation research is initially performed using unicellular model organisms, for which it is easier to perform large scale measurements of protein abundance and ribosomal densities [2, 17, 24, 33, 43]. In addition, in

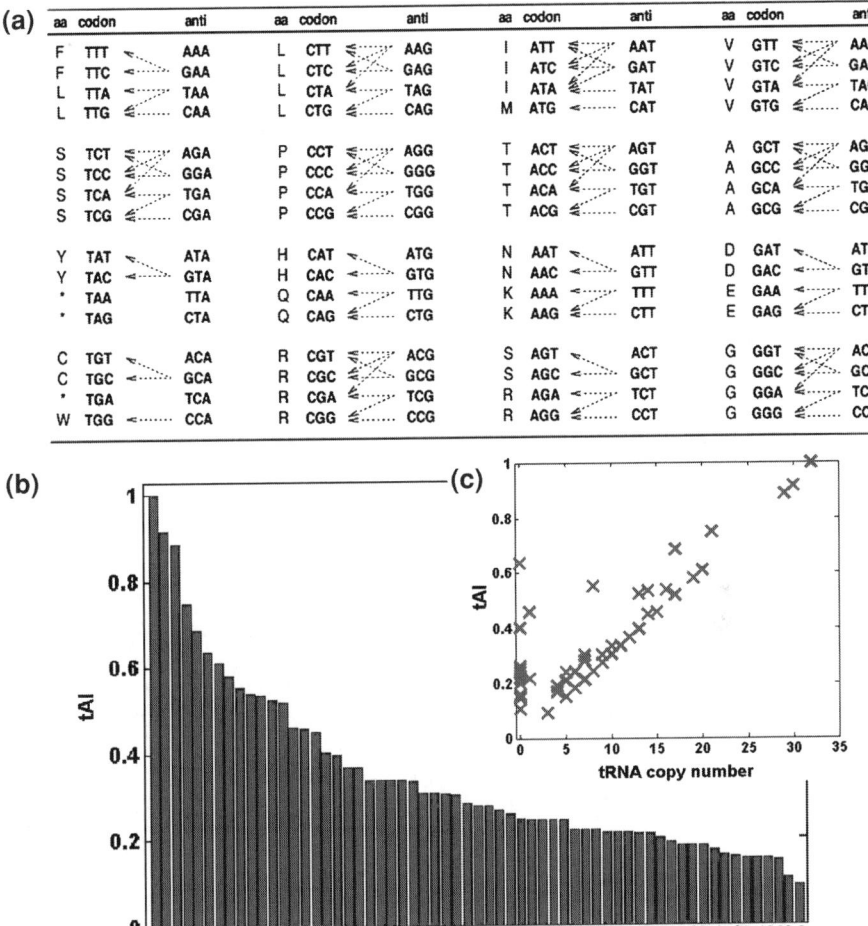

Fig. 2 a All possible interactions between anti-codons and codons (based on [12, 69, 72]). b. The tAI of all the human codons (based on [57]). c tAI versus genomic tRNA copy number for the human codons (based on [57])

these organisms the correlations between the actual tRNA levels and their widely used approximations is high ([11, 23, 27, 46] see more details in the next section); increasing the evaluation feasibility of translation models in these organisms.

The correlation between mRNA levels and protein abundance is rather low in all organisms. For example, in yeast [17] and human (Vogel, Abreu Rde et al.) the correlation is between 0.5 and 0.6. Thus, the measured mRNA levels of genes explain only around 30% of the variance in protein abundance in human (and yeast), and in general are usually a very poor approximation of protein abundance. The reasons for the gap between protein abundance measurements and mRNA measurements include: (1) disparate bias and noise in the measurement technologies of mRNA levels and protein abundance [37], (2) post transcriptional regulation [1], and (3) the half lives of proteins are orders of magnitude longer than the half lives of mRNA molecules [56].

Today, it is rather easy to perform large scale measurements of mRNA levels [6] while, for technical reasons, the technologies for performing large scale measurements of protein abundance lag behind. For example, at the time of writing this chapter, the GEO database, http://www.ncbi.nlm.nih.gov/geo/, includes hundreds of thousands of large scale measurements of mRNA levels, whilst there are only a few such large scale measurements of protein abundance (see, for example, [17, 37, 43, 63]). Thus, researchers from various fields are forced to use mRNA levels, the rather rough proxy of protein abundance, instead of the protein abundance itself. As we explain later, computational modeling of gene translation can reduce this gap between the desired measurements (protein abundance) and the available ones (mRNA levels or predictions of protein abundance).

3 Modeling Gene Translation

In this section we concentrate on translation elongation. There are two major approaches for modeling or predicting the efficiency of translation elongation.

The first and more traditional one is the non 'physical' approach. For example, this group of measures/predictors includes the Codon Adaptation Index (CAI, [52]) which is based on the frequency of optimal codons, i.e. codons that tend to appear in highly expressed genes. A newer measure belonging to this group is the tRNA Adaptation Index (tAI, [12]). This measure gauges the availability of the different tRNA molecules for each codon along an mRNA. Other non physical approaches include regressors and various machine learning techniques that are based on a combination of coding sequence features and various large scale measurements related to gene expression (see, for example, [22, 58, 63]).

In the next paragraph we will detail the *tAI*, which has been shown to have high performance levels, and is employed as a main feature in many of the more sophisticated predictors [5, 49, 58].

The *tAI* is based on the following observation: Since codon-anti-codon coupling is not unique due to wobble interactions, several anti-codons can recognize the same codon, with different efficiency weights (see Fig. 2).

Let n_i be the number of tRNA isoacceptors recognizing codon i. Let $tCGNij$ be the copy number of the jth tRNA that recognizes the ith codon, and let S_{ij} be a parameter corresponding to the efficiency of the codon-anticodon coupling between codon i and tRNA j. The S_{ij} are inferred optimizing the correlation between the tAI and gene expression measurements [12].

We define the absolute adaptiveness, W_i, for each codon i as [12, 60]:

$$W_i = \sum_{j=1}^{ni} (1 - S_{ij}) tCGN_{ij}$$

From W_i we obtain w_i, which is the *relative* adaptiveness value of codon i, by normalizing the W_i's values (dividing them by the maximal of all the 61 W_i).

The final tAI of a *gene*, g, is the following geometric mean:

$$tAIg = \left(\prod_{k=1}^{l_g} w_{i_{kg}}\right)^{1/l_g}$$

Where i_{kg} is the codon defined by the kth triplet on gene g; and lg is the length of the gene (excluding stop codons). Thus, the tAI of a gene is a number between 0 (extremely non-efficient codons) and 1 (utmost efficiency).

Figure 2b depicts the W_i (i.e. the tAI scores) of all the codons in human based on the data from [57]; note that the most efficient codon in human is AAC (one of the codons encoding the amino acid Asparagine) whilst the least efficient codon is CTA (one of the codons encoding the amino acid Leucine).

Figure 2c includes the tAI versus the tRNA copy number of human codons; as can be seen the correlation between the tAI of a codon and its tRNA copy number is high but not perfect (Spearman r = 0.73; p-value = $2.94*10^{-11}$).

The second approach is based on physical understanding of the process. Though theoretical physical models and simulations related to translation have been suggested more than 30 years ago [20, 39] only recently have such approaches been implemented on real large scale biological data.

Physical models take into account the dynamic and physical nature of the process. The most basic feature is the flow of ribosomes and the interactions between them [49, 53, 57, 74]. This feature can be modeled in a deterministic [74], or stochastic manner ([49]; Fig. 3a) in which the translation time of each codon is a random variable (e.g. with exponential distribution, [49]). The deterministic model is significantly computationally faster than the stochastic model, but is a cruder approximation of translation which is a stochastic phenomenon.

Recently, we have shown that stochastic ribosomal flow can be approximated by a non- stochastic, computationally efficient model, that we named *Ribosome Flow Model* [49].

The *Ribosome Flow Model* is illustrated in Fig. 3b. To approximate the process of ribosomal movement by a *deterministic* process, mRNA molecules are coarse-grained into sites of C codons; (in Fig. 3b $C = 3$); in practice we use $C = 25$, a value that is close to various geometrical properties of the ribosome, such as its footprint on the mRNA sequence and the length of the exit channel [1, 16, 24, 26, 57, 74].

Ribosomes arrive at the first site with initiation rate λ, but are only able to bind if this site is not occupied by another ribosome. The initiation rate is a function of physical features such as the number of available free ribosomes [57, 68, 73], the folding energy of the 5'UTRs [33, 60], and at the beginning of the coding sequence [29, 33, 42, 60], and the base pairing potential between the 5'UTR of the sequence and the ribosomal rRNA [44].

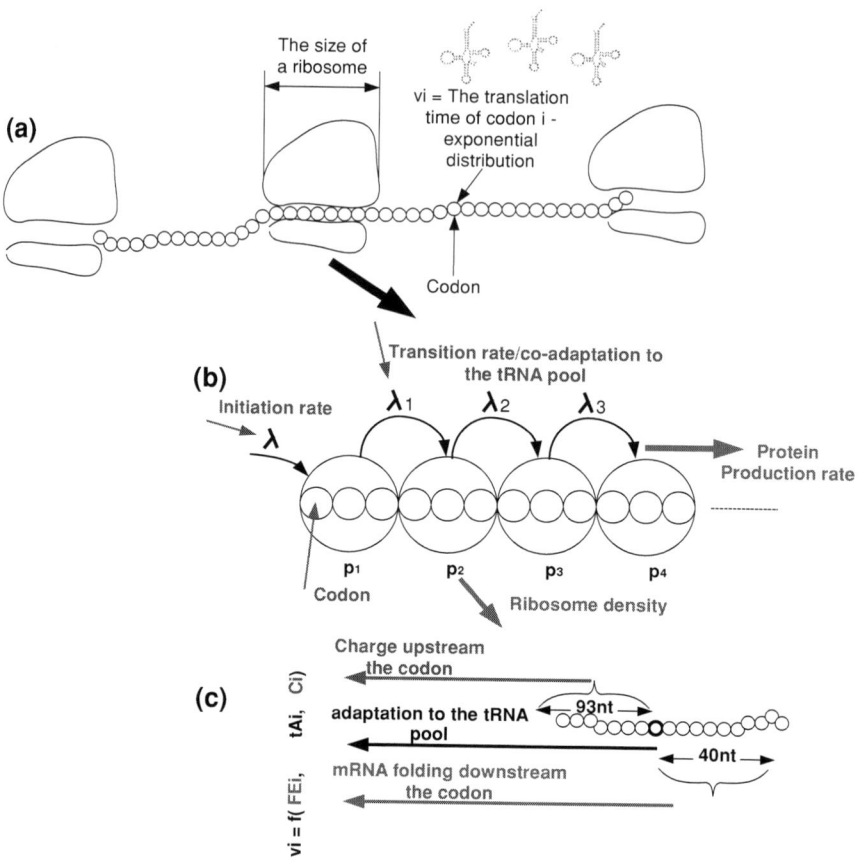

Fig. 3 Physical modeling of gene translation. **a**. Simple physical modeling of translation elongation. Each ribosome moves according to the translation speed of the current codon, but can be delayed as a result of ribosome collisions, caused by the downstream ribosome translating a slow codon. Ribosomal movement can be deterministic [74] or stochastic [53]. **b**. The ribosome flow model [49] (see details in the main text). **c**. The speed of codon translation is a function of the charge of the amino acids encoded in the codons upstream of the codon, the adaptation of the codon to the tRNA pool, and the folding energy of the mRNA sequence downstream from the codon

A ribosome that occupies the ith site moves, with rate λ_i, to the consecutive site provided the latter is not occupied by another ribosome. Transition rates are determined by the codon composition of each site and the tRNA pool of the organism. Briefly, the rate of a codon is proportional to the abundance of the tRNA species that recognize it, taking into account the affinity of the interactions between the tRNA species and the codons (Fig. 1, similarly to the tAI mentioned above).

Denoting the probability that the ith site is occupied at time t by $p_i(t)$, it follows that the rate of ribosome flow into/out of the system is given by: $\lambda[1 - p_1(t)]$ and

$\lambda_n p_n(t)$ respectively. Hence, the rate of ribosome flow from site i to site $i+1$ is given by: $\lambda_i p_i(t)[1-p_{i+1}(t)]$. Thus we get the following set of differential equations that describe the process of translation elongation:

$$\begin{cases} \frac{dp_1(t)}{dt} = \lambda[1-p_1(t)] - \lambda_1 p_1(t)[1-p_2(t)] \\ \frac{dp_i(t)}{dt} = \lambda_{i-1} p_{i-1}(t)[1-p_i(t)] - \lambda_i p_i(t)[1-p_{i+1}(t)] & 1 < i < n \\ \frac{dp_n(t)}{dt} = \lambda_{n-1} p_{n-1}(t)[1-p_n(t)] - \lambda_n p_n(t) \end{cases} \quad (1)$$

The correlation between the predictions of the ribosome flow model and the stochastic simulation of ribosome movement is very high (up to 0.96 in some cases) while the ribosomal flow model is computationally much faster (for example, implementing it on the genome of S. cerevisiae takes minutes versus several days for the stochastic simulation of ribosome movement [49]).

Recently, it was shown that the speed and density of ribosomes is determined not only by the co-adaptation of codons to the tRNA pool of the organism, but also by other features of the coding sequence. First, as the exit channel of the ribosome is negatively charged, interactions with positive amino acids may decelerate the translation speed; thus, the charge of the coding sequence upstream from the codon has effect on its translation rate [36, 59].

Second, the mRNA sequence tends to fold into secondary structures; thus during translation it is unfolded by the RNA-helicase [25]. It usually takes the helicase more time to unfold regions in the coding sequences with stronger folding; thus, the strength of the mRNA folding (folding energy) downstream from the codon is also a relevant determinant of the translation rate [59, 60]. These results suggest using computational models that consider these additional features (Fig. 3c [59]).

3.1 Measures of Translation Efficiency

Translation efficiency in our context, is the rate in which the transcript is translated by the ribosome to a protein (note that there are other definitions of translation efficiency, such as the precision of the translation). To accurately estimate translation efficiency one needs to consider the mRNA levels of the gene (in addition to its protein abundance), as those also have a positive effect on protein abundance.

There are several ways to measure translation efficiency, based on measurements of protein abundance, mRNA levels, or ribosomal densities. These measures have been used to evaluate the predictive power of translation efficiency computational predictors.

The first measure is the number of ribosomes on the mRNA sequence or, similarly, the ribosomal density which is the number of ribosomes on the transcript normalized by the length of the transcript [24, 47]. Assuming a constant elongation rate at steady state, more ribosomes on the transcript correspond to higher initiation and translation rates [47, 49].

The second measure is the ratio between protein abundance and mRNA levels [58, 60]. This ratio indicates the number of proteins generated from an mRNA molecule, when not considering the half lives of the mRNA molecule and protein (i.e. the rates of decay of proteins and mRNA molecules).

The third measure is the partial correlation of a predictor (e.g. one of the predictors defined above) with protein abundance when controlling for the mRNA levels of the protein [49, 60].

The last measure is the correlation between (1) the multiplication of the predicted translation rate with the mRNA levels, and (2) the protein abundance [49, 58]. As was shown in [49], assuming steady state, since protein abundance monotonically increases with this multiplication; a better predictor is the one that yields higher correlations under this procedure.

3.2 Large Scale Measurements Related to Translation Elongation

Data related to translation is a crucial factor in accelerating the development of predictive modeling of this process, and for improving its understanding. In this section we will survey the data sources that are available today. Specifically, these data are used for inferring various free parameters of the models and evaluating their performances.

The basic and essential data sources are large scale measurements of protein abundance of endogenous genes. Today, there are several such large scale measurements in various organisms including the model organisms *E. coli* [37, 56], *S. cerevisiae* [17, 43], and human [38, 63]. The main disadvantage of analyzing endogenous genes is the fact that these genes are shaped by evolution, thus, hindering the ability to understand the causality of the observed correlation between various variables related to translation efficiency (see, for example, [33]). Additionally as the different determinants of translation efficiency are correlated and are under selection, it is difficult to evaluate the contribution of each of them to the translation speed (see, for example, [60]). Recently, new datasets of protein abundance measurements of heterologous genes have emerged (see, for example, [33, 70]). Each of these libraries include many versions of proteins; each such version includes an amino acid sequence identical to the original gene but with a different set of codons (i.e. synonymous modifications). Thus, by analyzing such datasets it is possible to overcome the two aforementioned problems [33, 49, 70].

The heterologous genes datasets, on the other hand, may have two major problems. First, in order to derive general conclusions about translation, many such datasets of heterologous genes based on numerous amino acid sequences should be generated (significantly more than those existing today). Second, in many cases heterologous genes are expressed in extreme and unnatural levels, complicating the generalization of their analysis results to endogenous genes [47]. Thus, these datasets instigate cautious analysis as well.

Another important data source is the abundance of tRNA species. As we previously mentioned, the speed of translation elongation is based on the abundance of tRNA molecules. However, measuring the expression levels of tRNA is not trivial. These molecules have strong folding and thus, for example, do not hybridize strongly to conventional DNA chips.

Therefore, currently, there are only a few large scale measurements of tRNA abundance (see, for example, [8–10, 34, 45, 57]) most of which were generated by a special chip developed by Prof. Tao Pan. Opportunely, recent new approaches that are based on deep sequencing are emerging (see, for example, [34]).

In many organisms, it was shown that the copy number of tRNA genes (i.e. the number of times the tRNA is encoded in the genome of the organism) is strongly correlated with their expression levels [11, 23, 27, 46, 57, 66]. Specifically, in *S. cerevisiae* this correlation is up to $r = 0.91$ [46]. Thus, a common approximation in this field is to use the genomic tRNA copy number as a proxy to the tRNA expression levels (see, for example, [12, 40, 57]).

It was shown that the folding energy of mRNA sequences, plays an important role in their translation to proteins by affecting translation initiation [33, 60] (via affecting the attachment of the ribosome to the mRNA) and possibly translation elongation [59, 60]. Traditionally, the folding energies of mRNA have been inferred based on computational predictions (see, for example, [21, 59, 60, 75]). However, mRNA levels may change between different cellular conditions, and these softwares may miss such changes. Recently, a new technology for measuring the actual mRNA folding has been developed [30]. These new data can be used for improving the relevant predictions of translation efficiency under various conditions.

4 Potential Applications for Human Health and Individualized Medicine

In this section, we will discuss the potential applications of the computational models described above; emphasizing those related to human health and individualized medicine.

We begin with the prediction of protein abundance based on mRNA levels, features of the coding sequences, and, potentially, the abundance of tRNA species.

It has been shown that the computational models mentioned above can yield significant predictions of protein abundance in organisms such as *S. cerevisiae* [49, 58] and specifically in human [49, 66]. Today, there are hundreds of papers illustrating the diagnosis of various diseases based on gene expression (for example, see [18, 31, 50, 55]). However, as we mentioned, mRNA levels explain only 30% of the variance in protein abundance in humans [63]. Thus, diagnosis based on a predictor of protein abundance, which is trivially more correlative with protein abundance than mRNA levels, should have enhanced performances.

The second important application is understanding tumorigenesis and consequently, the diagnosis of cancer. Thousands of papers have been written about the regulatory mechanisms of tumorigenesis (see, for example, [4, 15]) and diagnosis or clinical predictions based on gene expression in cancerous cells (see, for example, [18, 31]). Modeling of gene translation, in particular translation elongation, may shed additional light on our comprehension of cancer evolution.

Indeed preliminary studies have demonstrated that there are significant regulatory changes in translation elongation in cancerous cells. For example, [45] compared genome-wide tRNA expression in cancerous versus non-cancerous breast cell lines, as well as tRNA expression in breast tumors versus normal breast tissues. They found that in cancer derived versus non-cancer derived cell lines, nuclear-encoded tRNAs increase by up to 3-fold. In tumors versus normal breast tissues, both the nuclear-encoded tRNAs increase up to 10-fold. This tRNA over-expression is selective and coordinates with the properties of cognate amino acids. Nuclear-encoded tRNAs exhibit distinct expression patterns in cancerous versus normal cells and tend to be over-expressed in cancerous cells. These results indicate that the expression levels of tRNAs may be used as a diagnostic tool of breast cancer. They also showed that tRNA expression levels are not geared towards optimal translation of house-keeping or cell line specific genes. Instead, tRNA isoacceptor expression levels may favor the translation of cancer-related genes having regulatory roles. Thus, their results suggest a functional consequence of tRNA over-expression in tumor cells. Specifically, tRNA over-expression may increase the translational efficiency of genes relevant to cancer development and progression.

In addition, Waldman et al. [65] performed a large scale study of mutations in the cancerous tumor suppressor gene *TP53*. *TP53* is known to be a key regulator in cancer, and more than half of human cancers exhibit mutations in this gene; however, it has also been shown that mutations in this gene may make the mutated gene oncogenic—i.e. a higher translation rate augments the survival of the cancerous gene. The analysis of Waldman et al. [65] has shown that translation efficiency in human cancer mutated *TP53* variants increases, and is correlated with the frequency of *TP53* mutations.

Furthermore, mutations with a known oncogenic effect significantly increase their translation efficiency compared with the other *TP53* mutations. Thus, the efficiency of translation elongation plays an important role in the selection of *TP53* cancerous mutations.

These two studies suggest that computational modeling of gene translation, based on tRNA expression in cancer, and various additional cellular and coding sequence features, should significantly improve our conception of tumorigenesis and may enable the development of diagnostic predictors of cancer. Patient specific variables that will be relevant in such analyses include: (1) the tRNA pool in the tumor cells—it may be altered due to chromosomal aberrations such as deletion and duplications of tRNA genes, (2) the coding sequences of cancerous genes—they may be modified due to synonymous or non-synonymous mutations that affect different determinants of translation efficiency (e.g. mRNA folding, charge, and adaptation to the tRNA pool), and (3) the un-translated regions of cancerous genes—a mutation in these regions may affect the initiation or termination rates.

The third essential application is improving association studies. The aim of association studies is to find genetic features that are relevant to different diseases. For many years researchers have believed that Single-Nucleotide Polymorphism (SNPs) that change the amino acid encoded in the gene (named non-synonymous SNPs) are the main cause of diseases. However, recently increasing evidence suggests that synonymous SNPs that affect other genomic features, such as translation efficiency, are also very relevant. One such example was reported recently by Kimchi-Sarfaty et al. [32], who showed that an SNP that instigates a slower codon, results in miss-folding of the resultant protein and a disease.

In addition, it was shown that SNPs that induce larger changes in the translation efficiency are under stronger purifying selection in recent human history [7, 64].Thus, computational modeling of gene translation can be used for evaluating the affect of SNPs on the translation rate, and can hence improve the predictions of association studies by incorporating this information (see, for example [14]). Similar concepts can be used when analyzing copy number variations in the human genome (see, for example, [48]).

The three patient specific variables mentioned in the case of cancer are also relevant here. SNPs or copy number variations may effect translation via changes in (1) the tRNA pool (SNPs or copy number variations related to tRNA genes), (2) the coding sequences (affecting elongation), and (3) the un-translated regions (affecting initiation or termination).

Finally, diagnosis and patient classification, based on protein abundance or SNPs is at the heart of customizing medical treatments to specific patient groups. The treatments for specific patient groups are tailored based on differences between their genomes; such differences usually correspond to specific SNPs and/or expression patterns. Thus, the three aforementioned points bear a strong relation to personalized medicine (Fig. 4).

5 Conclusions and Future Research

Computational gene translation is an emerging field with increasing impact on various disciplines including personalized medicine. In this chapter we reported the basic concepts of computational gene translation, the corresponding large scale measurements that are available, and the potential applications of this field.

Currently, the field includes many important open questions. For example, (1) How well can one predict protein abundance based on features such as the coding sequence, mRNA levels, and tRNA levels, (2) What is the contribution of modifications in translation features to translation efficiency (e.g. due to changes in the tRNA pool) and to the pathogenesis of different diseases (e.g. cancer), (3) What is the effect of SNPs that modify the translation efficiency of codons, on the

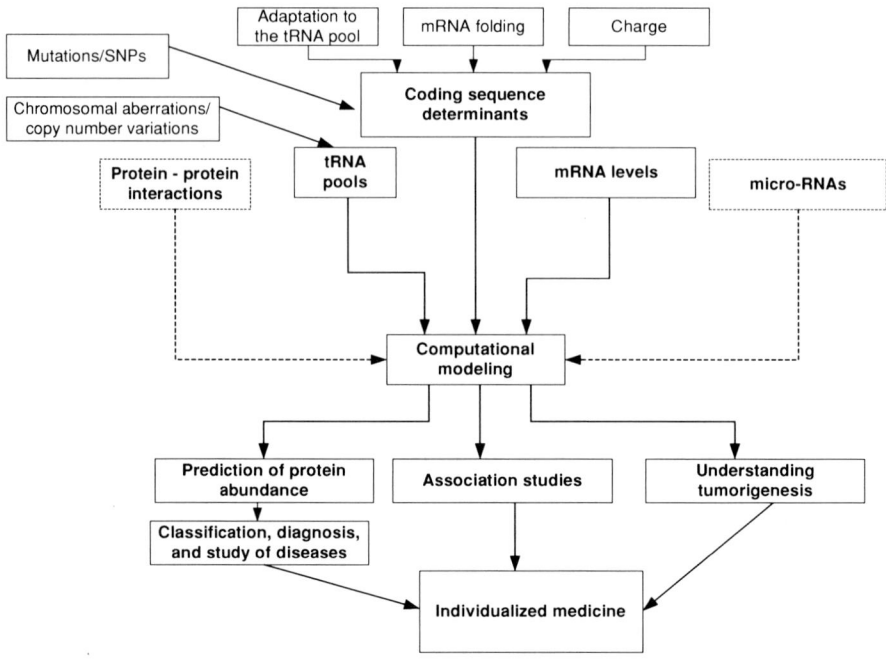

Fig. 4 The potential applications of predictive modeling of gene translation. Computational models that are based on the coding sequence determinants, the tRNA pool, and possibly additional cellular measurements such as micro-RNAs and protein–protein interactions are sensitive to mutations (and SNPs) or chromosomal aberrations (and copy number variations). Thus, they can be utilized for diagnosis, classification of disease, association studies, and tumorigenesis studies to improve individualized medicine

probability of contracting different diseases? (4) How to incorporate information related to translation efficiency (e.g. SNPs and tRNA pools of patients) in order to improve diagnosis and personalized medicine?

We believe that many of these open questions/issues will be solved in the following years with the emergence of more large scale measurements related to the process of gene translation, and with the development of more sophisticated computational approaches for analyzing and modeling gene translation.

Acknowledgements T.T. is supported by the Koshland center for basic research.

References

1. Alberts, B., Johnson, A., et al.: Molecular Biology of the Cell. Garland Science, New York (2002)
2. Arava, Y., Wang, Y., et al.: Genome-wide analysis of mRNA translation profiles in Saccharomyces cerevisiae. Proc. Natl. Acad. Sci. USA **100**(7), 3889–3894 (2003)

3. Bahir, I., Fromer, M., et al.: Viral adaptation to host: a proteome-based analysis of codon usage and amino acid preferences. Mol. Syst. Biol. **5**(311), 311 (2009)
4. Bartkova, J., Horejsi, Z., et al.: DNA damage response as a candidate anti-cancer barrier in early human tumorigenesis. Nature **434**(7035), 864–870 (2005)
5. Brockmann, R., Beyer, A., et al.: Posttranscriptional expression regulation: what determines translation rates? PLoS. Comput. Biol. **3**(3), e57 (2007)
6. Churchill, G.A.: Fundamentals of experimental design for cDNA microarrays. Nat. Genet. **32**, 490–495 (2002)
7. Comeron, J.M.: Weak selection and recent mutational changes influence polymorphic synonymous mutations in humans. Proc. Natl. Acad. Sci. USA **103**(18), 6940–6945 (2006)
8. Dittmar, K.A., Goodenbour, J.M., et al.: Tissue-Specific differences in human transfer RNA expression. PLoS. Genet. **2**(12), e221 (2006)
9. Dittmar, K.A., Mobley, E.M., et al.: Exploring the regulation of tRNA distribution on the genomic scale. J. Mol. Biol. **337**(1), 31–47 (2004)
10. Dittmar, K.A., Sorensen, M.A., et al.: Selective charging of tRNA isoacceptors induced by amino-acid starvation. EMBO Rep. **6**(2), 151–157 (2005)
11. Dong, H., Nilsson, L., et al.: Co-variation of tRNA abundance and codon usage in Escherichia coli at different growth rates. J. Mol. Biol. **260**(5), 649–663 (1996)
12. dos Reis, M., Savva, R., et al.: Solving the riddle of codon usage preferences: a test for translational selection. Nucleic Acids Res. **32**(17), 5036–5044 (2004)
13. Drummond, D.A., Wilke, C.O.: Mistranslation-Induced protein misfolding as a dominant constraint on coding-sequence evolution. **134**(2), 341–352 (2008)
14. Eskin, E.: Increasing power in association studies by using linkage disequilibrium structure and molecular function as prior information. Genome Res. **18**(4), 653–660 (2008)
15. Fearon, E.R., Vogelstein, B.: A genetic model for colorectal tumorigenesis. Cell **61**(5), 759–767 (1990)
16. Fredrick, K., Ibba, M.: How the sequence of a gene can tune its translation. Cell **141**(2), 227–229 (2010)
17. Ghaemmaghami, S., Huh, W.K., et al.: Global analysis of protein expression in yeast. Nature **425**(6959), 737–741 (2003)
18. Golub, T.R., Slonim, D.K., et al.: Molecular classification of cancer: class discovery and class prediction by gene expression monitoring. Science **286**(5439), 531–537 (1999)
19. Gustafsson, C., Govindarajan, S., et al.: Codon bias and heterologous protein expression. Trends Biotechnol. **22**(7), 346–353 (2004)
20. Heinrich, R., Rapoport, T.A.: Mathematical modelling of translation of mRNA in eucaryotes; steady state, time-dependent processes and application to reticulocytes. J. Theor. Biol. **86**(2), 279–313 (1980)
21. Hofacker, I.L.: Vienna RNA secondary structure server. Nucleic Acids Res. **31**(13), 3429–3431 (2003)
22. Huang, T., Wan, S., et al.: Analysis and prediction of translation rate based on sequence and functional features of the mRNA. PLoS One **6**(1), e16036 (2001)
23. Ikemura, T.: Correlation between the abundance of Escherichia coli transfer RNAs and the occurrence of the respective codons in its protein genes. J. Mol. Biol. **146**(1), 1–21 (1981)
24. Ingolia, N.T., Ghaemmaghami, S., et al.: Genome-wide analysis in vivo of translation with nucleotide resolution using ribosome profiling. Science **324**(5924), 218–223 (2009)
25. Jankowsky, E.: RNA helicases at work: binding and rearranging. Trends Biochem. Sci. **36**(1), 19–29 (2011)
26. Kaczanowska, M., Ryden-Aulin, M.: Ribosome biogenesis and the translation process in Escherichia coli. Microbiol. Mol. Biol. Rev. **71**(3), 477–494 (2007)
27. Kanaya, S., Yamada, Y., et al.: Studies of codon usage and tRNA genes of 18 unicellular organisms and quantification of Bacillus subtilis tRNAs: gene expression level and species-specific diversity of codon usage based on multivariate analysis. Gene **238**(1), 143–155 (1999)

28. Kapp, L.D., Lorsch, J.R.: The molecular mechanics of eukaryotic translation. Annu. Rev. Biochem. **73**(1), 657–704 (2004)
29. Kawaguchi, R., Bailey-Serres, J.: mRNA sequence features that contribute to translational regulation in Arabidopsis. Nucleic Acids Res. **33**(3), 955–965 (2005)
30. Kertesz, M., Wan, Y., et al.: Genome-wide measurement of RNA secondary structure in yeast. Nature **467**(7311), 103–107 (2010)
31. Khan, J., Wei, J.S., et al.: Classification and diagnostic prediction of cancers using gene expression profiling and artificial neural networks. Nat. Med. **7**(6), 673–679 (2001)
32. Kimchi-Sarfaty, C., Oh, J.M., et al.: A "Silent" polymorphism in the MDR1 gene changes substrate specificity. Science **315**(5811), 525–528 (2007)
33. Kudla, G., Murray, A.W., et al.: Coding-sequence determinants of gene expression in Escherichia coli. Science **324**(5924), 255–258 (2009)
34. Liao, J.Y., Ma, L.M., et al.: Deep sequencing of human nuclear and cytoplasmic small RNAs reveals an unexpectedly complex subcellular distribution of miRNAs and tRNA 3' trailers. PLoS One **5**(5), e10563 (2010)
35. Lithwick, G., Margalit, H.: Relative predicted protein levels of functionally associated proteins are conserved across organisms. Nucleic Acids Res. **33**(3), 1051–1057 (2005)
36. Lu, J., Deutsch, C.: Electrostatics in the Ribosomal Tunnel Modulate Chain Elongation Rates. J. Mol. Biol. **384**(1), 73–86 (2008)
37. Lu, P., Vogel, C., et al.: Absolute protein expression profiling estimates the relative contributions of transcriptional and translational regulation. Nat. Biotechnol. **25**(1), 117–124 (2007)
38. Lundberg, E., Fagerberg, L., et al.: Defining the transcriptome and proteome in three functionally different human cell lines. Mol. Syst. Biol. **6**(450), 450 (2010)
39. MacDonald, C.T., Gibbs, J.H., et al.: Kinetics of biopolymerization on nucleic acid templates. Biopolymers **6**(1), 1–5 (1968)
40. Man, O., Pilpel, Y.: Differential translation efficiency of orthologous genes is involved in phenotypic divergence of yeast species. Nat. Genet. **39**(3), 415–421 (2007)
41. Mattick, J.S., Makunin, I.V.: Non-coding RNA. Hum. Mol. Genet. **1**(15), R17–R29 (2006). 15 Spec. No. 1
42. Miyasaka, H.: The positive relationship between codon usage bias and translation initiation AUG context in Saccharomyces cerevisiae. Yeast **15**(8), 633–637 (1999)
43. Newman, J.R., Ghaemmaghami, S., et al.: Single-cell proteomic analysis of S. cerevisiae reveals the architecture of biological noise. Nature **441**(7095), 840–846 (2006)
44. Osada, Y., Saito, R., et al.: Analysis of base-pairing potentials between 16S rRNA and 5' UTR for translation initiation in various prokaryotes. Bioinformatics **15**(7–8), 578–581 (1999)
45. Pavon-Eternod, M., Gomes, S., et al.: tRNA over-expression in breast cancer and functional consequences. Nucleic Acids Res. **37**(21), 7268–7280 (2009)
46. Percudani, R., Pavesi, A., et al.: Transfer RNA gene redundancy and translational selection in Saccharomyces cerevisiae. J. Mol. Biol. **268**(2), 322–330 (1997)
47. Plotkin, J.B., Kudla, G.: Synonymous but not the same: the causes and consequences of codon bias. Nat. Rev. Genet. **12**(1), 32–42 (2010)
48. Redon, R., Ishikawa, S., et al.: Global variation in copy number in the human genome. Nature **444**(7118), 444–454 (2006)
49. Reuveni, S., Meilijson, I., et al.: Genome-Scale Analysis of Translation Elongation with a Ribosome Flow Model. RECOMB. Vancouver, BC (2011)
50. Scherzer, C.R., Eklund, A.C., et al.: Molecular markers of early Parkinson's disease based on gene expression in blood. Proc. Natl. Acad. Sci. USA **104**(3), 955–960 (2007)
51. Scholten, K.B., Kramer, D., et al.: Codon modification of T cell receptors allows enhanced functional expression in transgenic human T cells. Clin. Immunol. **119**(2), 135–145 (2006)
52. Sharp, P.M., Li, W.H.: The codon Adaptation Index–a measure of directional synonymous codon usage bias, and its potential applications. Nucleic Acids Res. **15**(3), 1281–1295 (1987)

53. Shaw, L.B., Zia, R.K., et al.: Totally asymmetric exclusion process with extended objects: a model for protein synthesis. Phys. Rev. E. Stat. Nonlin. Soft. Matter. Phys. **68**(2 pt 1), 021910 (2003)
54. Sonenberg, N., Hinnebusch, A.G.: Regulation of translation initiation in eukaryotes: mechanisms and biological targets. Cell **136**(4), 731–745 (2009)
55. Sotiriou, C., Neo, S.Y., et al.: Breast cancer classification and prognosis based on gene expression profiles from a population-based study. Proc. Natl. Acad. Sci. USA **100**(18), 10393–10398 (2003)
56. Taniguchi, Y., Choi, P.J., et al.: Quantifying E. coli proteome and transcriptome with single-molecule sensitivity in single cells. Science **329**(5991), 533–538 (2010)
57. Tuller, T., Carmi, A., et al.: An evolutionarily conserved mechanism for controlling the efficiency of protein translation. Cell **141**(2), 344–354 (2010)
58. Tuller, T., Kupiec, M., et al.: Determinants of protein abundance and translation efficiency in S. cerevisiae. PLoS Comput. Biol. **3**(12), e248 (2007)
59. Tuller, T., Veksler-Lublinsky, I., et al.: Composite Effects of Gene Determinants on the Translation Speed and Density of Ribosomes under-review (2011)
60. Tuller, T., Waldman, Y.Y., et al.: Translation efficiency is determined by both codon bias and folding energy. Proc. Natl. Acad. Sci. USA **107**(8), 3645–3650 (2010)
61. Uemura, S., Aitken, C.E., et al.: Real-time tRNA transit on single translating ribosomes at codon resolution. Nature **464**(7291), 1012–1017 (2010)
62. van den Berg, J.A., van der Laken, K.J., et al.: Kluyveromyces as a host for heterologous gene expression: expression and secretion of prochymosin. Biotechnology (NY) **8**(2), 135–139 (1990)
63. Vogel, C., Abreu Rde, S., et al.: Sequence signatures and mRNA concentration can explain two-thirds of protein abundance variation in a human cell line. Mol. Syst. Biol. **6**(400), 400 (2010)
64. Waldman, Y.Y., Tuller, T., et al.: Selection for Translation Efficiency on Synonymous Polymorphisms in Recent Human Evolution." under-review (2011)
65. Waldman, Y.Y., Tuller, T., et al.: TP53 cancerous mutations exhibit selection for translation efficiency. Cancer Res. **69**(22), 8807–8813 (2009)
66. Waldman, Y.Y., Tuller, T., et al.: Translation efficiency in humans: tissue specificity, global optimization and differences between developmental stages. Nucleic Acids Res. **38**(9), 2964–2974 (2010)
67. Warnecke, T., Hurst, L.D.: GroEL dependency affects codon usage–support for a critical role of misfolding in gene evolution. Mol. Syst. Biol. **6**(340), 340 (2010)
68. Warner, J.R.: The economics of ribosome biosynthesis in yeast. Trends Biochem. Sci. **24**(11), 437–440 (1999)
69. Watanabe, K., Osawa, S.: tRNA sequences and variation in the genetic code. In: Söll, D., RajBhandary, U. (eds.) tRNA Structure Biosynthesis and Function, pp. 225–250. AMS Press, Washington, DC (1995)
70. Welch, M., Govindarajan, S., et al.: Design parameters to control synthetic gene expression in Escherichia coli. PLoS One **4**(9), e7002 (2009)
71. Wenzel, S.C., Müller, R.: Recent developments towards the heterologous expression of complex bacterial natural product biosynthetic pathways. Curr. Opin. Biotechnol. **16**(6), 594–606 (2005)
72. Yokoyama, S., Nishimura, S.: Modified nucleosides and codon recognition. In: Söll, D., RajBhandary, U. (eds.) tRNA Structure Biosynthesis and Function, pp. 207–223. AMS Press, Washington, DC (1995)
73. Zenklusen, D., Larson, D.R., et al.: Single-RNA counting reveals alternative modes of gene expression in yeast. Nat. Struct. Mol. Biol. **15**(12), 1263–1271 (2008)
74. Zhang, S., Goldman, E., et al.: Clustering of low usage codons and ribosome movement. J. Theor. Biol. **170**(4), 339–354 (1994)
75. Zuker, M.: Mfold web server for nucleic acid folding and hybridization prediction. Nucleic Acids Res. **31**(13), 3406–3415 (2003)

Neural Network Modeling Approaches for Patient Specific Glycemic Forecasting

Scott M. Pappada and Brent D. Cameron

Abstract Prediction of glucose in patients with diabetes has been a major thrust of research in hopes to develop an artificial pancreas capable of automated and closed-loop glycemic control. Although it is less known, prediction of glucose in the critical care setting has also been the subject of considerable research endeavors. Successful prediction of glucose in both these patient populations requires the analysis of multiple factors and variables many of which are "patient specific" and vary from patient to patient. Thus, a well suited modeling technique for prediction of physiological glucose levels to needs to be adaptive and incorporate the effect of numerous dependent factors or variables to accurately forecast future glucose concentrations. Neural network models are a particularly well suited approach as they have the ability to learn and quantify the effect of various input factors/variables on a desired/predicted output. In this chapter, the application of neural network modeling to the prediction of glucose in diabetic and critical care patients that exhibit a lack of glucose control will be discussed, in addition to, the advantages of neural network modeling for patient specific glycemic forecasting with respect to other modeling techniques.

1 Introduction

Lack of glycemic (glucose) control in patients with diabetes is a well known phenomenon. This is especially difficult in patients with type 1 diabetes mellitus (DM1) as the body no longer produces insulin and subcutaneous insulin is required

---

S. M. Pappada · B. D. Cameron (✉)
Department of Bioengineering,
University of Toledo, Toledo, OH, USA
e-mail: brent.cameron@utoledo.edu

via either injection or continuous infusion of insulin via a pump system is required to manage glucose values. Optimum control of glucose concentration in patients with diabetes is the main goal of diabetes therapy and is attempted through the self monitoring of blood glucose (SMBG) values and adjustment of insulin based on these discrete sampled values. Research has been completed which indicates that intensive insulin therapy (IIT) is necessary to obtain adequate glycemic control in order to avoid long term complications associated with persistently elevated glucose levels [1–4].

What is less known, however, is the lack of glycemic control in critical care patients (e.g. trauma and cardio thoracic surgical patients in the intensive care unit). Persistently elevated glucose values in the critical care setting have been associated with increased morbidity [5–7] (e.g. deep sternal wound infection) and mortality [8–11]. Implementation of an intensive insulin infusion protocol, resulting in tight glycemic control (TGC), has been associated with a reduction of mortality and enhancement of patient outcomes in the critical care setting [12–17].

The optimization of glycemic control in either patient population can be enhanced with advanced knowledge of undesirable glycemic excursions. The prediction of high blood glucose values (hyperglycemia) and low blood glucose (hypoglycemia) would be extremely valuable and allow patients and clinicians to adjust insulin delivery in advance of these unwanted glucose values and mitigate and potentially avoid their occurrence. The enhancement of glycemic control resulting from the avoidance of these glycemic extremes would result in enhancement of outcomes in both diabetic [1–4] and critical care [12, 14, 17–19] patient populations.

Currently, patients with diabetes are recommended to monitor their blood glucose values 2–6 times daily. These SMBG values are discrete time measurements and do not provide insight to glycemic excursions when measurements were not taken (i.e. times between readings). In the critical care patient population, the current convention of care is to monitor a patient's blood glucose concentration every 1–4 h [20]. Based on these point-of-care (POC) glucose values, the patient is administered insulin according to the hospital's insulin delivery protocol. Even in the event that a patient is monitored every hour, there is a 60 min time span where the patient glucose concentration and glycemic excursions in response to the insulin infusion protocol are unknown. Studies have shown that significant variation can occur within this time period. In both patient populations, conventional glucose monitoring approaches are severely flawed or limited in that they are unable to provide reliable detection of many hyperglycemic and hypoglycemic glucose values experienced on an everyday basis.

Recent advances in glucose monitoring technology include the development and utilization of real-time continuous glucose monitoring (CGM) devices. Major medical device companies have developed and commercialized real-time CGM technologies. These companies (and their CGM technologies) include: Medtronic Diabetes® (Guardian RT™), Dexcom® (Seven PLUS™) and Abbott Diabetes Care® (Freestyle Navigator). These devices measure glucose concentration in the interstitial fluid every 1–5 min (depending on the technology). Such technologies are designed to provide more frequent and reliable measurement of glucose in order to detect hyperglycemia and hypoglycemia which would remain unrecognized using the aforementioned conventional glucose monitoring approaches. The ability to predict or forecast both short and long-term future glucose concentrations, especially when discrete SMBG or POC glucose values are not available would provide physicians and/or diabetics an excellent resource and tool for adjusting insulin therapy in order to optimize glucose control. Real-time CGM technologies provide a significant source of data for which to develop models for prediction of glucose which can be implemented in a therapeutic direction/guidance and perhaps eventually automation of glycemic control.

In this chapter, several modeling approaches will be discussed which have been historically implemented for prediction of glucose in both diabetic and critical care patient populations. The complexity of glycemic excursions will also be discussed in these populations and emphasis that successful prediction of glucose requires the analysis of multiple factors and variables will be justified. The majority of this chapter will focus on neural network modeling approaches that have implemented for "patient specific" glycemic forecasting. Neural network modeling (NNM) is a well suited approach for prediction of glucose as NNMs are adaptive and have the ability to learn and anticipate the effects that various and numerous variables and factors have on future glucose levels. Many of these variables are "patient specific" and vary from patient to patient, thus, a NNM is uniquely suited for "patient specific" glycemic forecasting as they possess the ability to learn and adapt to an individual patient overtime. The strengths and weaknesses of the discussed neural network modeling approaches implemented to date with respect to other modeling techniques for prediction of glucose will be highlighted.

2 Glycemic Excursions in Diabetic and Critical Care Patients: Multivariate and Complex Physiologic Systems

The major difficulty involving the successful treatment of diabetes is the appropriate dosing of insulin such that a therapeutic and normal range of glucose concentration (80–120 mg/dl) can be achieved. Multiple factors influence glycemic excursions in patients with diabetes. These factors include but are not limited to: insulin dosage, carbohydrate and nutritional intake, lifestyle (i.e. sleep-wake cycles and sleep quality, exercise, etc.), and emotional states (i.e. stress, depression, contentment, etc.) [21–31]. The effect of these various factors on subsequent glucose concentration is not fully understood, and may not be universal to all patients with diabetes or "patient specific" i.e. specific to an individual but vary from patient to patient.

Glycemic excursions experienced by patients with diabetes on a daily basis appear to be chaotic, however, previous research has identified patterns in glucose [30, 32–37]. Circadian rhythms in sleep and glucose regulation have been identified in previous research [30]. Research has also demonstrated patterns in other factors such as glucose tolerance, insulin activity, insulin sensitivity, insulin clearance, and hormone levels (e.g. cortisol, epinephrine, norepinephrine), which result in quantifiable patterns in glucose concentration [26, 30, 32–46]. The existence of rhythms and patterns related to insulin activity, and the resultant quantifiable patterns in glucose, provide a foundation and hypothetical construct for the development of modeling techniques for the prediction and ultimately control of physiological glucose levels.

These patterns identified in previous research in both diabetic and non-diabetic populations are also present in patients residing in critical care settings as well [26, 30–46]. Furthermore, patterns in glucose specific to critical care patients receiving intensive insulin therapy and insulin infusion have been discovered and documented [47–49]. A significant quantity of factors are detailed in patient medical records throughout the course of the patient's stay in the intensive care unit and many of these factors may also be indicators of directed influence on future glycemic trends. Such factors include but are not limited to: vital signs, lab results, ventilation data, pain levels, nutritional intake, medications, etc. For example, quantifiable vital signs, such as, increased heart rate or tachycardia have been correlated in the literature with increased glucose concentration [50]. Research has also substantiated that quantifiable changes in body temperature can be associated with low blood glucose concentration (hypoglycemia) [51], as well as, other factors that can directly influence glucose concentration such as daily inpatient use of oral or intravenous steroids, dextrose-containing intravenous fluids, and daily nutritional intake via tube feeds, inpatient total parenteral nutrition (TPN), etc. Perhaps the major difficulty facing caregivers in the intensive care unit is estimating the effect that these factors have on future glucose levels such that appropriate modification to intensive insulin therapy can be made to optimize glycemic control. The utilization of these records, in combination with glucose values collected throughout the patient's length of stay will be necessary components for the development of a successful predictive algorithm for accurate forecasting of glucose concentration in critical care patient populations.

3 Prediction and Control of Glucose Concentration in Diabetic and Critical Care Patients

As has been discussed, there are numerous factors which impact, or are indicators of future glucose concentration in diabetic and critical care patients. The ability to characterize the effect of such factors as related to glucose concentration is an extremely difficult task. Numerous modeling approaches for prediction and/or control of glucose in diabetic and critical care patients have been investigated to date. These models range in complexity, and many integrate several variables which impact glucose concentration. Modeling approaches which have been applied in both patient settings have included autoregressive modeling techniques

[52, 53], and model predictive control algorithms [54–56]. The goal of diabetes research and an emerging focus of critical care research have been in the development of technologies and algorithms for automated closed loop glycemic control. Many researchers have regarded the implementation of a classic control algorithm, such as the Proportional Integral Derivative (PID) controller, as perhaps the best suited technique and algorithm to support development of an artificial pancreas and to achieve closed loop glycemic control. In this section, applications which have been implemented using the PID controller will be discussed along with the relative strengths and weaknesses of this approach.

The use of PID controller based algorithms has been investigated for their potential in automating glycemic control in both diabetic [57–59] and critical care [60–62] patients. The utility the PID-based algorithm [59] is supported by the fact that the PID controller mimics the biphasic nature and function of pancreatic β cells [63, 64] (i.e. cells responsible for secretion of insulin in response to increases and changes in glucose concentration) in the human body. The output of the PID controller is the appropriate insulin dosage/response at any given time ($PID(t)$) required to maintain glucose at a normal/basal glucose value (G_B) and target range. Using the PID controller, optimal insulin delivery is calculated based on the tracking real-time values and the kinetics (i.e. rate of change) of glucose over time. The algorithm structure of the classic PID controller contains three key components which include proportional, integral, and derivative terms. The proportional component (K_p) of the PID algorithm is based on current/real-time glucose concentration. In terms of beta cell functionality the proportional component (K_p) is a key determining factor of the first phase of insulin secretion/delivery requirements in order to respond to hepatic glucose output from the liver. The integral component (K_I) of the PID algorithm gradually increases or decreases in response to glucose. From a closed-loop glycemic control perspective, K_I is the important factor in establishing the basal insulin delivery rate and requirements to maintain fasting plasma glucose at a set-point/target glucose value/range (G_B) irrespective of insulin sensitivity or rate of endogenous glucose appearance. The derivative component (K_D) determines insulin requirements in response to the rate of change in glucose which occurs over time $\left(\frac{dG}{dt}\right)$. Equation 1 contains the general equation for a PID control algorithm implemented for closed loop glycemic control.

$$PID(t) = K_p(G - G_B) + K_I \int (G - G_B) + K_D \frac{dG}{dt} \qquad (1)$$

While the PID control algorithm has been heavily investigated for its potential in supporting closed loop glycemic control, it can suffer from a number of important limitations which will restrict its ultimate clinical utility and acceptability. First and foremost, the algorithm is too simplistic and cannot adequately account for the effect of the various factors and variables (refer to Sect. 2) which impact future glycemic trends. For instance, patients with type 1 diabetes mellitus often have a degree of autonomic insufficiency in which the human body's physiologic counter-regulatory hormonal response is delayed or in some circumstances non-existent [65, 66]. For PID controllers, there is no input or ability to quantify the occurrence of lifestyle or emotional factors such as exercise or stress. These factors have known physiological implications which result in increased adrenal response and release of glucocorticoids, which then result in the elevation of glucose concentration. As has been described, the PID control algorithm would respond to this "perceived" elevation of glucose via delivery of insulin (an inappropriate response) which may lead to the occurrence of hypoglycemia. Given the degree of autonomic insufficiency that patients with diabetes commonly experience, this would be a very dangerous occurrence. Patients with diabetes often experience hypoglycemic unawareness, a condition in which the patient does not realize they are in a hypoglycemic state, which when combined with autonomic insufficiency would lead to prolonged occurrences of hypoglycemia [67–69] and a vicious cycle [70].

A further limitation of implementing the PID control algorithm is its inherent inability to appropriately account for nutritional intake. A PID control algorithm would not be able to distinguish meal content and insulin dosage needed to maintain a normal glycemic state. While an observed increase in glucose concentration following a meal would be accounted for, the overall insulin dosage to cover total nutritional intake could not be determined. A system which takes into account an approximation of carbohydrate/caloric intake could be more capable of maintaining tight glycemic control in response to a meal. A closed loop PID controller would therefore need to have at least a "meal button" to initiate to account for nutritional intake. Even in the presence of such a button, the meal content at specific meals and times of day would need to have a degree of consistency as well [37].

Regardless of the modeling approach or technique implemented for prediction and/or optimizing control of glucose, each requires the continuous monitoring of glucose to be effective, with minimal time delay, and clinically applicable. Many of the recent algorithms and models for prediction of glucose [52, 56] and automation of glucose control, including the PID control algorithm [57–59, 62, 64, 71], have made use of real-time continuous glucose monitoring (CGM) technologies which provide measurement of glucose concentration every 1–5 min. These devices measure *interstitial glucose* of which there is a demonstrated time-lag of 10–15 min with respect to actual blood glucose concentration [72]. The time delay is due to the time required for glucose to diffuse from the capillary to the tissue. Increases in this time-lag are observed during rapid changes in glucose and due to the magnitude of concentration differences in various tissues occurring at the time of rapid change [73]. Significant controversy has thus surrounded the clinical acceptability of algorithms which are based off interstitial CGM technologies due to this reported delay [74–77]. In order to address this significant limitation of CGM technologies, predictive models and algorithms need to employ large enough prediction horizons which negate/minimize such effects. For example, algorithms based off of CGM data which predict a 30 min glucose forecast, in reality only provide an indication of future glycemic excursions 15–20 min ahead of time.

The ability to predict/forecast glucose values using a horizon of ≥60 min would thus serve to mitigate and reduce the effects of the CGM time-lag and may be better suited for clinical and closed-loop glycemic control applications.

4 Neural Network Modeling for Prediction of Glucose

An artificial neural network is a mathematical or computational modeling technique. Neural network models (NNMs) contain a connected series of processing elements called neurons. Neurons exhibit complex global behavior which is determined by the connections existent between various processing elements and related parameters within the NNM architecture. The concept of neural network modeling is derived from knowledge of the central nervous system and specifically neurons (consisting of axons, dendrites, and synapses) which are the most basic information processing elements in neuroscience. Neural network modeling is desirable as they are adaptive technologies, which learn based on determining patterns existent in input data. Based on this learning process, weights within the NNM architecture are modified to minimize error in the model output and complete a specific task. Historically, neural network modeling has been used in a variety of applications. These applications include but are not limited to: time series prediction, pattern recognition, function approximation, regression analysis, classification, and data processing (e.g. filtering, and clustering).

As we have discussed, the existence of patterns in glycemic excursions and other parameters directly related to glucose in both diabetic [30, 32, 37, 44, 45] and critical care [33, 47] patient populations, indicate that predictability of glucose is possible. One of the major limitations of models/algorithms for prediction and control of glucose has been their inherent simplicity and the inability to incorporate other factors and variables which significantly impact future trends in glucose concentration. Most previously investigated modeling approaches have incorporated common and traditional variables which include discrete blood glucose monitoring results, insulin dosages, and nutritional intake. Due to the physiologic complexity of glycemic excursions and their dependency on other factors/variables such as daily activities (sleep [26, 28, 30, 40], exercise [21, 25], etc.), and emotional factors (stress, depression [31, 78], etc.) these "traditional" modeling approaches are not capable of successfully quantifying all responses in glucose. Due to the influence and effect of multiple factors and variables on glucose, an appropriate modeling approach for accurate glucose prediction would need to be highly sophisticated and have the ability to process and gauge the effect of a substantial quantity of factors. NNMs are "trained" such that a defined set of input variables produces the desired set of outputs. NNMs are trained by presentation of a significant quantity of teaching patterns (input data) such that weights within the NNM architecture are adapted/modified (according to a particular learning rule) to minimize the error between the NNM output and the desired set of outputs. NNMs are a particularly well suited for this application area as they have the ability to learn and quantify the effect of numerous input factors/variables which have an impact and effect on glucose.

In the past, NNMs have been investigated for their potential in prediction of glucose concentration [79–83]. These NNM applications have suffered from limitations of other mathematical modeling approaches in that they only utilized traditional inputs (blood glucose values, insulin dosages, exercise, and nutritional intake). It is also important to note that NNM accuracy relies on the availability and utilization of an appropriate training data set [84] which characterizes the system to be modeled. These past approaches also used only discrete blood glucose monitoring results, which are unable to effectively characterize trends and patterns present in glucose. Rather than conventional discrete glucose monitoring, continuous glucose monitoring (CGM) have the potential to be more effective in obtaining standardized training sets that are better suited for accurate glucose prediction. Only recently have NNMs began to utilize the availability of CGM technologies [85–89]. These models were designed in an attempt to overcome the aforementioned limitations of other approaches used for prediction and control of glucose. The next section will summarize some of these NNM approaches and their use for in vivo glucose prediction.

4.1 Neural Network Models for Glycemic Forecasting in DM1 Patients

In 2008, Pappada et al. reported on an initial investigation involving patients with type 1 diabetes mellitus (DM1) [88]. In this investigation, data was collected from 17 patients in an outpatient setting which included CGM data and other data documented through use of a pocket PC application which included traditional model inputs (SMBG values, insulin dosages, nutritional intake), and other unconventional inputs various activity (e.g. sleep-wake times) and emotional factors (e.g. stress). Data from these patients was used to train developed time-lagged feed forward neural network models (TLFFNNMs) for prediction of glucose.

TLFFNNMs are well suited for prediction as they contain memory structures to store historical values of neural network inputs. The developed models were designed with a 4 layer architecture which included an input layer, 2 hidden processing layers, and an output layer. The hidden processing layers are extremely important to the overall NNM effectiveness and functionality as they serve to confine input values to a defined range. This enables the neural network to process and interpret patterns in data more efficiently. For this model architecture, hidden processing elements (axons) which utilize a hyperbolic transfer function to effectively limit the range of model inputs to a range of ± 1 was implemented. Instead of model inputs, such as, glucose concentration ranging from 40 to 400 mg/dl, it was found that by normalizing these values to a more quantifiable range makes pattern recognition and processing in the NNM more efficient.

These models were trained using a well known and common modality known as back-propagation (BP). During BP training, processing elements within the NNM architecture (i.e. back-propagation axons) facilitate the overall training process. These axons derive a relative error at their input which is "back-propagated" to other processing elements preceding them in the NNM architecture. The overall NNM error for a given set of model weights (i.e. connections/synapses in the NNM architecture) is presented at the output of each axon in the NNM. Each of these axons is responsible for calculating the optimal weights which minimize the error between the desired response(i.e. glucose values being modeled) and the NNM output (i.e. predicted glucose values). Optimal weights for minimization of error in the predictive model are obtained via a well known algorithm known as gradient descent with momentum [90]. This algorithm calculates the optimal weights for minimization of total error in the neural network model via iteratively searching for the local minimum across the performance surface of model error as a function of different weight values. The weight values which correspond to the smallest error are thus the optimal solution (i.e. set of NNM weights) for the particular NNM. In the gradient descent with momentum algorithm, the momentum term gives the algorithm "inertia" and ensures that the algorithm continues towards (along the downward trajectory) the local minimum in its search for optimal NNM weights.

The TLFFNNMs were applied to retrospective prediction of glucose using data acquired from 10 patients not utilized for model development and training. Each of the NNMs were configured to output a trajectory of predicted CGM values across the prediction horizon defined in each model. For example, because CGM devices measure glucose every 5 min, a NNM configured with a prediction horizon of 60 min would output 12 predicted glucose values for each iteration. To provide an assessment of model performance, overall model error was calculated and expressed as mean absolute difference percent (MAD%) as calculated in Eqs. 2 and 3. Equation 2 is utilized for calculating the absolute difference percent (AD%) between each neural network predicted value and the corresponding actual CGM value. Equation 3 is then used to calculate the MAD%, which is defined as the mean of all obtained absolute difference percent values in the dataset.

Note: $AD\%(t)$ is the calculated AD% at time t, $NNet_{predict}(t)$ is the predicted neural network glucose value at time t, $CGM_{actual}(t)$ is the actual CGM data point at time t, and N is the number of data points in the dataset used for calculating the MAD%.

$$AD\%(t) = \frac{|NNet_{predict}(t) - CGM_{actual}(t)|}{CGM_{actual}(t)} * 100\% \qquad (2)$$

$$MAD\% = \frac{\sum_{i=1}^{N} AD\%(t)}{N} \qquad (3)$$

The initial results in the 2008 study were encouraging and indicated that prediction of glucose using a NNM approach was feasible. Furthermore, the

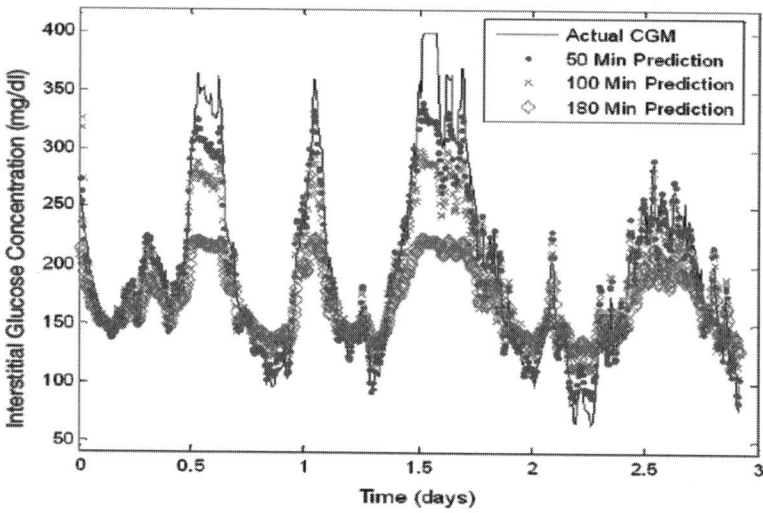

Fig. 1 TLFFNNM prediction of glucose in a DM1 patient using different model prediction horizons (reprinted with permission from [88])

approach demonstrated the potential to forecast glucose concentrations on the order of 50–180 min in the future, which was a significant extension to previously reported modeling approaches at that time. Overall model accuracy and the percentage glycemic extremes (normal, hypoglycemic, and hyperglycemic) predicted decreased as the model prediction horizon was increased. Model error ranged from 6.7 to 18.9% for prediction horizons of 50 and 180 min respectively. Overall, greater than 70% of the hyperglycemic glucose values (CGM values $\geq$180 mg/dl) for all model prediction horizons were able to be predicted. An observed limitation of this modeling approach was the inability to predict hypoglycemic glucose values (CGM values $\leq$70 mg/dl). This limitation was attributed to a lack of sufficient hypoglycemic data in the initial model training set. This limitation will be further discussed in the following section. Figure 1 demonstrates prediction of glucose for TLFFNNMs with variable length model prediction horizons. The decreases in model accuracy should be noted, which occurs as the model prediction horizon is increased.

In order to translate the NNM technique into an application which has a high degree of utility and applicability in the outpatient and clinical setting, research began to focus on the development of NNMs suitable for real-time glycemic forecasting. The initial TLFFNNMs [88] were fairly complex due to the existence of memory structures within the model architecture and the existence of two hidden processing layers. As the number of processing elements in a NNM increases, so does the number of NNM weights and consequently time required for training. For real-time prediction of glucose, NNM training and subsequent prediction of glucose must be completed within the sampling rate of the CGM device. In order to meet this criteria, a reduced complexity feed forward neural network

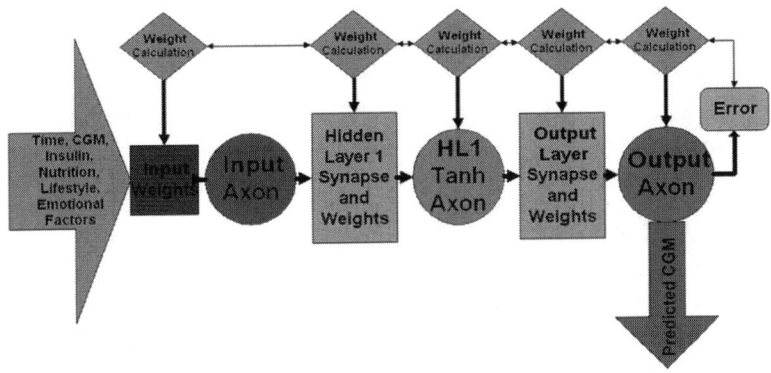

Fig. 2 Reduced complexity feed forward neural network model architecture for real-time prediction of glucose [86]

model (FFNNM) architecture was developed. This model architecture is similar to the original model architecture, but did not contain memory structures to store historical NNM inputs and consisted of only one hidden processing layer. This new reduced complexity NNM architecture is shown in Fig. 2.

This design was implemented for real-time prediction of glucose concentration using a predictive horizon (PH) of 75 min. For demonstration purposes, the neural network model was trained using the same 17 patient training dataset as implemented in the previously discussed investigation [88]. NNM functionality was integrated into a graphical user interface (GUI) based computer application in order to facilitate real-time training and prediction of glucose. For each iteration of the predictive computer application, two files containing the NNM input data and desired glycemic responses were utilized for real-time training and prediction. These files contained 800 vectors, of which 799 included historical values and the final vector was the current real-time input data and desired glycemic responses. Because the desired response is unknown in real-time and to maintain real-time model training, the rate of change in CGM data was calculated and the desired response implementing the 75 min PH was estimated. After 15 samples (i.e. number of samples in the model PH) occurred, the current real-time CGM was placed at the appropriate index of the desired response file to replace previous estimations with actual glycemic response. The NNM was then trained using the same back-propagation training technique which was previously described. Due to reduced model complexity of the NNM architecture, the developed computer application was able to effectively train and predict glucose in real-time within the 5 min sampling rate of the CGM device [86].

Further validation and demonstration of real-time NNM training and prediction of glucose was completed using data collected from 10 additional patients that were not included in the model training set. The predictive model error was determined with respect to actual CGM values via calculation of the root mean squared error (RMSE) and mean absolute difference percent (MAD%) techniques.

Due to the fact that previous investigative efforts demonstrated that hypoglycemic extremes were routinely overestimated via NNM model implementation because of lack of a significant quantity of hypoglycemic training data, RMSE and MAD% was also calculated at non-hypoglycemic extremes. To further assess model performance, the percentage of hypoglycemic (CGM values ≤ 70 mg/dl), normal (CGM values >70 and <180 mg/dl), and hyperglycemic (CGM ≥ 180 mg/dl) values predicted by the NNM were calculated.

In order to provide an indication of the clinical acceptability of model predictions, Clarke Error Grid Analysis (CEGA) of the model predictions with respect to reference CGM measurements was utilized. CEGA was established in 1987 and was originally used to assess patient estimates of blood glucose compared to those obtained using a "gold standard" reference glucose meter, but has since been applied to CGM data as well [91, 92]. Zone A contains predicted values within 20% of the reference concentration, and Zone B contains predictions outside 20% of the reference concentration, which would not have any adverse effect on treatment. Zones A and B therefore contain predicted values that can be classified as "clinically acceptable". Zone C contains points that lead to unnecessary treatments, and zone D contains points indicating a potentially dangerous failure to detect hypoglycemia. Zone E contains predicted values that would confuse treatment of hypoglycemia for hyperglycemia, and vice versa.

This model predicted 2.1, 88.6, and 72.6% of hypoglycemia, normal, and hyperglycemic glucose values, respectively. Furthermore, overall MAD% was calculated as 22.1% at all glycemic extremes and 18.1% at non-hypoglycemic glucose values (CGM values >70 mg/dl). Overall RMSE $\pm$ standard deviation (SD) in mg/dl was calculated as 43.9 $\pm$ 6.5 and 43.0 $\pm$ 6.4 for non-hypoglycemic glucose values. Predicted glucose values accurately followed trends in actual glycemic excursions. Clarke Error Grid Analysis (CEGA) indicated that 92.3% of predictions could be regarded as clinically acceptable and not lead to adverse therapeutic direction. CEGA also revealed that 62.3% of predictions fell within region A (region of highest accuracy holding predicted values within 20% of the reference values) of the error grid. For illustrative purposes, Fig. 3 includes model predictions for two DM1 patients generated by the computer program designed to simulate real-time prediction of glucose in an outpatient setting. For clarity, due to the significant quantity of predicted values used for model performance analysis (i.e. 15 predicted values for every real-time CGM value), the data was resampled to plot every 20th predicted value.

Model performance varied from patient to patient across the entire dataset. The percentage of predicted normal glucose values ranged from 64.9% to as high as 94.9%. The percentage of predicted hypoglycemic glucose values across the 10 patient dataset ranged from 0.0 to 11.1% whereas the percentage of hyperglycemia predicted ranged from 47.1 to 94.1%. Although the feasibility of real-time prediction of glucose using a model prediction horizon of 75 min was demonstrated, the variability of model performance across this multi-patient dataset demonstrates the challenges involved in developing a universal NNM. "Patient specific" factors such as insulin sensitivity, insulin resistance, sleep wake cycles, and other variables can

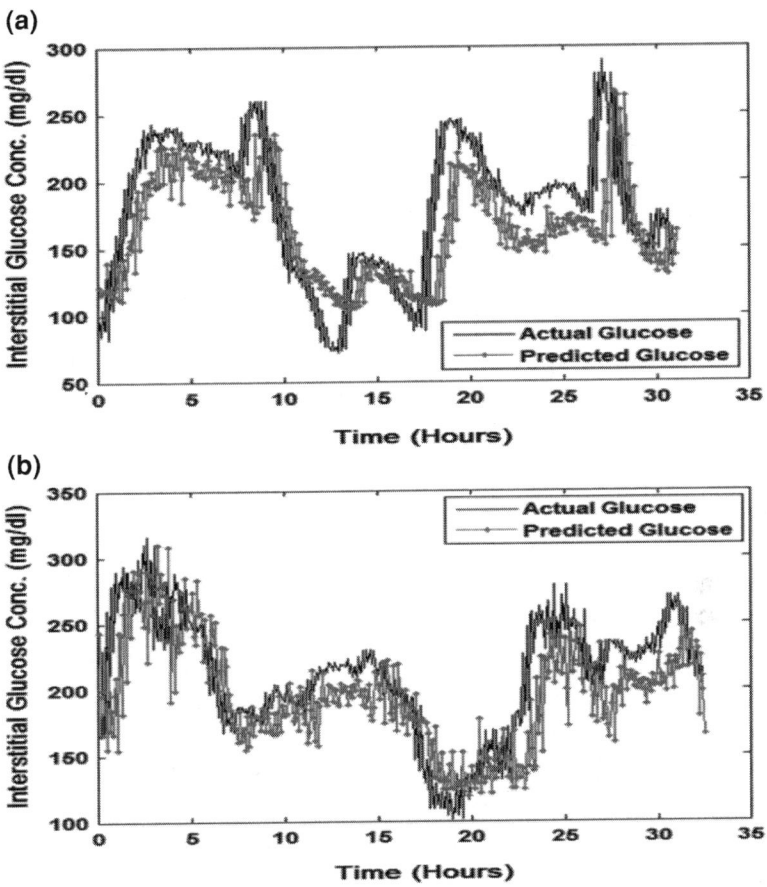

Fig. 3 Real-time prediction of glucose in 2 DM1 patients (**a** and **b**) using FFNNM architecture [86]

significantly affect future glycemic trends. Generating separate and individualized NNMs using data acquired from only a single patient may provide increased predictive accuracy. Thus, due to variability of these factors across patients, "patient specific" modeling may be well suited and incredibly advantageous.

4.2 Neural Network Models for Glycemic Forecasting in Critical Care Patients

Given the previous demonstrations of successfully predicting glucose concentration in diabetic patients [86, 88], similar NNM efforts have been pursued to investigate the applicability of performing glycemic forecasting in the critical care setting, where there is a similar and significant need. In this investigation,

19 critical care patients were subjected to CGM and routine documentation of their medical records throughout their length of stay in the intensive care unit (ICU). The length of stay of the patients varied from 2 to 16 days and was dependent upon the severity of injury/illness. All CGM and medical records collected in this investigation were utilized for the development and validation of the real-time predictive NNMs. The models developed for use in the critical care patient population were designed with the same FFNNM architecture implemented in patients with type 1 diabetes. In comparing model architectures, the major difference was in the NNM input structure.

For the critical care setting, there can be hundreds of potential inputs which may affect or be indicative of current and future glucose concentrations. Many of these inputs/variables are documented on a consistent basis every hour in the routine medical records during each patient's length of stay in the ICU. Thus, the optimal selection of inputs for the NNM can require advanced identification and processing algorithms to establish the best predictors for glucose. For a recent investigation, a customized computer program was developed to convert paper-based medical records into a usable electronic format to facilitate NNM development. Through this effort, there were 131 possible medical records which were documented for each patient's length of stay in the ICU. The use of all these inputs would result in a highly complex NNM requiring extensive computational resources and time which is not feasible for real-time operation. Therefore, in order to minimize and optimize the number of inputs and predictors, a genetic algorithm approach was used to determine which of these 131 inputs were optimal predictors of glucose concentration. Genetic algorithms are inspired by the principles of natural evolution, which include inheritance, mutation, selection, and crossover. Genetic algorithms are extremely useful techniques that have been utilized for a number of different applications [93–95] to optimize generating useful solutions and to handle search related problems. Thus, they are well suited to determine which medical records collected are useful predictors of glucose. The theory behind utilization of a genetic algorithm is that given an *x-block* of predictor data and *y-block* of data to be predicted, the key variables from the *x-block* can be identified in order to minimize the error in the *y-block* predictions. This is accomplished through cross-validation and regression to determine the root mean squared error of cross validation (RMSECV) obtained when a subset of variables from the *x-block* are utilized for prediction. This process is iterated to determine which variables from the *x-block* produce the lowest RMSECV. For the critical care investigation, the *x-block* included the documented medical records, CGM device sensor current, and CGM values categorized as glycemic states. The *y-block* was defined as CGM glucose concentration values measured every 5 min [85].

Through implementation of the genetic algorithm, the number of NNM inputs were reduced by 69.4% (i.e. from 131 to 40 inputs). Inputs to the NNM included medical records such as heart rate, temperature, intravenous dextrose solutions, insulin dosages, and POC blood glucose values. The inputs identified by the genetic algorithm were consistent with those indicated in literature, such as, the relationship that elevated glucose levels are often associated to increased heart rate or tachycardia [50]. Furthermore, research has substantiated that body temperature is an indicator of glucose concentration specifically the occurrence of hypoglycemia [51]. Other factors/inputs such as dextrose solutions (D5, D5W, etc.) which are infused in critical care patients contain dextrose (glucose) which would also correlate to an increase in glucose concentration.

The model used for this study was designed with a FFNNM architecture with a prediction horizon of 75 min, which is particularly well suited for the critical care setting for which POC blood glucose values are routinely sampled every hour in patients demonstrating lack of glucose control. Similar to the diabetes investigation, NNM functionality (i.e. real-time training/weight adaptation/optimization and prediction) was integrated into a computer application to simulate real-time prediction of glucose [86] in 5 patients which were not utilized for model training/development. In order to gauge model performance, MAD% of model predictions with respect to reference glucose concentration, the percentage of hypoglycemic (CGM values $\leq$70 mg/dl), normal (CGM>, and hyperglycemic glucose values, were calculated. As a further measurement of performance analysis, CEGA was implemented to assess the clinical acceptability of the model predictions.

The patient dataset used for performance analysis consisted of 10.5 h of data (random segments of approximately 2 h of data from each patient). Thus, this provided a considerable dataset to assess model performance which included 9,405 predicted glucose values (i.e. 15 predicted glucose values for each simulated real-time CGM value and iteration of NNM prediction). The MAD% was calculated as 18.0%. No hypoglycemia occurred in this 5 patient validation dataset, however, the NNM successfully predicted 77.7 and 80.0% of the normal and hyperglycemic glucose values respectively. CEGA of model predictions with respect to reference CGM values revealed that 99.9% of model predictions could be regarded as clinically acceptable. A greater majority (62.3%) of predicted glucose values fell within the region of highest accuracy (region A) of the Clarke Error Grid. Figure 4 includes model predictions across the patient dataset. For clarity, due to the large dataset, the data was resampled to plot every 20th predicted and reference CGM value in order to better demonstrate model performance.

Analysis of results indicated that model predictions accurately tracked the rate of change (ROC) in glucose concentration across the 75 min predictive horizon. To enhance model accuracy, a post-processing algorithm was generated to modify the NNM predicted output. This algorithm utilizes the offset existent between the first predicted value in the model output and real-time CGM values. This offset is then added to the first predicted value in the output and the ROC in the NNM output is weighted based on the glycemic threshold (i.e. quantifiable range of glucose of the current real-time CGM value). These weight values were greater at elevated glucose values then at low and normal glycemic extremes due to insulin infusion protocols and glycemic control obtained in the critical care patient population. Equation 4 was utilized to implement this algorithm to determine the post-processed predicted CGM value $PP_{CGM}(j)$ at the jth index of the NNM predicted output. At initiation, the first value in the predicted output is difference between the current real-time CGM value and first predicted value in neural network output. $ROC_{Pr}(j)$ is the rate of change at

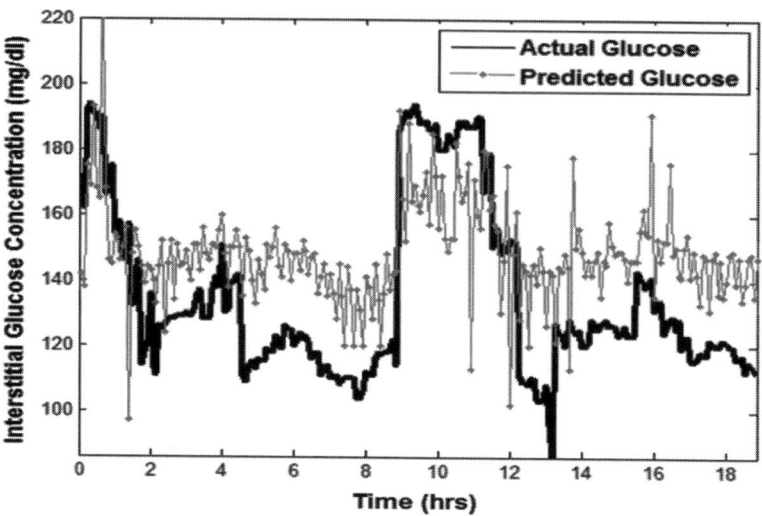

Fig. 4 Real-time prediction of glucose in critical care patients

the jth index of the predicted neural network output, $Samp_{rate}$ is the sampling rate of the CGM device (i.e. for this investigation $Samp_{rate}$ is 5 min), and W_i is a weight value (which varies between 0.2 and 0.9 [85]) for the rate of change based on the current real-time glycemic threshold.

$$PP_{CGM}(j) = CGM_{off}(j-1) + Samp_{rate} * W_i * ROC_{Pr}(j) \qquad (4)$$

Application of the post-processing algorithm model resulted in a significant increase in model performance. The overall MAD% was reduced to 7.1% (from 18.0%). Furthermore, the number of normal and hyperglycemic extremes predicted successfully was increased to 94.5 and 86.8% respectively. CEGA of model predictions following post-processing also indicated a substantial increase in performance as 93.2% of predictions fell in the region of highest accuracy in the error grid (Region A). Figure 5 contains the Clarke Error Grid containing model predictions which demonstrate a high degree of accuracy with most predictions falling within Region A of the error grid. CEGA further revealed that 99.9% of model predictions could be regarded as clinically acceptable. Figure 6 includes model predictions before and after application of the post processing algorithm for which improvements in model predictive accuracy following post-processing can be seen.

4.3 Neural Network Modeling for Patient Specific Glycemic Forecasting

Although real-time prediction of glucose in both diabetic and critical care patient populations is possible using "universal" or general NNMs (i.e. models generated

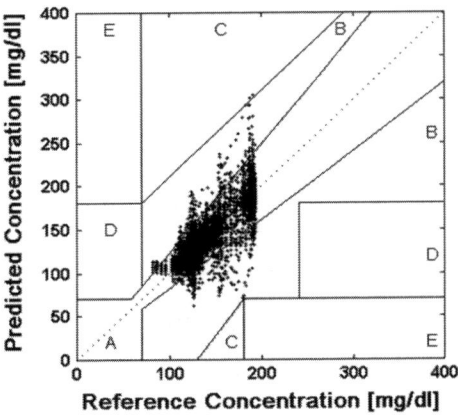

Fig. 5 Clarke Error Grid containing post-processed NNM predictions with respect to reference CGM values

using data from multiple patients), patient specific NNMs may provide increased accuracy as they are "individualized" and take into account various factors which are particular to a single patient. In the critical care investigation (described in Sect. 4.2), a 38 year old trauma patient admitted to the ICU had an extended length of stay of 16 days. During this extended length of stay in the ICU over 243.6 h of data was collected. This data consisted of 2,923 CGM values and provided a significant source of data to support development of a patient specific FFNNM for real-time prediction of glucose [87]. In order to compare performance of the patient specific model, a "universal"/general NNM was developed using a training set of data acquired from five critical care patients. This training set consisted of 515.7 h of data. This data consisted of 6,188 CGM values. The inputs to the FFNNMs included CGM and POC blood glucose values, as well as insulin delivery data.

Using the computer application for simulated real-time prediction of glucose, (as implemented in both investigations described in Sects. 4.1 and 4.2), data from the trauma patient not utilized for initial model development/training [containing 40.3 h (484 data points) of data] was used to test the performance of the patient specific and general NNMs. Figure 7 includes NNM predicted glucose values generated by the patient specific and "universal" NNMs (A and B respectively). Model performance of the patient specific NNM was superior to that of the "universal" model. The overall error (MAD%) of the predictions generated using the universal model was calculated as 15.9%. The patient specific model, therefore, generates more accurate predictions with a decrease in overall error of 8.0%. Furthermore, CEGA revealed that 95.1% of the predictions fell within region A of the error grid and 4.9% fell within region B of the error grid. For the "universal" NNM, 69.8% of the predictions fell within region A of the error grid and 30.2% fell within region B of the error grid. For both models, 100% of the predicted CGM values could be considered clinically acceptable as no predicted values were located within regions C, D, or E of the error grid.

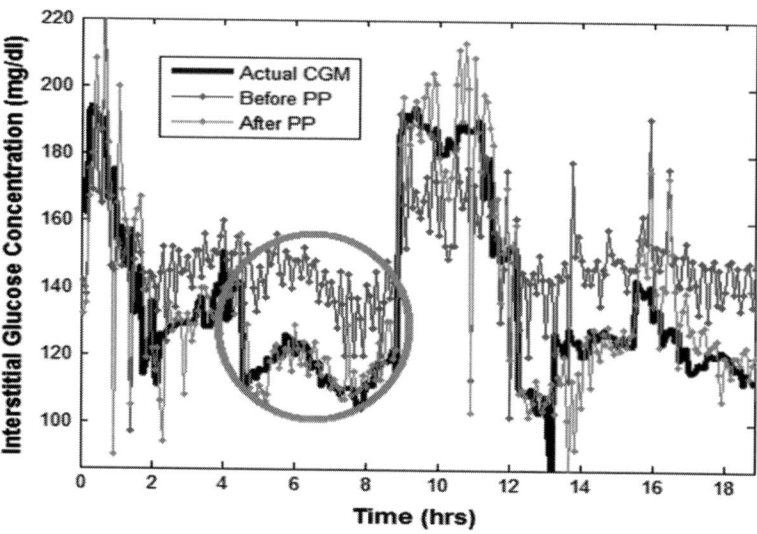

Fig. 6 Real-time prediction of glucose in critical care patients before and after application of the post-processing (PP) algorithm

For the patients with an extended length of stay in the ICU, the results of this investigation substantiate the position that a patient specific neural network model (generated and trained with data from a single patient) may provide improved model performance. This is likely attributed to the NNMs ability to adapt to factors such as insulin sensitivity which vary from patient to patient. Training a NNM using data from one patient results in an "optimized" model which is tailored to an individual patient.

5 Conclusion

To date, the feasibility of real-time prediction of glucose in both diabetic and critical care patient populations has been demonstrated [85–88]. Using such NNM approaches, real-time glycemic forecasting was demonstrated using prediction horizons of 75 min. If interstitial fluid based CGM technologies are used, model prediction horizons ≥60 min are desirable based on the inherent time lag associated with the measurement technology. This is necessary in order to optimize clinical utility and acceptability and to mitigate the time-lag existent between blood and interstitial glucose concentration [76] of 10–15 min. Furthermore, the reported NNM approach is unique because it integrates various inputs which, to date, have not been routinely implemented in predictive models. There are significant quantities of factors which are indicators of future glycemic excursions in both diabetic and critical care patient populations. NNMs are particularly well

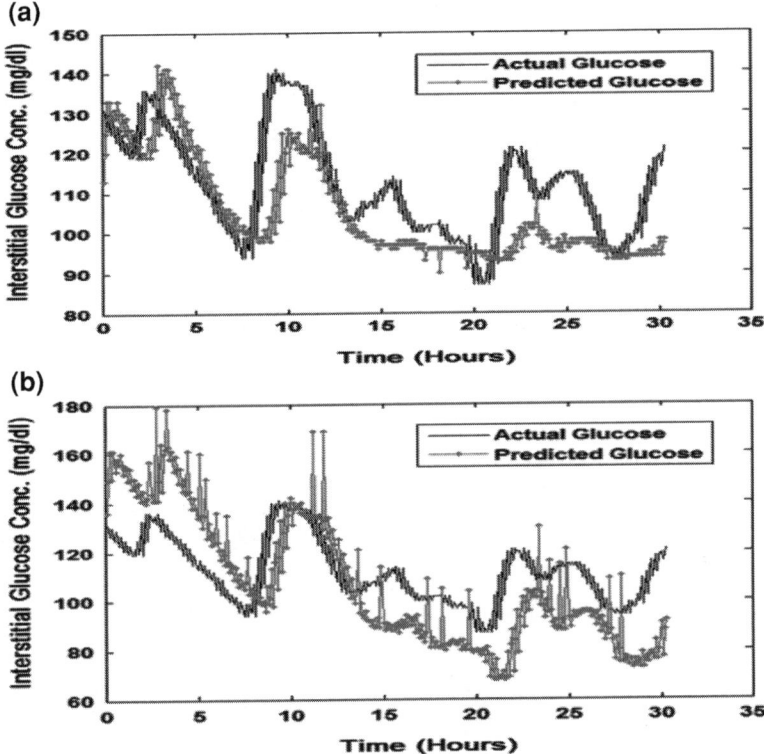

Fig. 7 Real-time prediction of glucose in a 38 year old trauma patient using patient specific (**a**) and "universal" NNMs (**b**)

suited to quantify the effect that these multiple factors can have on glucose concentration predictions. Furthermore, because NNMs are adaptive, they have the unique ability to be optimized for "patient specific" modeling where patterns in glucose and its related parameters are particular to a single patient. Further investigation is warranted to optimize NNM training sets for each patient population and to determine which of these factors are necessary components of the NNM input structure. The prediction of glucose concentration as provided by these NNM approaches would provide patients with diabetes and their healthcare providers a means for optimization of insulin therapy and glycemic control. This would in turn result in a reduction of complications (i.e. neuropathy, nephropathy, retinopathy) resultant of poor glycemic control [1–3] and increased patient safety and future quality of life.

CGM has only recently been investigated as a potentially useful clinical tool in the critical care setting [71, 96–98]. The presented NNM approach has also yielded promising results in such a setting and the 75 min model prediction horizon has the potential to provide valuable insight into future glycemic trends which will support optimization of glucose control and consequently enhancement of patient safety, outcome, and quality of life [12, 14, 18, 99, 100]. Such modeling approaches may one day be combined with other artificial intelligence methods to facilitate automation of blood glucose control. However, most importantly, the knowledge of real-time and future glucose values will provide critical care professionals more time to focus on higher level tasks and optimize the patient safety and care.

References

1. The Diabetes Control Complications Trial Research Group: The effect of intensive treatment of diabetes on the development and progression of long-term complications in insulin-dependent diabetes mellitus. N. Engl. J. Med. **329**, 977–986 (1993)
2. The Diabetes Control Complications Trial Research Group: Retinopathy and nephropathy in patients with type 1 diabetes four years after a trial of intensive therapy. N. Engl. J. Med. **342**, 381–389 (2000)
3. The Diabetes Control Complications Trial Research Group: Sustained effect on intensive treatment of type 1 diabetes Mellitus on development and progression of diabetic nephropathy. JAMA **290**, 2159–2167 (2003)
4. The Diabetes Control Complications Trial Research Group: Intensive diabetes treatment and cardiovascular disease in patients with type 1 diabetes. N. Engl. J. Med. **354**, 2643–2653 (2005)
5. Furnary, A.P., Zerr, K.J., Grunkemeier, G.L.: Continuous intravenous insulin infusion reduces the incidence of deep sternal wound infection in diabetic patients after cardiac surgical procedures. Ann. Thoracic. Surg. **67**, 352–360 (1999)
6. Grey, N., Perdrizet, G.: Reduction of nosocomial infections in the surgical intensive care unit by strict glycemic control. Endocrine. Prac. **10**(2), 46–52 (2004)
7. Zerr, K.J., Furnary, A.P., Grunkemeier, G.L., Bookin, S., Kanhere, V., Starr, A.: Glucose control lowers the risk of wound infection in diabetics after open heart operations. Ann. Thoracic. Surg. **63**, 356–361 (1997)
8. Doenst, T., Wijeysundera, D., Karkouti, K., Zechner, C., Maganti, M., Rao, V., Borger, M.A.: Hyperglycemia during cardiopulmonary bypass is an independent risk factor for mortality in patients undergoing cardiovascular surgery. J. Thorac. Cardiovasc. Surg. **130**(4), 1144–1152 (2005)
9. Jones, K.W., Cain, A.S., Mitchell, J.H., Millar, R.C., Rimmasch, H.L., French, T.K., Abbate, S.L., Roberts, C.A., Stevenson, S.R., Marshall, D., Lappe, D.L.: Hyperglycemia predicts mortality after CABG: postoperative hyperglycemia predicts dramatic increases in mortality after coronary artery bypass graft surgery. J. Diabetes Complicat. **22**, 365–370 (2008)
10. Laird, A.M., Miller, P.R., Preston, R., Kilgo, P.D., Meredith, J.W., Chang, M.C.: Relationship of early hyperglycemia to mortality in trauma patients. J. Trauma **56**(5), 1058–1062 (2004)
11. Umpierrez, G., Isaacs, S., Bazargan, N., You, X., Thaler, L., Kitabchi, A.: Hyperglycemia: an independent marker of in-hospital mortality in patients with undiagnosed diabetes. J. Clin. Endocrinol. Metab. **87**(3), 978–982 (2002)
12. Van den Berghe, G.: How does blood glucose control with insulin save lives in intensive care? J. Clin. Invest. **114**(9), 1187–1195 (2004)
13. Van Den Berghe, G., Wouters, P., Weekers, F.: Intensive insulin therapy in critically ill patients. N. Engl. J. Med. **345**, 1359–1367 (2001)

14. Van den Berghe, G., Wouters, P.J., Bouillon, R.: Outcome benefit of intensive insulin therapy in the critically ill: insulin dose versus glycemic control. Crit. Care Med. **31**, 359–366 (2003)
15. Van den Berghe, G., Wouters, P., Weekers, F., Verwaest, C., Bruyninckx, F., Schetz, M., Vlasselaers, D., Ferdinande, P., Lauwers, P., B, R.: Intensive insulin therapy in critically ill patients. N. Engl. J. Med. **345**, 1359 (2001)
16. van den Berghe, G., Bouillon, R.: Optimal control of glycemia among critically ill patients. JAMA **291**(10), 1198–1199 (2004)
17. Ingels, C., Debaveye, Y., Milants, I., Buelens, E., Peeraer, A., Devriendt, Y., Vanhoutte, T., Van Damme, A., Schetz, M., Wouters, P.J., Van den Berghe, G.: Strict blood glucose control with insulin during intensive care after cardiac surgery: impact on 4 years survival, dependency on medical care, and quality-of-life. Eur. Heart J. **27**(22), 2716–2724 (2006)
18. Collier, B., Diaz Jr, J., Forbes, R.: The impact of a normoglycemic management protocol on clinical outcomes in the trauma intensive care unit. J. Parenter Enteral Nutr. **29**(5), 353–358 (2005)
19. Finney, S.J., Zekveld, C., Elia, A., Evans, T.W.: Glucose control and mortality in critically ill patients. JAMA **290**(15), 2041–2047 (2003)
20. Clement, S., Braithwaite, S.S., Magee, M.F., Ahmann, A., Smith, E.P., HI, Schafer.R.G.: American diabetes association diabetes in hospitals writing committee: management of diabetes and hyperglycemia in hospitals. Diabetes Care **27**(2), 553–591 (2004)
21. Chipkin, S.R., Klugh, S.A., Chasan-Tabere, L.: Exercise and diabetes. Cardiol. Clin. **19**, 489–505 (2000)
22. Cox, D.J., Gonder-Fredrick, L., Kovatchev, B.P., Clarke, W.L.: The metabolic demands of driving for drivers with type 1 diabetes mellitus. Diab. Metab. Res. Rev. **18**, 381–385 (2002)
23. Dutour, A., Boiteau, V., Dadoun, F., Feissel, A., Atlan, C., Oliver, C.: Hormonal response to stress in brittle diabetes. Psychoneuroendocrinology **21**(6), 525–543 (1996)
24. Ficker, J.H., Dertinger, S.H., Siegfried, W., et al.: Obstructive sleep apnoea and diabetes mellitus: the role of cardiovascular autonomic neuropathy. Eur. Resp. J. **11**, 14–19 (1998)
25. Hargreaves, M., et al.: Effect of heat stress on glucose kinetics during exercise. J. App. Physiol. **81**(4), 1594–1597 (1996)
26. Jones, T.W., Porter, P., Sherwin, R.S., et al.: Decreased epinephrine responses to hypoglycemia during sleep. N. Engl. J. Med. **338**(23), 1657–1662 (1998)
27. Nomura, M., Fujimoto, K., Higashino, A., et al.: Stress and coping behavior in patients with diabetes mellitus. Acta Diabetologia **37**, 61–64 (2000)
28. Resnick, H.E., Redline, S., Shahar, E., et al.: Diabetes and sleep disturbances: findings from the sleep heart health study. Diabetes Care **26**, 702–709 (2003)
29. Trief, P.M., et al.: Impact of the work environment on glycemic control and adaptation to diabetes. Diabetes Care **22**, 569–574 (1999)
30. Van Cauter, E., Polonsky, K.S., Scheen, A.: Roles of circadian rhythmicity and sleep in human glucose regulation. Endocr. Rev. **18**, 716–738 (1997)
31. Winokur, A., Maislin, G., Phillips, J., Amsterdam, J.: Insulin resistance after oral glucose tolerance testing in patients with major depression. Am. J. Psychiatry **145**(3), 325–330 (1988)
32. Jarrett, R.: Rhythms in insulin and glucose. Endocrine Rhythms **12**, 247–258 (1979)
33. Simon, C., Brandenberger, G., Saini, J., Ehrhart, J., Follenius, M.: Ultradian oscillations of plasma glucose, insulin, and C-peptide in man during continuous enteral nutrition. J. Clin. Endocrinol. Metab. **64**, 669–674 (1987)
34. Simon, C., Brandenberger, G., Saini, J., Ehrhart, J., Follenius, M.: Slow oscillations of plasma glucose and insulin secretion rate are amplified during sleep in humans under continuous enteral nutrition. Sleep **17**, 333–338 (1994)
35. Tato, F., Tato, S., Beyer, J., Schrezenmeir, J.: Circadian variation of basal and postprandial insulin sensitivity in healthy individuals and patients with type 1 diabetes. Diabetes Res. Clin. Prac. **17**, 13–24 (1991)

36. Trumper, B., Reschke, K., Molling, J.: Circadian variation of insulin requirement in insulin dependent diabetes mellitus: the relationship between circadian change in insulin demand and diurnal patterns of growth hormone, cortisol, and glucagon during euglycemia. Horm. Metab. Res. **27**, 141–147 (1995)
37. Van Cauter, E., Shapiro, E., Tillil, H., Polonsky, K.: Circadian modulation of glucose and insulin responses to meals: relationship to cortisol rhythm. Am. J. Phys. **262**, E467–475 (1992)
38. Campbell, P.J., Gerich, J.E.: Occurrence of dawn phenomenon without change in insulin clearance in patients with insulin-dependent diabetes mellitus. Diabetes **35**, 749–752 (1986)
39. De Feo, P., Perriello, G., Ventura, M.M.: Studies on overnight insulin requirements and metabolic clearance rate of insulin in normal and diabetic man: relevance to the pathogenesis of the dawn phenomenon. Diabetologia **29**, 475–480 (1986)
40. Dodt, C., Breckling, U., Derad, I.: Plasma epinephrine and norepinephrine concentrations of healthy humans associated with nighttime sleep and morning arousal. Hypertension **30**, 71–76 (1997)
41. Dux, S., White, N.H., Skor, D.A.: Insulin clearance contributes to the variability of nocturnal insulin requirement in insulin-dependent diabetes mellitus. Diabetes **34**, 1260–1265 (1985)
42. Shapiro, E.T., Polonsky, K.S., Copinschi, G.: Nocturnal elevation of glucose levels during fasting in noninsulin-dependent diabetes. J. Clin. Endocrinol. Metab. **72**, 444–454 (1991)
43. Skor, D.A., White, N.H., Thomas, L.: Relative roles of insulin clearance and insulin sensitivity in the prebreakfast increase in insulin requirements in insulin-dependent diabetic patients. Diabetes **33**, 60–63 (1984)
44. Van Cauter, E., Desir, D., Decoster, C.: Nocturnal decrease in glucose tolerance during constant glucose infusion. J. Clin. Endocrinol. Metab. **69**, 604–611 (1989)
45. Van Cauter, E., Leproult, R., Kupfer, D.J.: Effects of gender and age on the levels and circadian rhythmicity of plasma cortisol. J. Clin. Endocrinol. Metab. **81**, 2468–2473 (1996)
46. Wu, M.S., Ho, L.T., Jap, T.S.: Diurnal variation of insulin clearance and sensitivity in normal man. Proc. Natl. Sci. Counc. Repub. China B **10**, 64–69 (1986)
47. Smith, S., Ovenson, K., Strauss, W., et al.: Ultradian variation of blood glucose in intensive care unit patients receiving insulin infusions. Diabetes Care **30**(10), 2503–2505 (2007)
48. Smith, S.M., Oveson, K.E., Strauss, W., Ahmann, A.J., Hagg, D.S.: Diurnal and other variations in blood glucose in ICU patients receiving insulin infusions. Crit Care **11**(Supp. 2), 133 (2007)
49. Smith, S., Hagg, D.: Glucose variance in ICU patients receiving insulin infusions. Chest **133**(5), 1288 (2008)
50. Filipovsky, J., Ducimetiere, P., Eschwege, E., Richard, J.L., Rosselin, G., Claude, J.R.: The relationship of blood pressure with glucose, insulin, heart rate, free fatty acids and plasma cortisol levels according to degree of obesity in middle-aged men. J. Hypertens. **14**(2), 229–235 (1996)
51. Molnar, G.W., Read, R.C.: Hypoglycemia and body temperature. JAMA **227**(8), 916–921 (1974)
52. Sparacino, G., Zanderigo, F., Corazza, S., Maran, A., Facchinetti, A., Cobelli, C.: Glucose concentration can be predicted ahead in time from continuous glucose monitoring sensor time-series. IEEE Trans. Biomed. Eng. **54**(5), 931–937 (2007)
53. Reifman, J., Rajaraman, S., Gribok, A., Ward, W.K.: Predictive modeling for improved management of glucose levels. J. Diabetes Sci. Technol. **1**(4), 479–486 (2007)
54. Plank, J., Blaha, J., Cordingley, J., et al.: Multicentric, randomized controlled trial to evaluate blood glucose control by the model predictive control algorithm versus routine glucose management protocols in intensive care unit patients. Diabetes Care **29**(2), 271–276 (2006)
55. Schlotthauer, G., Gamero, L.G., Torres, M.E., Nicolini, G.A.: Modeling, identification and nonlinear model predictive control of type I diabetic patient. Med. Eng. Phys. **28**(3), 240–250 (2006)

56. Schaller, H.C., Schaupp, L., Bodenlenz, M., Wilinska, M.E., Chassin, L.J., Wach, P., Vering, T., Hovorka, R., Pieber, T.R.: On-line adaptive algorithm with glucose prediction capacity for subcutaneous closed loop control of glucose: evaluation under fasting conditions in patients with type 1 diabetes. Diabet. Med. **23**(1), 90–93 (2006)
57. Steil, G., Rebrin, K., Mastrototaro, J.: Metabolic modelling and the closed-loop insulin delivery problem. Diabetes Res. Clin. Prac. **74**(2), S183–S186 (2006)
58. Steil, G.M., Clark, B., Kanderian, S., Rebrin, K.: Modeling insulin action for development of a closed-loop artificial pancreas. Diabetes Tech. Ther. **7**(1), 94–108 (2005)
59. Steil, G.M., Panteleon, A.E., Rebrin, K.: Closed-loop insulin delivery-the path to physiological glucose control. Adv. Drug. Deliv. **56**(2), 125–144 (2004)
60. Wintergerst, K.A., Deiss, D., Buckingham, B., Cantwell, M., Kache, S., Agarwal, S., Wilson, D.M., Steil, G.: Glucose control in pediatric intensive care unit patients using an insulin-glucose algorithm. Diabetes Technol. Ther. **9**(3), 211–222 (2007)
61. Chee, F., Fernando, T., van Heerden, P.: Closed-loop glucose control in critically ill patients using continuous glucose monitoring system (CGMS) in real time. IEEE Trans. Inf. Technol. Biomed. **7**(1), 43–53 (2003)
62. Chee, F., Fernando, T., Savkin, A., van Heerden, V.: Expert PID control system for blood glucose control in critically ill patients. IEEE Trans. Inf. Technol. Biomed. **7**(4), 419–425 (2003)
63. Ferrannini, E., Mari, A.: Beta cell function and its relation to insulin action in humans: a critical appraisal. Diabetologia **47**(5), 943–956 (2004)
64. Steil, G.M., Rebrin, K., Janowski, R., Darwin, C., Saad, M.F.: Modeling beta-cell insulin secretion—implications for closed-loop glucose homeostasis. Diabetes Technol. Ther. **5**(6), 953–964 (2003)
65. Segel, S.A., Paramore, D.S., Cryer, P.E.: Hypoglycemia-associated autonomic failure in advanced type 2 diabetes. Diabetes **51**(3), 724–733 (2002)
66. Dagogo-Jack, S.E., Craft, S., Cryer, P.E.: Hypoglycemia-associated autonomic failure in insulin dependent diabetes mellitus. J. Clin. Invest. **91**, 819–828 (1993)
67. Mokan, M., Mitrakou, A., Veneman, T., Ryan, C., Korytkowski, M., Cryer, P., Gerich, J.: Hypoglycemia unawareness in IDDM. Diabetes Care **17**(12), 1397–1403 (1994)
68. Gerich, J.E., Mokan, M., Veneman, T., Korytkowski, M., Mitrakou, A.: Hypoglycemia Unawareness. Endocr. Rev. **12**(4), 356–371 (1991)
69. Hoeldtke, R.D., Boden, G., Shuman, C.R., Owen, O.E.: Reduced epinephrine secretion and hypoglycemia unawareness in diabetic autonomic neuropathy. Ann. Intern. Med. **96**(4), 459–462 (1982). doi:10.1059/0003-4819-96-4-459
70. Cryer, P.E.: Iatrogenic hypoglycemia as a cause of hypoglycemia-associated autonomic failure in IDDM. A vicious cycle. Diabetes **41**(3), 255–260 (1992)
71. De Block, C., Manuel-y-Keenoy, B., Rogiers, P., Jorens, P., Van Gaal, L.: Glucose control and use of continuous glucose monitoring in the intensive care unit: a critical review. Curr. Diabetes Rev. **4**(3), 234–244 (2008)
72. Kovatchev, B.P., Shields, D., Breton, M.: Graphical and numerical evaluation of continuous glucose sensing time lag. Diabetes Technol. Ther. **11**(3), 139–143 (2009)
73. Van Der Valk, P.R., Van Der Schatte Olivier-Steding, I., Wientjes, K.J., et al.: Alternative site blood glucose measurement at the abdomen. Diabetes Care **25**, 2114–2115 (2002)
74. Rebrin, K., Steil, G.M.: Can interstitial glucose assessment replace blood glucose measurements? Diabetes Technol. Ther. **2**(3), 461–472 (2000)
75. Cengiz, E., Tamborlane, W.: A tale of two compartments: interstitial versus blood glucose monitoring. Diabetes Tech. Ther. **11**(s1), s11–s16 (2009)
76. Kovatchev, B.P., Shields, D., Breton, M.: Graphical and numerical evaluation of continuous glucose sensing time lag. Diabetes Tech. Ther. **11**(3), 139–143 (2009)
77. Wentholt, I.M.E., Hart, A.A.M., Hoekstra, J.B.L., Devries, J.H.: Relationship between interstitial and blood glucose in type 1 diabetes patients: delay and the push-pull phenomenon revisited. Diabetes Technol. Ther. **9**(2), 169–175 (2007)

78. Okamura, F., Tashiro, A., Utumi, A., Imai, T., Suchi, T., Tamura, D., Sato, Y., Suzuki, S., Hongo, M.: Insulin resistance in patients with depression and its changes during the clinical course of depression: minimal model analysis. Metabolism **49**(10), 1255–1260 (2000)
79. Prank, K., Jurgens, C., von zur Muhlen, A., Brabant, G.: Predictive neural networks for learning the time course of blood glucose levels from the complex interaction of counterregulatory hormones. Neural. Comput. **10**(4), 941–953 (1998)
80. Tresp, V., Briegel, T., Moody, J.: Neural network models for blood glucose metabolism of a diabetic. IEEE Trans. Neural. Netw. **10**(5), 1204–1212 (1999)
81. El-Jabali, A.: Neural network modeling and control of type 1 diabetes mellitus. Bioprocess Biosyst. Eng. **27**(2), 75–79 (2005)
82. Gogou, G., Maglaveras, N., Ambrosiadou, B.V., Goulis, D., Pappas, C.: A neural network approach in diabetes management by insulin administration. J. Med. Syst. **25**(2), 119–131 (2001)
83. Sandham, W.A., Nikoletou D., Hamilton D.J., Patterson K., Japp A., MacGregor C.: Blood glucose prediction for diabetes therapy using a recurrent artificial neural network. IX European signal processing conference (EUSIPCO) (1998)
84. Foody, G.M., McCulloch, M.B., Yates, W.B.: The effect of training set size and composition on artificial neural network classification. Int. J. Remote Sens. **16**, 1707–1723 (1995)
85. Pappada, S.M.: Prediction of glucose for enhancement of treatment and outcome: a neural network model approach. Doctoral dissertation, University of Toledo/OhioLINK (2010)
86. Pappada, S.M., Cameron, B.D., Rosman, P.M., Papadimos, T.J., Borst, M.J., Bourey, R.E.: Neural network based real-time prediction of glucose in patients with insulin dependent diabetes. Diabetes Technol. Ther. **13**(2), 135–141 (2011)
87. Pappada, S.M., Borst, M.J., Cameron, B.D., et al.: Development of a neural network model for predicting glucose levels in a surgical critical care setting. Patient Saf. Surgery **4**(15), 1–5 (2010)
88. Pappada, S.M., Cameron, B.D., Rosman, P.M.: Development of a neural network for prediction of glucose concentration in type 1 diabetes patients. J. Diabetes Sci. Technol. **2**(5), 793–801 (2008)
89. Perez-Gandia, C., Facchinetti, A., Sparacino, G., Cobelli, C., Gomez, E.J., Rigla, M., de Leiva, A., Hernando, M.E.: Artificial neural network algorithm for online glucose prediction from continuous glucose monitoring. Diabetes Technol. Ther. **12**(1), 81–88 (2010)
90. Qian, N.: On the momentum term in gradient descent learning algorithms. Neural Netw. **12**(1), 145–151 (1999)
91. Clarke, W.L.: The original clarke error grid analysis (EGA). Diabetes Technol. Ther. **7**(5), 776–779 (2005). doi:10.1089/dia.2005.7.776
92. Clarke, W.L., Cox, D., Gonder-Frederick, L.A., Carter, W., Pohl, S.L.: Evaluating clinical accuracy of systems for self-monitoring of blood glucose. Diabetes Care **10**(5), 622–628 (1987)
93. Renner, G., Ainko, E.: Genetic algorithms in computer aided design. Comput. Aided Des. **35**, 709–726 (2003)
94. McShane, M.J., Cameron, B.D., Cote, G.L., Spiegelman, C.H.: Improving complex near-IR calibrations using a new wavelength selection algorithm. Appl. Spectrosc. **53**(12), 1575–1581 (1999)
95. McShane, M.J.C.B., Cote, G.L., Motamadi, M., Spiegelman, C.H.: A novel peak-hopping stepwise feature selection method with application to Raman spectroscopy. Anal. Chem. Acta **388**, 251–264 (1999)
96. Goldberg, P.A., Siegel, M.D., Russell, R.R., Sherwin, R.S., Halickman, J.I., Cooper, D.A., Dziura, J.D., Inzucchi, S.E.: Experience with the continuous glucose monitoring system in a medical intensive care unit. Diabetes Technol. Ther. **6**(3), 339–347 (2004)
97. Holzinger, U., Warszawska, J., Kitzberger, R., Wewalka, M., Miehsler, W., Herkner, H., Madl, C.: Real-time continuous glucose monitoring in critically ill patients. Diabetes Care **33**(3), 467–472 (2010). doi:10.2337/dc09-1352

98. Price, G.C., Stevenson, K., Walsh, T.S.: Evaluation of a continuous glucose monitor in an unselected general intensive care population. Crit. Care Resusc. **10**(3), 209–216 (2008)
99. Bochicchio, G., Sung, J., Joshi, M.: Persistent hyperglycemia is predictive of outcome in critically ill trauma patients. J. Trauma **58**(5), 921–924 (2005)
100. Sung, J., Bochicchio, G., Joshi, M.: Admission hyperglycemia is predictive of outcome in critically ill trauma patients. J. Trauma **59**(1), 80–83 (2005)

Author Index

A
Alison C. Jones, 133
Amaya Pérez-del Palomar, 281
Amit Gefen, 441, 461
Andreas A. Linninger, 337
Angel Ginel, 281
Anup A. Gandhi, 75

B
Bar-Yoseph P. Z., 27
Brent D. Cameron, 505

C
Clayton J. Adam, 103

D
David Varssano, 461
Dime Vitanovski, 157
Dorin Comaniciu., 157
Douglas C. Fredericks, 75

E
Elin Diczfalusy, 357
Eran Peleg, 53

F
Fischer A., 27

H
Hadas Zur, 487
Hyungmin Kim, 413

I
Igor Sazonov, 241

J
Jean-François Pouget, 217
Johan Debayle, 217
Jose L. López-Villalobos, 281
Joseph D. Smucker, 75

K
Karin Wårdell, 357
Kiran H. Shivanna, 75

L
Laura Dubuis, 217
Liu D., 189

M
Manuel Doblaré, 281
Martyn P. Nash, 379
Mattias Åström, 357
Mauricio Reyes, 413
Miguel A González-Ballester, 281

N
Nicole A. DeVries, 75
Nicole A. Kallemeyn, 75
Nicole M. Grosland, 75
Nir Trabelsi, 3

O
Olfa Trabelsi, 281

P
Paige Little J., 103
Perumal Nithiarasu, 241
Philipp Jürgens, 413
Pierre Badel, 217
Pierre-Yves Rohan, 217
Podshivalov L., 27
Poul M. F. Nielsen, 379

R
Razvan Ioan Ionasec, 157
Roy Asher, 461
Ruth K. Wilcox, 133

S
Sarrawat Rehman, 133
Scott M. Pappada, 505

Serge Couzan, 217
Si Yong Ye, 241
Sigal Portnoy, 441
Stéphane Avril, 217
Swathi Kode, 75

T
Tamir Tuller, 487
Thiranja P. Babarenda Gamage, 379

V
Vijayaraghavan Rajagopal, 379
Vithanage N. Wijayathunga, 133

W
Wood N. B., 189

X
Xu X. Y, Sun N., 189

Z
Zohar Yosibash, 3

Printed by Publishers' Graphics LLC USA

2012